SPACE TELESCOPE SCIENCE INSTITUTE

SYMPOSIUM SERIES: 13
Series Editor S. Michael Fall, Space Telescope Science Institute

**SUPERNOVAE AND GAMMA-RAY BURSTS:
THE GREATEST EXPOSIONS SINCE THE BIG BANG**

Recent observations have uncovered that gamma-ray bursts of relatively long durations are at cosmological distances in star-forming galaxies. The detection of X-Ray, optical, and radio afterglows to gamma-ray bursts has literally revolutionized the understanding of these enigmatic events. Since the dramatic discovery that the supernova SN 1998bw coincided in position and time with a gamma-ray burst, the possibility was raised that these two types of spectacular explosions are related. This timely volume presents articles by a host of world experts who gathered together for an international conference at the Space Telescope Science Institute. This was the first meeting in which the communities of supernova researchers and gamma-ray burst researchers were brought together to share ideas. The contributions review the mechanisms for these explosive events, the possible connections between them, and their relevance for cosmology. Both observations and theoretical developments are covered. This book will be an invaluable source of information for both active researchers and graduate students in this exciting area of research.

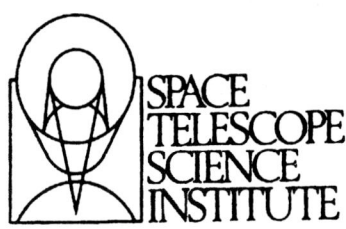

Other titles in the Space Telescope Science Institute Symposium Series.

1. Stellar Populations
   Edited by C. A. Norman, A. Renzini and M. Tosi 1987 0 521 33380 6
2. Quaser Absorption Lines
   Edited by C. Blades, C. A. Norman and D. Turnshek 1988 0 521 34561 8
3. The Formation and Evolution of Planetary Systems
   Edited by H. A. Weaver and L. Danly 1989 0 521 36633 X
4. Clusters of Galaxies
   Edited by W. R. Oegerle, M. J. Fitchet and L. Danly 1990 0 521 38462 1
5. Massive Stars in Starbursts
   Edited by C. Leitherer, N. R. Walborn, T. M. Heckman and C. A. Norman 1991 0 521 40465 7
6. Astrophysical Jets
   Edited by D. Burgarella, M. Livio and C. P. O'Dea 1993 0 521 44221 4
7. Extragalactic Background Radiation
   Edited by D. Calzetti, M. Livio and P. Madau 1995 0 521 49558 X
8. The Analysis of Emission Lines
   Edited by R. E. Williams and M. Livio 1995 0 521 48081 7
9. The Collision of Comet Shoemaker–Levy 9 and Jupiter
   Edited by K. S. Noll, H. A. Weaver and P. D. Feldman
   1996 0 521 56192 2
10. The Extragalactic Distance Scale
    Edited by M. Livio, M. Donahue and N. Panagia 1997 0 521 59164 2
11. The Hubble Deep Field
    Edited by M. Livio, S. M. Fall and P. Madau 1998 0 521 63097 5
12. Unsolved Problems in Stellar Evolution
    Edited by M. Livio 2000 0 521 78091 8

SUPERNOVAE AND GAMMA-RAY BURSTS
The greatest explosions since the Big Bang

Proceedings of the Space Telescope Science Institute Symposium,
held in Baltimore, Maryland
May 3–6, 1999

Edited by

MARIO LIVIO
Space Telescope Science Institute, Baltimore

NINO PANAGIA
Space Telescope Science Institute, Baltimore

KAILASH SAHU
Space Telescope Science Institute, Baltimore

Published for the
Space Telescope Science Institute

PUBLISHED BY THE PRESS SYNDICATE OF THE UNIVERSITY OF CAMBRIDGE
The Pitt Building, Trumpington Street, Cambridge, United Kingdom

CAMBRIDGE UNIVERSITY PRESS
The Edinburgh Building, Cambridge CB2 2RU, UK
40 West 20th Street, New York, NY 10011–4211, USA
10 Stamford Road, Oakleigh, VIC 3166, Australia
Ruiz de Alarcón 13, 28014 Madrid, Spain
Dock House, The Waterfront, Cape Town 8001, South Africa

http://www.cambridge.org

© Cambridge University Press 2001

This book is in copyright. Subject to statutory exception
and to the provisions of relevant collective licensing agreements,
no reproduction of any part may take place without
the written permission of Cambridge University Press.

First published 2001

Printed in the United Kingdom, at the University Press, Cambridge

Typeset by the authors (CRC)

A catalogue record for this book is available from the British Library

ISBN 0 521 79141 3 hardback

Contents

Participants	vii
Preface	x

Gamma-Ray Burst—Supernova relation
   B. Paczyński 1

Observations of Gamma-Ray Bursts
   G. Fishman 9

Fireballs
   T. Piran 17

Gamma ray mechanisms
   M. Rees 36

Prompt optical emission from gamma-ray bursts
   R. Kehoe, C. Akerlof, R. Balsano, S. Barthelmy, J. Bloch, P. Butterworth,
   D. Casperson, T. Cline, S. Fletcher, F. Frontera, G. Gisler, J. Heise,
   J. Hills, K. Hurley, B. Lee, S. Marshall, T. McKay, A. Pawl, L. Piro,
   B. Priedhorsky, J. Szymanski, and J. Wren 47

X-ray afterows of gamma-ray bursts
   L. Piro 67

The first year of optical-IR observations of SN1998bw
   I. Danziger, T. Augusteijn, J. Brewer, E. Cappellaro, V. Doublier,
   T. Galama, J. Gonzalez, O. Hainaut, B. Leibundgut, C. Lidman,
   P. Mazzali, K. Nomoto, F. Patat, J. Spyromilio, M. Turatto,
   J. Van Paradijs, P. Vreeswijk, and J. Walsh 79

X-ray emission of Supernova 1998bw in the error box of GRB980425
   E. Pian 85

Direct analysis of spectra of Type Ic supernovae
   D. Branch 96

The interaction of supernovae and gamma-ray bursts with their surroundings
   R. Chevalier 110

Magnetars, Soft Gamma-ray Repeaters and Gamma-ray Bursts
   A. Harding 121

Super-luminous supernova remnants
   Y.-H. Chu, C.-H. Chen, and S.-P. Lai 131

The properties of hypernovae: SNe Ic 1998bw, 1997ef, and SN IIn 1997cy
   K. Nomoto, P. Mazzali, T. Nakamura, K. Iwamoto, K. Maeda,
   T. Suzuki, M. Turatto, I. Danziger, and F. Patat 144

Collapsars, Gamma-Ray Bursts, and Supernovae
   S. Woosley, A. MacFadyen, and A. Heger 171

Pre-Supernova evolution of massive stars
   N. Panagia and G. Bono 184

Radio supernovae and GRB 980425
   K. Weiler, N. Panagia, R. Sramek, S. Van Dyk, M. Montes, and C. Lacey . . 198

Models for Ia Supernovae and evolutionary effects
   P. Höflich and I. Dominguez 218

Deflagration to detonation
   A. Khokhlov 239

Universality in SN Iae and the Phillips relation
   D. Arnett 250

Abundance from supernovae
   *F.-K. Thielemann, F. Brachwitz, C. Freiburghaus, S. Rosswog,
   K. Iwamoto, T. Nakamura, K. Nomoto, H. Umeda, K. Langanke,
   G. Martinez-Pinedo, D. Dean, W. Hix, and M. Strayer* 258

SNe, GRBs, and the global properties of the Universe
   B. Schmidt 287

How good are SNe Ia as standard candles?
   A. Sandage, G. Tammann, and A. Saha 304

Type Ia Supernovae and their implications for cosmology
   M. Livio 334

Conference summary: Supernovae and Gamma Ray Bursts
   J. Wheeler 356

Participants

Andrews, Thomas B.	
Araya-Gochez, Rafael	University of Costa Rica
Aretxaga, Itziar	Instituto Nacional de Astrofisica Optica y Electronica
Arnett, David	Steward Observatory
Baron, Edward	University of Oklahoma
Bauer, Franz	NRAO
Beckwith, Steve	Space Telescope Science Institute
Bloom, Joshua	Caltech
Boffi, Francesca	Space Telescope Science Institute
Bond, Howard	Space Telescope Science Institute
Bono, Giuseppe	Rome Astronomical Observatory
Branch, David	University of Oklahoma
Bulik, Tomasz	CAMK
Caraveo, Patrizia A.	Instituto di Fiscia Cosmica
Chevalier, Roger	University of Virginia
Chiu, Kuenley	University of California at Berkeley
Chu, You-Hua	University of Illinois
Danziger, John	Osservatorio di Trieste
De Salamanca, Isabel E.	Leiden Observatory
Della Valle, Massimo	Universita' di Padova
Diaz-Miller, Rose	Space Telescope Science Institute
Douvion, Thomas	CEA-Saclay Service d'Astrophysique
Duerbeck, Hilmar	Free University Brussels (VUB)
Dwarkadas, Vikram	University of Sydney
Ferguson, Harry	Space Telescope Science Institute
Fishman, Gerald	NASA/MSFC
Frail, Dale	NRAO, Very Large Aray
Fransson, Claes	Stockholm Observatory
Fruchter, Andy	Space Telescope Science Institute
Garnavich, Peter	Center for Astrophysics
Gehrels, Neil	NASA/GSFC
Giblin, Timothy	University of Alabama in Huntsville
Godon, Patrick	Space Telescope Science Institute
Goldhaber, Gerson	Lawrence Berkeley Laboratory
Graber, James	
Greyber, Howard	
Gull, Theodore	NASA/GSFC/LASP
Gursky, Herbert	Naval Research Laboratory
Hartmann, Dieter	Clemson University
Hauser, Michael	Space Telescope Science Institute
Heaton, Hal	JHU/APL & Space Telescope Science Institute
Hoeflich, Peter	University of Texas at Austin
Horvath, Jorge E.	Steward Observatory
Hurley, Kevin	University of California, Berkeley
Jha, Saurabh	Harvard-Smithsonian Center for Astrophysics
Kafka, Styliani	University of Delaware
Katz, Jonathan	Washington University
Kehoe, Robert	University of Michigan

Participants

Khokhlov, Alexei	Laboratory for Computational Physics and Fluid Dynamics
Kirshner, Robert	Harvard-Smithsonian Center for Astrophysics
Kulkarni, Shrinivas	California Institute of Technology
Lacey, Christina	NRL/NRC
Lamb, Don Q.	University of Chicago
Livio, Mario	Space Telescope Science Institute
Marani, Gabriela	NASA GSFC/NRC
Martin, Crystal	Space Telescope Science Institute
Mathews, Grant	University of Notre Dame
Mazzali, Paolo	University of Tokyo/Osservatorio Astronomico, Trieste
Meszaros, Peter	Penn State University
Milne, Peter	Naval Research Laboratory
Nomoto, Ken'ichi	University of Tokyo School of Science
Novick, Robert	Columbia University
Nugent, Peter	Lawrence Berkeley Laboratory
Oey, Sally	Spae Telescope Science Institute
Pacini, Franco	Arcetri Astrophysical Observatory
Paczynski, Bohdan	Princeton University Observatory
Palous, Jan	Astronomical Institute, Academy of Sciences of the Czech Republic
Panagia, Nino	Space Telescope Science Laboratory
Perlmutter, Saul	University of California, Berkeley
Petro, Larry	Space Telescope Science Laboratory
Pian, Elena	ITESRE-CNR
Piran, Tsvi	Columbia University
Piro, Luigi	Instituto Astrofisica Spaziale
Pun, Jason	GSFC/NOAO
Qui, Yulei	Beijing Astronomical Observatory
Ray, Alak	Tata Institute of Fundamental Research
Rees, Martin	Institute of Astronomy
Rhie, Sun	University of Notre Dame
Rhoads, James	Kitt Peak National Observatory
Ricker, George	MIT
Riess, Adam	University of California, Berkeley
Sahu, Kailash	Space Telescope Science Laboratory
Sandage, Allan	Carnegie Observatories
Schmidt, Brian	The Australian National University
Schreier, Ethan	Space Telescope Science Institute
Schwarzschild, Bert	Physics Today
Seitter, Waltraut	Muenster University
Smette, Alain	NASA/GSFC-NOAO
Stecker, Floyd W.	NASA/Goddard Space Flight Center
Tanvir, Nial	University of Hertfordshire
Theilemann, Friedrich	University of Basel
Turatto, Massimo	Osservatorio Astronomico di Padova
Urry, Meg	Space Telescope Science Laboratory
Wanatabe, Ken	USRA/LHEA; NASA/GSFC
Wang, Chih-Yueh	University of Virginia
Waxman, Eli	Department of Condensed Matter Physics

Weiler, Kurt	Naval Research Laboratory
Wheeler, Craig	University of Texas
Woosley, Stanford	University of California, Santa Cruz
Young, Timothy	University of Arizona

Preface

The Space Telescope Science Institute Symposium on "The Greatest Explosions Since the Big Bang" took place during 3–6 May 1999. An attempt was made to bring together for the first time researchers working on supernovae and on gamma-ray bursts. We strongly feel that a symbiosis between these two groups is obsolutely necessary for the understanding of these most energetic phenomena.

These proceedings respresent a part of the invited talks that were presented at the symposium. We thank the contributing authors for preparing their papers.

We thank Sharon Toolan of ST ScI for her help in preparing this volume for publication.

Mario Livio
Mino Panagia
Kailash Sahu
Space Telescope Science Institute
Baltimore, Maryland
May, 1999

Gamma-Ray Burst–Supernova relation

By BOHDAN PACZYŃSKI

Princeton University Observatory, Princeton, NJ 08544-1001; bp@astro.princeton.edu

There is growing evidence that long and hard gamma-ray bursts (GRBs), discovered at redshifts between 0.4 and 3.4, are related to some type of supernova (SN) explosions. The GRB ejecta are ultra-relativistic, and possibly beamed. There is a possibility that some SN ejecta are also beamed and/or relativistic. Prospects for further advances guided by expected and unexpected observational developments are very good. The prospects for developing a sound and quantitative GRB theory any time soon are rather modest, if histories of quasars, radio pulsars and supernovae are used for reference. However, the current progress in the understanding of GRB afterglows (which are relativistic) and remnants (which are non-relativistic) is likely to continue, as these appear to be simpler than the GRBs.

According to the current analysis of GRB 970508 the energy of gamma rays released by this event was about the same as the total energy of explosion. If correct, this result is difficult to reconcile with the internal shock models. It also implies that the global energy generation rate by GRBs is four orders of magnitude lower than the rate due to ordinary supernovae, which makes it very unlikely that the highly energetic supernova remnants were created by GRBs.

1. Introduction

The most dramatic recent breakthrough in our understanding of gamma-ray bursts (GRBs) was made by the *BeppoSAX* team, which discovered the first X-ray afterglow (Costa et al. 1997). That was quickly followed with the discovery of optical (van Paradijs et al. 1997) and radio (Frail et al. 1997) afterglows, and the determination of the first optical redshift (Metzger et al. 1997). By now about two dozen afterglows were detected, almost all within fraction of an arc second of very faint galaxies, with typical R-band magnitudes 24–26. Approximately ten redshift were measured. Gradually evidence emerged that GRBs appear to be associated with star forming regions (Paczyński 1998, Kulkarni et al. 1998, Galama et al. 1998). In several cases a direct association with a supernova (SN) appeared: GRB 980425–SN 1998bw (Galama et al. 1998), GRB 980326 (Bloom et al. 1999, Castro-Tirado & Gorosabel 1999), and GRB 970228 (Reichart 1999, Galama et al. 1999).

We should keep it in mind that all this exciting development is for the long duration GRBs, as these were the only type for which accurate coordinates became available within hours of the burst. The rest of this paper is about the long gamma-ray bursts only.

Until recently the most popular models of gamma-ray bursts (GRBs) were related to merging neutron stars, and neutron stars merging with stellar mass black holes. However, these would be located far away from star forming regions, and far away from parent dwarf galaxies. This does not seem to be the case for the location of GRB afterglows, and this is the reason why an association of bursts with explosions of massive stars became popular.

Throughout this paper I shall adopt popular assumptions and terminology. The bursts with strong high energy spectra require very large bulk Lorentz factors, $\Gamma > 300$, to reconcile their rapid variability with their huge luminosities and no evidence for spectral cut-off due to pair creation (Baring & Harding 1996). During its activity GRB's intensity varies rapidly. Several seconds or minutes after the beginning of the burst an afterglow becomes dominant, as recently shown by Burenin et al. (1999). The afterglows fade smoothly, usually as a broken power law of time, and they are almost certainly due to the interaction between the relativistic ejecta and ambient medium. Their emission is non-thermal, and thus it is fundamentally different form a thermal emission of a non-relativistic supernova. When the ejecta decelerate to non-relativistic expansion a GRB remnant is created, and at this stage it may resemble a supernova remnant.

2. Rates

I adopt Hubble constant $H_0 = 70$ km s^{-1} Mpc^{-1} throughout this paper.

According to Wijers et al. (1998) the energy generation rate due to GRBs is at present epoch (i.e. $z = 0$) equal

$$\epsilon_{\mathrm{GRB},0} \approx 10^{52} \text{ erg Gpc}^{-3} \text{ yr}^{-1} \;, \tag{2.1}$$

assuming that the GRB rate follows the star formation rate as a function of redshift. Note, that this number is independent of beaming of GRB emission. If there is beaming the energy per GRB is reduced, but the number of GRB explosions increases, so that the product, i.e. $\epsilon_{\mathrm{GRB},0}$ remains unchanged. Using a very different analysis Schmidt (1999) obtained the GRB energy generation rate about the same as Wijers et al. (1998).

The rate of all types of supernovae is approximately 1.5 per 10^{10} $L_{B,\odot}$ per century (van den Bergh & Tammann 1991). The mass density of the universe is probably $\Omega_m \approx 0.25$, and the average mass to blue light ratio is $M/L_B \approx 200$ M$_\odot$/L$_\odot$ (Bahcall et al. 1995). Therefore, the blue luminosity within one cubic gigaparsec is $\sim 1.6 \times 10^{17}$ L$_\odot$, and the local supernova rate is

$$n_{\mathrm{SN}} \approx 2.4 \times 10^5 \text{ Gpc}^{-3} \text{ yr}^{-1} \;. \tag{2.2}$$

Adopting 10^{51} erg of kinetic energy per supernova we obtain the overall energy generation rate (at $z = 0$)

$$\epsilon_{\mathrm{SN},0} \approx 2.4 \times 10^{56} \text{ erg Gpc}^{-3} \text{ yr}^{-1} \;. \tag{2.3}$$

It appears that global energy release rate is more than 4 orders of magnitude higher for supernovae than it is for gamma-ray bursts (Wijers et al. 1998, Schmidt 1999). Obviously, both rates are uncertain. It is possible that kinetic energy of GRB ejecta is considerably higher than their gamma ray output (Wijers et al. 1998 Kumar 1999). It is also possible that the actual supernova rate is much higher, as intrinsically faint explosions, like SN 1987A, are difficult to discover, yet they release about as much energy as ordinary SN Ia or SN II. While both, ϵ_{GRB} and ϵ_{SN}, may well be higher than the estimates given with the eqs. (2.1) and (2.3), it is likely that the ratio $\epsilon_{\mathrm{SN}}/\epsilon_{\mathrm{GRB}} \gg 1$. If this seemingly obvious conclusion is correct it has consequences for finding GRB remnants.

There is no generally accepted quantitative model of GRB emission at this time, and we may only guess what is the ratio of gamma-ray energy to kinetic energy of the ejecta. While it is common to think that this ratio is small (Wijers et al. 1998, Kumar 1999), it may just as well be much larger than unity, i.e. the kinetic energy may turn out to be much smaller than gamma-ray energy. This possibility follows from the recent analysis of the non-relativistic radio remnant of GRB 990508 by Frail, Waxman and Kulkarni (1999), who find that the total energy is only 5×10^{50} erg. At the same time Rhoads (1999b) finds that GRB 970508 was not strongly beamed, as its afterglow had unbroken power law decline for over 100 days. The total gamma-ray emission was at least 3×10^{50} erg for this burst (Rhoads 1999b). If these claims are correct then for this burst gamma-ray and kinetic energies were comparable, and this rules out the popular 'internal shock' models, which are very inefficient in generating gamma-rays (e.g. Kumar 1999).

Of course, GRB 970508 was not a typical gamma-ray burst. Its afterglow was the only one which first increased in luminosity for about 2 days, and later declined as unbroken power law for over 100 days. This is also the only event for which quantitative estimates were made for both: gamma-ray and kinetic energies. We have no direct information for the ratio of these two energy forms for any other burst.

3. GRB and SN remnants

The global energetics of supernovae and gamma-ray bursts has direct implications for the extra energetic supernova remnants. Recently, several suggestions were made that these may be remnants of gamma-ray burst explosions (Efremov et al. 1998, Loeb & Perna 1998, Wang 1999). However, if a typical GRB generates a factor f more energy than a typical supernova then the GRB rate must be lower than the supernova rate by a factor $10^4 f$, and correspondingly the number of GRB remnants must be vastly smaller than suggested by the number of very energetic remnants. Therefore, it is unlikely that the very energetic supernova remnants are related to gamma-ray bursts, unless GRBs generate vastly more kinetic energy than gamma-ray energy.

Let us suppose that the energetic remnants were caused by single explosions. We know that some rare supernovae are much more powerful than average. For example, SN 1998bw has released $\sim 20 \times 10^{51}$ erg (cf. Woosley et al. 1999, Iwamoto 1999, and references therein, but a much less energetic explosion has been proposed by Hoflich et al. 1999). It may well be that some stellar explosions are even more powerful than SN 1998bw. However, there is no obvious reason why the most powerful stellar explosions should be related to gamma-ray bursts. A classical GRB with a hard spectrum requires ejecta with the bulk Lorentz factor ~ 300, or more. Nobody knows how to generate outflow so highly ultra-relativistic, and it is not clear that the total energetics of the explosion has to be extraordinarily large, as a strongly beamed explosion may appear to be much more energetic that it really is. In other words: the ability to generate hard gamma-ray emission and the overall energetics of an explosion may be correlated with each other, or just as well the two may be uncorrelated. As long as we do not have a sound quantitative model there is no justification for either assumption; a semi-empirical approach may be more promising than theoretical speculations.

4. GRB and SN beaming

The possibility that highly relativistic GRB explosions may be jet-like was considered for a very long time and I do not know who was the first to make a suggestion. Some similarity between the GRBs and the blazars is so striking that a term 'micro-quasar' was suggested some years ago (Paczyński 1993). Similarities of these two classes of objects were recently analyzed by Dermer & Chiang (1998). If these are taken seriously a very strong beaming of GRBs follows, with a drastic reduction of the energetics compared to a spherical explosion. Recently, the breaks in the rate of decline of several afterglows were interpreted as evidence for beaming (Kulkarni et al. 1999, Stanek et al. 1999, Harrison et al. 1999). If GRB emission is confined to a very narrow beam they may not need much more energy than the 'standard' $\sim 10^{51}$ erg of an ordinary supernova. At this time there is no robust estimate of the degree and the possible range of GRB beaming (e.g. Rhoads 1997, 1999a,b).

More than a decade ago observations of a 'mystery spot' near SN 1987A were reported by Karovska et al. (1987) and Matcher et al. (1987). Piran & Nakamura (1987) suggested that this might have been a jet generated by the supernova. Not knowing about SN 1987A Cen (1998) suggested that supernovae may create relativistic jets, which may give rise to gamma-ray bursts. This idea gained some support when the new analysis of

SN 1987A data provided stronger evidence for the original 'mystery spot', and in addition provided evidence for a second spot on the opposite side of the supernova, suggesting relativistic jets (Nisenson & Papaliolios 1999). Evidence for a strong non-sphericity of SN 1998S was reported by Leonard et al. (1999). Hoflich et al. (1999) claim that the explosion of SN 1998bw was highly non-spherical.

Jets in supernovae became popular (e.g. Khokhlov et al. 1999, Cen 1999, Nagataki 1999), and often suggested to be associated with a beamed gamma-ray emission. A schematic picture may involve a quasi-spherical and non-relativistic supernova explosion with a narrow ultra-relativistic jet streaming along the rotation axis.

The possibility that some supernovae may generate jets is very interesting, and it should be possible to test it with the VLBA observations of very young radio supernovae. However, there is no reason to expect that all jets must generate gamma-ray bursts, as this would require all outflows to reach the huge Lorentz factor $\Gamma \sim 300$. It seems much more likely that there is a broad range of jet velocities, and only some are capable of GRB-like emission.

5. Hypernova

While the term 'hypernova' became popular recently, it was been sporadically used in the past (e.g. Wilkinson & Bruyn 1990). It does not have a clear, universally accepted meaning. The following are several examples.

1. Hypernova is just a name. The optical light curve of GRB 970508 was several hundred times brighter than any SN ever discovered. The absolute luminosity of several other afterglows, e.g. 990123, 971214, 990510, was higher by another factor ~ 100 (cf. Norris et al. 1999). So, rather than call it a super-super-nova, or a super-duper-nova, the term hypernova seems reasonable as a description of the phenomenon, with no implications for its nature.

2. Hypernova is a special type of a supernova explosion. At least some optical afterglows appear to be associated with star forming regions. Note that GRBs are many orders of magnitude less common than supernovae, and there may be an almost continuous transition from a typical massive SN to a typical GRB; the SN 1987A with its relativistic jet and GRB 980425–SN 1998bw may be examples of intermediate explosions. The link between the GRBs and the deaths of massive stars does not specify the mechanism for a GRB, and it is testable without a need for theoretical models. A question: 'are GRBs in star forming regions?' can be answered observationally. In this context a 'hypernova' is an explosion of a massive star, soon after its formation. Soon, means several million years, not a delayed explosion of the merging neutron star type.

3. Hypernova is a rotationally driven supernova. The idea that at least some supernovae explosions are driven by a rapid rotation of a compact core has been around for several decades (e.g. Ostriker & Gunn 1971). A qualitative reasoning proceeds as follows. A spherical collapse of a massive stellar core transforms $\sim 3 \times 10^{53}$ erg of gravitational energy into thermal energy of a hot neutron star, and 99.7% of that energy is lost in a powerful neutrino–anti-neutrino burst, with the remaining $\sim 10^{51}$ erg used to power a supernova explosion. If a pre-collapse core is rapidly rotating, than additional $\sim 3 \times 10^{53}$ erg may be stored in the rotation of the collapsed object. Some rotational energy is lost in gravitational radiation, but a large fraction cannot be readily disposed of. If an ultra strong magnetic field is generated by the differential rotation then it may act as the energy transmitter from the spinning relativistic object to the envelope, powering an explosion, perhaps in a form of a relativistic jet. The more rotation there is, the more jet-like explosion results, and the more relativistic the jet. This is just a speculation at

this time, recycled in dozens, perhaps hundreds of theoretical papers, with terms like a 'micro-quasar' (Paczyński 1993) or a 'failed supernova' (Woosley 1993) used at least as often as a 'hypernova' (Paczyński 1998).

6. Pessimistic conclusions

It is useful to put theoretical work on gamma-ray bursts in a broader perspective of other exotic objects and phenomena in order to asses the prospects for a short term progress.

There is almost universal agreement that GRB emission is non-thermal. Several important correlations were found for various GRB properties (e.g. Fenimore et al. 1995, Liang & Kargatis 1996, Beloborodov et al. 1998, Stern et al. 1999, Norris et al. 1999), but is not clear how to incorporate them in a theoretical model. This is not surprising. It is very difficult to prove which specific physical processes are responsible for the operation of a non-thermal source—consider current theories of quasars and radio pulsars. Well into the fourth decade of their development, and no serious ambiguity about the relevant distance scales, there are no generally accepted theories that account for either quasar or pulsar non-thermal emission. There is no reason why GRBs should be easier to understand.

For several decades there has been a consensus that Type Ia supernovae result from explosive carbon burning in white dwarfs close to the Chandrasekhar limit, while all other supernovae are related to core collapse of various massive stars. However, the detailed physics is so complicated that there is still no satisfactory and quantitative model that could describe the propagation of the nuclear burning front in SN Ia, without introducing free, adjustable parameters. There is also no agreement how $\sim 0.3\%$ of energy released in core collapse is channeled to drive the explosion of a SN II. As far as I can tell, if there were no observations of SN II it would be impossible to predict them from the first principles, even though hundreds of sophisticated papers were written on the subjects. The guidance provided by the observations of GRBs and their afterglows is less clear than it has been for supernovae. In my view there is no way to prove with theoretical models that either merging neutron stars or hypernova explosions should generate gamma-ray bursts. It is hard to believe that the puzzle of the central engine can be solved for GRBs more readily than for supernovae.

There is plenty of observational evidence that a huge diversity of rotating objects generates either bipolar outflows or jets—the phenomenon is obviously natural, as it appears so commonly in nature. Yet, there is no quantitative theory of the phenomenon that could explain (without ad hoc assumptions and ad hoc free parameters or free functions) what outflow velocities, or what rates of mass loss, should be associated with any particular object. The same applies to gamma-ray bursts and the current attempts to explain why their ejecta are likely to be beamed.

There is no theory that could predict the outflow velocity of any jet, but it seems natural to expect that only very specific conditions make it possible to reach the outflow with the Lorentz factor $\Gamma \sim 300$, as needed for HE bursts. There may be many more jets with more modest values of $\Gamma \sim 30$, 3, or non-relativistic at all. There is no direct evidence for a large Lorentz factor for the NHE bursts, which appear to have no photons above ~ 300 keV (Pendleton et al. 1997), and the pair creation argument does not apply to them. Perhaps the NHE GRBs are driven by non-relativistic explosions.

7. Optimistic conclusions

In spite of all theoretical problems there was a spectacular progress in our understanding of gamma-ray bursts. The statistics of GRB distribution obtained with BATSE on *Compton Gamma Ray Observatory* (Meegan et al. 1992, Paczyński 1995, and references therein) provided a very strong argument for a cosmological distance scale to the majority of GRBs. The obviously explosive nature of gamma-ray bursts provided the basis for the theoretical prediction of the afterglows as the products of interaction between GRB ejecta and ambient medium (Paczyński & Rhoads 1993, Katz 1994, Mészáros & Rees (1997). This prediction was confirmed with the discovery of afterglows with *BeppoSAX* (Costa et al. 1997), and soon provided the proof for the cosmological distance (Metzger et al. 1997). The observed distribution of the afterglows with respect to host galaxies indicated that GRBs are associated with star forming regions, and therefore with the explosions of massive stars, rather than with merging neutron stars (Paczyński 1998, Kulkarni et al. 1998, Galama et al. 1998). There is evidence that at least some bursts are directly associated with explosions of some supernovae (Galama et al. 1998, Bloom et al. 1999, Castro-Tirado & Gorosabel 1999, Reichart 1999, Galama et al. 1999).

There is every reason to expect more progress along similar lines: observations and their analysis providing more and more hints about the nature of the bursts. The following are some of the likely lines of progress in our understanding.

The new GRB instruments will provide hundreds of accurate positions within seconds of the burst's beginning, for long as well as for short bursts. We may expect that the distribution in distance will soon be known not only for the long HE bursts, but also for the NHE bursts and for the short bursts. It may well be that in several years some GRB will be the redshift record holder. If GRBs trace massive star formation rate, then they may become a new probe of the process in very dusty regions, or at very high redshifts.

While old GRB remnants may be difficult to distinguish from SN remnants, there is a possibility that a clear signature of the effect of non-thermal emission from a GRB and its afterglow may be detected in the interstellar medium (e.g. Perna et al. 1999, Draine 1999, Weth et al. 1999), and it may turn out to be a powerful new diagnostics for these events. The importance of the interstellar scintillation for the estimates of radio afterglow expansion has already proven to be an important research tool (Goodman 1997, Frail et al. 1997).

If GRBs are related to explosions of massive stars then we expect that circumstellar gas is left over from a strong stellar wind, as all massive stars appear to have winds. Currently there is mixed evidence from afterglow studies, with some events consistent with ambient gas density falling off as $1/r^2$, as expected of wind environment, while in some the ambient gas density appeared to be constant (Chevalier & Li 1999a,b). With many more afterglows followed with multi-band studies it will be possible to determine which environment is more common, and to make inferences about the nature of the exploding object.

At a cost much lower than any GRB space mission a super-super ROTSE or a super-super-LOTIS may be developed to follow up on the experience of ROTSE (Akerlof et al. 1999) and LOTIS (Williams et al. 1999). At a cost less than 10^6 it should be possible to implement an all sky optical monitoring system sensitive to optical flashes of ~ 1 minute duration, like the one discovered by ROTSE (Akerlof et al. 1999), detectable without any GRB trigger. There may be many more optical flashes than gamma-ray bursts if less extreme Lorentz factors are sufficient for generating optical flashes. Rather obviously, a major difficulty is not hardware but software.

We already know that some supernovae (SN 1998bw) eject some matter at a relativistic or sub-relativistic velocity (Waxman & Loeb 1999). There is a fairly strong case for a relativistic jet from the SN 1987A (Nisenson & Papaliolios 1999). We may expect (or at least hope) that other cases of relativistic motion will be discovered in other SN. For supernovae within ~ 100 Mpc it may be possible to detect anisotropy in their ejecta, perhaps even superluminal jets, using VLBA. If jets are detected in many cases it will be possible to study the distribution of jet velocities.

When the number of recorded supernova explosions will exceed 10^4 we shall know more about the high energy tail of their power distribution, and we may learn if there is a sharp maximum, or is there an extended tail, to the explosions in the 10^{53}–10^{54} erg range.

The ever more vigorous searches for distant (i.e. faint) supernovae will discover optical afterglows without a need for the GRB alert (Rhoads 1997). There may be a rich diversity of SN-like or afterglow-like events, perhaps even optical transients from merging neutron stars (Li & Paczyński 1998).

If the past can be used as a guide for the future than the most spectacular breakthroughs in the observations and understanding of gamma-ray bursts will be unexpected, just as the most recent *BeppoSAX* breakthrough was. An example may be the recent empirical finding of a very tight correlation between photon energy-dependent lags and peak luminosities of gamma-ray bursts (Norris et al. 1999).

REFERENCES

AKERLOF, C., ET AL. 1999, astro-ph/9903271.

BAHCALL, N. A., LUBIN, L. M., & DORMAN, V. 1995 *ApJ*, **447**, L81.

BARING, M. G. & HARDING, A. K. 1996 in *3rd Huntsville Symposium: Gamma-Ray Bursts* (eds. C. Kouveliotou, M. F. Briggs, & G. J. Fishman). AIP, p. 724.

BELOBORODOV, A. M., STERN, B., & SVENSSON, R. 1998 *ApJ*, **508**, L25.

BLOOM, J. S., ET AL. 1999, astro-ph/9905301.

BURENIN, R. A., ET AL. 1999, astro-ph/9902006.

CASTRO-TIRADO, A. & GOROSABEL, J. 1999, astro-ph/9906031.

CEN, R. 1998 *ApJ*, **507**, L131.

CEN, R. 1999, astro-ph/9904147.

CHEVALIER, R. A. & LI, ZHI-YUN 1999a *ApJ*, **520**, L29.

CHEVALIER, R. A. & LI, ZHI-YUN 1999b, astro-ph/9908272.

COSTA, E., ET AL. 1997 *Nature*, **387** 783.

DERMER, C. D. & CHIANG, J. 1998, astro-ph/9810222.

DRAINE, B. T. 1999, astro-ph/9907232.

EFREMOV, Y. N., ELMEGREEN, B. G., & HODGE, P. W. 1998 *ApJ* **501**, L163.

FENIMORE, E. E., ET AL. 1995, *ApJ*, **448**, L101.

FRAIL, D. A., ET AL. 1997 *Nature*, **389**, 261.

FRAIL, D. A., WAXMAN, E., & KULKARNI, S. R. 1999, in preparation.

GALAMA, T. J., ET AL. 1998 *Nature*, **395**, 670.

GALAMA, T. J., ET AL. 1999, astro-ph/9907264.

GOODMAN, J. 1997 *NewA*, **2**, 449.

HOFLICH, P., WHEELER, J. C., & WANG, L. 1999 *ApJ*, **521**, 179.

IWAMOTO, K. 1999 *ApJ*, **517**, L67.

KAROVSKA, M., NISENSON, P., & PAPALIOLIOS, C. 1987 *IAUC 4382*.

KATZ, J. I. 1994 *ApJ*, **422**, 248.

KHOKHLOV, A. M., ET AL. 1999, astro-ph/9904419.

KULKARNI, S. R., ET AL. 1998, astro-ph/9807001.

KULKARNI, S. R., ET AL. 1999, astro-ph/9902272.

KUMAR, P. 1999, astro-ph/9907096.

LEONARD, D. C., ET AL. 1999, astro-ph/9908040.

LI, L.-X. & PACZYŃSKI, B. 1998 *ApJ*, **507**, L59.

LIANG, E. & KARGATIS, V. 1996 *Nature*, **381**, 48.

LOEB, A. & PERNA, R. 1998 *ApJ*, **503**, L35.

MATCHER, S. J., MEIKLE, W. P. S., & MORGAN, B. L. 1987 *IAUC 4391*.

MEEGAN, C. A., ET AL. 1992 *Nature*, **355**, 143.

METZGER, M. R., ET AL. 1997 *Nature*, **387**, 878.

MÉSZÁROS, P. & REES, M. J. 1997 *ApJ*, **476**, 232.

NAGATAKI, S. 1999, astro-ph/9907109.

NISENSON, P., & PAPALIOLIOS, C. 1999 *ApJ*, **518**, L29.

NORRIS, J. P., MARANI, G. F., & BONNELL, J. T. 1999, astro-ph/9903233.

OSTRIKER, J. P. & GUNN, J. E. 1971 *ApJ*, **164**, L95.

PACZYŃSKI, B. 1993, in *Relativistic Astrophysics and Particle Cosmology*, (eds. C. W. Akerlof & M. A. Srednicki), p. 321. *Ann. NY Acad. Sci.*, **Vol. 688**.

PACZYŃSKI, B. 1995 *PASP*, **107**, 1167.

PACZYŃSKI, B. 1998 *ApJ*, **494**, L45.

PACZYŃSKI, B. & RHOADS, J. E. 1993 *ApJ*, **418**, L5.

PENDLETON, G. N., ET AL. 1997 *ApJ*, **489**, 175.

PERNA, R., RAYMOND, J., & LOEB, A. 1999, astro-ph/9904181.

PIRAN, T. & NAKAMURA, T. 1987 *Nature*, **330**, 28.

REICHART, D. E. 1999 *ApJ*, **521**, L111.

RHOADS, J. E. 1997 *ApJ*, **487**, L1.

RHOADS, J. E. 1999a, astro-ph/9903399.

RHOADS, J. E. 1999b, astro-ph/9903400.

SCHMIDT, M. 1999, astro-ph/9908206.

STANEK, K. Z., ET AL. 1999 *ApJ*, **522**, L39.

STERN, B., POUTANEN, J., & SVENSSON, R. 1999 *ApJ*, **510**, 312.

VAN DEN BERGH, S. & TAMMAN, G. A. 1991 *ARA&A*, **29**, 363.

VAN PARADIJS, J., ET AL. 1997 *Nature*, **386**, 686.

WANG, Q. D. 1999 *ApJ*, **517**, L27.

WAXMAN, E. & LOEB, A. 1999 *ApJ*, **515**, 721.

WETH, C., MÉSZÁROS, P., KALLMAN, T., & REES, M. J. 1999, astro-ph/9908243.

WIJERS, R. A. M. J., BLOOM, J. S., BAGLA, J. S., & NATARAJAN, P. 1998 *MNRAS*, **294**, L13.

WILKINSON, P. N. & BRUYN, A. G. 1990 *MNRAS*, **242**, 529.

WILLIAMS, G. G., ET AL. 1999, astro-ph/9902190.

WOOSLEY, S. E. 1993, *ApJ*, **405**, 273.

WOOSLEY, S. E., EASTMAN, R. G., & SCHMIDT, B. P. 1999 *ApJ*, **516**, 788.

Observations of Gamma-Ray Bursts

By GERALD J. FISHMAN

Space Science Department, Code SD-50 NASA-Marshall Space Flight Center Huntsville, AL 35812

Gamma-ray bursts (GRBs) are now recognized as the most luminous known objects in the Universe. Their brief, random appearance in the gamma-ray region had made their study difficult since their discovery, over thirty-five years ago. The recent discoveries of GRB afterglow radiation in other wavelengths and observations of their faint host galaxies have provided the long-sought breakthrough in the direct determination of their distance and luminosity scales.

The observed time profiles of GRBs are very diverse and their durations cover a wide range. Their general spectral characteristics are summarized, primarily from data obtained from the BATSE experiment on the *Compton Gamma-Ray Observatory*. With over 2500 GRBs now observed, these temporal and spectral signatures, as well as population studies and global properties of GRBs, can now be described with greater accuracy than previously possible.

1. Introduction

This paper describes some of the distinguishing features of GRBs, with particular emphasis on recent observations with the BATSE experiment on the *Compton Gamma-Ray Observatory* (Figure 1). BATSE has been in operation since its launch by the Space Shuttle Atlantis in April 1991 and it is planned to continue operation for at least another five years.

About once per day, a burst of gamma-rays appears from a random direction on the sky. Often, the burst outshines all other sources of gamma-rays in the sky, combined. The serendipitous discovery of GRBs resulted from the cold-war space program of the 1960s. The *Vela* series of spacecraft were developed by the Los Alamos National Laboratory to detect clandestine nuclear explosions above the atmosphere. The orbits of these spacecraft were greatly elongated in order to simultaneously view a large fraction of the earth. Occasional bursts of gamma-rays were seen coming from random directions in space. It was several years later that the observed GRBs were determined to be of cosmic origin (Klebesadel, Strong, & Olson 1973). In the first years after their discovery, GRBs were observed at the rate of about 10 to 15 per year.

Now, with more complete coverage and larger detectors, GRBs are detected at the rate of over 300 per year. At this sensitivity, the all-sky burst rate is about 800 per year, corrected for sky exposure.

The field of gamma-ray bursts has undergone a dramatic change since BATSE. This has resulted primarily from more sensitive observations of the gamma-ray burst intensity and sky distributions (Meegan et al. 1992). Prior to these observations, the source of gamma-ray bursts were considered by most workers in the field to be relatively nearby neutron stars in the Galactic plane (cf. Higdon & Lingenfelter 1990). Another revolution in GRB research is now underway, made possible by the rapid, precise locations from the *BeppoSAX* spacecraft (Piro et al. 1999).

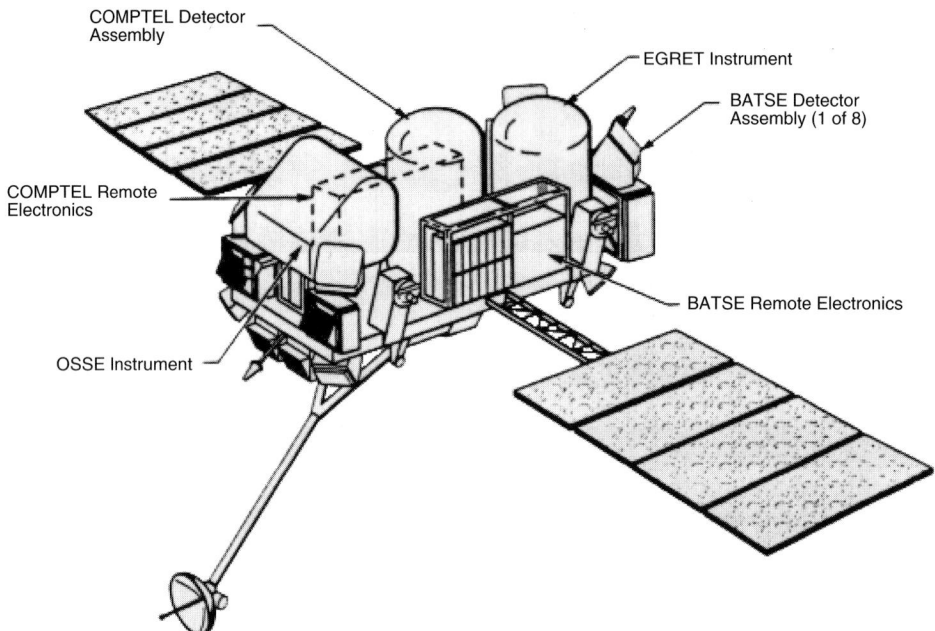

FIGURE 1. The *Compton Gamma-Ray Observatory*, showing the placement of the BATSE detector modules in the corners.

2. Temporal characteristics

The most striking feature of the time profiles of gamma-ray bursts is the diversity of their time structures and the wide range of their durations. Coupled with this diversity is the difficulty of placing many GRBs into well-defined types, based on their time profiles. Many bursts have multiple characteristics and many other bursts are too weak to classify. Some burst profiles are chaotic and spiky with large fluctuations on all time scales (Figure 2), while others show rather simple structures with few peaks (Figure 3). No periodic structures have been seen from gamma-ray bursts. There is, however, one general characteristic of the profiles: at higher energies, the overall burst durations are shorter and sub-pulses within a burst tend to have shorter rise-times and fall-times (sharper spikes). Most bursts also show an asymmetry, with the leading edges being of shorter duration than trailing edges. This feature is also evident when pulses from large numbers of GRBs are superposed (Mitrofanov et al. 1998). There are many bursts which have similar shaped sub-pulses within the burst. Numerous examples of GRB characteristics may be found in the latest BATSE GRB catalog (Paciesas et al. 1999). These characteristics are also summarized in a recent review (Fishman & Meegan 1995).

The wide diversity of GRB profiles has been known since the earliest observations. While several burst morphologies (e.g. smooth or spiky) are easy to identify, there are numerous gradations of these, as well as many complex forms. Examples of GRBs with well-defined, separated episodes of emission are also seen (Figure 4) but not accounted for by most emission models. Attempts to quantify these structures have been largely unsuccessful.

The range of the duration of gamma-ray bursts spans over five decades, from a few milliseconds to over a thousand seconds. The double peaked distribution of the duration, first noted many years ago (cf. Kouveliotou et al. 1993), is now much more evident with

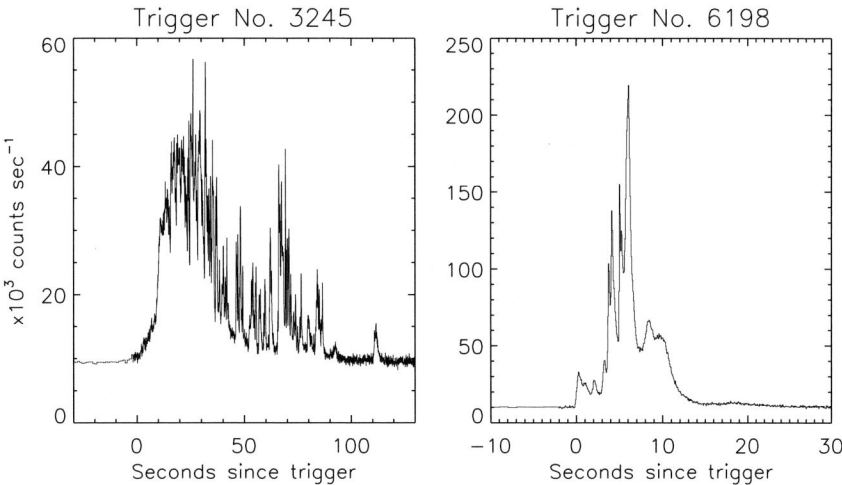

FIGURE 2. Gamma-Ray Bursts with complex pulse structures.

the large number of bursts now available for study. These two peaks in the duration distribution occur at ~ 2 s and ~ 34 s.

3. Spectral characteristics

Most of the power from gamma-ray bursts is emitted above 50 keV. In general, bursts have a rather simple, smooth continuum spectrum. The high-energy photon and νF_ν spectra of GRB990123 are shown in Figure 5, from data taken from *Compton Observatory* experiments covering a broad energy range (Briggs 1999a). The general spectral form is well-described by what has become known as the Band function (Band et al. 1993) both in the time-integrated spectra and within shorter intervals within bursts. The low-energy characteristics (below 20 keV) of GRBs have been examined in detail in several recent papers (Preece et al. 1998, Strohmayer et al. 1998). This is a crucial region to test emission models, such as the synchrotron shock model and the thick Comptonization model. Joint fits of spectral data over a wide energy range from *BeppoSAX* and BATSE data should become available in the near future.

The high-energy spectra of GRBs that are seen to extend above 10 MeV are best fit by a power-law with a spectral index of ~ 2, but the number of GRBs for which these data are available are limited (cf. Schaefer et al. 1998). Spectra within the BATSE range from 20 keV to ~ 2 MeV have been fit to numerous GRBs (Mallozzi et al. 1995). In these spectra, both the low and high-energy ends of the broad spectral distribution are fit by a power-law within the BATSE bandpass. The energy at which the νF_ν spectrum is a maximum is referred to as the E_p energy. The E_p distribution for a large number of GRBs is shown in Figure 6. There are many GRB spectra that are seen to cut-off much more steeply and do not have detectable emission above 300 keV (Pendleton et al. 1997). On the other hand, about a half-dozen strong GRBs have been seen by the EGRET experiment on the Compton Observatory with emission above 50 MeV, extending to 18 GeV (Hurley et al. 1994). These high-energy photons are seen long after the lower energy emission is observed; they also contain a significant fraction of the total energy of the GRB.

Spectral softening is usually (but not always) seen throughout a GRB and also in sub-pulses within a burst. This is illustrated in Figure 8, where a sub-peak is seen to have

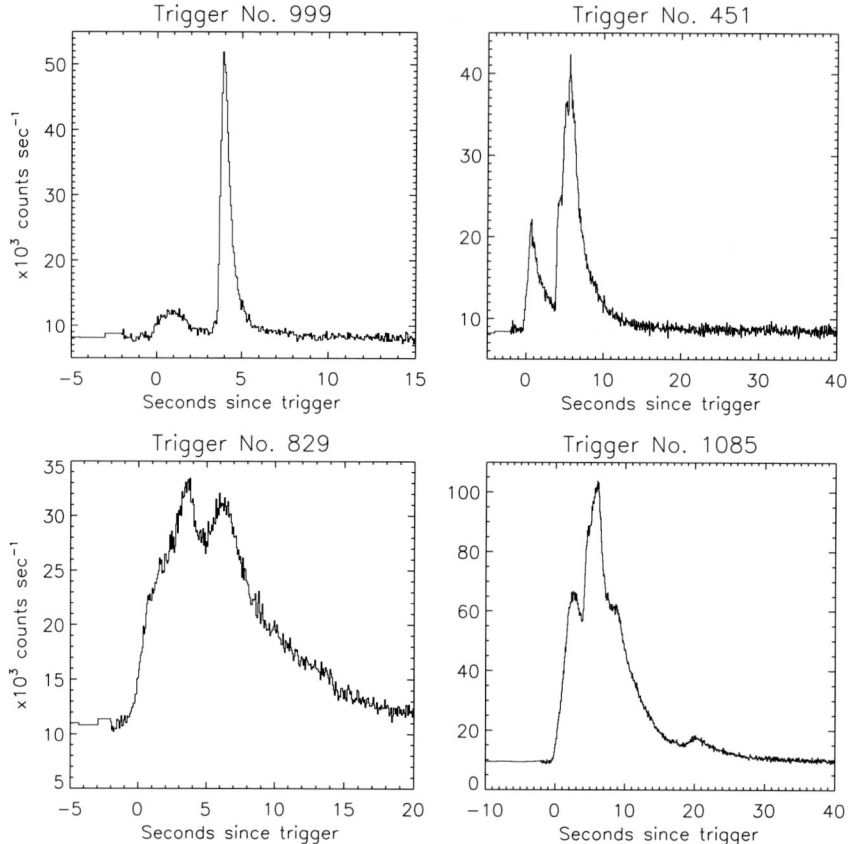

FIGURE 3. Examples of gamma-ray bursts which show very little temporal structure, as opposed to those shown in Figure 2.

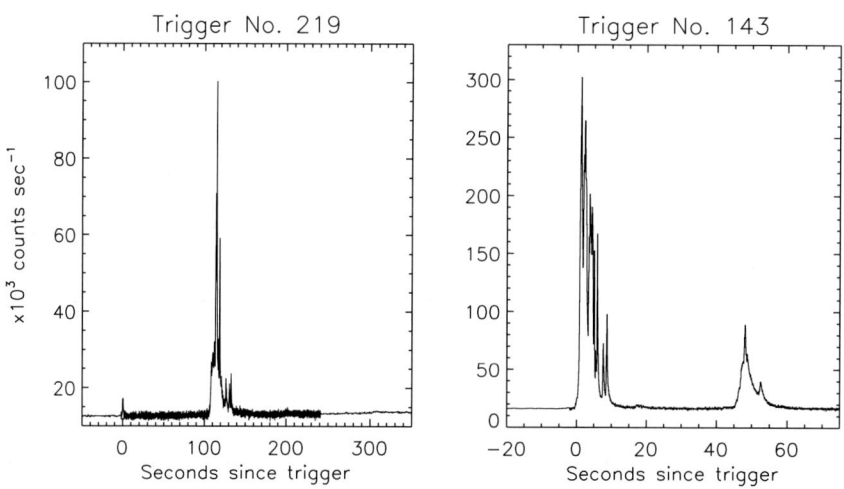

FIGURE 4. Two examples of GRBs with well-separated pulses.

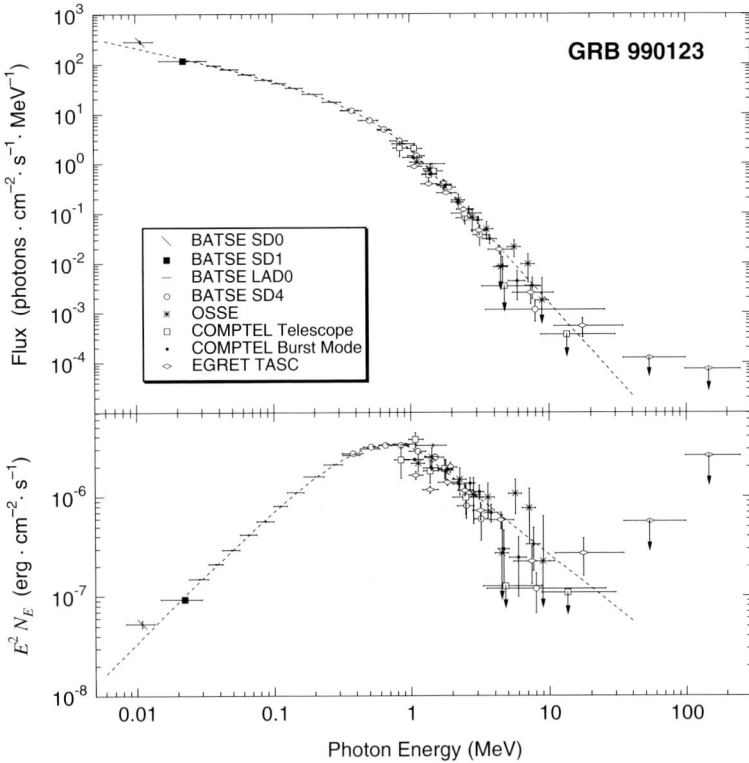

FIGURE 5. Broad band high-energy spectrum of GRB990123 (taken from Briggs 1999a).

soft emission (< 300 keV), for an extended duration, ∼ 8 s, following the hard peak, which only lasts for ∼ 2 s. In this particular case, the peaks of both the hard and soft emission occur simultaneously. However, in many cases the peaks are offset, with the softer peak delayed by up to a few seconds. Spectral evolution within GRBs is another GRB property that must be considered in burst emission models but is usually not, in contrast to the well-modeled spectral evolution of the GRB afterglow emission.

The reported observation of spectral line features in gamma-ray bursts, and their interpretation as cyclotron lines produced in the intense magnetic fields of neutron stars (cf. Harding 1991) had been a primary reason for the early association of gamma-ray bursts with galactic neutron stars. A search for line features (either absorption or emission features) with the BATSE detectors onboard the Compton Observatory has thus far been unable to confirm the earlier reports of spectral line features from gamma-ray bursts (Briggs 1999b).

4. Intensity distributions and sky distributions

Since the launch of the Compton Observatory, burst locations have been available for a large sample of weak bursts, which has never before been possible. BATSE determines directions to burst sources by comparing the count rates of individual detectors, whose response varies approximately as the cosine of the angle from the detector normal. The systematic error of these locations is presently about two degrees for the stronger bursts.

If the sources were distributed homogeneously in Euclidean space, i.e., the density and luminosity function are independent of position throughout the volume of space

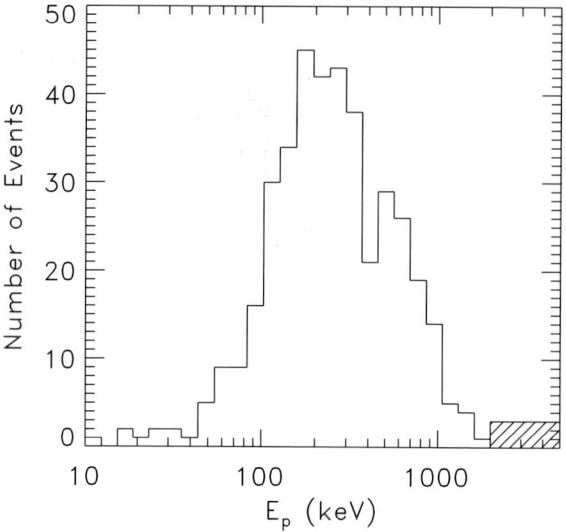

FIGURE 6. The BATSE E_p distribution (Mallozzi et al. 1995).

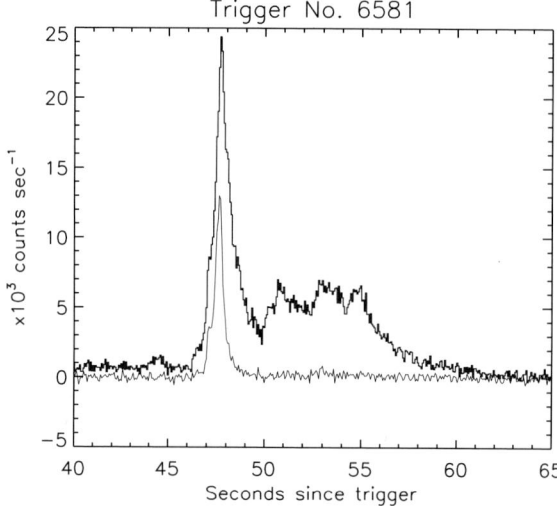

FIGURE 7. A GRB showing the common hard-to-soft spectral evolution.

observed, then the integral intensity distribution will be $N(>P) = P^{-3/2}$, where P is the observed peak intensity of the burst and N is the number of bursts observed with a peak intensity above the value P. Observations from BATSE and other spacecraft experiments show that there is a significant deviation of the observed intensity distribution from this $-3/2$ power-law. This clearly indicates an inhomogeneity in the region of space being sampled. Figure 8 shows a recent sky distribution of 2483 GRBs observed with BATSE. The statistics are now such that the effect of earth-blockage is becoming apparent on the map. This shows as a slight deficiency along the celestial equator. Corrections for this blockage yield a map with a distribution consistent with isotropy.

8.3 Years of BATSE Observations: 2483 GRBs

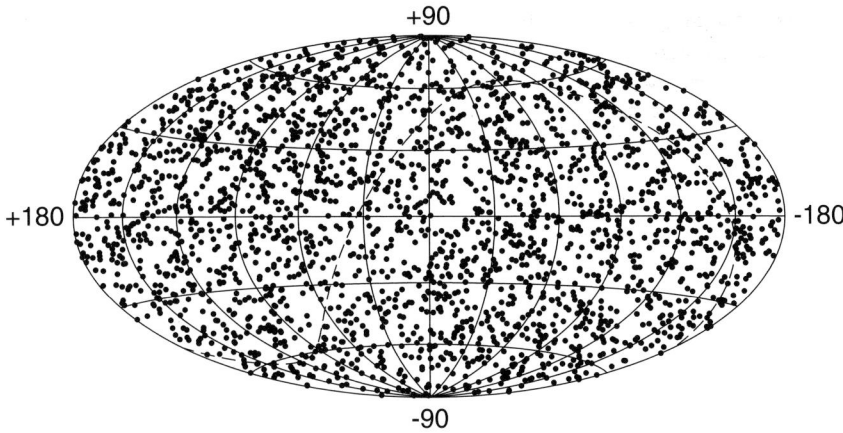

FIGURE 8. The sky distribution of 2483 gamma-ray bursts.

5. Rapid follow-up observations of GRBs

Ever since the initial discovery of GRBs, there has been a quest to discover a counterpart to a gamma-ray burst in any other wavelength region before, during, or after the gamma-ray event. These searches have taken many forms, including searches for statistical associations of known objects with bursts with poorly known locations as well as searches of archival plates and other data bases for transient or unusual objects within the error boxes of well-determined burst locations. Attempts have also been made to obtain a chance observation of a counterpart with wide-field patrol cameras. There have been several claims of candidate objects as counterparts, but these have generally been discounted because of low statistical significance, controversial instrumental effects or the spurious, one-time occurrence of a claimed counterpart which could not be independently confirmed by a second instrument.

The study of gamma-ray bursts has been revolutionized by the discovery of extremely well localized X-ray, optical, and radio transients (Costa et al. 1997; van Paradijs et al. 1997; Frail et al. 1997), made possible by precise, rapid locations of GRBs from the Italian-Dutch satellite, *BeppoSAX*. Follow-up of the optical transients (OTs) has shown that GRBs come from cosmological distances (Metzger et al. 1997). The host galaxies that contain the x-ray and optical transient have generally been between 22nd and 25th magnitude in the visible for those cases in which a host galaxy has been detected, which includes a majority of the well-localized bursts. The corresponding redshifts of the host galaxies are measures to be in the range $z \sim 0.4\text{--}2$.

The combination of the BATSE quick-alert system (Barthelmy 1998, Kippen et al. 1998) with the more precise follow-up location of *BeppoSAX*, allowed for the first simultaneous optical observation of a burst with the gamma-ray emission to be made (Akerlof et al. 1999). The intense gamma-ray burst, GRB990123, was viewed by a robotic set of CCD cameras, each only ~ 10 cm in diameter. The optical emission reached a relatively bright 9th magnitude during the burst, but not coincident with the peak of the gamma-ray emission. Subsequent observations by this camera and other larger telescopes, clearly showed a decay over several decades in time after the outburst. These results open the prospect for many observations of bursts very quickly after their occurrence.

6. Conclusions

In the past few years, gamma-ray burst research has undergone a tremendous revitalization, prompted by new observations. Although many possible models still exist, the boundary conditions of these models have been narrowed considerably. There has be a resurgence in theoretical studies and modeling of GRBs, as evidenced by the recent scientific literature and at conferences such as this one.

Gamma-ray bursts will allow us to extreme study physical conditions that are not encountered in any other known environments. Also, since they are the most luminous objects known in the Universe, they will allow more distant observations of early star formation and cosmology in regions unattainable by other means.

The author is grateful to T. Giblin for his help in preparing figures and the manuscript.

REFERENCES

AKERLOF, K., ET AL. 1999 *Nature*, **398**, 400.

BAND, D., MATTESON, J., FORD, L., SCHAEFER, B., PALMER, D., ET AL. 1993 *ApJ*, **413**, 281.

BARTHELMY, S., ET AL. 1998 in *Gamma-Ray Bursts*, AIP Conference Proceedings 428 (eds. C. A. Meegan, R. D. Preece, T. M. Koshut). p. 139.

BRIGGS, M. S., PENDLETON, G., KIPPEN, M., ET AL. 1999 *ApJ*, **524**, 82.

BRIGGS, M. S. 1999, to appear in *Gamma-Ray Bursts: The First Three Minutes*, ASP Conference Series 190, (eds. J. Poutanen and R. Svensson). (astro-ph 9910362).

COSTA, A., ET AL. 1997 *Nature*, **387**, 783.

FISHMAN, G. & MEEGAN, C. 1995 *Ann. Rev. Astron. Astrophys.*, **33**, 415.

FRAIL, D., ET AL. 1997 *Nature*, **389**, 261.

HARDING, A. K. 1991 *Phys. Reports*, **206**, 327.

HIGDON, J. & LINGENFELTER, R. 1990 *Ann. Rev. Astron. & Astrophys.*, **28**, 401.

HURLEY, K., DINGUS, B., MUKHERJEE, R., KOUVELIOTOU, C., MEEGAN, C., ET AL. 1994 *Nature*, **372**, 652.

KIPPEN, R. M., ET AL. 1998, in *Gamma-Ray Bursts*, AIP Conference Proceedings 428, (eds. C. A. Meegan, R. D. Preece, T. M. Koshut). p. 119.

KLEBESADEL, R., STRONG, I., & OLSON, R. 1973 *ApJ*, **182**, L85.

KOUVELIOTOU, C., MEEGAN, C., FISHMAN, G., ET AL. 1993 *ApJ*, **413**, L101.

MALLOZZI, R., PACIESAS, W. S., PENDLETON, G. N., BRIGGS, M. S., PREECE, R. D., ET AL. 1995 *ApJ*, **454**, 597.

MEEGAN, C., FISHMAN, G., WILSON, R., PACIESAS, W., PENDLETON, G., ET AL. 1992 *Nature*, **355**, 143.

METZGER, M., ET AL. 1997 *Nature*, **387**, 878.

MITROFANOV, I., ET AL, 1998 *ApJ*, **504**, 925.

PACIESAS, W., ET AL. 1999 *ApJS*, **122**, 465.

PENDLETON, G., ET AL. 1999 *ApJ*, **489**, 175.

PIRO, L., ET AL. 1999 *ApJ*, **514**, L73.

PREECE, R., BRIGGS, M., MALLOZZI, M., ET AL. 1998 *ApJ*, **506**, L23.

SCHAEFER, B., ET AL. 1998 *ApJ*, **492**, 696.

STROHMAYER, T., FENIMORE, E., MURIKAMI, T., & YOSHIDA, A. 1998 *ApJ*, **500**, 873.

VAN PARADIJS, J., ET AL. 1996 *Nature*, **466**, 768.

Fireballs

By TSVI PIRAN

Racah Institute for Physics, The Hebrew University, Jerusalem, 91904, Israel†
Physics Department, New York University, New York, NY 10003
Physics Department, Columbia University, New York, NY 10027

I compare the predictions of the Relativistic Fireball Model (specifically the internal-external shocks model) with recent observations of Gamma-Ray Bursts (GRBs) and their afterglows. These observations have demonstrated directly that the afterglow expands relativistically. The prompt optical emission seen in GRB 990123 has shown that the early stages of the GRBs and the afterglow involve ultra-relativistic motion, with Lorentz factors exceeding 100. The breaks in the light curves of GRB 990123 and GRB 990510 and the peculiar light curves of GRB 980519 and GRB 980326 disclose that these GRBs are beamed. I examine these recent developments and discuss their implications for the models of the source. I argue that even though the observations reflect only the conditions at the outer regions far from the central engine, they already imply that GRBs signal the birth of stellar mass black holes. I also discuss the "energy crisis" and show that various puzzles related to the estimated very large energy seen in GRBs can be explained by a "hot spots" model.

1. Introduction

Gamma-Ray Bursts (GRBs) are short and intense bursts of MeV range γ-rays. During the last decade observational progress has revolutionized our understanding of GRBs. BATSE on Compton-GRO have found that GRBs are distributed isotropically, revealing their cosmological origin. More recently, BeppoSAX discovered X-ray afterglow (Costa et al. 1997). This enabled accurate position determination and the discovery of optical (van Paradijs et al. 1997) and radio (Frail et al. 1997) afterglows and host galaxies. Remarkably, the afterglow is a simple phenomenon that can be analyzed using a rather simple model. The resulting information tells us a lot about the properties of GRBs.

In this talk I confront the predictions of the Relativistic Fireball model with recent observations. This model recently had several successes in predicting the afterglow, the prompt optical flash and the beaming break in the light curve. The fireball model deals with the "outer" radiating regions. It doesn't deal directly with the "inner engine"—the source of the relativistic ejecta that powers the whole phenomenon. I review the implications of the current observations as interpreted within the fireball model for models of the source, and I discuss some of the possible sources.

2. The relativistic internal-external shocks fireball model

The nonthermal spectrum of practically all GRBs indicates that the observed emission emerges from an optically thin region. However, a simple estimate of the size of the source (using $c\delta t$ as implied by the observed variability) shows the source must be rather small. Using a cosmological distance and the observed flux, we obtain a simple estimate of the number density of photons above 500 keV and find that the source must be extremely optically thick to pair creation (Piran & Shemi 1993; Fenimore, Epstein & Ho, 1993; Woods & Loeb 1995; Piran 1997). Such a source cannot emit nonthermal emission. This is the Compactness problem.

† Permanent Address

The simplest way to overcome this problem is to say that the source is moving ultra-relativistically towards us (Ruderman 1975; Goodman 1986; Krolik & Pier, 1989). This has lead to the relativistic fireball model. According to this model a compact source produces a relativistic wind. The observed GRBs and their afterglow are produced as this relativistic ejecta is slowed down by relativistic shocks. Relativistic electrons accelerated in these relativistic shocks emit the observed gamma-rays via synchrotron or synchrotron self-Compton emission. Both the energy density of these electrons and of the magnetic fields should be close to equipartition for efficient emission. To bypass causality and the compactness limits the shocks must be extremely relativistic with $\gamma \geq 100$.

This suggests that GRBs involve three stages:

- First a source produces a relativistic energy flow. The observed fluctuations in the GRB light curves and the huge energy released indicates that the source is compact. This "inner engine" is hidden and it is not observed directly. This makes it difficult to constrain GRB models and leaves only circumstantial evidence on the nature of the sources.

- The energy is transferred relativistically from the compact source to distances larger than $\sim 10^{13}$ cm, where the system is optically thin. This is most likely in the form of a relativistic particle flow, but the possibility of a Poynting Flux should also be considered. A relativistic hydrodynamics flow can be generated and accelerated easily by deposition of a large amount of energy in a small region with a low baryonic load. This is the origin of the fireball model (Goodman, 1986; Paczyński 1986). As I discuss shortly, the flow must be highly irregular to produce internal shocks.

- The relativistic ejecta is slowed down and the shocks that form convert the kinetic energy into internal energy of accelerated particles, which in turn emit the observed gamma-rays.

Paczyński and Rhoads (1993) and (independently) Katz (1994) early on realized that the long term interaction of the relativistic ejecta with the surrounding matter will produce a low frequency afterglow. Paczyński and Rhoads (1993) discuss the radio afterglow, and Katz (1994) discusses the optical afterglow. Later, Mészáros and Rees (1997a) and (independently) Vietri (1997) developed detailed models of this afterglow. In all these studies, it was thought that both the GRB and the afterglow are produced by external shocks. This afterglow would have been the long time extrapolation of the GRB and it should scale with the GRB properties. Alternatively, Sari and Piran (1997a), from the fact that internal shocks can extract only a fraction of the total energy (Mochkovitch et al. 1995; Kobayashi et al. 1997; Daigne & Mochkovitch, 1998), inferred that the remaining energy will be extracted later via external shock giving rise to additional emission at different wavelengths that will follow the bursts. Thus, GRBs have a fourth stage:

- The relativistic flow, which has been slowed down but has not been stopped, is slowed further by the surrounding material producing the afterglow. This phase is regular and can be modelled rather well with the adiabatic Blandford & McKee (1976) adiabatic self-similar solution.

The overall internal-external model is summarized in Fig. 1.

External shocks arise due to the interaction of the relativistic matter either with the surrounding matter (Mészáros, & Rees, 1992), like the ISM, or with a circumstellar wind that took place during an earlier epoch. These shocks are the relativistic analogues of SNRs. Like in SNRs, these shocks are collisionless. External shocks become effective at $\min(R_\gamma, R_\Delta)$, where $R_\gamma \equiv l/\gamma_0^{2/3}$ and $R_\Delta \equiv l^{3/4}\Delta^{1/4}$. γ and Δ are the Lorentz factor and the width of the shell (in the observer frame) and the subscript 0 indicates, here and elsewhere, the initial value (Sari & Piran, 1995). The Sedov length,

The Internal-External Scenario

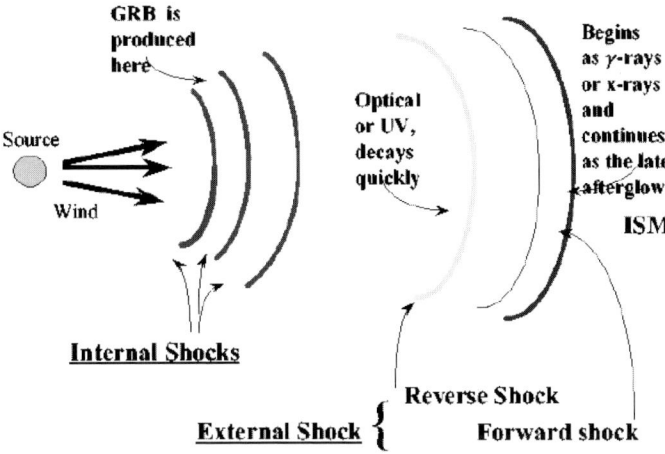

FIGURE 1. The internal-external shocks model: from the source to the GRB.

$l \equiv \left(E_0/[(4\pi/3)n_{ism}m_p c^2]\right)^{1/3}$, is the radius within which the rest mass energy of the external material, whose density is n_{ism}, equals the initial energy of the ejecta, E_0. Typical values are: $l \sim 10^{18}$cm and $R_\gamma \sim R_\Delta \sim 10^{15}$–$10^{16}$ cm.

Sari and Piran (1997) and Fenimore, Madras and Nayakshin (1996) have shown that external shocks cannot produce efficiently the observed highly variable temporal structure seen in GRB.† Thus, GRBs are produced by the only other alternative, internal shocks (Narayan, Paczyński, & Piran, 1992; Paczyński & Xu, 1993; Rees & Mészáros, 1994). These shocks arise in an irregular flow when faster shells overtake slower ones (see Fig. 2). If the flow varies on a scale length $\delta\ell$ then internal shocks would take place at $R_{\rm int} \sim \delta\ell\gamma_0^2$. The length $\delta\ell = c\delta t$ can be inferred from the observed temporal variability $\delta t \leq 1$ sec, indicating that these shocks take place at $\sim 10^{13}$ cm. If the shocks arise earlier the system is still optically thick and the radiation does not escape. The observed GRB time scales reflect the time scales of the "inner engine" (Kobayashi et al. 1997). The GRB duration corresponds to the time during which the "inner engine" is active.

Recently Kumar (1999) has shown that the efficiency of γ-ray generation via internal shocks is only a few percent. This efficiency is the product of three factors. The first is the efficiency of conversion of kinetic energy into thermal energy by the relativistic shocks. This factor is the same as calculated earlier (Mochkovitch et al. 1995; Kobayashi et al. 1997; Daigne & Mochkovitch, 1998). The two other factors involve radiative efficiency and early neutrino losses within the fireball. It should be stressed that when comparing external and internal shocks, only the first factor should be compared, because the second and third are of the same order in both models. Hence it is simply wrong to compare, as some have done, the overall internal shocks efficiency calculated by Kumar to the efficiency of conversion of kinetic energy to thermal energy in external shocks.

Within the external shocks model, late subpulses, which are produced when the Lorentz factor is lower and at larger radii, are expected to last longer. The lack of any correlation

† Dermer and Mitman (1999) suggest that within external shocks with a very inhomogeneous medium (see Fig. 3), subpulses close to the line of sight could produce the observed peaks. But these subpulses cannot produce the large amplitude variability observed. Additionally, there is no explanation why the expected correlation between arrival time and subpulses duration is not seen.

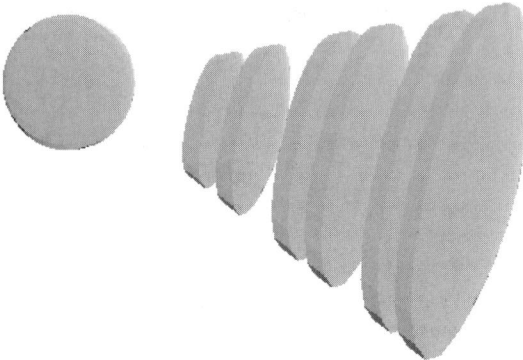

FIGURE 2. Internal shocks: Faster shells catch up with slower ones and collide, converting some of their kinetic energy to internal energy.

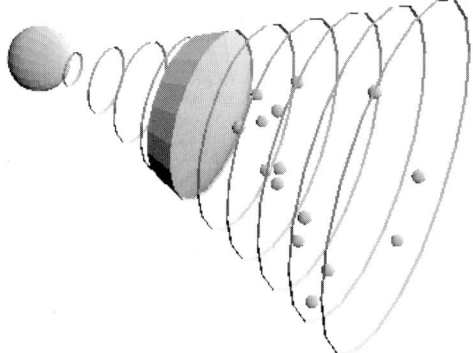

FIGURE 3. To produce variability in external shocks the relativistic ejecta must encounters interstellar bubbles. However, most of the shell passes between the bubbles and its kinetic energy is lost.

between the width of subpulses and their time of arrival is inconsistent with this model (Fenimore et al. 1998). The lack of a direct scaling between the GRB and the afterglow is another evidence for the internal shocks model. This is a very important clue to the nature of the source. External shocks can be produced by an explosive event in which the "inner engine" releases all of its energy at once. The observed temporal structure could have been produced in this case within the shocks due to irregularities in the surrounding material (see Fig. 3), but this process would have been extremely inefficient (Sari & Piran, 1997a). This rules out the possibility of an explosive "inner engine."

If GRBs arise as an effect of internal shocks and the afterglow is produced by external shocks (Sari & Piran, 1997a) we don't expect direct scaling between the two. In this "Internal-External" model the GRB and the afterglow are produced by two different processes. The recent afterglow observations provide a significant evidence for this picture.

3. Afterglow observations

The Italian-Dutch satellite, BeppoSAX, discovered an X-ray afterglow on February 28th, 1997 (Costa et al. 1997). By now X-ray afterglows have been observed from two dozen bursts. As BeppoSAX can trigger only on long bursts, we do not know if short bursts are also followed by afterglow. Optical (van Paradijs et al. 1997) and radio (Frail et al. 1997) afterglows were discovered using the accurate position obtained by BeppoSAX in about half of the cases in which an X-ray afterglow was seen. It is not clear what determines whether optical and radio afterglow are observed.

- The energy involved can be estimated from the late phases of the optical (Galama et al. 1998a; Granot et al. 1999a; Wijers & Galama, 1999; Vreeswijk et al. 1999) and the radio (Waxman 1999) afterglows. The overall energy emitted in the afterglow is of order 10^{50}–10^{52}. Quite generally it is only a fraction of the energy emitted during the GRB.
- The afterglow light curve decays, in most cases, as a single power law in time:† $F_\nu \propto t^{-\alpha}$, with $\alpha \sim 1.2$. In two cases (GRB 980326 and GRB 980519) $\alpha \sim 2$, and in two cases GRB 990123 and GRB 990510 there is a clear break from a shallow decay $\alpha \sim 1.1$–1.2 to a much faster decline. I show later that these features indicate that these GRBs are narrowly beamed (I also clarify in that section the confusion between relativistic and geometric beaming).
- Prompt optical emission from GRB 990123 was observed by Akerloff et al. (1999). This emission peaked with a ninth magnitude signal which lagged 70 seconds after the gamma-ray peak and coincided with the prompt X-ray peak.

The afterglow provided additional direct confirmation of the fireball model. The radio afterglow of GRB 970508 showed significant flickering during the first few weeks. This flickering decreased and eventually stopped after about a month. Goodman (1997) quickly suggested that the flickering be due to scintillation. Initially the source is small and it is within the scintillation regime. As the source expands the scintillation stops. Using this idea, Frail et al. (1997) estimated that the size of the afterglow of GRB 970508 was $\sim 10^{17}$ cm one month after the burst. Even before GRB 970508, Katz and Piran (1997) suggested that synchrotron self-absorption would result in a rising spectrum in radio frequencies for which the source is optically thick, and using the observed flux and an estimate of the temperature of the emitting regions, one could estimate the size of the emitting region. In this way Katz and Piran (1997) obtain a size of $\sim 10^{17}$ cm after one month. The agreement of the two independent estimates is reassuring. These observations imply that the fireball is expanding relativistically, and provide for the first time a confirmation of the notion of relativistic motion in GRBs.

4. Synchrotron spectrum and afterglow observations

The generic emission process for both the GRB and the afterglow is synchrotron. It is generally assumed that the emitting electrons have a power law energy distribution: $N(E) \propto E^{-p}$. A typical value that fits both the GRB and afterglow observations is $p \sim 2.5$.‡ With $p = 2.5$ the distribution diverges at low energies and there must be a low energy cutoff which is determined by the energy density available: $E_{\min} = [(p+2)e_e/(p+1)n_e]$, where n_e and e_e are the electrons' density and their energy density. The largest number of electrons is around $E_{\min}$, and hence this is also the characteristic electron energy. We denote by ν_m the synchrotron frequency of an electron with this energy. This is the "typical" synchrotron frequency. The electrons' energy density, as well as the magnetic field energy density, are characterized as fractions ϵ_e and ϵ_B of the total internal energy (Sari, Narayan, & Piran, 1996).

† Sari, Piran and Narayan (1998) have introduced in the astro-ph version the notation $F_\nu \propto t^{-\alpha} \nu^\beta$. This notation was changed to $t^{-\beta} \nu^\alpha$ in the *ApJ* version and in my review (Piran, 1999). However, the astro-ph notation caught so well that I return to this original notation here and elsewhere.

‡ One can expect that this number will be universal and won't vary from one burst to another.

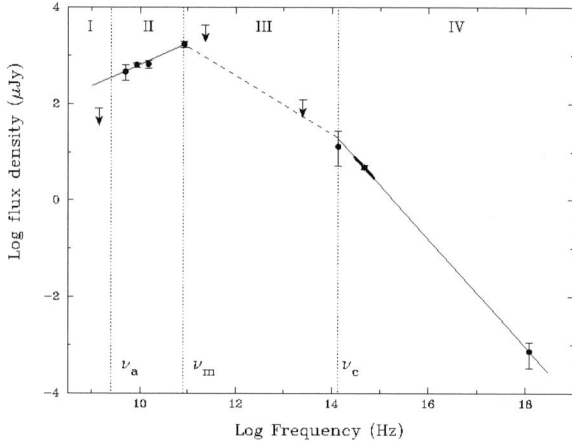

FIGURE 4. The X-ray to radio spectrum of GRB 970508 on May 21.0 UT (12.1 days after the event). The fit to the low-frequency part $\alpha_{4.86-86\text{GHz}} = 0.44 \pm 0.07$ is shown, as well as the extrapolation from X-rays to the optical (solid lines). The local optical slope (2.1–5.0 days after the event) is indicated by the thick solid line. Also indicated is the extrapolation $F_\nu \propto \nu^{-0.6}$ (dashed line). Indicated are the rough estimates of the break frequencies ν_a, ν_m and ν_c for May 21.0 UT from Galama et al. (1998a).

The low (but not extremely low) frequency spectrum is given by the low energy synchrotron tail: $F_\nu \propto \nu^{-1/3}$. At intermediate frequencies the spectrum depends on the whether the "typical" electrons are cooling within the hydrodynamic time t_{hyd}. In a fast cooling system the cooling time of an electron with $E_{\min}$ is shorter than t_{hyd}. We define the cooling frequency, ν_c, as the synchrotron frequency of an electron that cools during the local hydrodynamic time scale: $E_c/P_{\nu_c}(E_c) = t_{\text{hyd}}$. For fast cooling $\nu_c < \nu_m$ and $F_\nu \propto \nu^{-1/2}$ whereas for slow cooling $\nu_m < \nu_c$ and $F_\nu \propto \nu^{-(p-1)/2}$. The highest peak of the spectrum is always dominated by emission from fast cooling electrons with: $F_\nu \propto \nu^{-p/2}$. At very low frequencies, usually at radio frequency, the system may become optically thick to synchrotron self absorption. We denote by ν_{sa} the self absorption frequency for which $\tau(\nu_{sa}) = 1$. Unlike the common discussion of synchrotron self absorption in textbooks (which assumes $\nu_{sa} > \nu_m$ and obtains $F_\nu \propto \nu^{5/2}$), here $\nu_{sa} \ll \nu_m$. In this case $F_{\nu_{sa}} \propto \nu^2$, just like the usual Wien part of a black body spectrum. Combined we have $F_\nu \propto \nu^\beta$ with:

$$\beta = \begin{cases} 2 & \text{for } \nu < \nu_{sa} \text{ - self absorption;} \\ 1/3 & \text{for } \nu_{sa} < \nu < \min(\nu_m, \nu_c); \\ -1/2 & \text{for } \nu_c < \nu < \nu_m \text{ - fast cooling;} \\ -(p-1)/2 & \text{for } \nu_m < \nu < \nu_c \text{ - slow cooling;} \\ -p/2 & \text{for } \max(\nu_m, \nu_c) < \nu. \end{cases} \quad (4.1)$$

The resulting spectrum is a combination of four power laws,†, with three of the four slopes fixed and one depending on whether the electrons are fast cooling or not. A comparison of the theoretical model and the observation is shown in Fig. 4 (Galama et al. 1998a). The agreement is remarkable for such a simple model.

In reality the emitting region is inhomogeneous and different parts are moving with a different Lorentz factors. A specific example (Granot et al. 1999a) of integration over an inhomogeneous Blandford-McKee solution including different viewing angles results

† The self absorption region may split into two parts, and a fifth power law may arise due to inhomogeneities in the emitting medium (Granot et al. 1999b—see below).

in a spectrum which is basically similar to the one shown above, but with the sharp corner replaced by smooth curves. Inhomogeneities in the emitting region may also complicate the self absorption approximation and lead to a fifth region with a steep spectrum (Granot et al. 1999b) in the fast cooling phase. Mészáros and Rees (1999) discuss other implications of inhomogeneities in the emitting regions.

From the simple synchrotron model one expects generically that the low frequency (say X-ray, for the GRB phase) slope will always be less steep than 1/3. Cohen et al. (1997) find that this is satisfied in several strong bursts that have a well determined spectrum. Preece et al. (1998) have found that a few percent of the bursts have a steeper low energy slope, suggesting that some modification to the simple synchrotron model may be needed (see Mészáros, & Rees, 1999; Granot et al. 1999b for possible modifications). On the other hand, at high frequency, one should consider the effects of inverse Compton scattering which might dominate over synchrotron cooling (Sari & Piran 1997b). Ghisellini and Celotti (1999) even suggest that numerous pairs are produced and their inverse Compton emission dominates the observed gamma-ray emission during the GRB. Recently Milagrito has detected very high energy emission (at energies higher than 10 GeV) accompanying GRB 970417a (Nemety, 1999). Such emission could have arisen from inverse Compton emission.

Sari, Piran and Narayan (1998) used the scaling laws of the Blandford-McKee solution (which assumes adiabatic evolution and propagation into a constant density surrounding medium) and the scaling laws of a radiative solution to obtain specific light curves. A nice feature that arises is a simple relation between α, β and p. For the spherical adiabatic case we have:

$$\alpha = \begin{cases} 3\beta/2 = 3(p-1)/4 & \text{for } \nu < \nu_c, \\ (3\beta - 1)/2 = (3p-2)/4 & \text{for } \nu > \nu_c. \end{cases} \quad (4.2)$$

These relations are satisfied, for example, for the afterglow of GRB 970508 for which $\alpha = 1.12$ and $\beta = 1.14$ corresponding to $p \sim 2.4$.

5. The early afterglow and the prompt optical flash

Radio flickering and the rising radio spectrum of the afterglow of GRB 970508 have shown that this afterglow was expanding relativistically one month after the burst. However, at this stage it had slowed down significantly and its Lorentz factor was "only" 5 or less. This is not the extreme ultra-relativistic motion expected during the burst itself. Can we determine what is the initial Lorentz factor? This important question could provide another clue to the nature of the relativistic flux and could distinguish between different source models. An extremely high Lorentz factor (of order 10^5 or so) would indicate a Poynting flux (Usov, 1992; Thompson, 1994). While a lower Lorentz factor does not rule out a Poynting flux, it is an indication in favor of a baryonic flow.

There are already some limits on the initial Lorentz factor. The Compactness problem provides a lower limit of ~ 100. However, as this was the motivation for introducing relativistic motion in the first place, this cannot be considered as an independent test. Modeling of the internal shocks emission suggests an upper limit of ~ 1000 (Sari & Piran 1997b). However, we would like a direct measurement. For this our best bet is to turn to the very early afterglow—to the initial stages of the interaction of the ejecta with the ISM. This phase is almost simultaneous with the GRB itself (Sari, 1997a). Thus, its detection poses an observational challenge in obtaining quickly an accurate position and following it up with a rapid response.

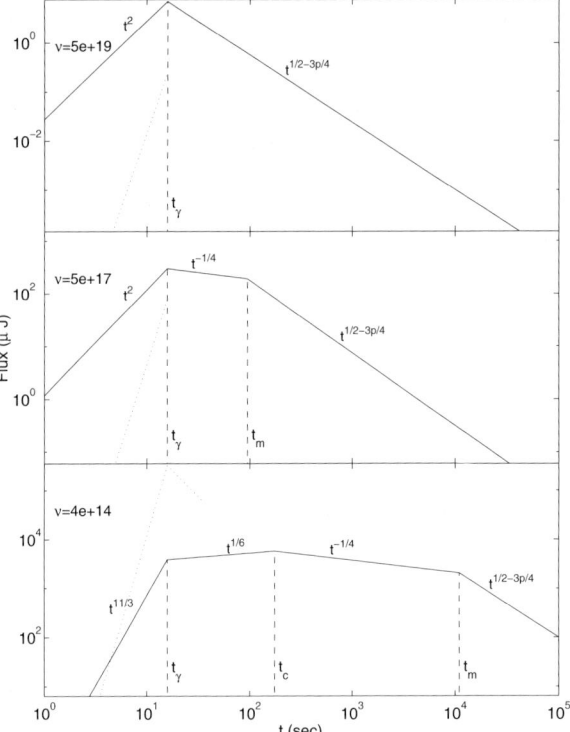

FIGURE 5. The early gamma-ray, X-ray and optical afterglow. The dashed curve on the optical range is due to the reverse shock emission (Sari & Piran, 1999a,b).

Unlike the late adiabatic afterglow, the early afterglow is most likely radiative. Namely, the energy carried away by the emitted radiation is significant compared to the total energy and these losses influence the dynamics. A radiative system must be fast cooling (if the "typical" electron does not cool there cannot be efficient energy losses).

The early afterglow peaks in the soft gamma-rays or hard X-rays (Sari & Piran 1999b). This emission could be viewed as a superposition of a longer, smoother and somewhat softer component on top of the variable, hard internal shocks signal (Sari, 1997a). This signal should be scaled and compared with the late afterglow as they arise from basically the same phenomenon. Indeed in many cases BeppoSAX' late time X-ray data extrapolates nicely with a $t^{-1.3}$ power law decay to the initial observations. It should not necessarily correlate directly with the GRB as they are produced by a different phenomenon. Moreover, as Kumar and Piran (1999) pointed out, the emitting regions of the GRB and the afterglow could have different angular sizes which could lead to drastically different fluxes, even if comparable amounts of energy is emitted. This explains why ROTSE's team (Kehoe et al. 1999) does not find a correlation between the prompt optical flash and the GRB.

A new component, a prompt optical flash, that appeared in GRB 990123 is shown in Fig. 5. Early Gamma-rays and X-rays and late optical afterglow emission arises from the forward shock: i.e. from shocked ISM material. During the early interaction of the ejecta with the ISM, before the solution has settled down to the Blandford-McKee self similar solution, there is also a reverse shock that heats the ejecta.† The emission from the reverse shock peaks at lower frequencies: $\nu_{m_{\rm rev}} \approx \nu_{m_{\rm for}}/\gamma^2$. If $\nu_{m_{\rm for}}$ is in the X-ray and $\gamma_0 \sim 100$ we expect $\nu_{m_{\rm rev}}$ to be in the optical or UV. Comparable amounts of energy are generated by the forward and the reverse shocks, and this energy is comparable to the energy of the GRB itself. A strong 5th magnitude optical flash would have been produced if the fluence of a moderately strong GRBs, 10^{-5} erg s^{-1} cm^2 would have been released on a time scale of 10 seconds in the optical band. Even a small fraction of this would be easily observed.

While the synchrotron frequency of the reverse shock is much lower than the synchrotron frequency of the forward one, both have the same cooling frequency (both have the same magnetic field and the same bulk Lorentz factor). As the forward shock must be radiative: $\nu_{c_{\rm for}} < \nu_{m_{\rm for}}$. If $\nu_{c_{\rm rev}} = \nu_{c_{\rm for}} \gg \nu_{m_{\rm rev}}$, the reverse shock might not be radiating efficiently and this might lower the observed signal. Even with these factors, one can hardly avoid a strong optical emission from the reverse shock. Additional effects such as inverse Compton scattering and self absorption can somewhat reduce this flux, but even so a signal stronger than 15th magnitude is expected.

Sari and Piran (1999a) presented this detailed prediction of a strong optical flash at the Rome meeting that took place in October 1998. The possibility of strong optical emission was noticed by Mészáros & Rees (1997a) in two of several models they examined. This prediction was almost in contradiction with the LOTIS upper limits on several bursts (see e.g. Williams et al. 1999). Less than two months later, on January 23, 1999, ROTSE (Akerloff et al. 1999) was triggered by the GCN network (Barthelmy et al. 1994) and detected a 9th magnitude optical signal accompanying GRB 990123. A careful examination of the different light curves of GRB 990123 (see Fig. 6) reveals that the peak optical emission was offset by at least 25 seconds from the peak gamma-ray emission. In fact, there is no temporal correlation between the gamma-rays and the optical photons. Moreover, the soft X-ray signal peaks at the time of the optical peak, and quite generally the spectrum evolves from hard to soft. All this is in complete agreement with the picture presented above. The early afterglow slightly lags after the GRB. The forward shock emission peaks in X-rays while the reverse shock emission peaks at the optical or UV band. The optical light curve shows an initial phase of rapid decline $\sim t^{-2}$, again in a complete agreement with the prediction of the reverse shock emission (Sari & Piran, 1999a; 1999b). Later on this turns to the common $t^{-1.1}$ decay seen in other bursts (see Fig. 7).

These observations provide us with three independent estimates of the Lorentz factor during the early afterglow (Sari & Piran, 1999a):
- The time of the optical flash peak, ~ 70 s $\sim l/(c\gamma^{8/3})$. For $l \sim 10^{18}$ cm we have: $\gamma_0 \sim 200$.
- The initial decay like t^{-2} suggests that the typical synchrotron frequency was below the optical early on. This suggests $\gamma_0 \sim 200$.
- The initial decline of the X-ray suggests that already initially the typical synchrotron frequency was below the 1.5–10 keV band. Using the initial ratio $\nu_{m_{\rm for}}/\nu_{m_{\rm rev}}$ we find that $\gamma_0 \sim 70$.

Radio emission was also observed from GRB 990123 (Kulkarni et al. 1999b; Galama et al. 1999). This radio emission peaked around one day with marginal detections prior to and later than that (see Fig. 8). This radio emission is also produced by the reverse shock. The expected radio light curve (see Fig. 8) was calculated using the optical data

---

† This notation is somewhat confusing as this reverse shock propagates backwards only in the fluid's frame. It is propagating relativistically toward the observer in the observer frame.

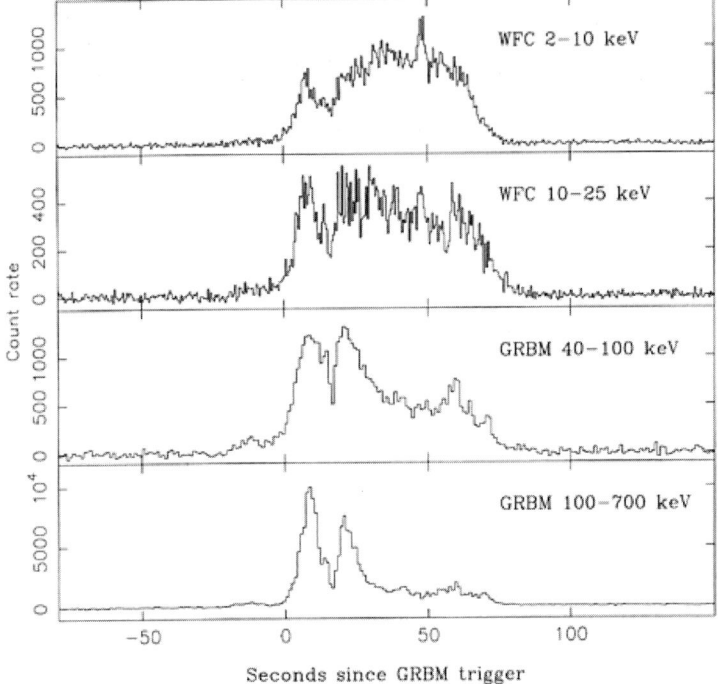

FIGURE 6. The light curves of GRB 990123 in different bands (Costa, 1999): from top to bottom: 2–13 keV; 13–40 keV; 40–100 keV and 100–700 keV. The increasing X-ray flux at late time is seen clearly. Note that BeppoSAX's trigger is ~ 25 sec after BATSE's trigger.

(Sari & Piran 1999c). The fit to the observations is almost too nice. The radio emission is suppressed early on due to synchrotron self absorption. This enables us to estimate the perpendicular size of the system, γct, at one day after the burst to be $\sim 10^{15}$ cm!

6. Beaming, jets and flying pancakes

If we assume isotropic emission, then the energies emitted by GRB 971214, at $z = 3.418$ and by GRB 990123 at $z = 1.65$ are $\sim 3 \times 10^{53}$ ergs and 4×10^{54} ergs respectively.† The first is comparable to the binding energy of a neutron star, the second is greater than a solar rest mass energy. Clearly, these values are problematic for most GRB models.

6.1. Theory

If the GRB emission is beamed into some angle, θ, the overall energy would be lower by a factor $\theta^2/4$ (the overall event rate will be larger by the inverse factor). There has been some confusion between such "geometric" beaming and the relativistic beaming that arises due to the relativistic motion. It is worthwhile to discuss this issue first. The radiation from a source moving with a Lorentz factor γ towards the observer is beamed into a narrow cone with an opening angle γ^{-1}. Typical values during the GRB and the afterglow are $\gamma \sim 200$ and $\gamma \sim 2$–10 respectively, corresponding to a relativistic beaming of 10^{-2} rad and 0.1–0.5 rad. These are the maximal (smallest angle) beaming possible.

† These values depend, of course, on the cosmological model assumed.

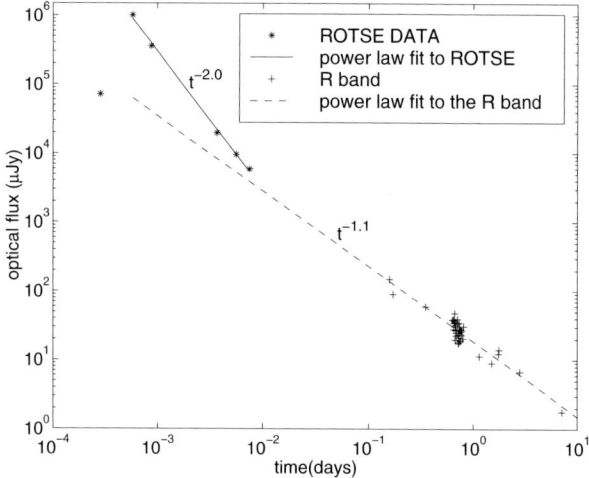

FIGURE 7. The early optical light curve of GRB 990123 (Sari & Piran 1999c).

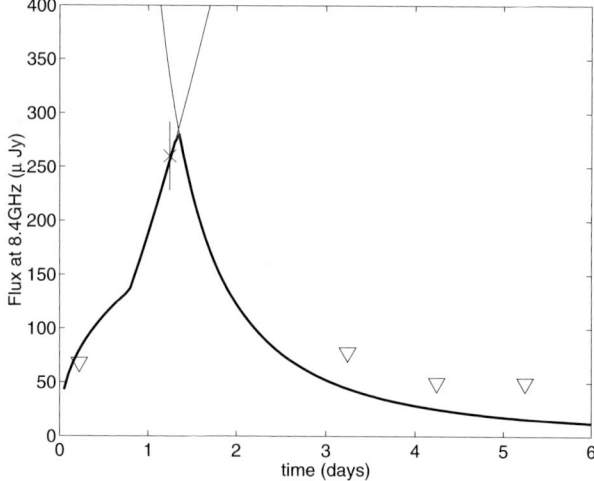

FIGURE 8. Observations and theoretical light curve for the early radio signal of GRB 990123 (Sari & Piran, 1999c).

However, this is just a lower limit to the beaming angle. If the source has an opening angle $\theta > \gamma^{-1}$, then the beaming is determined by θ and not by γ^{-1}. Observers with a viewing angle up to θ from the center can see the source. However, each observer sees a local patch whose size is only γ^{-1}. There are $(\theta\gamma)^2$ such patches. As all these patches are causally disconnected, different observers that are more than γ^{-1} apart will observe different emitting regions and may record a different time profile and a different spectrum from the same burst. The causally connected regions grow as γ decreases. Thus, during the afterglow there are fewer and fewer such regions and when $\gamma^{-1} \sim \theta$ there is only one. We will return to this issue later, when I discuss the implications of this phenomenon to the implied energy emission and efficiency of GRBs.

What distinguishes the dynamics of a spherical ejecta from a nonspherical one? Initially there is little difference. The proper time required for the matter to reach a radial distance R is $R/c\gamma$. The maximal sideways expansion is therefore R/γ. Hence the angular size of a causally connected region is γ^{-1}. As long as $\gamma^{-1} < \theta_0$, the initial opening angle, the matter simply does not have enough time to expand sideways and to know that it is not a part of a spherical shell (Piran, 1995). However, once $\gamma^{-1} \approx \theta_0$ the matter suddenly "discovers" its nonspherical structure and begins to expand sideways. As the matter at the front is constantly shocked to relativistic energies, Sari, Piran and Halpern (1999) expect it to expand with the speed of light: $\theta \sim \gamma^{-1}$. Rhoads (1999) assumes that this sideways expansion is at the sound speed which results in $\theta \sim \gamma^{-1}/\sqrt{3}$. This sideways expansion is so rapid that it dominates completely the radial expansion. For an expansion into a homogeneous ejecta this yields $\gamma \propto R^{-3/2} \exp[-[3/(2\gamma_o\theta_o)](R^{3/2}/R_o^{3/2} - 1)]$, which is valid before, during and after the transition. Additionally, the radiation is now beamed into a larger cone since $\gamma^{-1} > \theta_0$. Both effects reduce the observed emission and will cause a break in the light curve, roughly by an additional factor† of t^{-1}. Within the adiabatic synchrotron model we have new relations between α, β and p:

$$\alpha_{beam} = \begin{cases} 2\beta + 1 = p & \text{for } \nu < \nu_c, \\ 2\beta = p & \text{for } \nu > \nu_c. \end{cases} \quad (6.3)$$

6.2. Observations

The first, longest and best observed light curves from GRB 970228 and GRB 970508 show a single power law decay with no indication for a beaming break. This has lead to the early impression that GRBs are not beamed. However, other bursts were different. GRB 980519 was unique among GRB afterglows with its most rapidly fading $t^{-2.05\pm0.04}$ in optical as well as in X-rays (Halpern et al. 1999). The optical spectrum of this burst shows: $\beta = 1.15 \pm 0.15$ (see Fig. 9). These values are in perfect agreement with an expanding beamed emission with $p \approx 2.2$. As transition is not seen in the light curve, the sideways spreading must have begun before the first optical observation, namely, less than 8.5 hours after the burst. Using the detector's time equation, we find that the corresponding opening angle must have been rather small: $\theta < 0.05$ radians, leading to a beaming factor of 500 or larger!

GRB 980326 was another burst with a rapid decline. Groot et al. (1998) derived a temporal decay slope of $\alpha = 2.1 \pm 0.13$ and a spectral slope of $\beta = 0.66 \pm 0.7$ in the optical band, suggesting once more beaming. As Groot et al. (1998) noted, the large uncertainty in the spectral index allows in this case also a spherical expansion interpretation (with somewhat unusual values $p = 4.2$ or $p = 5.2$). However, this measured temporal decay was dependent upon a report of a host galaxy detection at $R = 25.5 \pm 0.5$, which was included as a constant term. The detection of a host has since been determined to be spurious; better data show no constant component to a limiting magnitude of $R = 27.3$ (Bloom et al. 1999). If the last detection is interpreted as a different phenomenon (Bloom et al. 1999) then the remaining points show a rapid decline—in agreement with a spreading beam interpretation.

GRB 990123 provided the first direct evidence for a beaming break (Kulkarni et al. 1999a). The prompt optical flash, which we interpret to arise from the reverse shock, decayed like t^{-2} and disappeared quickly. The intermediate optical afterglow showed a power law decay with $t^{-1.1\pm0.03}$. The decay and the spectrum fit well an electron distribution with $p = 2.5$. This behavior continued from the first late observation (about 3.5 hours after the burst) until about 2.04 ± 0.46 days after the burst. Then the optical emission began to decline faster. The simplest explanation is that we have observed the transition from a spherical like phase to a sideways expanding phase. The transition took

† The beaming break takes place at $\gamma \approx \theta_0^{-1}$. If the hydrodynamic break is only at $\gamma \approx (\sqrt{3}\theta_0)^{-1}$ then two successive breaks will take place (Panaitescu, & Mészáros, 1998).

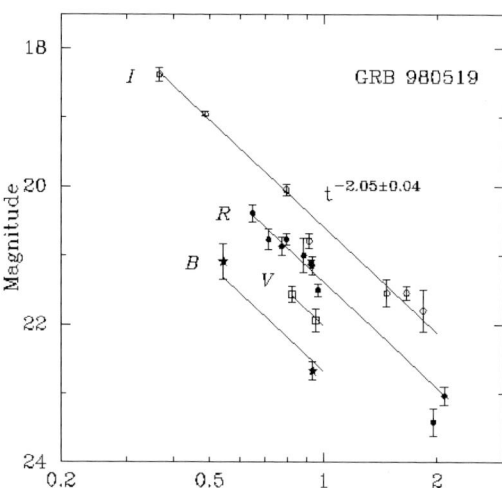

FIGURE 9. An optical light curve for GRB 980519. The fast decline indicate a spreading beam (Halpern et al. 1999).

place at ~ 2 days, corresponding to $\theta_0 \sim 0.1$ radians. This implies a beaming factor of about 100, reducing the energy of the burst to 3×10^{52} ergs.

Finally, just last May, GRB 990510 displayed a beautiful transition from $t^{-0.82 \pm 0.02}$ to $t^{-2.18 \pm 0.05}$ at 1.2 ± 0.08 days after the burst (Harrison et al. 1999; Stanek et al. 1999). The isotropic energy of this burst is 2.7×10^{53} ergs. With a beaming factor of 300 this becomes a "modest" 10^{51} ergs.

6.3. On jets, bullets or flying pancakes

Probably because of the analogy with AGNs in which beautiful jets are observed, the terminology jets has been used to describe beamed emission from GRBs. Unfortunately this terminology is somewhat misleading. A jet is long and narrow. It corresponds to continuous activity. The jets seen in AGNs reflect activity on scales of 10^7 to 10^8 years, much longer than the typical time scales of the inner black holes. In GRBs we observe a transient phenomenon and the length of the ejecta (in the direction along its motion) is much shorter. At first one may think that the term bullet would be more appropriate. A bullet is not so extended in length and it clearly represents a transient phenomenon. However, a more detailed consideration reveals that the perpendicular size of the ejecta is larger than its length. The angular size is $R\theta$ and the "length" is $\Delta \sim cT$. Typical values are $\Delta \sim 10^{12}$ cm (for $T = 30$ sec) and $\theta \sim 0.1$. Thus, quite generally, $R\theta > \Delta$. This is not even a bullet. This is a pancake flying at relativistic velocity perpendicular to its flat direction (see Fig. 2). One may wonder if relativistic contraction has played some tricks on us. When we look at the ejecta in its own rest frame, we find that it is longer by a factor γ, so initially it is actually a bullet. However, even at this early phase the ejecta expands sideways proportionally to R (unless it is continuously collimated) and even in its own frame it looks like a pancake.

7. On the nature of the "inner engine"

One can divide the various clues to the nature of the "inner engine" into three groups. The first one is based on general features of GRBs. The second one includes constraints that arise from the fireball model. The third one includes indications that arise from the hosts of GRBs.

We turn now to summarize the evidence concerning the "inner engine." We concentrate first on evidence that follows from the "internal shocks model":

The first group includes:

- **Energy:** The implied isotropic energy is: $\sim 10^{51}$–10^{54} ergs. Even with beaming (which we discuss later) this is a significant fraction of the binding energy of a solar mass compact object.
- **Rate:** The observed GRB rate corresponds approximately to one burst per 10^7 years per galaxy (see e.g. Schmidt 1999 for a recent summary). The actual rate may be higher by a factor of $4/\theta^2$, approximately a hundred, if most GRBs are beamed.
- **Luminosity Function:** The GRB luminosity function is wide. Both the peak luminosity and the total energy varies by more than a factor of a hundred.
- **The Afterglow Luminosity Function:** The observed X-Ray afterglow luminosity function is narrower than the GRB luminosity function. The observed R magnitude is even narrower (Kumar & Piran, 1999). Note that this result is based on a small number of events as there are only half a dozen bursts with measured redshifts.

The second group include constraints that follow from the internal shocks model:

- **Irregular flow:** To produce internal shocks the "inner engine" must produce a long-lasting and irregular wind. Single explosions won't work.
- **Variability:** The variability time scale, δt, suggests a solar mass compact object.
- **Duration:** The duration, T, suggests a prolonged activity which is much longer than the source's gravitational time scale.
- **Relativistic Flow:** The central engine must produce efficiently a relativistic flow. The baryonic load is less than about 10^{-5} M$_\odot$.
- **Beaming:** The emitted flow is sometimes highly nonspherical with an opening angle of a few degrees. Since the late afterglow is less beamed than the GRB, this suggests a search for "orphan" afterglows that are not accompanied by GRBs. This also suggests some similarity to AGNs.

Finally, we have observational evidence that follows from the positions of the afterglows and from the properties of their host galaxies.

- **Host Galaxies:** Host galaxies have been detected for most GRBs with optical afterglows. The long standing no-host problem has disappeared. Most bursts are located at the central regions, indicating that the progenitors have not escaped from their host galaxies as would have been the case in long-lived binary neutron stars in dwarf galaxies (Narayan, Paczyński, & Piran, 1992).
- **Association with star forming regions:** There is some evidence that host galaxies of GRBs are star forming galaxies (Hogg & Fruchter, 1999; see however Djorgovski 1999). This would favor short-lived progenitors: collapsing massive stars or short-lived† (10^4–10^5 years) binary neutron stars (Tutukov & Yungelson 1994).
- **GRB 980425, SN1999bw and the GRB-SN association:** The error box of GRB 990425 contains the bright supernova SN1999bw which is at $z = 0.085$ (Galama et al. 1998b). This has lead to an ongoing debate on the association of GRBs with type Ic supernovae (Wang & Wheeler 1998; Kippen et al. 1998, Graziani et al. 1998; Wheeler, 1999). Support to the possibility of an association between the two was recently received with the observation of the afterglow of GRB 980326, in which the spectrum at a late

---

† The lifetime of the observed galactic neutron star binaries is 10^8–10^9 years. On the other hand, there is a strong observational bias against detection of short-lived neutron star binaries.

epoch, whose optical point falls high above the extrapolated afterglow light curve, is compatible with a supernova signal (Bloom et al. 1999).

This evidence, and more specifically, the energy and the time scale considerations suggest that GRBs are powered by accretion of a massive (~ 0.1 $M_\odot$) accretion disk onto a compact object, most likely a black hole. Such a massive accretion disk must form simultaneously with the black hole, from matter that was slowed down by centrifugal forces. Thus GRBs signal the formation of black holes. The gravitational energy of a 0.1 $M_\odot$ mass can supply the energy required for the process, the accretion time would determine the overall duration, while the variability would be determined by the gravitational time scale of the central object or by the hydrodynamic time scale of the accretion disk. Both are rather short. Mészáros and Rees (1997b) and Katz (1997) suggest that the energy extraction takes place via an electromagnetic process such as the Blandford-Znajek (1977) mechanism. This mechanism extracts the rotational energy of the black hole—a larger reservoir than the gravitational energy of the accretion disk. As this effect involves electromagnetic fields it is likely that, using this effect, it is easier to produce a clean relativistic flow, possibly Poynting flux.

Several routes can lead to a black hole—accretion torus system:

• **Binary neutron star merger:** Binary neutron star mergers (Eichler et al. 1989) have been considered the canonical cosmological GRB sources for some time. These mergers are known to take place at a rate of one per $\sim 10^6$–10^7 years per galaxy (Narayan, Piran, & Shemi, 1991). This rate is comparable to the rate of GRBs (Piran, Narayan, & Shemi, 1992). The final outcome of such a merger (Davies et al. 1994; Ruffet & Janka, 1999; Rosswog et al. 1999) is a ~ 2.4 $M_\odot$ black hole surrounded by a 0.1–0.2 $M_\odot$ thick accretion disk, which could power the burst (Narayan, Paczyński, & Piran 1992).

• **Neutron star–Black hole merger:** This is a simple variation on the previous theme. One expects here that the neutron star will be torn apart by the black hole, again leaving a massive disk (possibly slightly more massive than in the binary neutron star case), which will power the burst. If the mass of the black hole is ~ 10 $M_\odot$, we expect slightly different time scales and different behavior between the neutron star binary and this case. While black hole–neutron star binaries are expected to be as common as neutron star–neutron star binaries (Narayan, Piran, & Shemi 1991) or even more common (Bethe & Brown 1998), none was observed so far.

• **White dwarf–Neutron star merger:** A white dwarf orbiting a neutron star will be pulled inwards via gravitational radiation emission. If the mass ratio is small it will become unstable† when it reaches the Roche limit and it will dump its mass into an accretion disk within its hydrodynamic timescale, a few seconds, producing a solar mass torus surrounding a neutron star. Once accretion from the disk begins, this neutron star will turn into a black hole. Depending on the viscosity and on neutrino losses, the accretion rate may be high enough so that the accretion time would be of order few seconds which will be the duration of the GRB.

• **"Failed Supernova," Collapsar or Hypernova:** These are all different names for a collapsing star that produces a GRB. Both Woosley's (1993) failed supernova model and Paczyński's (1998) hypernova model assume that the rapidly rotating collapse produces a rotating massive black hole surrounded by a thick torus that accretes on it.

It is interesting to note that these sources can be arranged as a sequence in terms of their maximal timescale with neutron star binaries the shortest, followed by black hole–neutron star merger, a white dwarf–neutron star merger and a failed supernova.

---

† The expected mass ratio in a white dwarf–black hole binary is small and we expect these systems to be stable.

The same sequence also arranges the sources from less to more baryons surrounding the source.

8. Energetics

The most basic and fundamental issue in any problem is energetics. Here we ask: Can solar mass compact sources generate the required energy? Put in other words: Is there an "energy crisis"? Assuming isotropic emission, one finds that the energy of the most energetic bursts is higher than 10^{54} ergs. The recently observed evidence for beaming with an opening angle of a few degrees reduces the energy estimate by a factor of a hundred (Kulkarni et al. 1999; Sari, Piran & Halpren, 1999; Harrison et al. 1999, Stanek et al. 1999). Internal shocks can convert under reasonable conditions $\sim 10\%$ of the kinetic energy into thermal energy (Mochkovitch et al. 1995; Kobayashi et al. 1997; Daigne & Mochkovitch, 1998; Kumar 1999). External shocks are even less efficient. Then there is the question of radiative efficiency (Sari, Narayan & Piran, 1996; Sari & Piran, 1997b; Kumar, 1999) and of additional energy losses, which may reduce the overall efficiency to ~ 0.01 (Kumar 1999) and increase the total energy budget back to $\sim 10^{54}$ ergs (in extreme cases). Even before all that, one should wonder what was the efficiency of the source in accelerating the relativistic flow?

A related issue is the question of the energy distribution between the GRB and its afterglow. According to the internal-external shocks model a comparable amount of energy should be released in the GRB and in the early afterglow. Actually the early afterglow should be more powerful than the GRB by a factor of a few. However, early afterglow observations show the opposite. The prompt X-ray emission (which could be interpreted as an early afterglow) is at most 10% of the GRB energy. The impressive optical flash from the reverse shock of GRB 990123 was only 1% of the gamma-ray energy. Has anything gone wrong?

It is possible that even after making the beaming correction we still overestimate the GRB energy. This could happen if the GRB luminosity is dominated by hot spots (Kumar & Piran 1999) whose angular size γ^{-1} is much smaller that the overall geometrical beaming θ_0. The size of a causally connected region of a shell of radius R moving with Lorentz factor γ is $\sim R/\gamma$. Because of relativistic beaming this is also the size of the region visible to a distant observer. During the GRB the angular size of these regions (< 0.01 rad) is significantly smaller than the inferred angular width of the ejecta, $\theta_0 \sim$ a few degrees. There are therefore $(\gamma\theta_0)^2$ causally disconnected regions within this cone. Thus the observed γ-ray luminosity seen by different observers from the same burst could fluctuate strongly due to small scale inhomogeneities in the emitting regions. Therefore, one would over-estimate the energy release in γ-rays in cases in which a hot spot has been observed.† At a later time, when the Lorentz factor of the ejecta has become smaller and the size of causally connected regions is larger, the dispersion of the afterglow luminosity seen along different lines of sight should be smaller as well. The emission at this stage yields a better estimate of the overall total energy involved.

Kumar and Piran (1999) present a simple model of inhomogeneities in GRBs which results in a wide luminosity function consistent with the observed $\log N - \log S$ distribution. This model also explains naturally the decrease in the width of the luminosity function of the X-ray and optical afterglow (relative to the width of the GRB luminos-

† Note that as BeppoSAX can detect only rather strong bursts the BeppoSAX sample might be biased towards cases in which such a hot spot has been seen.

ity function). It also explains the lack of correlation between the γ-ray fluence and the afterglow flux.

9. Conclusion

The relativistic synchrotron model provides an excellent model for GRB afterglow. It appears that we understand the afterglow phenomenon. As many have remarked, this is simpler and easier to analyze than the GRB itself. Unlike more pessimistic opinions expressed at this Conference (Paczyński, 1999), I think that we are also on the right track towards a viable model (the internal shocks model) for the γ-ray emission process.

As for the "inner engine" that drives the GRBs, here I share the pessimistic point of view. It is unlikely that we could prove with theoretical models that either merging neutron stars or hypernova explosions should generate GRBs (Paczyński, 1999). Still, I have tried to show here how our understanding of the emitting regions directs us towards a generic central engine—a (most likely newborn) black hole surrounded by a compact, fairly massive, accretion disk. As this central engine can arise in several scenarios we will have to resort to circumstantial evidence to distinguish between these different models.

I thank Re'em Sari for a wonderful and productive collaboration and J. Granot, J. Katz, S. Kobayashi, P. Kumar, R. Narayan and F. K. Thielemann, for many helpful discussions. This research was supported by the US-Israel BSF grant, by a grant from the Israeli Space Agency and by NASA grant NAG5-3516. I thank Columbia University, NYU and Basel University for their hospitality while this research was done.

REFERENCES

AKERLOFF, C. W., ET AL. 1999, *Nature*, **398**, 400.
BARTHELMY, S. D., ET AL. 1994, in *Proceedings of the Second Huntsville Workshop*, eds. G. Fishman, J. Brainerd, K. Hurley, 307, 643.
BETHE, H. A. & BROWN, G. E. 1998, *ApJ*, **506**, 780.
BLANDFORD, R. D. & MCKEE, C. F. 1976, *Phys. of Fluids*, **19**, 1130.
BLANDFORD, R. D. & ZNAJEK, R. L. 1977, *MNRAS*, **179**, 433.
BLOOM, J. S., ET AL. 1999, sumbitted to *Nature*, astro-ph/9905301.
COHEN, E., ET AL. 1997, *ApJ*, **480**, 330.
COSTA, E., ET AL. 1997, *Nature*, **387**, 783.
COSTA, E. 1999, talk given at the Santa Barbara workshop, March 1999.
DAIGNE, F. & MOCHKOVITCH, R. 1998, *MNRAS*, **296**, 275.
DAVIES, M. B., BENZ, W., PIRAN, T., & THIELEMANN, F. K. 1994, *ApJ*, **431**, 742.
DERMER, C. D. & MITMAN, K. E. 1999, *ApJ*, **513**, L656.
DJORGOVSKI, S. G. 1999, talk given at the Santa Barbara workshop, March 1999.
EICHLER, D., LIVIO, M., PIRAN, T., & SCHRAMM, D. N. 1989, *Nature*, **340**, 126.
FENIMORE, E. E., EPSTEIN, R. I., & HO, C. H. 1993, *A&AS*, **97**, 59.
FENIMORE, E. E., MADRAS, C., & NAYAKSHIN, S. 1996, *ApJ*, **473**, 998.
FENIMORE, E. E., RAMIREZ, E., & SUMMER, M. C. 1998, in *Gamma-Ray Bursts 4th Huntsville Symposium* (eds. C. Meegan, R. Preece & T. Koshut). AIP Conf. Proc. 428. AIP.
FRAIL, D. A., ET AL. 1997, *Nature*, **389**, 261.
GALAMA, T. J., ET AL. 1998a, *ApJ*, **500**, L1008.
GALAMA, T. J., ET AL. 1998b, *Nature*, **395**, 670.
GALAMA, T. J., ET AL. 1999, *Nature*, **398**, 394.

GHISELLINI, G. & CELOTTI, A. 1999, *ApJ*, **511**, L93.
GOODMAN, J. 1986, *ApJ*, **308**, L47.
GOODMAN, J. 1997, *New Astronomy*, **2**, 449.
GRANOT, J., PIRAN, T., & SARI, R. 1999a, *ApJ*, **513**, 679.
GRANOT, J., PIRAN, T., & SARI, R. 1999b, in preparation.
GRAZIANI, C., LAMB, D. Q., & MARION, G. H. 1998, astro-ph/9810374.
GROOT, P. J., ET AL. 1998, *ApJ*, **502**, L123.
HALPERN, J. P., KEMP, J., PIRAN, T., & BERSHADY, M. A. 1999, *ApJ*, 1999, **517**, L105.
HARRISON, F. A., ET AL. 1999, *ApJ*, **523**, L121.
HOGG, D. W. & FRUCHTER, A. 1999, *ApJ*, **520**, 54.
KATZ, J. I. 1994, *ApJ*, **422**, 248.
KATZ, J. I. 1997, *ApJ*, **490**, 633.
KATZ, J. I. & PIRAN, T. 1997, *ApJ*, **490**, 772.
KEHOE, R.. ET AL. 1999, this volume, astro-ph/9909219.
KIPPEN, R. M., ET AL. 1998, *ApJ*, **506**, L27.
KOBAYASHI, S., PIRAN, T., & SARI, R. 1997, *ApJ*, **490**, 92.
KROLIK, J. H. & PIER, E. A. 1991, *ApJ*, **373**, 277.
KULKARNI, S. R., ET AL. 1999a, *Nature*, **393**, 35.
KULKARNI, S. R., ET AL. 1999b, *ApJ*, **522**, L97.
KUMAR, P. 1999, *ApJL*, in press, astro-ph/9907096.
KUMAR, P. & PIRAN, T. 1999, *ApJ*, in press, astro-ph/9909014.
MÉSZÁROS, P. & REES, M. J. 1992, *MNRAS*, **258**, 41.
MÉSZÁROS, P. & REES, M. J. 1997a, *ApJ*, **476**, 232.
MÉSZÁROS, P. & REES, M. J. 1997b, *ApJ*, **482**, L29.
MÉSZÁROS, P. & REES, M. J. 1999, astro-ph/9908126.
MOCHKOVITCH, R., MAITIA, V., & MARQUES, R. 1995, in: *Towards the Source of Gamma-Ray Bursts, Proceedings of 29th ESLAB Symposium* (eds. K. Bennett & C. Winkler), p. 531.
NARAYAN, R., PACZYŃSKI, B., & PIRAN, T. 1992, *ApJ*, **395**, L83.
NARAYAN, R., PIRAN, T., & SHEMI, A. 1991, *ApJ*, **379**, L1.
NEMETY, P. 1999, private communication.
PACZYŃSKI, B. 1986, *ApJ*, **308**, L43.
PACZYŃSKI, B. & RHOADS, J. 1993, *ApJ*, **418**, L5.
PACZYŃSKI, B. & XU, G. 1994, *ApJ*, **427**, 709.
PACZYŃSKI, B. 1998, *ApJ*, **494**, L45.
PACZYŃSKI, B. 1999, this volume, astro-ph/9909048.
PANAITESCU, A. & MÉSZÁROS, P. 1998, *ApJ*, **492**, 683.
PIRAN, T. 1995, in *Gamma-Ray Bursts–The Second Huntsville Meeting*.
PIRAN, T. 1997, in *Some Unsolved Problems in Astrophysics* (eds. J. N. Bahcall & J. P. Ostriker). Princeton University Press.
PIRAN, T. 1999, *Physics Reports*, **314**, 575.
PIRAN, T., NARAYAN, R., & SHEMI, A. 1992, in AIP Conference Proceedings **265**, *Gamma-Ray Bursts, Huntsville, Alabama, 1991*, (eds. W. S. Paciesas & G. J. Fishman). p. 149. AIP.
PIRAN, T. & SHEMI, A. 1993, *ApJ*, **403**, L67.
PREECE, R. D., ET AL. 1998, *ApJ* **506**, L23.
REES, M. J. & MÉSZÁROS, P. 1994, *ApJ*, **430**, L93.
RHOADS, J. E. 1999, *ApJ*, in press, astro/ph-9933099.

Rosswog, S., et al. 1999, *A&A*, **341**, 499.
Ruderman, M. 1975, *Ann. N.Y. Acad. Sci.*, **262**, 164.
Ruffet, M. & Janka, H.-Th. 1999, *A&A*, **344**, 573.
Sari, R. 1997a, *ApJ*, **489**, L37.
Sari, R., Narayan, R., & Piran, T. 1996, *ApJ*, **473**, 204.
Sari, R. & Piran, T. 1995, *ApJ*, **455**, L143.
Sari, R. & Piran, T. 1997, *ApJ*, **485**, 270.
Sari, R. & Piran, T. 1997b, *MNRAS*, **287**, 110.
Sari, R., Piran, T., & Narayan, R. 1998, *ApJ*, **497**, L41.
Sari, R. & Piran, T. 1999a, *A&A* in press, astro-ph/9901105.
Sari, R. & Piran, T. 1999b, *ApJ* **520**, 641.
Sari, R. & Piran, T. 1999c, *ApJ*, **517**, L109.
Sari, R., Piran, T., & Halpern, J. 1999, *ApJ*, **519**, L17.
Schmidt, M. 1999, astro-ph/9908190.
Stanek, A. Z., et al. 1999, *ApJ*, **522**, L39.
Thompson, C. 1994, *MNRAS*, **270**, 480.
Tutukov, A. V. & Yungelson, L. R. 1994, *MNRAS*, **268**, 871.
Usov, V. V. 1992, *Nature*, **357**, 472.
van Paradijs, J., et. al. 1997, *Nature*, **386**, 686.
Vietri, M. 1997, *ApJ*, **478**, L9.
Vreeswijk. P. M., et al. 1999, *ApJ*, in press, astro-ph/9904286.
Wang, L. & Wheeler, J. C. 1998, *ApJ*, **504**, L87.
Waxman, E. 1999, talk given at the Santa Barbara workshop, March 1999.
Wheeler, J. C. 1999, this volume.
Wijers, R. A. M. J. & Galama, T. J. *ApJ*, **523**, 177.
Williams G. G., et al. 1999, *ApJ*, **519**, L25.
Woods, E. & Loeb, A. 1995, *ApJ*, **383**, 292.
Woosley, S. E. 1993, *ApJ*, **405**, 273.

Gamma ray mechanisms

By MARTIN J. REES

Institute of Astronomy, University of Cambridge, Madingley Road, Cambridge, CB3 OHA, U.K.

Gamma-ray bursts, an enigma for more than 25 years, are now coming into focus. They involve extraordinary power outputs, and highly relativistic dynamics. The most plausible progenitors, ranging from NS-NS mergers to collapsars (sometimes called "hypernovae") eventually lead to the formation of a black hole with a debris torus around it, the extractable energy being up to 10^{54} ergs. Magnetic fields may exceed 10^{15} G; if so, the most efficient extraction of energy is likely to be electromagnetic. Details of the afterglow may be easier to understand than the initial trigger. Bursts at very high redshift can be astronomically-important as probes of the distant universe.

1. Introduction

Astrophysics is an observation-led subject: theorists generally play a subsidiary role— certainly a more modest one than their counterparts in, for instance, particle physics. But in the case of gamma ray bursts the lag between gathering data and making sense of it has been specially embarrassing, even by astrophysical standards. Until two years ago, there was absolutely no consensus on what, or even where, the bursts are. Owing primarily to the impetus of the Italian/Dutch Beppo-SAX satellite, there is now general agreement that the bursts (or at least a substantial subset of them) are at high redshifts. In this paper I shall discuss the physics of the energy production and its conversion into an intense burst and a prolonged afterglow, emphasizing the role of strong magnetic fields. I hope this will be complementary to the survey by Piran and the more specific discussion by Woosley.

Although theorists took decades to "home in" on plausible models it is amusing to recall that it did not take long for them to become enthusiastically engaged. At the "Texas conference on Relativistic Astrophysics" held in December 1974, less than two years after the classic paper by Klebesadel et al. (1973), Ruderman (1975), gave a review of models and theories. He presented an exotic menu of alternatives that had already appeared in the literature, involving supernovae, neutron stars, flare stars, antimatter effects, relativistic dust, white holes, and some even more bizarre options. He noted also the tendency (still often apparent; not only among astrophysicists!), for theorists to "strive strenuously to fit new phenomena into their chosen specialties."

In April 1995 an interesting debate took place on the location of GRBs, in which the two main protagonists were Don Lamb and Bohdan Paczyński (a written version of the argument appears in Lamb (1995) and Paczyński (1995)). It was held in the Washington Museum of Natural History, to commemorate the 75th anniversary of the famous debate that took place there between Shapley and Curtis on whether some of the so called "nebulae" were stellar systems (i.e. other galaxies) beyond our Milky Way. I had the privilege of moderating this debate, perhaps because I was one of the few people who had explored both options (cf. Podsiadlowski et al. 1995).

There was an agreement among all participants on the kind of new evidence that could settle the issue. Most valuable of all would be firm identification with objects detectable in other wavebands.

The controversies in the Shapley-Curtis debate were settled within a few years; our knowledge of extragalactic astronomy thereby made a forward leap, and astronomers moved on to address more detailed issues. The GRB distances were actually settled even more quickly and decisively: the crucial step, as is now well known, was the detection of gradually-fading afterglows within some of the arc-minute-scale error boxes that the BeppoSax satellite was able to supply within a few hours of the burst.

2. What is the trigger?

The photon *luminosity*, for the few-second duration of a typical burst, is of course colossal: it exceeds by many thousands the most extreme output from any active galactic nucleus (thought to involve supermassive black holes), and is 14 orders of magnitude above the Eddington limit for a stellar-mass object. The total *energy*, however, is not out of line with some other phenomena encountered in astrophysics—indeed it is reminiscent of the energy released in the core of a supernova, the big difference being that the primary sudden event (with a timescale of seconds) is not smothered by a stellar envelope, as in a normal supernova, but manifests itself in hard radiation that escapes more promptly.

Unless they are beamed into less than one percent of the solid angle, the triggers for GRBs are thousands of times rarer than supernovae. A widely discussed possibility is coalescence of binary neutron stars (see, for example, Narayan, Paczyński and Piran 1992). Systems such as the famous binary pulsar will eventually coalesce, when gravitational radiation drives them together. When a neutron star (NS) binary coalesces, the rapidly-spinning merged system would be too massive (for most presumed equations of state) to form a single NS; on the other hand, the total angular momentum is probably too large to be swallowed immediately by a black hole. The expected outcome, after a few milliseconds, would therefore be a spinning black hole (BH), orbited by a torus of neutron-density matter.

Other types of progenitor have been suggested—e.g. a NS-BH merger, where the neutron star is tidally disrupted before being swallowed by the hole; the merger of a white dwarf with a black hole; or a category labeled as hypernovae or collapsars, where the collapsing core is too massive to become a neutron star, but has too much angular momentum to collapse quietly into a black hole (as in a so called "failed supernova"). The simple point that I wish to stress, however, is that a BH surrounded by a neutron-density torus is a common feature of all these models; moreover the overall energetics of these various progenitors differ by at most an order of magnitude, the spread reflecting the differing spin energy in the hole and the different masses left behind in an orbiting torus. (There has been some confusion on this point in the literature, through failure to appreciate that the dominant energy from a NS-NS event comes after a black hole forms, rather than during the precursor stage that Narayan et al. (1992) discussed.) How might such a system generate relativistic outflow or a release of electromagnetic energy?

3. Energy from a black hole and debris torus?

Two large reservoirs of energy are in principle available: the binding energy of the orbiting debris, and the spin energy of the black hole. The first can provide up to 42% of the rest mass energy of the torus, for a maximally rotating black hole: the second can provide up to 29% (for a maximal spin rate) of the mass of the black hole itself.

How can the energy be transformed into outflowing relativistic plasma after such a coalescence event? There seem to be two options. The first is that some of the energy released as thermal neutrinos is reconverted, via collisions outside the dense core, into electron-positron pairs or photons.

The second option (which allows higher efficiency) is that strong magnetic fields anchored in the dense matter convert the rotational energy of the system into a Poynting-dominated outflow, rather as in pulsars. Let us consider these two options in turn.

(i) Neutrinos could give rise to a relativistic pair-dominate wind if they converted into pairs in a region of low baryon density (e.g. along the rotation axis, away from the equatorial plane of the torus). The $\nu\bar{\nu} \to e^+e^-$ process can tap the thermal energy of the torus produced by viscous dissipation. For this mechanism to be efficient, the neutrinos must escape before being advected into the hole; on the other hand, the efficiency of conversion into pairs (which scales with the square of the neutrino density) is low if the neutrino production is too gradual. Typical estimates suggest a limit of $\lesssim 10^{51}$ erg (Ruffert, 1997, Ruffert et al. 1997, Ruffert & Janka 1999, Popham, Woosley & Fryer, 1998), except perhaps in the "collapsar" or failed SN Ib case where Popham et al. (1998) estimate 2×10^{53} ergs for optimum parameters. These models are discussed fully in Woosley's contribution. If the pair-dominated plasma were collimated into a solid angle Ω_j then of course the apparent "isotropized" energy would be larger by a factor $(4\pi/\Omega_j)$, but unless Ω_j is $\lesssim 10^{-2} - 10^{-3}$ this may fail to reach the apparent isotropized energy of the most luminous bursts.

(ii) An alternative way to tap the torus energy is via magnetic fields threading the torus (Paczyński, 1991; Narayan, Paczyński & Piran, 1992; Mészáros & Rees, 1997b; Katz & Piran 1997). Even before the BH forms, a NS-NS merging system might lead to winding up of the fields and dissipation in the last stages before the merger (Mészáros & Rees, 1992; Vietri, 1997). Similar amplification may occur in the degenerate tori discussed by Woosley and his collaborators.

Simple scaling from the familiar results of pulsar theory tell us that fields of order 10^{15} G, are needed to carry away the rotational or gravitational energy in the time scales of tens of seconds (Usov, 1994, Thompson, 1994). If the magnetic fields do not thread the BH, then a Poynting outflow can at most carry the gravitational binding energy of the torus. This is between 0.06 and 0.42 of the rest mass energy of the torus, depending on the spin of the hole. The torus mass in a NS-NS merger is $M_t \sim 0.1$ M$_\odot$ (Ruffert et al. 1997), and for an NS-BH or WD-BH merger it may be $M_t \sim 1$ M$_\odot$ (Paczyński, 1998, Fryer & Woosley, 1998).

The extractable energy could amount to several times $10^{53}\epsilon(M_t/\rm M_\odot)$ ergs, where ϵ is the efficiency in converting gravitational into MHD jet energy. Tori masses even higher than ~ 1 M$_\odot$ may occur in scenarios involving massive supernovae. Conditions for the efficient escape of a high-Γ jet may, however, be less propitious if the "engine" is surrounded by an extensive envelope.

If magnetic fields of comparable strength thread the BH, its rotational energy offers an extra (and even larger) source of energy that can in principle be extracted via the B-Z mechanism (Mészáros & Rees, 1997b). For a maximally rotating BH, this is $0.29 M_{\rm bh} c^2$ ergs, multiplied, of course, by some efficiency factor. A near-maximally rotating black hole is guaranteed in a NS-NS merger. The central BH will have a mass of about 2.5 M$_\odot$; the NS-BH merger and hypernova models may not produce quite such rapidly-spinning holes, but the hole masses are larger, so the expected rotational energy should be comparable. Spinning holes can thus power a jet of up to $\sim 1.5 \times 10^{54}$ ergs. Even allowing for low total efficiency (say 30%), a system powered by the torus binding energy would only require a modest beaming of the γ-rays by a factor $(4\pi/\Omega_j) \sim 20$, or no beaming if the jet is powered by the B-Z mechanism, to produce the equivalent of an isotropic energy of up to 10^{54} ergs. The fields of $\sim 10^{15}$ G required for efficient electromagnetic extraction of energy are not significantly higher than those directly inferred in "magnetars" (cf. Kouveliotou, 1999).

Even in the collapsar model, magnetic extraction of energy seems likely to be more efficient than relativistic pairs generated by neutrinos. (It is, however, harder to quantify than the latter, and therefore has not been included in the simulations by Woosley, Janka and their collaborators.)

Nevertheless, I think these models would be found to supply the requisite energy with less "tuning" of the viscosity if magnetic fields and Poynting flux could be allowed for.

4. The gamma-ray emission mechanism

Well-known arguments connected with opacity, variability timescales and so forth (see, for instance, Piran (1997) and these proceedings) require highly relativistic outflow. Best-guess numbers are Lorentz factors Γ in the range 10^2 to 10^3, allowing rapidly-variable emission to occur at radii in the range 10^{14} to 10^{16} cms. The entrained baryonic mass would need to be below 10^{-4} $M_\odot$ to allow these high relativistic expansion speeds.

Because the emitting region must be several powers of ten larger than the compact object that acts as "trigger," there is a further physical requirement: the original energy outflowing in a magnetized wind would, after expansion, be transformed into bulk kinetic energy (with associated internal cooling). This energy cannot be efficiently radiated as gamma rays unless it is re-randomized. This requires relativistic shocks. Impact on an external medium would randomize half of the initial energy merely by reducing the expansion Lorentz factor by a factor of 2. Alternatively, there may be internal shocks within the outflow: for instance, if the Lorentz factor in an outflowing wind varied by a factor more than 2, then the shocks that developed when fast material overtakes slower material would be internally relativistic (Rees & Mészáros 1994; see also references given in Piran's paper).

In an unsteady outflow, if Γ were to vary by a factor of $\gtrsim 2$ on a timescale δt, internal shocks would develop at a distance $\Gamma^2 c\delta t$, and randomize much of the energy. For instance, if Γ ranged between 500 and 2000, on a timescale of 1 second, efficient dissipation would occur at 3×10^{16} cms.

There is a general consensus that the longer complex bursts must involve internal shocks, though simple sharp pulses could arise from an external shock interaction (the latter would in effect be the precursor of the afterglow). An external shock moving into a smooth medium would obviously give a burst with a simple time-profile. A blobby external medium could give features, but only if the covering factor of blobs is low, implying modest efficiency.

Even if the bursts were caused by a completely standardized set of objects, their appearance would be likely to depend drastically on orientation relative to the line of sight. Along any given line of sight, the time-structure would be determined partly by the advance of jet material into the external medium, but probably even more by internal shocks within the jet, which themselves depend on the evolution of the torus, from its formation to its eventual swallowing or dispersal.

The radiation processes for the gamma rays are probably no more than synchrotron radiation. This would imply the presence of magnetic fields where the shocks occur. If the outflow from the central trigger is Poynting-dominated, then a field of 10^{15} G at (say) 10^7 cm would imply a comoving field of $10^7(\Gamma/100)^{-1}$ G out at 10^{13} cm—strong enough to ensure rapid cooling of shocked relativistic electrons. (Note, conversely, that even if magnetic fields were not important near the central trigger, they must be present, with about the same amount of flux that Poynting-dominated models require, at the location of the actual gamma-ray emission.)

We are a long way from modelling what triggers gamma ray bursts. If we had a precise description of the dynamics, along with the baryon content, magnetic field, and Lorentz factor of the outflow, we could maybe predict the gross time-structure. But we could not predict the intensity or spectrum of the emitted radiation—still less answer key questions about the emission in other wavebands—without also having an adequate theory for particle acceleration in relativistic shocks. We need the answers from plasma physicists to the following poorly-understood questions:

(i) Do ultra-relativistic shocks yield power laws? The answer probably depends on the ion/positron ratio, and on the relative orientation of the shock front and the magnetic field (e.g. Gallant et al. 1992).

(ii) In ion-electron plasmas, what fraction of the energy goes into the electrons?

(iii) Even if the shocked particles establish a power law, there must be a low-energy break in the spectrum at an energy that is in itself relativistic. But will this energy, for the electrons, be $\Gamma m_p c^2$, or (or even, if the positive charges are heavy ions like Fe, $\Gamma m_{Fe} c^2$)?

(iv) Can ions be accelerated up to the theoretical maximum where the gyroradius becomes the scale of the system? If so, the burst events could be the origin of the highest energy cosmic rays (as discussed in Waxman's paper).

(v) Do magnetic fields get amplified in shocks? (This is relevant to the magnetic field in the swept-up external matter outside the contact discontinuity, and determines how sharp the external shock actually is.)

5. Intrinsic time scales

A question which has remained largely unanswered so far is what determines the characteristic duration of bursts, which can extend to tens, or even hundreds, of seconds. This is of course very long in comparison with the dynamical or orbital time scale for the "triggers," which is measured in milliseconds. While bursts lasting hundreds of seconds can easily be derived from a very short, impulsive energy input, this is generally unable to account for a large fraction of bursts which show complicated light curves. This hints at the desirability for a "central engine" lasting much longer than a typical dynamical time scale.

Observationally (Kouveliotou, et al., 1993) the short ($\lesssim 2$ s) and long ($\gtrsim 2$ s) bursts appear to represent two distinct subclasses, and one early proposal to explain this was that accretion induced collapse (AIC) of a white dwarf (WD) into a NS plus debris might be a candidate for the long bursts, while NS-NS mergers could provide the short ones (Katz & Canel, 1996). As indicated by Ruffert et al. (1997), $\nu\bar{\nu}$ annihilation will generally tend to produce short bursts $\lesssim 1$ s in NS-NS systems, requiring collimation by 10^{-1}–10^{-2}, while Popham, Woosley & Fryer (1999) argued that in and WD/He-BH systems longer $\nu\bar{\nu}$ bursts may be possible.

An acceptable model requires that the surrounding torus should not completely drain into the hole, or be otherwise dispersed, on too short a time scale. In a torus that was massive and/or thin enough to be self-gravitating, bar-mode gravitational instabilities could lead to further redistribution of angular momentum and/or to energy loss by gravitational radiation within only a few orbits. Whether a torus of given mass is dynamically unstable depends on its thickness and stratification, which in turn depends on internal viscous dissipation and neutrino cooling.

The disruption of a neutron star (or any analogous process) is almost certain to lead to a situation where violent instabilities redistribute mass and angular momentum within a few dynamical time scales (i.e. in much less than a second). A key issue for gamma ray burst models involving coalescing binaries is the nature of the surviving debris after these violent processes are over: what is the maximum mass of a remnant disc/torus which is immune to very violent instabilities, and which can therefore in principle survive for long enough to power the bursts? If the torus results from the disruption of a compact binary, then the *residual* mass left over after violent instabilities on a dynamical timescale have done their work is the relevant M_t in the above expressions (in Section 3) for the extractable energy of the torus. In collapsar models, the torus is not created suddenly, but is replenished by infall from the degenerate stellar core on a timescale ~ 10 sec. In these latter models, the long durations arise naturally, since they do not require a low viscosity (and long residence time) in the relativistic torus.

If the trigger is to liberate its energy over a period 10–100 sec via Poynting flux—either through a relativistic wind "spun off" the torus or via the B-Z mechanism—the required field is a few times 10^{15} G. A weaker field would extract inadequate power; on the other hand, if the large-scale field were even stronger, then the energy would be dumped too fast to account for the longer complex bursts. It is not obvious why the fields cannot become even higher. Note that the virial limit is $B_v \sim 10^{17} =$ G.

Kluzniak and Ruderman (1998) note that, starting with 10^{12} G, it only takes of order a second for simple winding to amplify the field to 10^{15} G; amplification in a newly-formed torus could well occur more rapidly, for instance via convective instabilities, as in a newly formed neutron star (cf. Thompson & Duncan, 1993, Thompson, 1994). Kluzniak and Ruderman suggest, however, that the amplification may be self-limiting because magnetic stresses would then be strong enough for flares to break out. A magnetic field configuration capable of powering the bursts is likely to have a large scale structure. Flares and instabilities occurring on the characteristic millisecond) dynamical time scale would cause substantial irregularity or intermittency in the overall outflow that would manifest itself in internal shocks. There is thus no problem in principle in accounting for sporadic large-amplitude variability, on all time scales down to a millisecond, even in the most long-lived bursts. Note also that it only takes a residual torus of 10^{-3} $M_\odot$ to confine a field of 10^{15} G, which can extract energy from the black hole via the B-Z mechanism.

6. How much beaming?

Computer simulations of compact object mergers and black hole formation can address the fate of the bulk of the matter, but there are some key questions that they cannot yet tackle. In particular, high resolution of the outer layers is needed because even a tiny mass fraction of baryons loading down the outflow severely limits the attainable Lorentz factor—for instance a Poynting flux of 10^{53} ergs could not accelerate an outflow to $\Gamma > 100$ if it had to drag more than $\sim 10^{-4}$ solar masses of baryons with it. Further 2D numerical simulations of the merger and collapse scenarios are under way largely using Newtonian dynamics, and the numerical difficulties are daunting. There may well be a broad spread of Lorentz factors in the outflow—close to the rotation axis Γ may be very high; at larger angles away from the axis, there may be an increasing degree of entrainment, with a corresponding decrease in Γ. Even if the outflow is not narrowly collimated, some beaming is expected because energy would be channeled preferentially along the rotation axis. Moreover, we would expect baryon contamination to be lowest near the axis, because angular momentum flings material away from the axis, and any gravitationally-bound material with low angular momentum falls into the hole. In hypernovae, the envelope is rotating only slowly and thus would not initially have a marked centrifugal funnel; even 10^{53} ergs would not suffice to blow out more than a narrow cone of the original envelope with a Lorentz factor or more than 100. So in these models the gamma rays would be restricted to a narrow beam, even though outflow with a more moderate Lorentz factor (relevant to the afterglow) could be spread over a wider range of angles. A wide variety of burst phenomenology could be attributable to a standard type of event being viewed from different orientations.

Two further effects render the computational task of simulating jets even more challenging, The first stems from the likelihood that any entrained matter would be a mixture of protons and neutrons (neutrons, being unconstrained by magnetic fields, could also drift into a jet from the denser walls at its boundary). If a streaming velocity builds up between ions and neutrons (i.e. if they have different Lorentz factors in the outflow) then interactions can lead to dissipation even in a steady jet where there are no shocks (Derishev et al. 1999).

A second possibility (Mészáros & Rees 1998a,b) is that entrained ions in a relativistic jet could become concentrated in dense filaments confined by the magnetic field. As already mentioned, the comoving field strength, even out at 10^{13} cm, is of order 10^6 G. Trapped filaments of iron-rich thermal, with density up to $10^{19}\,\mathrm{cm}^{-3}$ and with kT of order a keV, could be confined by such fields. Such filaments must of course have a small volume-filling factor: otherwise they would load down the jet too much. However, in these strong fields the gyroradii would be so small that filaments could survive against thermal conduction and other diffusion processes even if their dimensions (transverse to the field) were less than 100 cm. Such thin filaments can provide a large covering factor even while filling a tiny fraction of the volume. If they were moving relativistically outwards, they could contribute ultra-blueshift spectral features—for instance, K-edges of Fe could be shifted up to hundreds of keV.

7. Brief comments on the afterglows

Astrophysicists understand supernova *remnants* reasonably well, despite continuing uncertainty about the initiating explosion; likewise, we may hope to understand the afterglows of gamma ray bursts (see Hartmann (1999) for a recent review) despite the uncertainties about the "trigger" that I have already emphasized. The simplest hypothesis is that the afterglow is due to a relativistic expanding blast wave. The complex time-structure of some bursts suggests that the central trigger may continue for up to 100 seconds. However, at much later times all memory of the initial time-structure would be lost: essentially all that matters is how much energy and momentum has been injected, its distribution in angle, and the mass fractions in shells with different Lorentz factors.

The discovery of afterglows has not only extended observations to longer time scales and other wavebands, making the identification of counterparts possible, but also provided confirmation for much of the earlier work on the fireball shock model of GRB, in which the γ-ray emission arises at radii of 10^{13}–10^{15} cm (Rees & Mészáros, 1992, 1994; Mészáros & Rees 1993; Paczyński & Xu, 1994; Katz, 1994; Sari & Piran, 1995). In particular, this model led to the prediction of the quantitative nature of the signatures of afterglows, in substantial agreement with subsequent observations (Mészáros & Rees, 1997a; Costa et al., 1997; Vietri, 1997a, Tavani, 1997; Waxman, 1997; Reichart, 1997; Wijers et al., 1997).

The simplest spherical afterglow model—where a relativistic blast wave decelerates as it runs into ambient matter, leading to a radiative output with a calculable spectrum, and a characteristic power law decay—has been remarkably successful at explaining the gross features of the GRB970228, GRB970508 and other afterglows (e.g. Wijers, et al., 1997). The gamma-rays we receive come only from material whose motion is directed within one degree of our line of sight. They therefore provide no information about the ejecta in other directions: the outflow could be isotropic, or concentrated in a cone of any angle substantially larger than one degree (provided that the line of sight lay inside the cone). At observer times of more than a week, the blast wave would however be decelerated to a moderate Lorentz factor, irrespective of the initial value. The beaming and aberration effects are thereafter less extreme, so we observe afterglow emission not just from material moving almost directly towards us, but from a wider range of angles.

The afterglow is thus a probe for the geometry of the ejecta—at late stages, if the outflow is beamed, we expect a spherically-symmetric assumption to be inadequate; the deviations from the predictions of such a model would then tell us about the ejection in directions away from our line of sight. It is quite possible, for instance, that there is relativistic outflow with lower Γ (heavier loading of baryons) in other directions (e.g. Wijers, Rees, & Mészáros, 1997); this slower matter could even carry most of the energy (Paczyński, 1998). Rhoads (1997) noted that if the energy were channeled into a solid angle Ω_j, one expects a faster decay of Γ after it drops below $\Omega_j^{-1/2}$. A simple calculation using the usual scaling laws leads then to a steepening of the flux power law in time. Anisotropy in the burst outflow and emission affects the light curve at the time when the inverse of the bulk Lorentz factor equals the opening angle of the outflow. If the critical Lorentz factor is less than 3 or so (i.e. the opening angle exceeds $20°$) such a transition might however be masked by the transition from ultrarelativistic to mildly relativistic flow, so quite generically it would difficult to limit the late-time afterglow opening angle in this way if it exceeds $20°$.

The beaming angle for the gamma ray emission, requiring Γ to be $\gtrsim 100$, could be far smaller than for the overall relativistic outflow, and is much harder to constrain directly.

If the gamma rays were much more narrowly beamed than the optical afterglow there should be many "homeless" afterglows, i.e. ones without a GRB preceding them. The rate of GRB with peak fluxes above 1 ph cm^{-2} s^{-1} as determined by BATSE is about 300/yr, i.e. 0.01/sq. deg/yr. According to Wijers et al. (1998) this flux corresponds to a redshift of 3. The transient sky at faint magnitudes is poorly known, but there are two major efforts under way to find supernovae down to about $R = 23$ (Garnavich et al. 1998, Perlmutter et al. 1998). These searches have by now covered a few tens of "square degree years" of exposure and would be sensitive to afterglows of the brightness levels thus far observed. It therefore appears that the afterglow rate is not more than a few times 0.1/sq. deg/yr. Since the magnitude limit of these searches allows detection of optical counterparts of GRB brighter than 1 ph cm^{-2} s^{-1} it is fair to conclude that the ratio of homeless afterglows to GRB is unlikely to exceed ~ 20. It then follows that $\Omega_\gamma > 0.05 \Omega_{\rm opt}$, which combined with our limit to $\Omega_{\rm opt}$ yields $\Omega_\gamma > 0.02$. The true rate of events that give rise to GRB is therefore at most 600 times the observed GRB rate, and the opening angle of the ultrarelativistic, gamma-ray emitting material is no less than $5°$.

Obviously, the above calculation is only sketchy and should be taken as an order of magnitude estimate at present. However, it should improve (cf. Rhoads 1997) as more afterglows are detected and the modeling gets more precise.

8. Conclusions and prospects

There are two key questions regarding the "trigger." First, does it involve a black hole orbited by a dense torus (which I've advocated as a "best buy")? Second, if so, can we decide between the various alternative ways of forming it: NS-NS, NS-BH or collapsar/hypernova?

The locations should help to settle the second question. This is because a collapsar/hypernova would be expected to lie in a region of recent star formation; on the other hand, a neutron star binary could take hundreds of millions of years to spiral together, and could by then (especially if given a "kick velocity" on formation) have moved many kiloparsecs from its point of origin (Bloom, Sigurdsson & Pols, 1999). There is also already tentative evidence that some detected afterglows arise in relatively dense gaseous environments—e.g. by evidence for dust in GRB970508 (Reichart, 1998) and the absence of an optical afterglow and strong soft X-ray absorption in GRB970828 (Groot et al., 1997; Murakami et al. 1997). On the other hand, fits to the observational data on GRB970508 and GRB971214 suggest external densities in the range of 0.04–0.4 cm^{-1}, which would be more typical of a tenuous interstellar medium (Wijers & Galama 1998).

We must nonetheless remain open minded about other possibilities. For instance, we may be wrong in supposing that the central object becomes dormant after the gamma-ray burst itself. It could be that the accretion-induced collapse of a white dwarf, or (for some equations of state) the merger of two neutron stars, could give rise to a rapidly-spinning pulsar, temporarily stabilized by rapid rotation. The afterglow could then, at least in part, be due to a pulsar's continuing power output (cf. Usov 1994). It could also be that mergers of unequal mass neutron stars, or neutron stars with other compact companions, lead to the delayed formation of a black hole. Such events might also lead to repeating episodes of accretion and orbit separation, or to the eventual explosion of a neutron star which has dropped below the critical mass, all of which would provide a longer time scale, episodic energy output.

And there could be more subclasses of classical GRB than just short ones and long ones. There is for instance the apparent coincidence of GRB980425 with the SN Ib/Ic 1998bw (Galama et al., 1998). Much progress has been made in understanding how gamma-rays can arise in fireballs produced by brief events depositing a large amount of energy in a small volume, and in deriving the generic properties of the long wavelength afterglows that follow from this. There still remain a number of mysteries, especially concerning the identity of their progenitors, the nature of the triggering mechanism, the transport of the energy and the time scales involved. Finally, it is worth noting that gamma-ray bursts, even if we do not understand them, may still be useful as powerful beacons for probing the high redshift ($z > 5$) universe.

Even if their total energy is reduced by beaming to a "modest" $\sim 10^{52}$ ergs in photons, they are the most extreme phenomena that we know about in high energy astrophysics. The modeling of the burst itself—the trigger, the formation of the ultrarelativistic outflow, and the radiation processes—is a formidable challenge to theorists and to computational techniques. It is, also, a formidable challenge for observers, in their quest for detecting minute details in extremely faint and distant sources. And if the class of models that we have advocated here turns out to be irrelevant, the explanation of gamma-ray bursts will surely turn out to be even more remarkable and fascinating, perhaps implicating magnetic fields even stronger than 10^{15} G, or matter even more exotic than in neutron stars.

I am especially grateful to Peter Mészáros and Ralph Wijers for extended collaboration on this subject, and to Josh Bloom, Max Ruffert and Stan Woosley for discussions. This research has been supported by the Royal Society.

REFERENCES

BLANDFORD, R. D. & ZNAJEK, R. L. 1977 *MNRAS*, **179**, 433.
BLOOM, J. S., SIGURDSSON, S., & POLS, O. 1998 *MNRAS*, **305**, 763.
DERISHEV, E. V., KOCHAROVSKY, V. V., & KOCHAROVSKY, V. V. 1999, preprint.
FRYER, C. & WOOSLEY, S. 1998 *ApJ*, 502, L9.
GALAMA, T., ET AL. 1998 *Nature*, **395**, 670.
GALLANT, Y. A., HOSHINO, M., LANGDON, A. B., ARONS, J., & MAX, C. E. 1992 *ApJ*, 391, 73.
GARNAVICH, P., ET AL. 1998 *ApJ*, **493**, L53.
GROOT, P., ET AL. 1997 in *Gamma-Ray Bursts*, (eds. C. Meegan, R. Preece, & T. Koshut). AIP. p. 557.
HARTMANN, D. 1999 *Proc. Nat. Acad. Sci.*, **96**, 4753.
KATZ, J. 1994 *ApJ*, **422**, 248.
KATZ, J. & CANEL, L. M. 1996 *ApJ*, **471**, 915.
KATZ, J. & PIRAN, T. 1997 *ApJ*, **490**, 772.
KLEBESADEL, R. W., STRONG, I. B., & OLSEN, R. A. 1973 *ApJ*, **182**, L85.
KLUŹNIAK, W. & RUDERMAN, M. 1998, preprint.
KOUVELIOTOU, C. 1999 *PNAS*, **96**, 5351.
KOUVELIOTOU, C., ET AL. 1993 *ApJ*, **413**, L101.
LAMB, D. 1995 *PASP* **107**, 1152.
MÉSZÁROS, P. & REES, M. J. 1992, *ApJ*, **397**, 570.
MÉSZÁROS, P. & REES, M. J. 1993, *ApJ*, **405**, 278.
MÉSZÁROS, P. & REES, M. J. 1997a, *ApJ*, **476**, 232.
MÉSZÁROS, P. & REES, M. J. 1997b, *ApJ*, **482**, L29.
MÉSZÁROS, P. & REES, M. J. 1998a, *ApJ*, **502**, L105.
MÉSZÁROS, P. & REES, M. J. 1998b, *MNRAS*, **299**, L10.
MURAKAMI, T., ET AL. 1997, in *Gamma-Ray Bursts* (eds. C. Meegan, R. Preece & T. Koshut). (AIP). p. 435.
NARAYAN, R., PACZYŃSKI, B., & PIRAN, T. 1992 *ApJ*, **395**, L83.
PACZYŃSKI, B. 1991 *Acta. Astron*, **41**, 257.
PACZYŃSKI, B. 1995 *PASP*, **107**, 1167.
PACZYŃSKI, B. 1998 *ApJ*, **494**, L45.
PACZYŃSKI, B. & XU, G. 1994 *ApJ*, **427**, 708.
PERLMUTTER, S., ET AL. 1998 *Nature*, **391**, 51.
PIRAN, T. 1997 in *Unsolved Problems in Astrophysics*, (eds. J. N. Bahcall & J. P. Ostriker). Princeton UP.
PODSIADLOWSKI, P., REES, M. J., & RUDERMAN, M. 1995 *MNRAS*, **273**, 755.
POPHAM, R., WOOSLEY, S., & FRYER, C. 1999 *ApJ*, **518**, 356.
REES, M. J. & MÉSZÁROS, P. 1992 *MNRAS*, **258**, 41P.
REES, M. J., & MÉSZÁROS, P. 1994, *ApJ*, **430**, L93.
REICHART, D. 1997 *ApJ*, **485**, L57.
REICHART, D. 1998 *ApJ*, **495**, L99.

RHOADS, J. E. 1997 *ApJ*, **487**, L1.
RUDERMAN, M. 1975 *Ann. N. Y. Acad. Sci.*, **262**, 164.
RUFFERT, M. 1997 *A& A*, **317**, 793.
RUFFERT, M. JANKA, H.-T., TAKAHASHI, K., & SCHAEFER, G. 1997 *A& A*, **319**, 122.
RUFFERT, M. & JANKA, H.-T. 1999 *A& A*, **344**, 573.
SARI, R. & PIRAN, T. 1995 *ApJ*, **455**, L143.
TAVANI, M. 1997 *ApJ*, **483**, L87.
THOMPSON, C. 1994 *MNRAS*, **270**, 480.
THOMPSON, C. & DUNCAN, R. C. 1993 *ApJ*, **408**, 194.
USOV, V. V. 1994 *MNRAS*, **267**, 1035.
VIETRI, M. 1997a *ApJ*, **478**, L9.
WAXMAN, E. 1997 *ApJ*, **489**, L33.
WIJERS, R. A. M. J., BLOOM, J. S., BAGLA, J. S., & NATARAJAN, P. 1998 *MNRAS*, **294**, L13.
WIJERS, R. A. M. J. & GALAMA, T. 1998 *ApJ*, in press (astro-ph/9805341).
WIJERS, R. A. M. J., REES, M. J., & MÉSZÁROS, P. 1997 *MNRAS*, **288**, L51.
WOOSLEY, S., EASTMAN, R., & SCHMIDT, B. 1999 *ApJ*, **516**, 788.

Prompt optical emission from gamma-ray bursts

By ROBERT KEHOE,[1] CARL AKERLOF,[1]
RICHARD BALSANO,[2] SCOTT BARTHELMY,[3,4]
JEFF BLOCH,[2] PAUL BUTTERWORTH,[3,5]
DON CASPERSON,[2] TOM CLINE,[3]
SANDRA FLETCHER,[2] FILLIPPO FRONTERA,[6]
GALEN GISLER,[2] JOHN HEISE,[7] JACK HILLS,[2]
KEVIN HURLEY,[8] BRIAN LEE,[1]
STUART MARSHALL,[9] TIM MCKAY,[1]
ANDREW PAWL,[1] LUIGI PIRO,[10]
BILL PRIEDHORSKY,[2] JOHN SZYMANSKI,[2]
AND JIM WREN[2]

[1]University of Michigan, Ann Arbor, MI 48109

[2]Los Alamos National Laboratory, Los Alamos, NM 87545

[3]NASA/Goddard Space Flight Center, Greenbelt, MD 20771

[4]Universities Space Research Association, Seabrook, MD 20706

[5]Raytheon Systems, Lanham, MD 20706

[6]Università degli Studi di Ferrara, Ferrara, Italy

[7]Space Research Organization, Utrecht, The Netherlands

[8]Space Sciences Laboratory, University of California, Berkeley, CA 94720-7450

[9]Lawrence Livermore National Laboratory, Livermore, CA 94550

[10]Instituto Astrofisica Spaziale, Rome, Italy

The Robotic Optical Transient Search Experiment (ROTSE) seeks to measure contemporaneous and early afterglow optical emission from gamma-ray bursts (GRBs). The ROTSE-I telescope array has been fully automated and responding to burst alerts from the GRB Coordinates Network since March 1998, taking prompt optical data for 30 bursts in its first year. We will briefly review observations of GRB990123 which revealed the first detection of an optical burst occurring during the gamma-ray emission, reaching 9th magnitude at its peak. In addition, we present here preliminary optical results for seven other gamma-ray bursts. No other optical counterparts were seen in this analysis, and the best limiting sensitivities are $m_V > 13.0$ at 14.7 seconds after the gamma-ray rise, and $m_V > 16.4$ at 62 minutes. These are the most stringent limits obtained for GRB optical counterpart brightness in the first hour after the burst. This analysis suggests that there is not a strong correlation between optical flux and gamma-ray emission.

1. Introduction

1.1. Gamma-ray observations

Fast, intense bursts of cosmic gamma-rays and energetic X-rays were first observed about 30 years ago (Klebesadel et al. 1973). Since that time, satellite missions have determined several characteristics of these events. They are generally very brief but are otherwise extremely diverse in their gamma-ray temporal variations. Durations range from 0.005 to 100s of seconds, and intensity fluctuations are as short as 0.3 ms (Hurley et al. 1984). They are often instantaneously the brightest gamma-ray sources in the sky. Studies of the over 2000 currently recorded bursts indicate a thoroughly isotropic distribution with no detectable concentration towards the galactic plane and no angular correlations with other astrophysical structures (Meegan et al. 1998).

1.2. Counterparts at other wavelengths

Given the lack of a spatial or temporal pattern, it has been extremely difficult to comprehend the physical mechanisms from the gamma-ray observations alone. Since the mid-70s, there have been many attempts to detect counterparts at other wavelengths, but they were unsuccessful until 1997. The difficulty arose from the brevity of bursts, the lack of arc-minute localizations and theoretical prejudices concerning the burst progenitor. Currently, the two main GRB efforts utilize the BATSE detectors (Fishman et al. 1989) on-board the *Compton Gamma-Ray Observatory*, and the GRBM (Feroci et al. 1997) and WFC (Jager et al. 1997) on the *BeppoSAX* satellite, and they have addressed the observational limitations very differently. BATSE's advantage is its near complete coverage of the sky, which allows observation of about 300 bursts per year, and its unique ability to provide rough coordinates very rapidly ($\sim$ 5 seconds). These localizations are distributed in the form of triggers over the GRB Coordinates Network (GCN) (Barthelmy et al. 1998, Barthelmy et al. 1995). *BeppoSAX*, on the other hand, is able to deliver positions accurate to a few arcminutes in a few hours for about a dozen bursts per year.

The *BeppoSAX* positions are accurate enough for follow-up with conventional, small field-of-view telescopes which have detected optical counterparts for twelve GRBs. As a result, we now know that at least some GRBs are at cosmological distance (e.g. Metzger et al. 1997, Kulkarni et al. 1998) and, if their emission is isotropic, release a significant fraction of $M_\odot c^2$. These GRBs, at least, are associated with galaxies (Hogg & Fruchter 1999). Studying this afterglow period of a few hours to days after the burst has generally revealed a slow, roughly power-law decay of optical emission with time.

These observations, however, are mute concerning the details of the burst itself, and there are several limitations in the current sample. In particular, the number of such events is small. There is also a bias which arises from the requirement of an observable X-ray counterpart. In addition, because the *BeppoSAX* GRBM uses a 1 second integration period, no counterparts at any wavelength have been identified for the class of short ($\sim$ 0.1 second) bursts. Short bursts are more likely to have high energy emission, and they occupy a region of the hardness-ratio vs. duration space well-separated from long bursts (Kouveliotou et al. 1993, Kouveliotou et al. 1995). This may imply a different type of progenitor for short bursts. Lastly, these observations occur hours after any gamma-ray emission, so that despite a growing understanding of afterglows, the burst origin remains a mystery.

1.3. Prompt, unbiased optical detections

Studying early optical emission in an unbiased way has several advantages. First, observations at early times may elucidate details of shock development, the burst environment and beaming. Second, by looking for optical emission unbiased by selections based on burst duration or fluence, we can probe more thoroughly their range of behavior.

While we do not know the actual mechanism by which a GRB occurs, there is a general picture of the development of the cataclysm aftermath. A highly relativistic expanding fireball is created in which shells develop within the outflow with a spread in velocities. Gamma-rays emerge from interactions among these shells (Rees & Meszaros 1994, Paczynski & Xu 1994). The chaotic time histories of the gamma-rays favor variability from a central engine such as in the internal shock models (Fenimore et al. 1996). As the relativistic shell propagates into the interstellar medium, its deceleration produces a forward shock wave (Rees & Meszaros 1992, Meszaros & Rees 1993) and possibly a reverse shock. The afterglow is believed to arise from the forward shock, while significant early optical emission may arise from the reverse shock. In this scenario, comparisons of simultaneous optical and gamma-ray emission comment on the presence and progress of the external shocks. For instance, the relative timing of optical and gamma-ray emission indicates the Lorentz factors involved (Sari & Piran 1999) as well as the process by which the shells responsible for the external shocks arise.

In addition, we can learn about the environment of the source at the time of the burst. The detectability of optical emission alone demonstrates that the local environment of the burst is not opaque to optical photons.

Since the relativistic shell is initially moving at very high bulk Lorentz factors, $\Gamma > 100$, there will be a beaming angle which is expected to vary roughly as $1/\Gamma$. If the optical photons arise from a more slowly moving region than the gamma-rays, they will be more isotropically emitted. This can occur, for instance, in the case of later optical emission from the decelerating relativistic shell. As a result, the correlation between this optical emission and the gamma-rays will be weak and we could observe a bright optical counterpart to a dim or absent gamma-ray burst (Rhoads 1999). In addition, the energy release of a GRB might not be isotropic, being restricted to a jet of angular width, θ. This restriction will dominate the optical emission once $\Gamma \sim 1/\theta$. After some time, however, there will be lateral spreading of the jet which will further increase the isotropic distribution of the later optical light (Rhoads 1999).

Any detection of optical emission from short bursts could reveal their relation to the longer bursts. The short bursts may have a different redshift distribution than that observed for long bursts. The two populations might arise from very different sources, such as neutron star mergers and hypernovae. If so, they should have distinct distributions within galaxies, and their local environments might be quite different. The fireball mechanism may be entirely different for short bursts. If so, prompt optical observations should help illuminate their properties.

2. The ROTSE Project

2.1. Challenges

To probe the nature of GRBs, the ROTSE project seeks to detect their prompt optical emission. In particular, we wish to detect optical photons coincident with a burst and observe optical afterglows to a few hours afterwards. Ultimately, we wish to do this for a sample of GRBs as free of detection biases as possible. We have a secondary mission to obtain and provide a stream of arcsecond-level positions for many GRBs for more sensitive follow-up.

To do this, several technical challenges must be met. First, the gamma-ray emission lasts a few seconds or less so we need to respond to triggers for gamma-ray bursts in real-time. Second, they vary rapidly during their gamma-ray emission and might be expected to do the same in the optical. Therefore, we must frequently image GRB positions to measure short time-scale variation. All of this requires a fully automated system. Third, the field-of-view of the instrument must match the positional accuracy of the trigger which, for BATSE coordinates, is 5 to 20 degrees. While this is far larger than the field-of-view of conventional optical telescopes, it is achievable in a moderately sensitive ($m_V \sim 15$) configuration. Finally, for analysis purposes, we must be able to distinguish rare signals from a variety of backgrounds.

We have developed a compact, flexible design consisting of fully automated mini-observatories. In the rest of Section 2, we discuss the technical details of our experiment, Section 3 outlines the operation of ROTSE-I, Section 4 reviews the observations of GRB990123, Section 5 presents results on several other GRBs, and Section 6 presents an interpretation of those results.

2.2. Mini observatories

The ROTSE telescopes are sited in northern New Mexico inside enclosures providing for the control and protection of the hardware. These enclosures possess an automatable enclosure cover ('clamshell'). In general, they are instrumented with weather sensors to detect rain, clouds, temperature, humidity, and excessive wind. In addition, lightning is a serious hazard at the site, so surge suppressors must protect all electrical lines to the outside world. Uninterruptable power supplies (UPSs) perform some power-line filtering and provide about 10 minutes of power to gracefully shut down the system in case of power failure. Each enclosure's internal network and external connection runs on 100 Mbps ethernet. At the moment, the site itself is limited to a 10 Mbps connection.

Within the enclosures, there are computers and a custom control box for operation and monitoring of the various devices. The control box provides power to the telescope mount, clamshell, and the weather monitoring devices, as well as communication to the weather devices and clamshell. One of the keys to performing our experiment is the utilization of fast, inexpensive PCs. The division of labor among the computers in the enclosures varies, but in general there is a main computer on which our data acquisition system runs. This includes monitoring the weather devices, incoming triggers and system errors, control of the clamshell, and observation scheduling. Operation of the mount and data processing may also occur from this computer. Each camera is interfaced to an auxiliary PC via an ISA card interface.

Our data storage needs are handled by the Los Alamos Computing Division Mass Store System. This storage system has a several Petabyte capacity, and provides crucial random access to our large data set via a quick and convenient interface.

2.3. Software components

To achieve prompt response times and maximum livetime, each instrument must be automated, and it must operate in real-time. We chose Linux based on its stability, capability and cost, as well as the availability of drivers and other software parts necessary for the experiment. Although Linux was not inherently designed for real-time application, we can tolerate 0.1 second latencies in responsiveness which is well within the operating system's capabilities.

We have produced a small suite of programs as diagrammed in Figure 1. The overall structure of the system consists of a central switchboard process which channels *commands* from user input processes to hardware control programs via shared data structures. This switchboard also relays *status* information from the hardware control and monitoring programs (i.e. camera, mount, clamshell, GCN monitor, weather monitor, and watchdog) back to the users. The two user processes are an astronomical scheduler for automation and a modified UNIX shell for manual control. Aside from small portions directly interfacing to specific hardware, we have designed a simple, general structure for easy porting to newer systems as we develop them. A large effort has been made to produce a responsive system, so we have taken maximum advantage of Linux's interrupting and multitasking capabilities.

In order to be sure of the absolute timing of events, the main computer is synchronized to public servers using the Network Time Protocol (NTP), and the camera computers

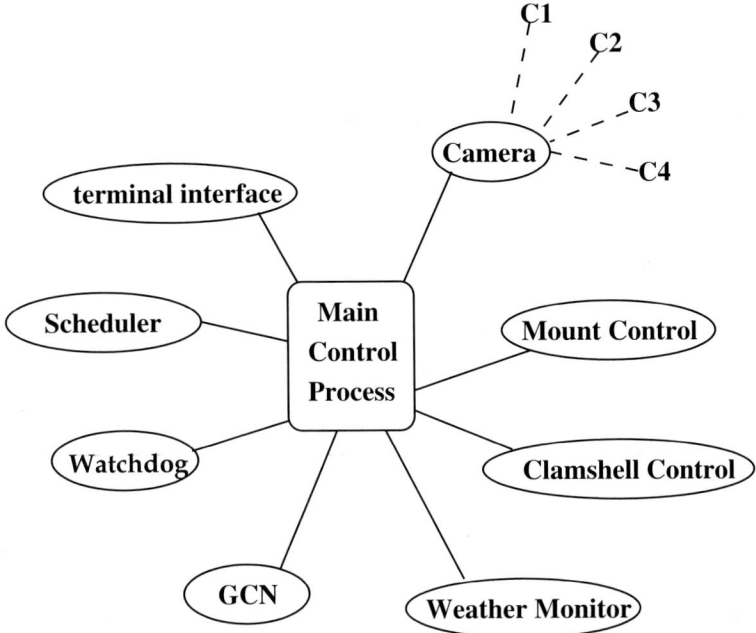

FIGURE 1. Scheme of ROTSE data-acquisition system. A central switchboard process running on the control computer channels commands from a user interface and scheduler to hardware control processes via shared data channels (solid lines). The instantaneous status of these processes and other monitors is continuously returned. Four additional PCs are used to control the four CCD cameras, with communication with the main computer proceeding over network connections (dashed lines).

are synchronized to the main computer also via NTP. This configuration maintains the main computer within 10 ms of UTC and the camera computers within about 1 ms of the main computer and each other.

2.4. Telescopes

We have developed a two-tiered program to cover the initial outburst and afterglow periods. ROTSE-I is fast, wide field-of-view, and moderately sensitive. It consists of a 2×2 array of small telescopes co-mounted on a rapid-slewing platform (see Figure 2). Each telescope is instrumented with a thermoelectrically cooled CCD camera employing a 14-bit Thompson TH7899M chip with 2048×2048 14 micron pixels. Read noise is $\sim 25 e^-$, and readout is limited by the ISA interface to take about 7 seconds. The optics of each telescope consist of a Canon FD 200 mm $f/1.8$ telephoto lens. We have equipped each with a focus-ring clamp positioned by a micrometer for accurate manual adjustment. Our sensitivity to faint point sources is maximized by the match of the optical point spread function to the pixel size. The plate scale for ROTSE-I images is $14.4''$/pixel. To further improve sensitivity, the cameras are operated without filters, and the peak response is in the R, V and I bands. Each telescope is sensitive to 14th magnitude in a 5 second exposure, and the array covers a $16.4° \times 16.4°$ field-of-view. The mount is capable of slewing to any point in the sky in less than 3 seconds. As shown in Figure 3, ROTSE-I is capable of seeing optical counterparts as dim as 14th magnitude by 10 seconds after a burst, and longer exposures achieve 16th magnitude sensitivity.

FIGURE 2. The ROTSE-I telephoto array. Four Canon lenses are each mounted on CCD cameras on a compact, fast mount. The telescope sits on the roof of the ROTSE-I enclosure.

The second stage of the experiment brings significant improvements in sensitivity to faint objects. Each of the two existing ROTSE-II telescopes consists of a wide-field modified Cassegrain optical tube instrumented with the same cameras as ROTSE-I and a mount with an average slew time of 15 seconds. We have started building an additional set of eight similar telescopes, called ROTSE-III, with improved optics, back-illuminated CCD chips, and substantially faster slew times. All of these consist of $f/1.9$ optical assemblies with 45 cm apertures and $1.9°$ field-of-view. The plate scale is $3.4"$/pixel. With these telescopes, we will study optical bursts for those few, prompt, accurate localizations from HETE2 (Ricker et al. 1990), and we will observe optical afterglows by scanning the neighborhood of BATSE burst locations. We will also search for non-triggered fast-fading optical transients which might have a similar physical origin. The estimated sensitivity of these instruments is shown in Figure 3.

3. ROTSE-I operations

3.1. Observation scheduling

The astronomical scheduler is responsible for starting and stopping a night's run, designing observing sequences during the night, and scheduling darks for image noise correction. There are currently two main observing modes. Most of the time is spent in a lower priority sky patrol. Given the ROTSE-I field-of-view, 206 frames cover the celestial sphere with reasonable overlaps. We observe all fields with elevation $> 20°$ in two successive images taken twice nightly. These images were 25 second exposures until December 1998, and generally 80 seconds since then. These data are valuable for untriggered transient

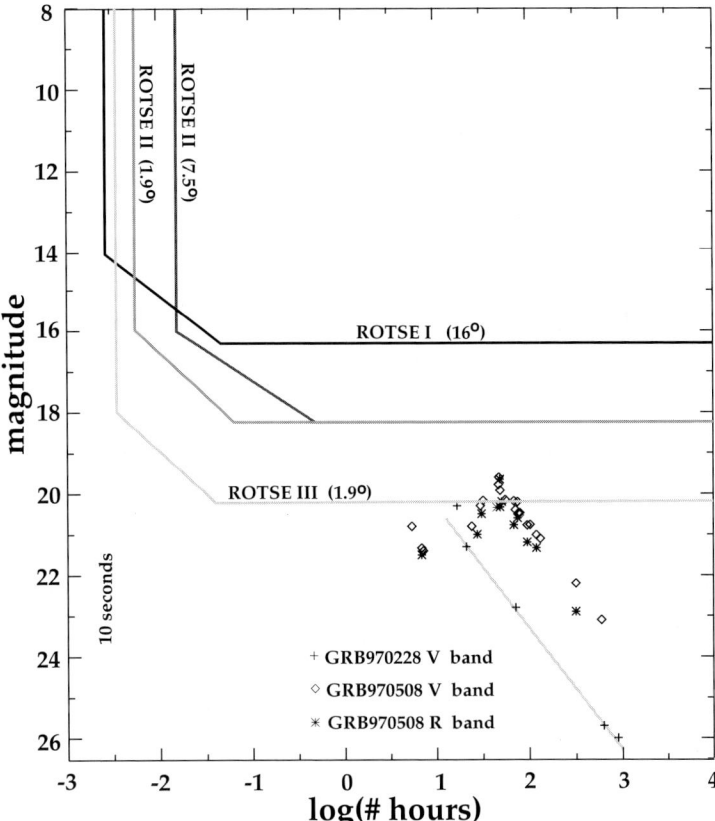

FIGURE 3. Sensitivity of ROTSE instruments vs. delay time. ROTSE-I's limiting sensitivity is 14th to 16th magnitude depending on exposure length, ROTSE-II goes approximately two magnitudes deeper, and ROTSE-III goes four magnitudes deeper. The field-of-view of each configuration is given in parentheses (7.5° for ROTSE-II is with tiling). Afterglow points for the first two optical counterparts are shown for comparison (see Galama et al. 1998, Fruchter et al. 1998, Pedersen et al. 1998a).

studies and calibrations. They also provide precursor images for GRB fields and permit studies of the optical transient background.

About once per week, an observable trigger is received via GCN, and in these instances we interrupt any ongoing sky patrol observations for the higher priority alert. A response is scheduled which depends on the trigger coordinates and type. Different trigger types arrive with different transient position errors as well as different delays from the initial event detection. About half of all triggers received correspond to classic GRBs. In general, a series of exposures with increasing durations is taken as the response progresses. Until December 1998 we employed exposure lengths of 5, 25 and 125 seconds, then changed briefly to 5, 75, and 200 seconds, and since January we have used 5, 20, and 80 seconds. If we are observing the burst within seconds of its rise, we take short exposures initially. Longer delays begin with longer exposures. If the trigger is of a type with a large position error, then we also 'tile' around the given position (32° × 32°) at specific points in the sequence. In this case, several direct-pointing images are taken and then a pair of images is taken in each of the four corners around the trigger coordinates.

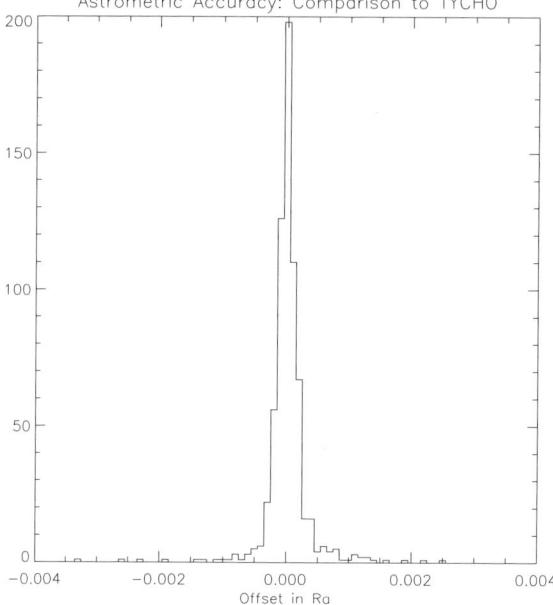

FIGURE 4. RA residuals in degrees for calibrated objects identified in ROTSE-I images compared to Hipparcos coordinates. Centroids are accurate to 1.4 arcecond (i.e. 0.1 pixel).

We then return to the direct pointing with longer exposures and begin the sequence again.

3.2. *Online data processing*

Every observing night, multiple raw darks are taken for each exposure length and averaged to produce a reference dark. Flats are produced by dark-subtracting and median-averaging ~ 60 sky patrol frames. For the most part, the flat variation is dominated by vignetting which amounts to a 60% loss at the frame corners. The process of making flats and darks also generates diagnostics which are regularly examined for signs of hardware problems. These correction frames are applied to the rest of the data to compensate for CCD noise and photometric response variations. After the correction procedure, images are reduced to lists of objects using SExtractor (Bertin & Arnouts 1996) which provides rough photometry and cluster shape information. Due to processing and data transfer limitations, only the triggered data and some of the sky patrol data can currently be processed online. Once a night's observing is done, the data is moved automatically to mass storage.

3.3. *Astrometry and photometry*

We currently perform our final astrometric and photometric calibrations offline by comparing our raw object lists to Hipparcos data (Høg et al. 1998). Photometry is established by comparing raw ROTSE magnitudes to V-band measures and color correcting based on B–V. Astrometry is determined by triangle-matching approximately 1000 catalog stars to each image, and determining warp corrections via a third order polynomial fit. As shown in Figure 4, our astrometric errors are 1.4 arcsec, and Figure 5 indicates our photometric errors to be as good as 0.02 magnitude for bright stars.

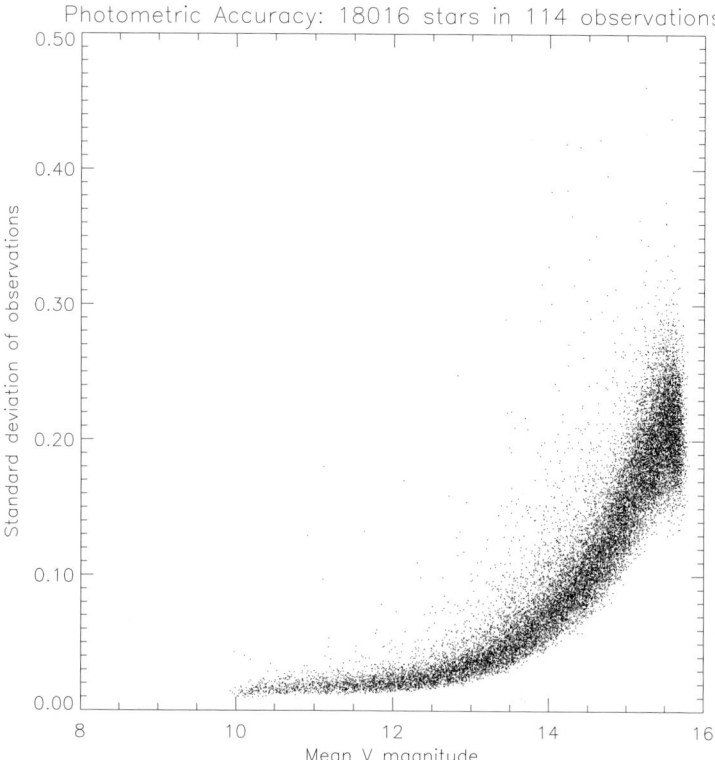

FIGURE 5. Magnitude residuals for calibrated objects identified in ROTSE-I images compared to Hipparcos photometry. Our final photometric errors are ~ 0.02 magnitude for bright stars.

3.4. Run summary

ROTSE-I operations achieved robust automation in March 1998, and during the first 12 months observed approximately 75% of all nights. The downtime resulted from very bad weather and from occasional hardware and software failures. In a typical night, the entire visible sky is imaged to 15th magnitude four times. In the first year, every field north of declination $-30°$ was observed between 200 and 900 times. The data stream is approximately 8 Gb/day, and the total amount of data generated is currently > 2 Tb. In that time, ROTSE-I responded to 49 physically interesting triggers. Of these, 30 were from classic gamma-ray bursts, 13 were from soft gamma-ray repeater events (10 of SGR1900+14), and 6 were X-ray transients. Response times for the subsample of prompt GRB and SGR triggers are shown in Figure 6.

4. Contemporaneous optical emission from GRB990123

At approximately 9:47 UTC on Jan 23, 1999, BATSE and *Beppo-SAX* triggered on an intense GRB. After 4 seconds, ROTSE received the estimated GCN coordinates (see Figure 7), and scheduled an observational sequence which began the first exposure after another 6 seconds. One hundred minutes and 200 images later, the ROTSE trigger response was complete. In about 4 hours, the *Beppo-SAX* position, accurate to $0.1°$, became available (Piro et al. 1999), leading to the detection of the optical afterglow (Odewahn et al. 1999). It also permitted quick identification of a burst of optical photons in the ROTSE images spatially coincident with the optical afterglow (see Figure 8). As

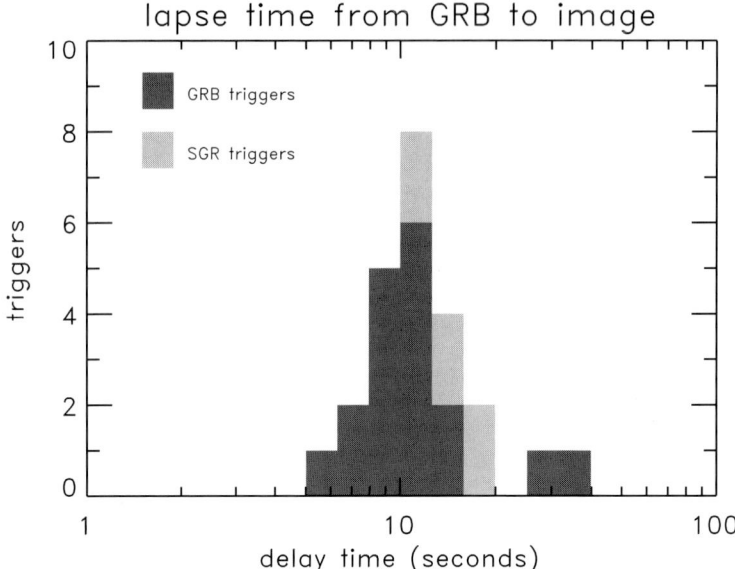

FIGURE 6. Delays for prompt GRB triggers. Times refer to period between the gamma-ray rise and the opening of the shutter for the first ROTSE-I exposure. The typical delay time is 10 seconds.

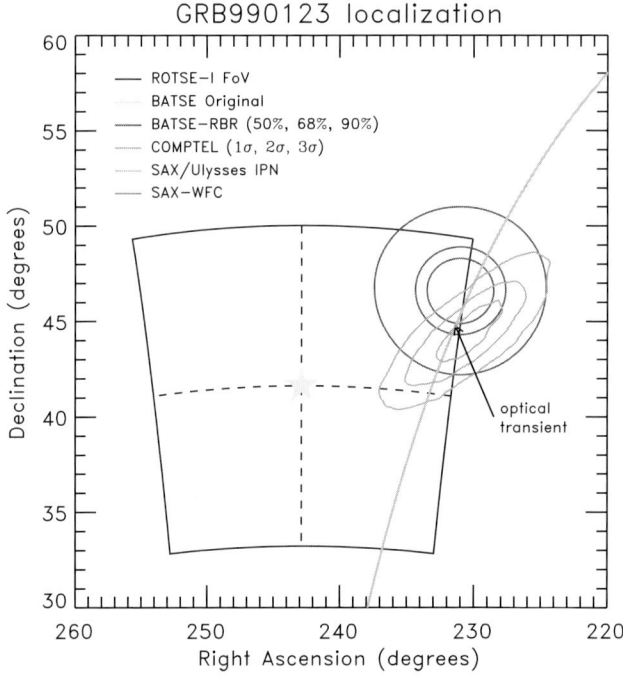

FIGURE 7. Various localizations of GRB990123 superimposed on ROTSE-I field-of-view. The optical counterpart was within 0.1° of the frame edge.

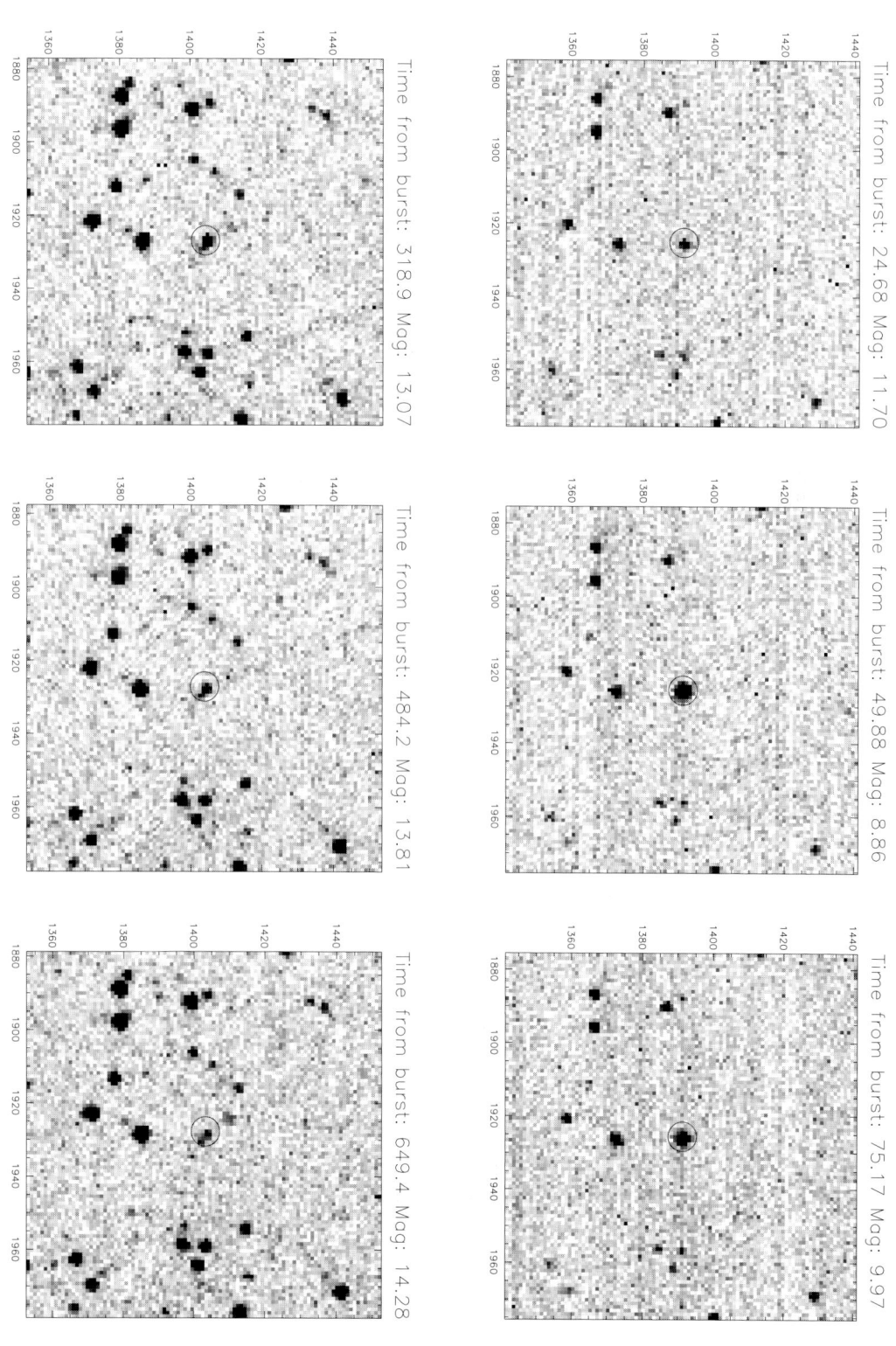

FIGURE 8. 100 × 100 pixel subimage (24 arcminutes across) surrounding GRB990123 optical counterpart. The top row shows 5 second exposures, while the bottom row shows 75 second exposures.

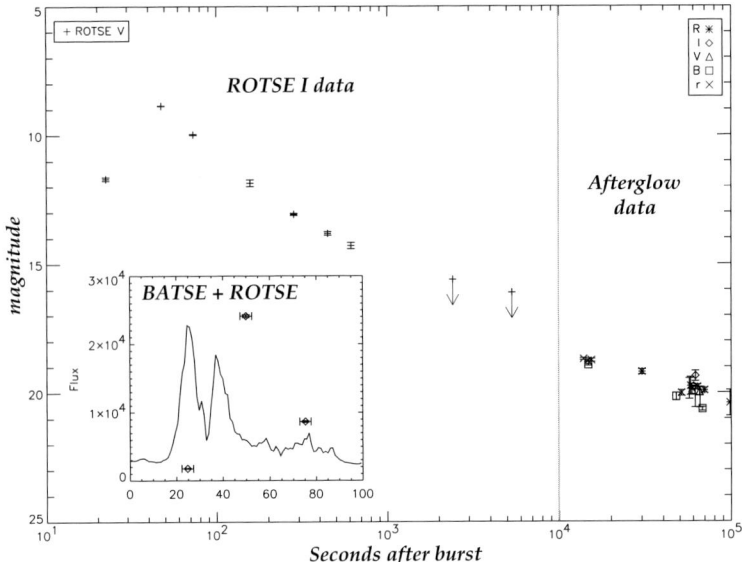

FIGURE 9. Lightcurve from ROTSE-I observations with some early afterglow points. The inset superimposes the first three ROTSE points on the BATSE gamma-ray lightcurve. The vertical line dividing the ROTSE data from the afterglow observations indicates approximately when the earliest arcminute positions can be made available.

shown in Figure 9, this 'optical burst' was surprisingly bright, reaching 9th magnitude at its peak 50 seconds after the gamma-ray rise (Akerlof et al. 1999).

The observations of GRB990123 demonstrate several things about gamma-ray bursts. First, some subset of GRBs do exhibit optical bursts with fluctuations as violent as the gamma-ray variations (see inset in Figure 9). This observation and the optical brightness of the event imply that optically absorbing material local to the event was minimal.

The fact that the optical and gamma-ray emission were both intense and similarly short-lived suggests that the two are connected. On the other hand, while the gaps between the optical observations do not permit an exact location of the optical peak, it clearly did not occur when the gamma-rays were at maximum, thereby suggesting the gamma-rays and optical emission came from different processes in the burst. In particular, the observations are consistent with coming from a reverse shock (Sari & Piran 1999).

When considering the measured redshift, $z = 1.6$, of this GRB (Kelson et al. 1999, Hjorth et al. 1999), it becomes evident that this is a truly colossal event. In optical light, it is the most luminous object ever recorded, having a peak absolute magnitude of -36.4. Assuming isotropy, over $M_\odot c^2$ was released in gamma-rays. Such an energy output is large enough to cause great difficulty for most GRB models, which typically provide only about 1% of the inferred energy (see Janka & Ruffert 1996, Ruffert et al. 1997 and Meszaros & Rees 1997 and references within). This has led to speculation that the emission is not isotropic, suggesting a beaming scenario (Kulkarni et al. 1999).

date	trigger #	loc. source	coverage (%)	dur. (sec.)	rel. fluence (%)
980329a	6665	SAX	100	55	32
980401	6672	IPN	100	37	8
980420	6694	IPN	85–100	40	8
980527	6788	BATSE	86	0.1	1
980627	6880	IPN	60	14	1
981121	7219	IPN	68–100	60	7
981223	7277	IPN	100	60	13

TABLE 1. Characteristics of seven bursts responded to by ROTSE-I. The second column gives the BATSE trigger number. The third column specificizes the origin of the best localization. The fourth column indicates the coverage of the GRB probability in percent. The fifth column indicates the duration in seconds. The last column indicates the fluence of the burst as a percent of GRB990123.

5. What about other bursts?

5.1. Strategies

The gamma-ray fluence of GRB990123 is about 100 times that of a median BATSE burst. If there is a strong linear correlation between gamma-ray fluence and optical flux, ROTSE-I should be sensitive enough to find optical counterparts to roughly half of the GRBs observed. Our ongoing analysis of seven earlier bursts (see Table 1) might then be expected to reveal more optical counterparts.

We have simplified the analysis by choosing those with the smallest position errors—six of the seven possess square-degree level localizations or better, while one (GRB980527) has a BATSE statistical error of about 1.1°. This results in an enormous reduction in background (> 200×). Aside from the one *BeppoSAX* position for GRB980329a, the more accurate positions arise from the use of the gamma-ray detectors on-board Ulysses (Hurley et al. 1999). By comparing timing information between two widely spaced detectors, thin annuli are generated which are several degrees long but only about 0.1° wide (Hurley et al. 1998). The intersection of the BATSE position probability distribution with such an 'Interplanetary Network' (IPN) timing annulus produces an IPN arc. The downside to restricting ourselves to using these localizations is that they cannot be obtained for GRBs on the faint end of the BATSE fluence distribution.

We are currently using several analysis strategies to check the consistency of our methods. The simplest is a lightcurve analysis which looks for sources varying by > 2 magnitudes in a trigger response. We also match our object lists to more complete catalogs such as USNO (Monet et al. 1998) to identify any new objects. The most sensitive method we employ is image differencing. A template for our trigger response images is constructed from precursor sky patrol images and subtracted from the triggered data. If we do not have a very precise localization such as from an X-ray, optical, or radio counterpart, new or varying objects are only considered bona fide optical counterpart candidates if they appear in at least two successive images. This is to remove the background arising from ghosts, cosmic rays, satellite glints, etc. which show up in nearly every image with the field-of-view of ROTSE-I. If we have an IPN arc, we search through the unconstrained IPN annulus in these images for interesting objects. If no source is found, limits are obtained whenever coverage exceeds 50% of the IPN arc. Our results take the form of average magnitudes, $\langle m_V \rangle_x$, during an exposure length, x, vs. the time, t_+, after the start of the burst. Limits refer to the faintest $\langle m_V \rangle_x$ to which we are > 50% efficient at finding objects after our analysis selection. Unless specifically noted, they do not

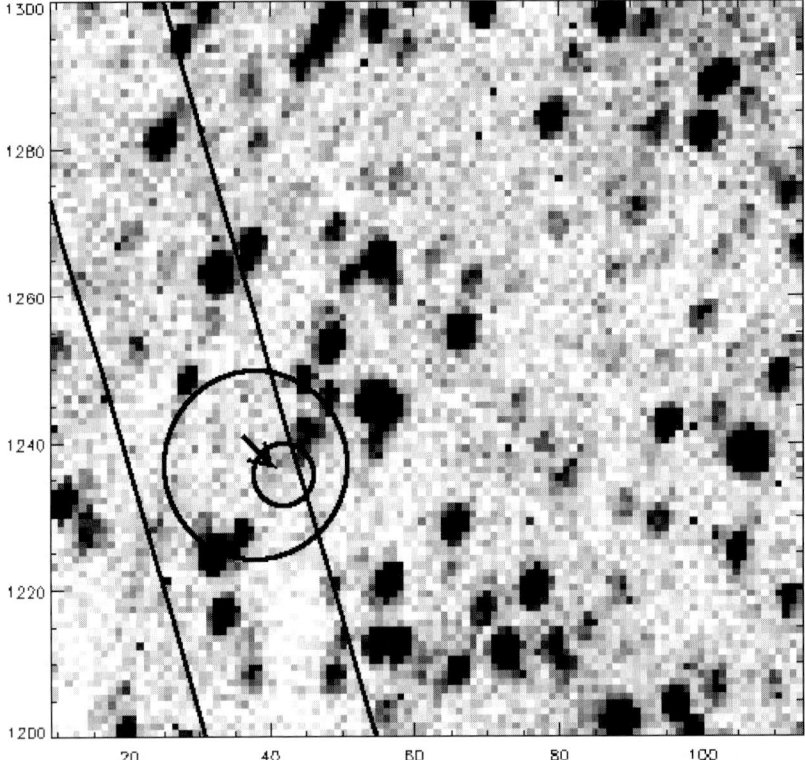

FIGURE 10. Subimage surrounding GRB980329a radio counterpart. Three images were co-added, giving an effective exposure length of 375 seconds. Circles denote the *BeppoSAX* localizations (Frontera et . 1998).

correspond to long integration times (i.e. hundreds of seconds) obtained from co-adding multiple images. The results discussed here are preliminary, and a final, more robust analysis will exploit the full sensitivity of the telescope.

5.2. *GRB980329a*

On March 29, 1998, ROTSE-I successfully responded to its first trigger in automation. The first exposure was begun 11.5 seconds after the gamma-ray rise, and the complete response was finished approximately one hour later. Unfortunately, the sky was fairly cloudy until the last three frames were taken, and these are somewhat hazy. There were, however, optical (Djorgovski et al. 1998, Palazzi et al. 1998, Pedersen et al. 1998b), X-ray (Frontera et al. 1998) and radio (Taylor et al. 1998) counterparts observed hours later with the result that the burst location is known precisely. Despite the poor quality of the early data, some images are clear enough in the immediate region of the burst to see any 9–11 magnitude objects. Since we know exactly where the source is in this case, we use the following reference stars in to estimate the sensitivity of the image in the region: SAO 59687 ($\alpha = 105.297$, $\delta = 39.177$, $m_V = 8.24$), SAO 59692 ($\alpha = 105.308$, $\delta = 38.859$, $m_V = 9.63$), SAO 59708 ($\alpha = 105.585$, $\delta = 39.169$, $m_V = 8.65$) and GSC 958 ($\alpha = 105.583$, $\delta = 38.883$, $m_V = 10.09$). To be conservative, we take the sensitivity to be equal to the magnitude of the dimmest of these objects that can be reliably observed in a given image. The last three images were co-added to produce an image (see Figure 10) sensitive to about 14.8 magnitude as obtained from comparison

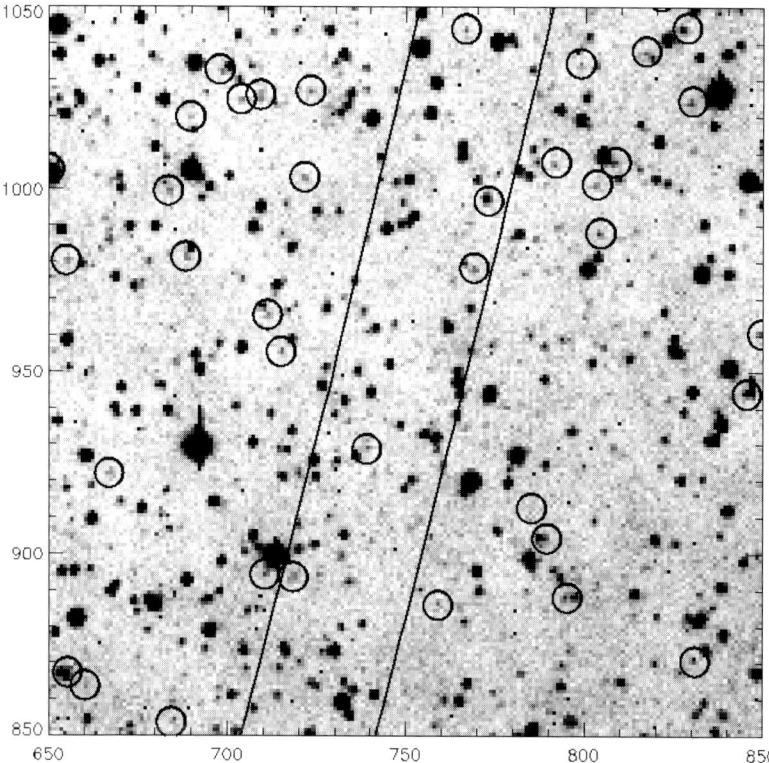

FIGURE 11. Subimage from GRB980401 data with portion of overlapping IPN arc (delineated with solid lines). Circles indicate catalog objects with magnitudes between 16.0 and 16.4.

with the USNO catalog. No object is detected at the known source location in our images. The limits are summarized in Table 2.

5.3. GRB980401

Our third trigger, GRB980401, arrived during focusing tests, so a manual response was performed starting 2477 seconds after we received the trigger. This burst has an IPN localization which is completely contained in one camera. The last four 125 second exposures were co-added into one exposure. While the total exposed time is 500 seconds, the manual nature of the trigger spreads the integration time over about 900 seconds. The resultant image is sensitive to $\langle m_V \rangle_{500} = 16.4$ at $t_+ = 62$ minutes and shows no unknown objects when compared to the USNO catalog (see Figure 11). This is the most stringent optical upper limit on a GRB in the first hour timeframe.

5.4. GRB980420

Although conditions for GRB980420 were very good, the camera containing most of the final IPN arc in the initial images was out of focus, thereby greatly reducing the image sensitivity. In tiled and later response images, in-focus cameras overlap the IPN arc resulting in a greater sensitivity. No optical source was observed which varied by more than 2 magnitudes, with an analysis sensitive to 14th magnitude objects in the later images.

5.5. GRB980527

While much more difficult, we have begun analyzing bursts for which only BATSE positions are available. Such an analysis requires image differencing to work efficiently. One such burst is GRB980527. The ROTSE-I array covers 100% of the statistical and 86% of the combined statistical and systematic error region, where we have used the systematic error parameterization found in Briggs et al. (1997). The first observation starts at $t_+ = 12.6$ seconds. Our preliminary analysis is sensitive to $\langle m_V \rangle_5 = 13.0$ and $\langle m_V \rangle_{25} = 14.2$. No candidates were found in the early non-tiled images. It should be remembered that this is the only short burst in this sample. The limit of 13.0 at 15 seconds is the best limit obtained by any experiment so soon after a burst.

5.6. GRB980627

GRB980627 was fairly dim in gamma-rays and the IPN arc was located very far from the original trigger localization. As a result, a majority of the probability distribution is only covered in our tiled images. No objects were found which varied by more than 2 magnitudes in this data. The images are sensitive to approximately magnitude 12.5, irrespective of exposure length.

5.7. GRB981121

We responded to a GRB on November 21, 1998 which was fairly intense in gamma-rays. The data were taken under good conditions, although our shutters were occasionally not opening completely due to very cold weather. Despite this problem, we cover most of the IPN arc for this burst. No unidentified objects were seen with a sensitivity to $\langle m_V \rangle_5 = 12.8$ in early exposures and $\langle m_V \rangle_{25} = 14.3$ in later exposures.

5.8. GRB981223

GRB981223 was another burst which was bright in gamma-rays. Our trigger response was prompt and the weather was clear. No unidentified objects were seen with limits in the range 12.4 to 13.5.

6. Results

The preliminary limits placed on the early optical emission for these seven bursts are shown in Figure 12, and a summary of the limit results is given in Table 2. We are able to place a constraint on the overall power-law decline of optical emission from GRB980329a to be shallower than -2.0 with respect to the afterglow points. The best limits are currently $\langle m_V \rangle_5 > 13.0$ at 14.7 seconds for GRB980527, and $\langle m_V \rangle_{500} > 16.4$ at 62 minutes for GRB980401. Given the measurements presented, we can conclude that bright optical counterparts (i.e. $m_V \sim 10$) are uncommon.

Now that the optical signature has been seen in one case, the natural question is whether other bursts behave like GRB990123. One way to address this question is to compare optical to gamma-ray levels. To bring all bursts onto some common footing, we correct for their fluence by defining:

$$\mu \equiv m_V - 2.5 \log(f/f_{GRB990123}), \qquad (6.1)$$

where f is the fluence measured in BATSE channels 2 and 3 to avoid systematics due to problems in spectral fitting the other channels (Briggs 1999).

Several issues of optical extinction arise in our comparison. First, galactic extinction varies significantly over the IPN arcs preventing us from quoting an accurate value in most of the bursts we analyzed. We note, however, that for almost all of these bursts it

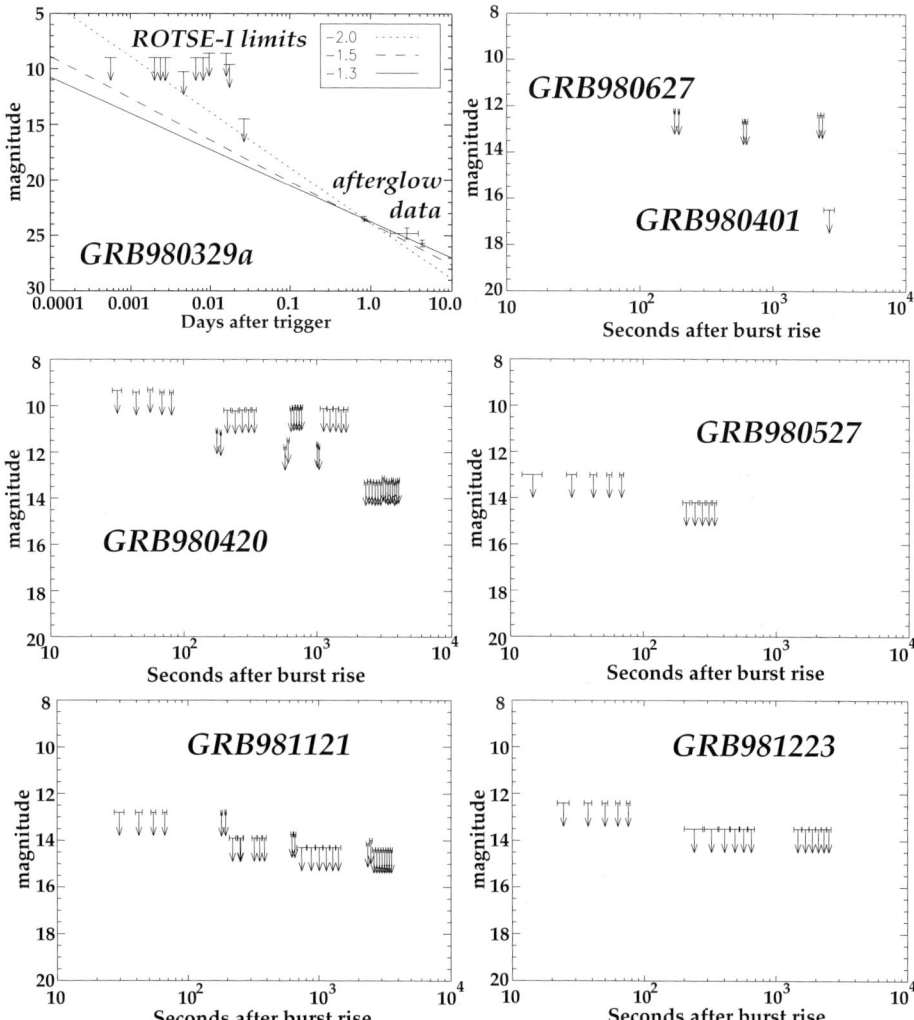

FIGURE 12. Magnitude limits for seven GRB data sets. Each plot shows apparent V band magnitude vs. time after the gamma-ray rise. The time history for GRB980329a has three R-band afterglow observations (Djorgovski et al. 1998, Palazzi et al. 1998, Pedersen et al. 1998b) superimposed, along with three power law decays passing through the earliest optical detection.

is much less than 1 magnitude at their most probable location. Since GRB990123 has a similar low value (= 0.04), its effect on our comparison should be minimal. The one exception is GRB980420 which may have over 2 magnitudes of absorption. We ignore the effects of extinction at the source because there is no measure of it in most of these cases, aside from GRB980329a and GRB990123. A statistical argument has been made, however, that most GRBs are not heavily obscured at the source (Frail et al. 1999).

The limits are replotted as μ vs. t_+ in Figure 13. If the scaled optical emission was much higher than in GRB990123 around $t_+ = 50$ seconds or around $t_+ = 300$ seconds, we would have seen it. In several dimmer bursts, however, we observed no such behavior. While we cannot rule out beaming, there is no evidence for it in this analysis.

date	t_1 (s)	t_{exp1}	$m_V(t_1)$	t_2 (s)	t_{exp2}	$m_V(t_2)$	t_3 (s)	t_{exp3}	$m_V(t_3)$
980329a	51.0	5	9.0	415	25	10.1	2231	375	14.8
980401	–	–	–	–	–	–	3726	500	16.4
980420	31.5	5	9.4	178	5	11.1	2324	125	13.5
980527	14.7	5	13.0	208	25	14.2	–	–	–
980627	–	–	–	180	5	12.2	602	25	12.7
981121	29.7	5	12.8	219	25	13.9	742	125	14.3
981223	24.4	5	12.4	238	25	13.5	–	–	–

TABLE 2. Summary of limits for seven bursts responded to by ROTSE-I. The time an image was taken, the exposure length, and the limiting sensitivity are given. Each time corresponds to the middle of an exposure. Multiple times are listed when the sensitivity significantly improves.

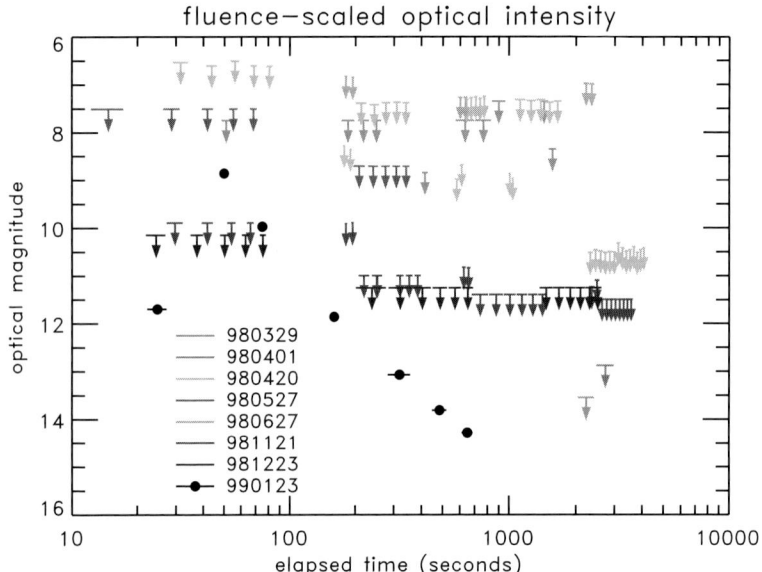

FIGURE 13. Rescaled limits, μ, for seven GRB data sets vs. t_+. Relative to GRB990123, significantly higher optical emission would have been seen during the first few minutes. In two cases, the optical emission is at least one magnitude fainter than GRB990123.

More importantly, in two cases, GRB981121 and GRB981223, the optical emission around one minute is at least 1 magnitude fainter than in GRB990123. The main implication of this analysis is that there does not appear to be a strong correlation of optical flux with gamma-ray fluence. The inherent dispersion to any actual correlation must be larger than one magnitude to explain the results from GRB981121 and GRB981223, in particular. A significant correlation, however, is expected in models of reverse-shock development (Sari & Piran 1999).

There are several caveats and cautions to this analysis. The most important is that there is a large diversity of GRB behavior. Our results are based on a handful of events, and aside from GRB980527, we are only looking at long bursts. Another limitation arises from the IPN source of final positions, which discriminates against dim bursts. Therefore, our results may not be representative of GRBs in general, and more GRB triggers and further analysis are necessary.

7. Conclusions

We have observed a prompt optical burst during the gamma-ray emission of GRB990123. It is as violent as the burst of gamma-rays, but it displays a different temporal behavior. This difference is consistent with the expected signature of a reverse shock from the explosion.

Preliminary studies of seven other bursts reveal several further points about GRBs. No optical counterparts were identified, and from this we can conclude that bright optical bursts (i.e. $m_V \sim 10$) are uncommon. When using fluence as an estimator of total energy output, no bursts with optical flux much greater than GRB990123 have been observed. In two cases, the scaled optical emission around 1 minute is at least 1 magnitude dimmer than for GRB990123. While not conclusive, the non-detection of another optical burst suggests that there is not a strong correlation between gamma-ray and optical emission.

ROTSE is supported by NASA under $SR\&T$ grant NAG5-5101, the NSF under grants AST-9703282 and AST-9970818, the Research Corporation, the University of Michigan, and the Planetary Society. Work performed at LANL is supported by the DOE under contract W-7405-ENG-36. Work performed at LLNL is supported by the DOE under contract W-7405-ENG-48.

REFERENCES

AKERLOF, C. ET AL. 1999 *Nature*, **398**, 400.

BARTHELMY, S., ET AL. 1995 *Astrophysics and Space Science*, **231**, 235.

BARTHELMY, S., ET AL. 1998 GCN: The GRB Coordinates Network. http://gcn.gsfc.nasa.gov/gcn/.

BERTIN, E. & ARNOUTS, S. 1996 *ApJS*, **117**, 393.

BRIGGS, M., ET AL. 1998 in *Gamma-Ray Bursts 4th Huntsville Symposium* (eds. C. Meegan, R. Preece, and T. Koshut), p. 104.

BRIGGS, M. 1999 private communication.

DJORGOVSKI, S., ET AL. 1998 *GCN Circ.* **41**.

FENIMORE, E., MADRAS, C., & NAYAKCHIN, S. 1996 *ApJ*, **473**, 988.

FEROCI, M., ET AL. 1997 in *EUV, X-ray, and Gamma-Ray Instrumentation for Astronomy VIII*, Proc. SPIE Vol. 3114. p. 186.

FISHMAN, G., ET AL. 1989 in *Proc. of the Gamma Ray Observatory Science Workshop*, (ed. W. Neil Johnson), p. 2.

FRAIL, D., ET AL. 1999 these proceedings.

FRONTERA, F., ET AL. 1998 *IAU Circ.*, **6853**.

FRUCHTER, A., ET AL. 1998 in *Gamma-Ray Bursts 4th Huntsville Symposium* (eds. C. Meegan, R. Preece, and T. Koshut), p. 509.

GALAMA, T., ET AL. 1998 in *Gamma-Ray Bursts 4th Huntsville Symposium*, (eds. C. Meegan, R. Preece, and T. Koshut), p. 478.

HJORTH, J., ET AL. 1999 *GCN Circ.*, **219**.

HØG, E., ET AL. 1998 *A&A*, **335**, L65.

HOGG, D. & FRUCHTER, A. 1999 *ApJ*, **520(1)**, 54.

HURLEY, K., ET AL. 1984 in *Gamma-Ray Bursts* (eds. E. Liang and V. Petrosian), p. 3.

HURLEY, K., ET AL. 1998 GCN/IPN Notices. http://gcn.gsfc.nasa.gov/gcn/ipn.html.

HURLEY, K., ET AL. 1999 *ApJS*, **120**, 399.

JAGER, R., ET AL. 1997 *A&AS*, **125**, 557.

JANKA, H. & RUFFERT, M. 1996 *A&A*, **307**, L33.

Kelson, D., et al. 1999 *IAU Circ.*, **7096**.
Klebesadel, R., Strong, I., & Olson, R. 1973 *ApJ*, **182(2)**, L85.
Kouveliotou, C., et al. 1993 *ApJ*, **413**, L101.
Kouveliotou, C., et al. 1996 in *Gamma-Ray Bursts 3rd Huntsville Symposium* (eds. C. Kouveliotou, M. Briggs, and G. Fishman), p. 42.
Kulkarni, S., et al. 1998 *Nature*, **393**, 35.
Kulkarni, S., et al. 1999 *Nature*, **398**, 389.
Meegan, C., et al. 1998 in *Gamma-Ray Bursts 4th Huntsville Symposium* (eds. C. Meegan, R. Preece, and T. Koshut), p. 3.
Meszaros, P. & Rees, M. 1993 *ApJ*, **405**, 278.
Meszaros, P. & Rees, M. 1997 *ApJ*, **482**, L29.
Metzger, et al. 1997 *IAU Circ.*, **6676**.
Monet, D., et al. 1998 *A Catalog of Astrometric Standards*. US Naval Observatory.
Odewahn, S., Bloom, J., & Kulkarni, S. 1999 *GCN Circ.*, **201**.
Palazzi, E., et al. 1998 *GCN Circ.*, **48**.
Paczynski, B. & Xu, G. 1994 *ApJ*, **427**, 708.
Pedersen, H., et al. 1998 in *Gamma-Ray Bursts 4th Huntsville Symposium* (eds. C. Meegan, R. Preece, and T. Koshut), p. 530.
Pedersen, H., et al. 1998 *GCN Circ.*, **52**.
Piro, L., et al. 1999 *GCN Circ.*, **199**.
Rees, M. & Meszaros, P. 1992 *MNRAS*, **258**, 41.
Rees, M. & Meszaros, P. 1994 *ApJ*, **430(2)**, L93.
Rhoads, J. 1999 astro-ph/9903399.
Ricker, et al. 1990 in *Gamma-Ray Bursts Observations, Analyses and Theories* (eds. C. Ho, R. Epstein, and E. Fenimore).
Ruffert, M., et al. 1997 *A&A*, **319**, 122.
Sari, R. & Piran, T. 1999 *ApJ*, **517(2)**, L109.
Taylor, G., et al. 1998 *GCN Circ.*, **40**.

X-ray afterglows of gamma-ray bursts

By LUIGI PIRO

Istituto Astrofisica Spaziale, C.N.R., Via Fosso del Cavaliere, 00133 Roma, Italy

Gamma-ray bursts (GRBs) have been one of the greatest mysteries of high energy astrophysics since their discovery in 1969. The fast and precise localization of the burst provided by BeppoSAX led to the discovery that GRBs continue to glow in X-rays and longer wavelengths for days after the initial burst. Observations of afterglows of gamma-ray bursts are now providing a wealth of new information on these mysterious objects that, during the few seconds of their explosion, outshine any other object in the Universe. Along with their distance-scale, we are gathering information that should disclose the origin of these explosions and their relationship with the environment of the host galaxy in the early phases of the Universe. In this review I will focus primarily on observations of X-ray afterglows.

1. Introduction

Gamma-Ray Bursts (GRBs) were discovered in 1969 (Klebesadel et al. 1973) by the Vela satellites, deployed by the USA to verify the compliance of USSR to the nuclear test ban treaty. In the following 28 years, thousands of events have been observed by several satellites, leading to a good characterization of the global properties of this phenomenon. A big step in this area was achieved with BATSE (Fishman et al. 1994). The isotropical distribution of the events in the sky (Fishman & Meegan 1995) was suggestive of an extragalactic origin, but a direct measurement of the distance in a single object was not available. What was lacking was a *fast AND precise* position, where the *Holy Grail* of GRB scientists, i.e. the *counterpart*, could have been searched for at all wavelengths with more chances to catch it. This was achieved in 1996, with BeppoSAX observations of GRBs.

2. Gamma-ray bursts in the Afterglow Era

2.1. *The BeppoSAX mission*

The BeppoSAX satellite (Piro, Scarsi & Butler 1995, Boella et al. 1997a) was launched on April 30, 1996 from Cape Canaveral by an Atlas-Centaur rocket in an equatorial low-background orbit (3.9° inclination, height 600 km). The spacecraft covers a complete orbit in about 97 minutes. Once per orbit it passes over the ground station in Malindi (Kenya) and downloads all the data recorded onboard during the orbit, and uplinks all the telecommands needed for the operations in the next orbit. The entire radio-contact of the satellite with the ground station lasts about 10 minutes each orbit. Telecommands and data are relayed through an Intelsat satellite to the Science Operation Center in Rome where they are analyzed by Duty Scientists 24 hours a day.

A schematic of the scientific payload of the satellite is shown in Figure 1, where the individual instruments are visible. They are basically divided in *Narrow Field Instruments* (NFI, with a field of view of about 1°, looking at the same direction in the sky), namely:

• one Low Energy Concentrator Spectrometer (LECS, 0.1–10 keV) (Parmar et al. 1997);

• three Medium Energy Concentrator Spectrometers (MECS, 2–10 keV) (Boella et al. 1997b);

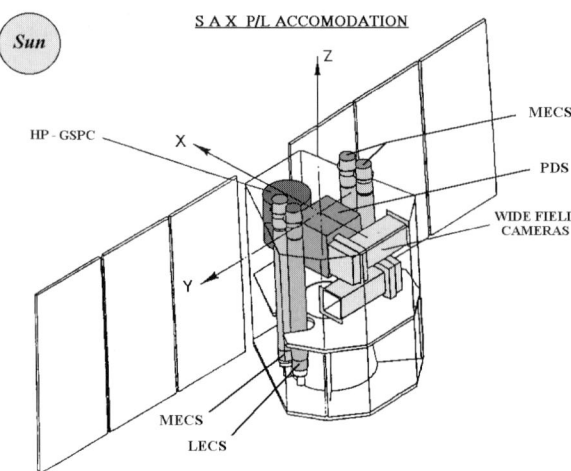

FIGURE 1. Schematics of the BeppoSAX satellite with the scientific instruments indicated. The GRBM is a part of the Phoswich Detection System, located in the core of the satellite.

- one High Pressure Gas Scintillation Proportional Counter (HPGSPC, 4–100 keV) (Manzo et al. 1997);
- one Phoswich Detection System (PDS, 15–300 keV) (Frontera et al. 1997);

and *Wide Field Instruments*, looking at directions 90° from the NFI field of view:

- two Wide Field Cameras (WFC, 2–30 keV, field of view at zero response of $40° \times 40°$) pointed at two opposite directions (Jager et al. 1997);
- one Gamma-Ray Burst Monitor (GRBM, 40–700 keV, open field of view, including the WFC field of view) (Frontera et al. 1997, Feroci et al. 1997).

The BeppoSAX instrumentation is therefore well matched *both* for wide band (more than three decades of energy, from 0.1 to 300 keV) pointed observations of celestial sources with its good sensitivity, spatial and energy resolution, *and* for wide field monitoring of transient X/gamma events with the WFC and GRBM. The ground operation system has been designed to react promptly to X-ray transient sources discovered by the wide field instruments with fast follow-up observations with the more sensitive NFI (sensitivity $\sim 5 \times 10^{-14}$ erg cm^{-2} s^{-1}).

The complete set of BeppoSAX instruments is used to observe GRBs. Firstly, the wide field devices are used for a fast and accurate location (Piro et al. 1998a). The limited positional capabilities of gamma-ray instrumentation have been overcome by coupling the Gamma-Ray Burst Monitor (GRBM), which provides the temporal signature of the event (the so-called *trigger*), with a Wide Field X-ray Camera (WFC), which provides a localization within few arcmin in the X-ray band. It was indeed known that GRBs emit a small fraction ($\sim 5\%$) of their luminosity in the X-ray band (Yoshida et al. 1989). Therefore, whenever a GRB occurs within the field of view of one of the two WFCs, covering about 3% of the sky each, it can be promptly recognized by the simultaneous appearance of a burst in the light curves of both instruments (Fig. 2) and localized by the WFC to a level of accuracy (about 3 arcmin, see Fig. 3) good enough to define an appropriate target for the NFI. This can be done in about 1 to 2 hours after the detection of the GRB onboard, because the event can occur within in any part of the orbit. At the same time an alert message with the coordinates of the GRB is sent to ground- and space-based telescopes working at other wavelengths while the operations to follow up the GRB with BeppoSAX start. The GRB location is usually acquired with the NFI within a few hours.

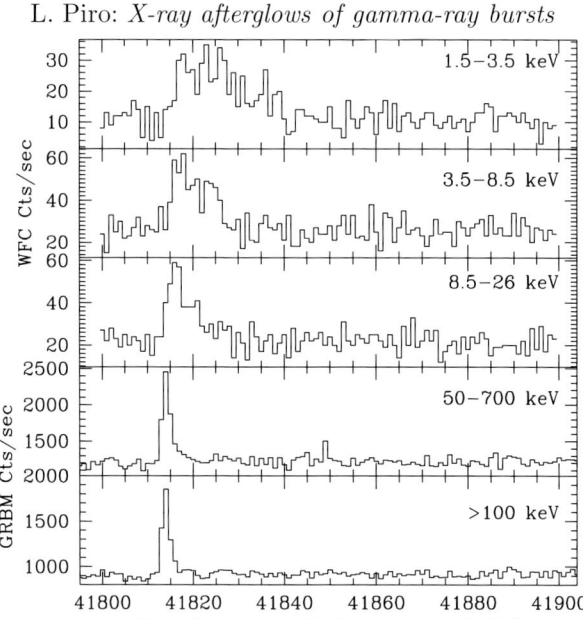

FIGURE 2. BeppoSAX GRBM and WFC light curves of GRB960720 (from Piro et al. 1998a)

2.2. The first afterglow: GB970228

Before the advent of BeppoSAX, GRB astronomy proceeded on a statistical approach and no one had an idea of how a GRB emitter could evolve soon after the event. The capabilities for a prompt reaction to GRBs became operative on December 1996, after an off-line analysis of a GRB (GRB960720: Piro et al. 1998a) had demonstrated the designed capability of the mission. The first opportunity occurred on January 11, 1997: GRB970111. The field was searched with the NFI 16 hours after the GRB. The possible association of one of the faint sources found in the error box with the GRB was still under scrutiny (Feroci et al. 1998), when on February 28, 1997, another event, GRB970228, was detected by the BeppoSAX GRBM and WFC. The NFI were pointed to the GRB location 8 hours after the burst. A previously unknown X-ray source was detected in the field of view of the LECS and MECS instruments with a flux in the 2–10 keV energy range of 3×10^{-12} erg cm^{-2} s^{-1}. The new source appeared to be fading away during the observation. On March 3 we made another observation that confirmed that the source was quickly decaying: at that time its flux was a factor of about 20 lower than the first observation's (Figure 4). This was the first detection of an "afterglow" of a GRB (Costa et al. 1997).

The flux of the source appeared to decrease following a power law dependence on time ($\sim t^{-\alpha}$) with index $\alpha = (1.3 \pm 0.1)$. Further X-ray observation with the X-ray satellites ASCA and ROSAT detected the source about one week later with a flux consistent with the same time decline law (Yoshida et al. 1997, Frontera et al. 1998a). This kind of temporal behavior agrees with the general predictions of the fireball models for GRBs (e.g. Mészáros & Rees 1997). A backward extrapolation of this power law decay (Fig. 5) is consistent with the X-ray flux measured during the burst, suggesting that the afterglow started soon after the GRB. Another important result came from the spectral analysis of the X-ray afterglow. It excluded a black body emission, therefore arguing against a model in which the radiation comes from the cooling of the surface of a neutron star (Frontera et al. 1998b).

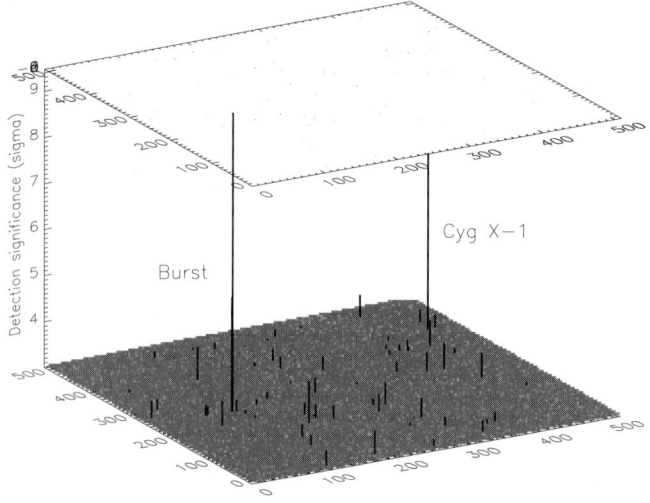

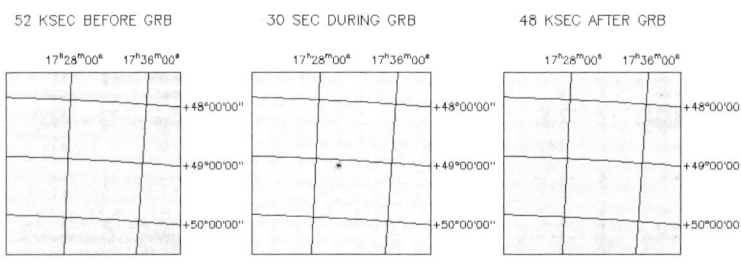

FIGURE 3. **a)** The $40° \times 40°$ image in the 2–26 keV range of the WFC integrated on a period of 15 s on the burst. The y axis gives the significance of the detection (σ). The source close to the edge is Cyg X-1. **b)** Images of the field centered on GB960720 in time sequence. The first and last images were obtained integrating over $\sim 50,000$ s before and after the burst. No source was detected. The second image is a 30 s long shot which shows the sudden presence of GB960720 (from Piro et al. 1998a).

While the X-ray monitoring of GRB970228 was going on, an observational campaign of the same object was simultaneously started with the most important optical telescopes. This campaign led to the discovery (van Paradijs et al. 1997) of an optical transient associated with the X-ray afterglow. As in the X-ray domain, the optical flux of the source showed a decrease well described by a power law with index -1.12 (e.g. Garcia et al. 1998), again in agreement with the general predictions of the fireball model. The images taken with the Hubble Space Telescope (HST) (Sahu et al. 1997, Fruchter et al. 1997) showed the presence of a nebulosity around the optical transient. However the nebulosity was very weak, and it was not possible to clarify whether it was associated with the host galaxy (extragalactic origin) or with a transient diffuse emission representing the residual of the explosion (galactic origin).

2.3. *The first measurement of redshift: GRB970508*

On 8 May 1997 the second breakthrough arrived with GRB970508 (Piro et al. 1998b), detected just few minutes before the satellite was passing over the ground station in Malindi. This opportunity, and the experience gained from previous events, allowed us

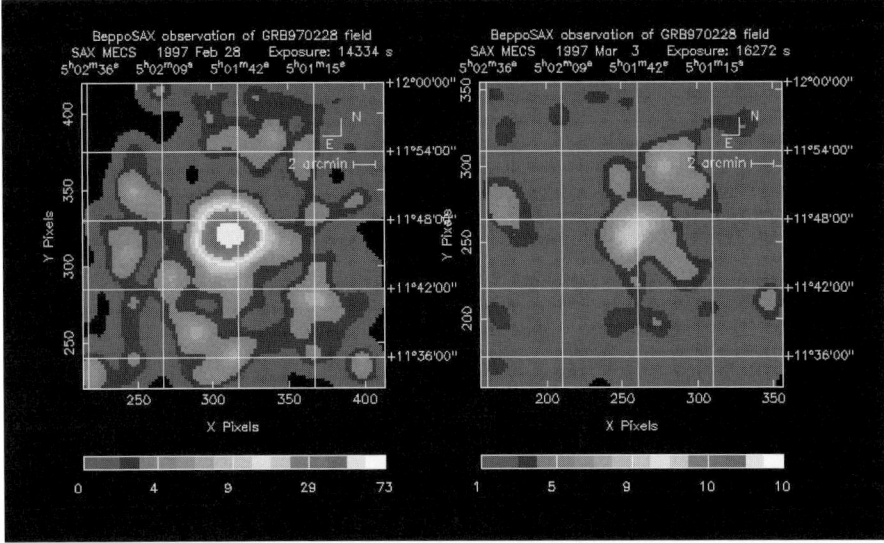

FIGURE 4. BeppoSAX MECS images of the GRB970228 afterglow, 8 hours after the GRB (left) and 3 days after the GRB (right) (from Costa et al. 1998)

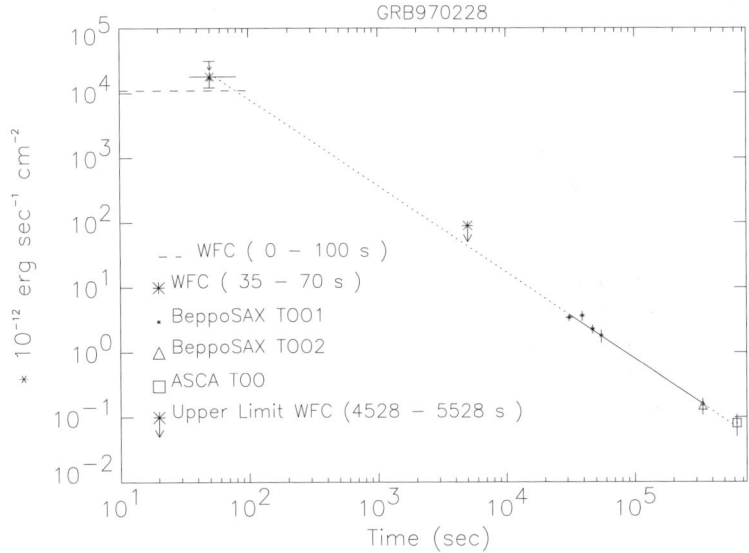

FIGURE 5. BeppoSAX decay curve of the GRB970228 afterglow in the 2–10 keV energy range, obtained with the WFC and the NFI. The result of the ASCA observation is also shown at the bottom right(from Costa et al. 1998).

to point the BeppoSAX NFI on source 5.7 hours after the burst, while optical observations started 4 hours after the burst.

The early detection of the optical transient (Bond 1997) and its relatively bright magnitude allowed a spectroscopic measurement of its optical spectrum with the Keck telescope (Metzger et al. 1997). The spectrum revealed the presence of FeII and MgII absorption lines at a redshift of $z = 0.835$, attributed to the presence of a galaxy between us and the GRB, and therefore demonstrated that GRB970508 was at a cosmological distance.

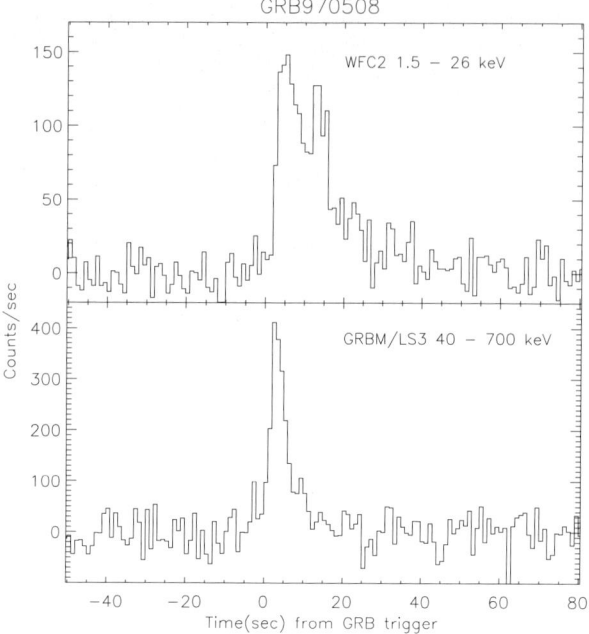

FIGURE 6. WFC and GRBM light curves of GRB970508 (from Piro et al. 1998b).

2.4. *GRBs are long-lasting phenomena ... after all!*

The BeppoSAX observation of GB970508 has also changed our view of the GRB *phenomenon*. The old concept of a brief or sudden release of luminosity concentrated in few seconds does not stand with the new information provided by BeppoSAX. Indeed the name *afterglow* attributed to the X-ray emission observed after the event is somewhat misleading. This is clear when one considers the energy produced in the afterglow phase, which turns out to be comparable to that of the GRB. In order to compute the energy emitted in the afterglow phase it is necessary to integrate in time its luminosity and, therefore, it is crucial to know *when* the afterglow starts. A detailed analysis of the data of the WFC of GB970508 (Piro et al. 1998b) shows that the X-ray emission is present (Fig. 7) even when the signal of the γ-ray light curve disappears into the noise at ~ 30 s (Fig. 6), and remains visible for at least 2000 seconds, when the X-ray flux drops below the sensitivity of the WFC (Fig. 7). The conclusion is, then, that the afterglow phase starts immediately after (or even before) the prompt emission settles down. The energy emitted in the afterglow in X-rays turns out to be a substantial fraction (~ 40–50%) of the energy produced by the GRB.

Furthermore, the light curve shows a rebursting event starting 1 day after the initial burst, an evidence that indicates that the source of the energy can re-ignite on long time scales (Piro et al. 1998b).

3. The largest explosions since the Big Bang from the edge of the Universe

The GRB observed by BeppoSAX on December 14, 1997 has, on the one hand, consolidated the extragalactic origin of GRBs, and on the other hand has underlined the problem of the energy budget and, ultimately, of the nature of the "central engine."

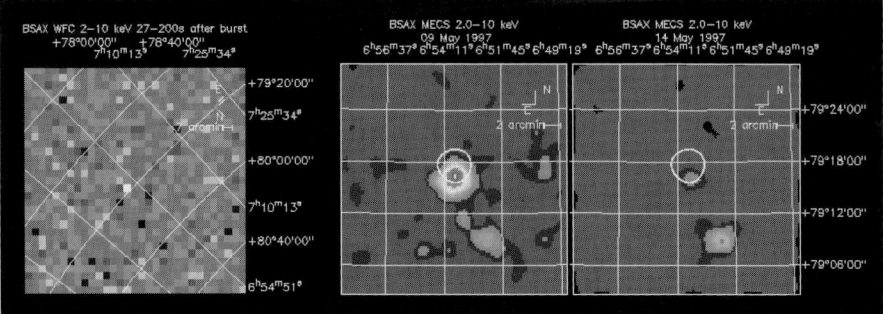

FIGURE 7. The X-ray afterglow of GB970508 (from Piro et al. 1998b). Time sequence of images of the field of GB970508 observed by the WFC2 (left image, 27–200 s after the burst), MECS(2+3) on May 9 (center image, 6 hours after the GRB), and MECS(2+3) on May 14 (after 6 days). The WFC2 show the presence of the afterglow that was then detected by the LECS and MECS (1SAXJ0653.8+7916 visible in the 99% error circle of the WFC). Note the decrease in intensity between the two MECS observations, as compared to 1RXSJ0653.8+7916, the source in the lower right corner.

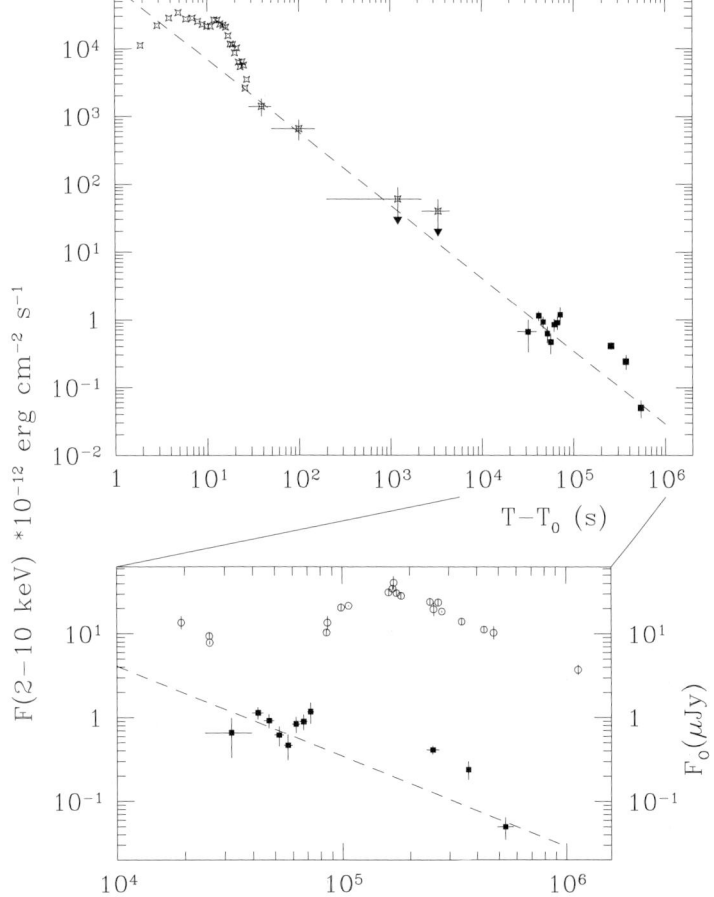

FIGURE 8. Top panel: Decay law of the GRB970508 afterglow as detected by the BeppoSAX WFC and NFI. The WFC provided the data up to 5.000 seconds after the burst. Bottom panel: enlargement of the X-ray decay law and comparison with the simultaneous time history of the optical transient (open circles) (from Piro et al. 1998b).

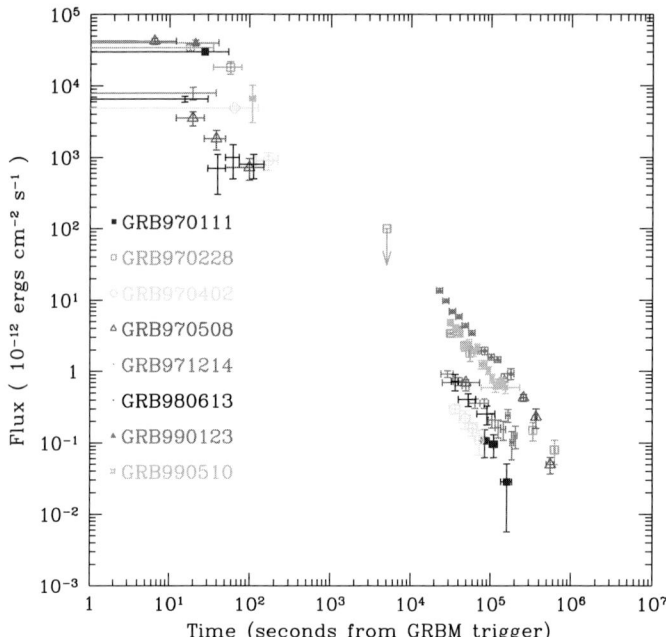

FIGURE 9. X-ray (2–10 keV) light curves of a sample of afterglows by BeppoSAX.

The chain of steps leading to the identification of the counterpart of GRB971214 (Dal Fiume et al. 1999) and its distance was the same of previous BeppoSAX observations. With a redshift $z = 3.42$ (Kulkarni et al. 1998a) this GRB and its host galaxy are at a distance that corresponds to a look-back time of about 85% of the present age of the Universe. At this distance, the luminosity would be about 3×10^{53} erg s^{-1}, were the emission isotropic. This was the highest luminosity ever observed from any celestial source. Initially the huge luminosity did not appear to be compatible with the energy available in the coalescence of neutron-star mergers (Kulkarni et al. 1998a) unless beaming were invoked. Other alternative energetic models are based on the death of extremely massive stars, leading to an explosion orders of magnitude more energetic than a supernova, hence named *hypernova* (Paczynski 1998). However it was shown (Mészáros & Rees 1998a) that all these progenitors, whether Neutron Star–Neutron Star mergers or hypernovae, eventually go through the formation of a same Black Hole/torus system, from which the energy is extracted to form the GRB. The radiation physics and energy of all mergers and hypernovae are then, to order of magnitude the same, and still compatible with the luminosity observed in GRB971214.

The energy problem became much more severe with GRB990123, another one of the BeppoSAX GRBs (Heise et al. 1999, Piro et al. 1999). It was one of the brightest GRBs ever observed, ranking in the 0.3% top of the BATSE flux distribution. Its distance ($z = 1.6$, Kulkarni et al. 1999) would imply a total energy of 1.6×10^{54} erg, assuming isotropical emission. This corresponds to ~ 2 $M_\odot c^2$, at the limit of all models of mergers (Mészáros & Rees 1998a).

This piece of evidence is lending support to the idea that, at least in some cases, the emission is collimated. This would reduce the energy budget by $\sim \theta^2/4\pi$, where θ is the opening angle of the jet. A typical feature of a jet expansion (vs. spherical expansion) would be the presence of an achromatic (i.e. energy-independent) break in the light curve that appears when the relativistic beaming angle $1/\Gamma$ becomes $\approx \theta$

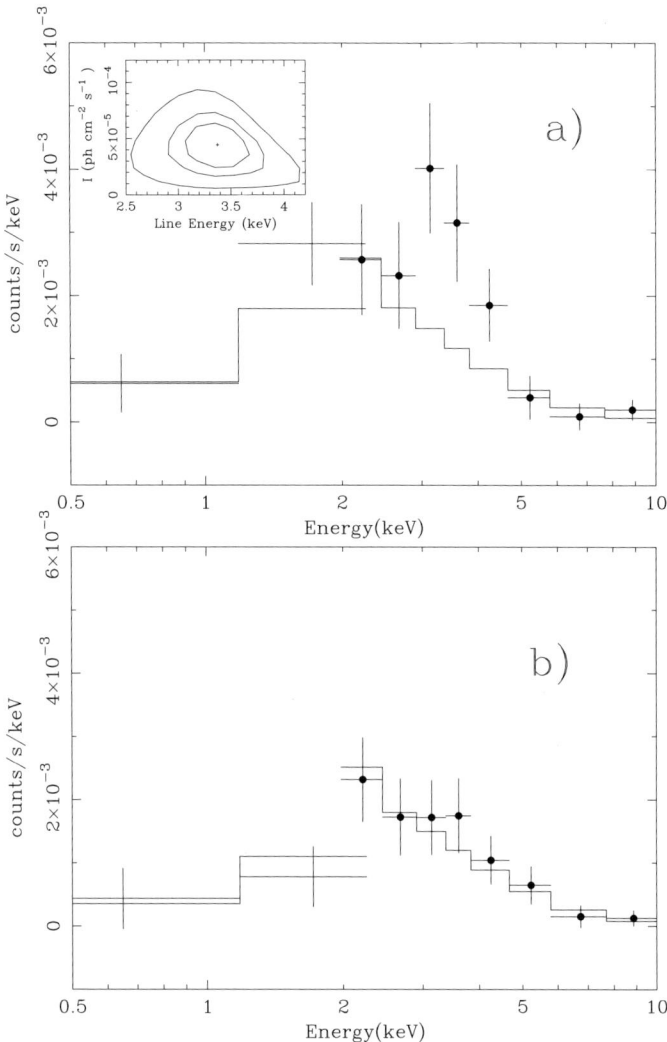

FIGURE 10. Spectra (in detector counts) of the X-ray afterglow of GRB970508 taken in the periods 0.2–0.6 day, (panel a) and 0.6–1 day (panel b) after the burst. The continuous line represents the best fit power law. Note the feature around 3.5 keV in the first part (contour plots at 68%, 90%, 99% in the inset) that, at a redshift of 0.83, corresponds to the energy of the iron line complex (6.4–6.8 keV). The line disappears in the second part of the observation (panel b) (from Piro et al. 1999).

(e.g. Sari et al. 1999, Rhoads 1997). The presence of such a break has been claimed in GRB990123 (Kulkarni et al. 1999) and in another more recent GRB, GRB990510 (Harrison et al. 1999). With an angle $\theta \approx 10°$, the total energy would be reduced by $\approx 10^3$, which falls within the limits of current models. So far evidence for an achromatic break is limited to the optical range and an independent measurement confirming its occurrence in different regions of the spectrum is lacking or not conclusive (e.g. in X-rays Kuulkers et al. 1999).

4. The nature of the progenitor

Information on the nature of the progenitor can be drawn from the GRB environment. In the case of a hypernova, the progenitor massive star has a very short lifetime ($\approx 10^6$ years) and therefore GRBs should be preferentially hosted in regions near the center of star-forming galaxies. On the contrary, NS-NS coalescence occurs on much longer time scales and the kick velocity given to the system by two consecutive supernova explosions should bring a substantial fraction of these systems away from the parent galaxy. So far the angular displacement of 5 optical counterparts indicates that GRBs are located within their host galaxies (Bloom et al. 1999a), favoring the association with star-forming regions. We note also that those events are not located in the very center of their galaxies, which excludes an association of GRB with AGN activity.

Another diagnostics of the progenitor nature is based on spectral measurements of broad and narrow features imprinted by a dusty, gas-rich environment expected in the hypernova scenario (e.g. Perna & Loeb 1998, Mészáros & Rees 1998b, Bottcher et al. 1998). The absence of an optical transient in about 50% of well localized BeppoSAX GRBs (18 as of May 99), in which instead an X-ray afterglow has been found in almost all the cases, may be explained by heavy absorption by dust in the optical range, which leaves almost unaffected the X-rays (Owens et al. 1998).

An exciting possibility is opened by the possible detection of X-ray iron line features in two different GRBs, one by BeppoSAX (GRB970508 Piro et al. 1999; Fig. 10) and the other by ASCA (GRB970828 Yoshida et al. 1999), associated with rebursting on time scales of the order of a day. It should be remarked that the presence of rebursting appears to be an uncommon feature of X-ray afterglows, whose temporal behavior is very well described by power laws (Fig. 9) at least until 2–3 days, when the X-ray flux goes below the sensitivity of current X-ray instruments. Both the temporal and spectral features betray the presence of a dense ($n \sim 10^{10}$ cm^{-3}) medium of ≈ 1 M$_\odot$ near the site of the explosion ($\approx 10^{16}$ cm) (Piro et al. 1999). Such a medium should have been pre-ejected before the GRB explosion, but the large value of the density excludes stellar winds. A possible, intriguing explanation is that the shell is the result of a SN explosion preceding the GRB (Piro et al. 1999, Vietri et al. 1999, Vietri & Stella 1998).

Other evidences argue in favor of a GRB-SN association. In the BeppoSAX error box of GRB980425 (Pian et al. 1999) two groups (Kulkarni et al. 1998, Galama et al. 1998) found a supernova (SN1998bw) that had exploded at about the same time as the GRB. The probability of a chance coincidence of the two events is $\approx 10^{-4}$. Since the majority of GRBs are not associated with SN (e.g. Graziani et al. 1999), this event (if the association is correct) should apparently represent an uncommon kind of GRB. However, it is also possible that the two families are indeed associated: this scenario would require that the GRBs be emitted by collimated jets. The majority of GRBs and afterglows we see are beamed towards us, so that the contribution of the supernova to the total emission is negligible. The case of SN1998bw was, then, peculiar in that the jet producing the GRB was collimated away from our line of sight, allowing the detection of the (isotropic) SN emission at an early phase. This scenario also explains why GRB980425 was not particularly bright, notwithstanding its redshift ($z = 0.0085$), much lower than the typical value of the other GRBs ($z \approx 1$). Since the afterglow decays as a power law, it is possible that at late times the emission of the SN becomes detectable. Evidence of such emission has been claimed in at least two cases (GRB990326: Bloom et al. 1999b, GRB970228: Reichart et al. 1999).

5. Conclusions

Several bits of evidence supporting the association of GRBs to star-forming regions have been gathered so far. The potential perspectives of this link are extremely exciting. Since GRBs are the most powerful and distant sources of ionizing photons, we can think of using them as probes of heavy elements and star/galaxy formation in the early Universe. A direct proof of this association is still missing, but the near future appears very promising in this respect. BeppoSAX is discovering and localizing GRBs and X-ray afterglows at a pace of 1 per month. Other satellites have also successfully set up procedures for rapid GRB localization (BATSE, XTE, ASCA and IPN). The launch of HETE2, foreseen to be in early 2000, will substantially increase the number of well-localized GRBs. Furthermore, in the near future, big X-ray satellites, like Chandra, XMM, and ASTRO-E will allow detailed spectral studies of X-ray afterglows and provide (Chandra) arcsec position of X-ray counterparts and, possibly, a direct redshift determination.

The BeppoSAX results presented here were obtained through the joint effort of all the components of the BeppoSAX Team. BeppoSAX is a major program of the Italian Space Agency (ASI) with participation of the Netherlands agency for aerospace programs (NIVR).

REFERENCES

BLOOM, J. S., ET AL. 1999a *ApJ*, in press.
BLOOM, J., ET AL. 1999b *Nature*, in press.
BOELLA, G., ET AL. 1997a *A&AS* **122** 299.
BOELLA, G., ET AL. 1997b *A&AS* **122**, 327.
BOND, H. 1997 *IAU Circular n. 6654*.
BOTTCHER, M., DERMER, C. D., CRIDER, A. W., & LIANG, E. D. 1998 *A&A* **343**, 111.
COSTA, E., ET AL. 1997 *Nature* **387**, 783.
DAL FIUME, D., ET AL. 1999 *A&A*, in press.
FEROCI, M., ET AL. 1997 *Proc. SPIE Conference* **3114**, 186 (astro-ph/9708168).
FEROCI, M., ET AL. 1998 *A&A* 332, L29.
FISHMAN, G. J., ET AL. 1994 *ApJS* **92**, 229.
FISHMAN, G. J. & MEEGAN, C. A. 1995 *Annual Review Astron. Astrophys.* **33**, 415.
FRONTERA, F., ET AL. 1997 *A&AS* **122**, 357.
FRONTERA, F., ET AL. 1998a *A&A* **334**, L69.
FRONTERA, F., ET AL. 1998b *ApJ* **493**, L67.
FRONTERA, F., ET AL. 1999 *ApJS*, in press.
FRUCHTER, A., ET AL. 1997 *IAU Circular n. 6747*.
GALAMA, T., ET AL. 1998 *Nature* **395**, 670.
GRAZIANI, C., LAMB, D., & MARION, G. H. 1999 *A&AS* **138**, 469; *Proc. of Gamma-Ray Bursts in the Afterglow Era* (eds. F. Frontera & L. Piro).
GARCIA, M. R., ET AL. 1998 *ApJ* **500**, L105.
HARRISON, F. A., ET AL. 1999 *ApJ* **523**, L21.
HEISE, J., ET AL. 1999, submitted to *Nature*.
KLEBESADEL, R. W., ET AL. 1973 *ApJ* **182**, L85.
JAGER, R., ET AL. 1997 *A&AS* **125**, 557.
KULKARNI, S. R., ET AL. 1998a *Nature* **393**, 35.
KULKARNI, S. R., ET AL. 1998 *Nature* **395**, 663.

Kulkarni, S., et al. 1999 *Nature* **398**, 389.
Kuulkers, E., et al. 1999 *ApJ*, submitted.
Manzo, G., et al. 1997 *A&AS* **122**, 341.
Mészáros, P. & Rees, M. J. 1997 *ApJ* **476**, 319.
Mészáros, P. & Rees, M. J. 1998a *New Astronomy*, (astro-ph/9808106)
Mészáros, P. & Rees, M. J. 1998b *M.N.R.A.S.* **299**, L10.
Metzger, M. R., et al. 1997 *Nature* **387**, 87.
Owens, A., et al. 1998 *A&A* **339**, L37.
Paczynski, B. 1998 *ApJ* **494**, L45.
Parmar, A. N., et al. 1997 *A&AS* **122**, 309.
Perna, R. & Loeb, A. 1998 *ApJ* **501**, 467.
Pian, E., et al. 1999 *A&AS* **138**, 463; *Proc. of Gamma-Ray Bursts in the Afterglow Era* (eds. F. Frontera & L. Piro).
Piro, L., Scarsi, L., & Butler, R. C. 1995 in *X-Ray and EUV/FUV Spectroscopy and Polarimetry* (ed. S. Fineschi). SPIE 2517, 169–181.
Piro, L., et al. 1998a *A&A* **329**, 906.
Piro, L., et al. 1998b *A&A* **331**, L41.
Piro, L., et al. 1999 *GCN* **199**, 203.
Piro, L., et al. 1999 *ApJ* **514**, L73.
Reichart, D., et al. 1999 *ApJ*, in press.
Rhoads, J. E. 1997 *ApJ* **478**, L1.
Sahu, K. C., et al. 1997 *Nature* **387**, 476.
Sari, R., Piran, T, Halpern, J. P. 1999 *ApJ* **497**, L17.
van Paradijs, J., et al. 1997 *Nature* **386**, 686.
Vietri, M., et al. 1999 *M.N.R.A.S.*, in press.
Vietri, M. & Stella, L. 1998 *ApJ* 507, L45.
Yoshida, A., et al. 1989 *Publs. Astron. Soc. Japan* 41, 509.
Yoshida, A., et al. 1997, in *IAU Circular n. 6593*.
Yoshida, A., et al. 1999 *A&AS* **138**, 433; *Proc. of the Gamma-Ray Bursts in the Afterglow Era* (eds. F. Frontera & L. Piro).

The first year of optical-IR observations of SN1998bw

By I. J. DANZIGER,[1] T. AUGUSTEIJN,[2] J. BREWER,[2]
E. CAPPELLARO,[3] V. DOUBLIER,[2] T. GALAMA,[4]
J. F. GONZALEZ,[2] O. HAINAUT,[2] B. LEIBUNDGUT,[5]
C. LIDMAN,[2] P. MAZZALI,[1] K. NOMOTO,[6]
F. PATAT,[2] J. SPYROMILIO,[5] M. TURATTO,[3]
J. VAN PARADIJS,[4] P. M. VREESWIJK,[4]
AND J. WALSH[5]

[1] Osservatorio Astronomico di Trieste, via G. B. Tiepolo 11, I-34131 Trieste, Italy

[2] European Southern Observatory, A. de Cordova 3107, Casilla 19001 Santiago, Chile

[3] Osservatorio Astronomico di Padova, via Osservatorio 5, I-35122, Italy

[4] Astronomical Institute "Anton Pannekoek"/CHEAF, Kruislaan 403, 1098 SJ Amsterdam, The Netherlands

[5] European Southern Observatory, Karl Schwarzschild Strasse 2, D-85748 Garching bei München, Germany

[6] Dept. of Astronomy, University of Tokyo, Tokyo 113-0033, Japan

We describe and attempt to interpret the various sets of observations, mostly spectroscopic, made at ESO, La Silla for the past year from the time of the outburst of the associated GRB980425. These observations include optical spectra taken in all phases photospheric through nebular, infrared spectra, and spectrophotometric linear polarization. Because of the increasing attention to the possibility of asymmetric explosions, our results are discussed in the framework of an attempt to identify observable signatures of asymmetry.

1. Introduction

The reasons for associating GRB980425 with SN1998bw in the first instance have to do with the near coincidence in position and time and the statistical significance of these does not merit further discussion here. What should be also stressed in considering this association is the fact that SN1998bw is one of the brightest supernovae ever detected in both optical and radio wavelengths. Unfortunately the X-ray positions are not sufficiently accurate to allow us to make a similar statement about X-ray luminosities. To supernova optical spectroscopists the spectrum was also a puzzle, its almost unique appearance being dictated mostly by the very large velocities of expansion of the envelope. This resulted in contradictory reports of its classification when the first spectra became available. Even now to classify it as a Type Ic, the possibly most appropriate category of currently used classes, underscores the limitations of too few or too restrictive boxes into which supernovae are placed. In any case one can state with some confidence that it is not a Type II of any sub-category because there has never been any sign of hydrogen in emission or absorption.

The three Nature papers (Galama et al. 1998, Iwamoto et al. 1998, Kulkarni et al. 1998) have presented various aspects of the early observations and papers by Pian et al. (1999 and this conference) discuss the X-ray observations. The gamma-ray characteristics have also been discussed by Pian (1999) with a more extensive paper by Frontera et al. (1999).

2. Spectroscopy

It may be instructive to compare and contrast spectra of SN1998bw with other Type I spectra obtained at similar phases both early (photospheric) and late (nebular). Near maximum light the peak emission in a Type Ia (SN1992A) occurs near 4000 Å while in SN1998bw there is a major broad peak at 5200 Å, while it is clear that absorption features are much broader in SN1998bw. While the Type Ib (SN1983N) has a less broad feature at 5200 Å the overall continuum has a different shape revealing more flux than SN1998bw extending to 7000 Å. Again SN1998bw has much broader absorption features and lines of HeI are not apparent as they are in Type Ib spectra. The early spectrum of the Type Ic (SN1994I) again shows a broad feature near 5200 Å, a continuum shape somewhat similar to that of SN1998bw, but again much narrower absorption features. The broad feature at 5200 Å is so isolated and peaked in SN1998bw that it has been incorrectly identified as an emission line feature. Subsequent modelling has shown that as a consequence of the very high velocities in the expanding envelope the redward moving photons tend to accumulate in this wavelength region owing to a lack of overlapping line opacity that would allow them to move even further redwards.

In the nebular phase a comparison of Type Ia (SN1992A) with SN1998bw shows that the latter has developed very strong emission features ascribed to [OI]6300,63 and the CaII triplet blend near 8500 Å, neither of which are present in the former. Also the ratio of various emission features ascribed to [FeII,III] in the region 4500–5500 Å are very different in the two supernovae. The spectrum of the Type Ib (SN1990B) shows the [OI]6300 feature but the CaII8500 is more pronounced and the emission features 4500–5500 Å less pronounced than in SN1998bw. A similar remark is justified in the comparison of the Type Ic (SN1990I) with SN1998bw. Thus these comparisons justify the remark that SN1998bw resembles a Type Ic supernova more than any other, but that the velocities are clearly significantly higher in SN1998bw.

That the velocities were very high has been shown by measurements of the SiII 6355 line and presented by Iwamoto et al. (1998). The velocity measured at maximum absorption in the line was 28000 km s^{-1} at day 8 after the explosion, and it is not difficult to intuit that extrapolated backwards to day 1 this velocity could be as high as 60000 km s^{-1}. In addition there is material moving at much higher velocities than this as evidenced by the absorption profile extending bluewards from this wavelength of maximum absorption. While it is impossible because of the complexity of the spectrum to estimate at what velocity one still sees absorption, it is clear that relativistic velocities in the outer part of the envelope are to be expected. One might think that the CaII triplet which seems to be present at a very early stage would lend itself to a better estimate of limiting velocities. However care is required before doing this for 2 reasons. Firstly there may be significant absorption due to the OI7774 triplet, and secondly the P Cygni feature ascribed to the CaII triplet has a temporal behaviour that has not been understood. Whereas the peak emission of such a feature should be always near the rest wavelength, in SN1998bw it appears to be moving in time from a considerably shorter wavelength. This second effect may prove to be a result of the OI7774 feature interacting with the CaII triplet feature. It hardly needs to be emphasized that successfully decoding this complex profile provides a means of determining how much oxygen existed in outer layers of the progenitor star.

3. Models

Model fits to the light curves and photospheric spectra of SN1998bw have been presented by Iwamoto et al. (1998). A family of models have been constructed to fit the

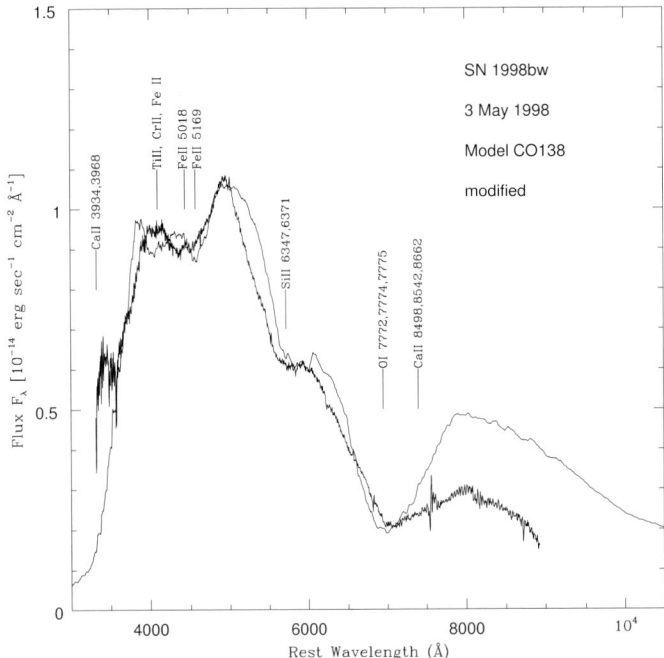

FIGURE 1. The observed spectrum of SN 1998bw on 3 May 1998 (bold line) compared to a synthetic spectrum obtained with our Monte Carlo code and using the explosion model CO138 (Iwamoto et al. 1998) but with a modified outer density structure, with $\rho \propto n^{-6}$ at $v > 30,000$ km s^{-1}. The original model had $\rho \propto n^{-8}$. The spectrum computed with a modified density profile reproduces the observed line blending much better because of the increased high-velocity line absorption.

light curve around the maximum peak whose characteristics are governed by the photon diffusion time in the envelope. This photon diffusion time is a function of the product of different powers of the ejecta mass and the explosion energy. Thus from the light curve alone one cannot obtain a unique value of these two essential quantities. Therefore the spectra have been used to obtain the best model fit requiring a match of the overall spectrum and the velocity widths in particular. We display in Figure 1 a spectrum of the supernova 7 days before maximum light together with a model fit the characteristics of which are described in the caption. (This model exemplifies how the remarkable peak near 5000 Å is formed without recourse to emission lines. Modelling of successive spectra shows how the peak also moves redwards as a function of time). Since this velocity is also a function of different powers of the ejecta mass and the energy of the explosion, through the equation for kinetic energy, one can then find unique solutions for both mass and energy.

The net outcome of this work resulted in the following parameters for the best fitting model:
- Mass of exploding star—13.8 M$_\odot$
- Mass of radioactive ^{56}Ni—0.7 M$_\odot$
- Mass of ejecta—10.9 M$_\odot$
- Explosion energy—$3 \cdot 10^{52}$ ergs
- Mass of remnant—2.9 M$_\odot$ (possibly a black hole?)
- Mass of progenitor—30–40 M$_\odot$

4. Infrared spectra

Three infrared spectra have been obtained with SOFI at the NTT during the interval from 8 days to 51 days after maximum light on 10 May, 1998. A fuller discussion of these spectra will be given elsewhere. However we draw attention to a strong P Cygni feature which on day 8 would correspond to HeI10830 with a velocity of 18000 km s^{-1}. This is higher than the velocity of SiII at the same phase which was 13000 km s^{-1}. Such a difference might be expected if the helium were confined to an outer layer of the envelope. That this might well be the case is evidenced by the fact that in the subsequent 2 spectra the velocity of the absorption has not changed. This is precisely what would be predicted to happen if, on or before day 8, this helium layer had become optically thin in the line so that the effective "photosphere" lay well below this level. Within this absorption profile there is a secondary absorption at shorter wavelengths which still requires identification. The later 2 spectra show other features growing in strength. One obvious candidate which can be confirmed with modelling is the NaI 11381,404 (Multiplet 3) transition whose lower level corresponds to the upper level for the NaID lines transition.

Why are other helium lines at shorter wavelengths not detected? There are 2 reasons for this. First because of the special nature of the metastable lower level of HeI10830 transition, it is overpopulated with respect to a simple thermal equilibrium situation, and intrinsically stronger than all the HeI lines at shorter wavelengths. In addition these lines at shorter wavelengths are smeared out and lose their identity owing to the redward photon transfer at the very high velocities because of the significantly greater density of all lines at shorter wavelengths. What remains to be tested with models is whether a star with an outer layer of helium whose mass remains to be determined, will evolve differently and therefore whether a significant change in the parameters given above will be required.

5. Polarization

Spectrophotometric linear polarization measurements were made at 2 phases, 8 days following the GRB event and 23 days. These were accomplished by means of Wollaston prisms mounted in EFOSC2 at the ESO 3.6 meter telescope. The degree of polarization measured on the earlier date is 0.8 percent averaged over the range 4500–6500 Å, and is 0.4 percent for the same range on the later date. Since nearby stars do not have polarizations greater than 0.2 percent, one may conclude that the polarization is not due to the interstellar medium in our Galaxy. The fact that it varies and tends to decrease strongly suggests that a major part is due to intrinsic polarization and not to the ISM in the parent galaxy of the supernova.

The recent monograph by Gnedin and Silant'ev (1997) presents analytic solutions for polarization resulting from extended electron scattering atmospheres of different shapes. That we obtain a finite degree of polarization means that the envelope is not spherically symmetric. Both oblate and prolate figures are demonstrated to show polarization (a few percent linearly dependent on the scattering optical depth) with moderately small axial ratios (2.5/1–4/1), and particularly if viewed at the most favourable angles, namely orthogonal to the axis of symmetry. Therefore the modest polarization recorded for SN1998bw could be envisaged to be a result of a modest asymmetry, or a viewing angle intermediate between the axis of symmetry and the orthogonal direction, or both. Unfortunately we cannot with this data alone decode the shape and the viewing angle but some degree of asymmetry is apparent. This is not the first case of net polarization being recorded for Type Ib,c supernovae.

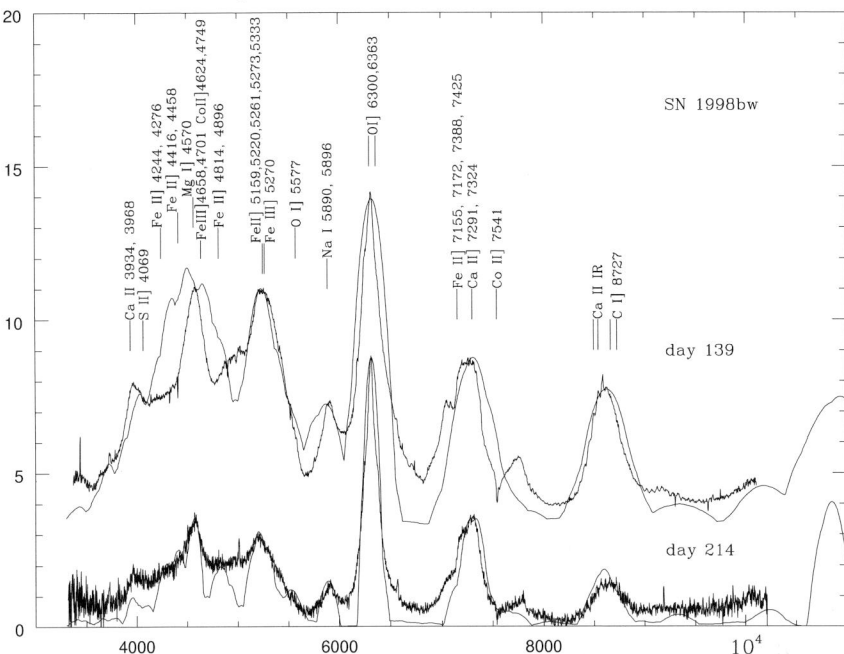

FIGURE 2. Nebular spectra of SN 1998bw on 12 Sept 1998 (rest frame epoch 139 days) and 26 Nov 1998 (rest frame epoch 214 days) are compared to synthetic spectra obtained with a NLTE nebular model based on the deposition of gamma-rays from ^{56}Co decay in a nebula of uniform density. The ^{56}Ni mass is 0.65 M$_\odot$, and the outer nebular velocity is 11,000 km s^{-1} for both epochs, but the O mass and the total mass increase from 3.5 and 5.2 M$_\odot$, respectively, on day 139, to 5.0 and 7.2 M$_\odot$, respectively, on day 214. The average electron density in the nebula is $\log n_e = 7.47$ cm^{-3} on day 139 and $\log n_e = 7.06$ cm^{-3} on day 214. The filling factor used is 0.1 for both models.

6. Nebular Spectra

In Figure 2 we show two spectra taken in the nebular phase at days 139 and 214. Strong emission lines now dominate, and the strongest, clearly identified with [OI]6300 is characterized by a sharply peaked profile, the underlying reasons for which still require an explanation. Other lines whose identity is certain even if other transitions contribute to a blend are NaID5889,95, the CaII triplet and [CaII]7291,7324, and MgI]4571. The overlying spectra result from models whose characteristics are given in the caption. These models represent a first attempt to quantify what we can understand about the spectrum. Readers should be aware that the models invoke a uniform single density which must be a gross approximation to an envelope with a density and probable temperature gradient. This being so, it has not been possible so far to take account of critical densities for particular lines of particular ions. In the observed envelope it is therefore possible, because of critical density effects and varying temperature, that the relative contributions of different transitions from different ions to the integrated line emission can vary significantly through the envelope.

The mismatch of the model spectra with the observed one near 4500 Å on day 139 suggests a number of possibilities. One is that the [FeII] spectrum has been quenched by prevailing densities higher than the critical density. This leaves open the possibility of a significant contribution from FeII and FeIII transitions if densities are sufficiently high. The critical density of [OI]6363 is approximately 3 times lower than that of [OI]6300 and so the weaker line could also be quenched relative to the former if the density fell in the appropriate range. A visual inspection of the spectrum of day 139 suggests that [OI]6300 is narrower than other features but this is only partially borne out by the model fit. It will be important in the future to obtain models that give a good fit to all features because in this way one can examine the interesting question whether lines of different elements have different profiles and are therefore expanding in a different configuration. The observed spectra show a temporal variation, and therefore it is important that there be consistency in the best-fit models required for each epoch.

7. Asymmetries

It is appropriate to examine whether any or all of these observations can give any indication of an asymmetrical explosion and expansion of the envelope. There are a number of reasons related to the energetics of the gamma-ray burst and the early radio emission which make asymmetry attractive. In addition, the dominant role of supernovae in explaining GRBs would be enhanced if asymmetries and beaming were significant.

Unfortunately, as we have shown above, the polarization measurements can be interpreted as evidence for an asymmetrical shape of the envelope, but we cannot decode this information into quantitative information about actual shape and orientation.

Absorption line profiles in the photospheric phase offer another opportunity. An inspection of the SiII6355 line in the early spectra reveals significant variations of the profile which seem real given the quality of the spectra. However defining an absorption profile emanating from an expanding envelope is extremely difficult both because P Cygni features may be present and the continuum is virtually impossible to define accurately because of the complex overlap of lines. It is made even worse by the fact that this line in particular lies in a region of a very steeply sloping continuum. Our best hope here might be to wait until modelling gives a much better accounting for all spectral features, both lines and continuum.

Lastly there is the nebular spectrum where progress at the moment is hindered by secure identification of some emission features. There is at the moment a slight hint that the [OI]6300 profile may differ from others such as those due to [FeIII]. If this impression proves to be correct and differing density and temperature conditions cannot reasonably explain such an effect, it might be possible to invoke an asymmetric explosion mechanism in which oxygen from the outer part of the progenitor is being thrown off in an asymmetric way as a result of rotation (?) while iron is not, and the explosion is being viewed from some angle that does not reveal the maximum component of the oxygen velocity. Although this at the moment is speculative, observational criteria must be sought to answer the challenge posed by, for example, the asymmetric models proposed by Popham et al. (1999) and Woosley (this conference and references therein).

REFERENCES

FRONTERA, F., ET AL. 1999 *ApJ*, in press.
GALAMA, T. J., ET AL. 1998 *Nature*, **395**, 670.
GNEDIN, YU. N. & SILANT'EV, N. A. 1997 *Astrophys. Space Phys.*, **10**, 1.
IWAMOTO, K., ET AL. 1998 *Nature*, **395**, 672.
KULKARNI, S., ET AL. 1998 *Nature*, **395**, 663.
PIAN, E., ET AL. 1999 *A&A Suppl.*, in press.
POPHAM, R., WOOSLEY, S. E., FRYER, C. 1999 *ApJ*, **518**, 356.

X-ray emission of Supernova 1998bw in the error box of GRB980425

By ELENA PIAN

Istituto di Tecnologie e Studio delle Radiazioni Extraterrestri, Via Gobetti 101, I-40129 Bologna, Italy

The spatial and temporal coincidence of a GRB and a supernova explosion (1998bw) on 25 April 1998 has raised conjectures on the physical connection between the two phenomena, and in general on the association of GRBs with supernovae, at least with the most powerful among them (hypernovae or collapsars). In fact, multiwavelength observations of SN 1998bw have revealed unusual characteristics: extremely high energy output at radio and optical wavelengths, and relativistic expansion of the outgoing shock. The X-ray emission of SN 1998bw, monitored by *BeppoSAX* starting ~ 10 hours after the GRB detection, was remarkably prompt (within one day of supernova detonation), but exhibited spectral and temporal properties similar to those of other supernovae detected in X-rays.

1. Introduction

The discovery of a supernova within the $8'$ radius error circle of the GRB980425 has been regarded as a major puzzle within the thick mystery of GRBs. The GRB980425, which has been detected by the *BeppoSAX* Gamma Ray Burst Monitor (GRBM, 40–700 keV) and by BATSE (Kippen 1998), and rapidly localized by the *BeppoSAX* Wide Field Cameras (WFC, 2–26 keV) Unit 2, appears as a relatively weak burst, characterized by a single, non structured peak of longer duration in the 2–26 keV range (52 seconds) than in the 40–700 keV range (31 seconds). In Figure 1 (left panel) are reported the temporal profiles in both energy ranges. The spectrum of the GRB rapidly softens with time (Figure 1, right panel; see also Frontera et al. 1999).

Weak intensity, a single peak, a soft and fastly evolving spectrum, and a ~ 5 second temporal delay of the X-rays with respect to the γ-rays appear to be the main characteristics of this GRB. However, these features are common to other GRBs, and therefore cannot be considered as an obvious suggestion that GRB980425 is peculiar.

It came as a surprise for the teams involved in GRB search and follow-up at longer wavelengths, and subsequently for the whole astronomical community, that in the error box of GRB980425 a supernova was detected, 1998bw, at the optical (Galama et al. 1998; Galama et al. 1999a), and radio wavelengths (Kulkarni et al. 1998a), 17 hours and 3 days after the GRB event, respectively. The inferred time of supernova explosion is consistent with the GRB occurrence to within $+0.7/-2$ days (Iwamoto et al. 1998).

SN 1998bw lies in a spiral arm of the galaxy ESO 184-G82, at a redshift $z = 0.0085$ (Tinney et al. 1998). Its radio luminosity ($\sim 10^{38}$ erg s^{-1} at peak) is the largest ever measured for any supernova, and the optical one ($\sim 10^{43}$ erg s^{-1} at peak) ranks among the highest supernova luminosities.

SN 1998bw stands out not only for its positional and temporal coincidence with GRB980425 and for its unusual radio and optical luminosity, but also for the properties of the radio light curves (Kulkarni et al. 1998a; Wieringa et al. 1999) and for the broad optical spectral lines, which indicate high photospheric velocities. Based on its optical spectrum, SN 1998bw was classified as a peculiar Type Ib before maximum light (Sadler et al. 1998) and Type Ic at later epochs (Iwamoto et al. 1998; Galama et al. 1998; Patat & Piemonte 1998).

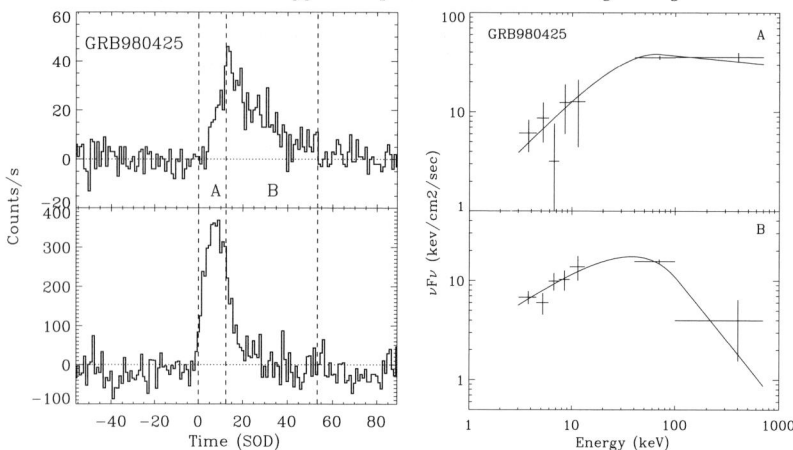

FIGURE 1. Left panel: *BeppoSAX* WFC (top) and GRBM (bottom) light curves of GRB980425. Time is in seconds; the onset of the GRB, indicated by the zero abscissa, corresponds to 1998 April 25.9091. The typical 1-σ uncertainty associated with the individual flux points is ~ 4 counts s^{-1} for the WFC data and ~ 40 counts s^{-1} for the GRBM data. The vertical dashed lines divide the burst duration in two time intervals denoted by "A" (12 s) and "B" (40 s), over which the signal has been integrated to construct 2–700 keV spectra. These are reported in the right panel, along with the Band law fitting curves (Band et al. 1993).

2. *BeppoSAX* Target of Opportunity observations of the GRB980425 error box

Following the GRB event, the field of GRB980425 was promptly acquired by the *BeppoSAX* Narrow Field Instruments (NFI; these include the LECS, MECS, HPGSPC, and PDS detectors. See Butler & Scarsi 1990, and Boella et al. 1997 for a description of the *BeppoSAX* mission), and observations in the energy range 0.1–300 keV started 10 hours after the GRB detection (26–27 April 1998). Two previously unknown sources have been detected within the WFC error box by the MECS in the 2–10 keV energy range (Pian et al. 1999a; Pian et al. 1999b).

The brighter source, S1, is consistent with the position of SN 1998bw, while the fainter one, S2, is not (Figure 2). The LECS data have a significantly lower signal-to-noise ratio than the MECS data and the HPGSPC and PDS instruments yielded no detection above the background, therefore we will briefly report here only the MECS results and refer to a paper of imminent submission for details about both LECS and MECS data (Pian et al. 1999b).

Note that the coordinates of sources S1 and S2 distributed by Pian et al. (1998) have been revised in November 1998, to take into account a systematic error due to the non-optimal spacecraft attitude during the April and May 1998 observations (see Piro et al. 1998a). Figure 2 illustrates the updated, correct location of the sources.

The following NFI pointings, one week (2–3 May 1998) and six months later (10–12 November 1998), have shown that neither source exhibits the behavior expected for an X-ray afterglow. Source S1 did not exhibit significant variability in one week, and was still detected, a factor of ~ 2 fainter, six months later (see Figure 3). Source S2 exhibits marginally significant variability between 26–28 April and 2–3 May 1998. It is not detected in November 1998, but its upper limit is consistent with the April-May flux level (see Figures 3 and 4a).

The variability of source S1 and its positional consistency with SN 1998bw suggest that S1 is the X-ray counterpart of the supernova. This is the earliest detection of X-ray

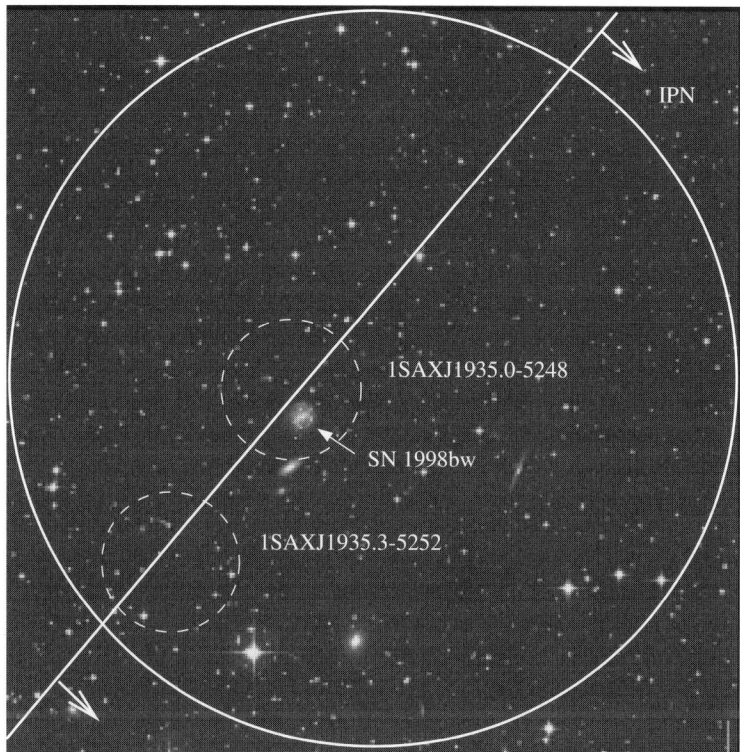

FIGURE 2. Digitized Sky Survey image ($16' \times 16'$) of the WFC error box of GRB980425 (large circle of radius $8'$). The left boundary of the IPN annulus is indicated as well as the error boxes of the two NFI X-ray sources (dashed circles of radius $1\rlap{.}'5$) and the position of the SN 1998bw. The two X-ray sources S1 (1SAXJ1935.0–5248, at the revised coordinates (J2000) $\alpha = 19^{\rm h}\ 35^{\rm m}\ 05.9^{\rm s}$, $\delta = -52°50'03''$) and S2 (1SAXJ1935.3–5252, at the revised coordinates $\alpha = 19^{\rm h}\ 35^{\rm m}\ 22.9^{\rm s}$, $\delta = -52°53'49''$), and SN 1998bw are consistent with the WFC and IPN locations. (From Galama et al. 1999a, reprinted with permission of *Astronomy & Astrophysics*.)

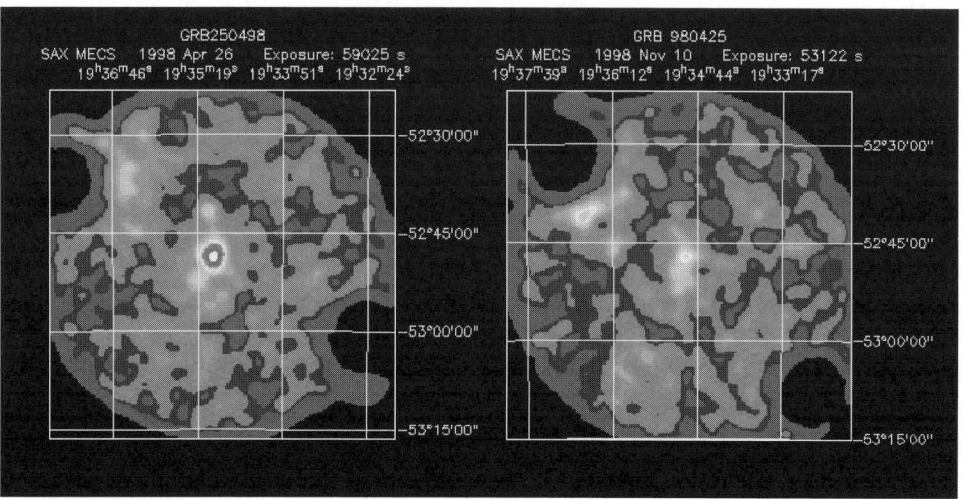

FIGURE 3. *BeppoSAX* MECS images of GRB980425 in April 1998 (left) and November 1998 (right). The data have been smoothed with a Gaussian function of $1\rlap{.}'5$ FWHM. Source S1 is visible in the image center; source S2 is toward the South-East, at $\sim 4'$ away from S1.

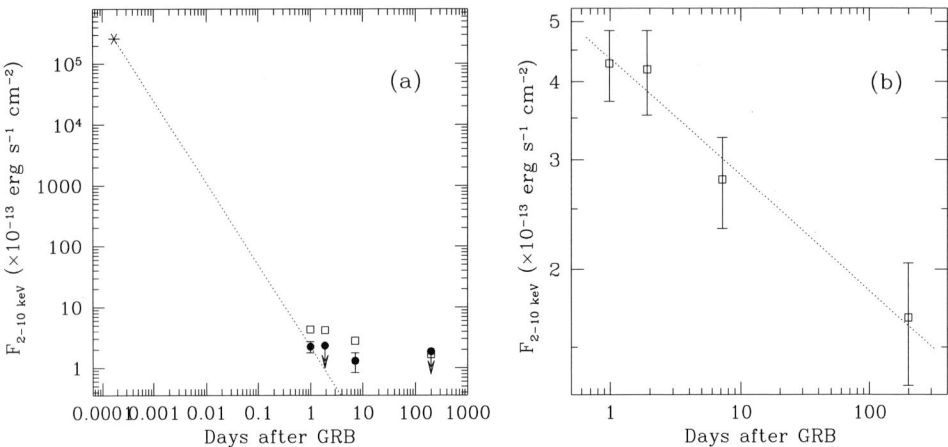

FIGURE 4. (a) *BeppoSAX* MECS light curves in the 2–10 keV band of the X-ray sources S1 (open squares) and S2 (filled circles) detected in the GRB980425 field. The WFC early measurement in the same band is also shown (star). The zero point for the abscissa is 1998 April 25.9091. Uncertainties associated with the WFC point and with the NFI measurements of S1, being equal to or smaller than the symbol size, have been omitted. The dotted line represents the power-law $f(t) \propto t^{-p}$ of index $p \simeq 1.3$ connecting the WFC measurement with the first NFI measurement of source S2. The extrapolation of the line to the time of the third observation falls below the lower bound of the S2 flux measurement but it is marginally consistent with it (the excess with respect to the power-law is $\sim 2.5\text{-}\sigma$). (b) Same as (a) for source S1 only. The fit to the temporal decay with a power-law of index ~ 0.2 is shown as a dotted line.

supernova emission, and the first detection of medium energy X-rays from a Type I supernova (the only other case of X-ray bright supernova is the Type Ic SN 1994I, detected in the soft X-rays by ROSAT, Immler et al. 1998a).

At the distance of SN 1998bw, the luminosity observed in the range 2–10 keV, $\sim 2\text{--}5 \times 10^{40}$ erg s^{-1}, is compatible with that of other supernovae detected in the same energy band, all Type II (see Table 1). However, the supernova X-ray luminosity could suffer from host galaxy contamination, which might be significant at these energies (see Fabbiano 1989). Similarly, the observed variation of a factor of two in six months is only a lower limit to the supernova X-ray variability amplitude. A power-law $f(t) \propto t^{-p}$ with index $p \sim 0.2$ provides an acceptable fit of the light curve (Fig. 4b), and is approximately consistent with the behavior observed for other supernovae (Kohmura et al. 1994; Houck et al. 1998) and with predictions based on interaction of energetic electrons with the circumstellar medium (Chevalier & Fransson 1994; Li & Chevalier 1999).

The prompt X-ray emission observed for SN 1998bw requires that the circumstellar medium is highly ionized (probably by the powerful explosion), to allow the X-rays to escape so soon after the explosion (see Zimmermann et al. 1994), and also very dense, as inferred also from the large radio output (Kulkarni et al. 1998a; Wieringa et al. 1999).

The spectrum of S1 in the 2–10 keV energy range is well fitted by a power-law $F_\nu \propto \nu^{-\alpha}$ of index $\alpha = 0.5 \pm 0.2$ (1-σ), or by a thermal bremsstrahlung model with temperature ~ 15 keV (see Pian et al. 1999b for details on the spectral fits). Both are consistent with spectral slopes and temperatures found for other supernovae detected in X-rays (e.g. Kohmura et al. 1994; Leising et al. 1994; Dotani et al. 1987).

The mildly relativistic conditions of the expanding shock of SN 1998bw (Kulkarni et al. 1998a) might suggest that the mechanism responsible for the X-ray emission is synchrotron radiation of very energetic electrons, or inverse Compton scattering of relativistic electrons (which produce the radio spectrum via synchrotron) off optical/UV

SN	Type	Host Galaxy	Distance (Mpc)	Date of optical max.	B mag at optical max.	Δt[†]	Satellite[‡]	X-ray Lum.[¶] (10^{39} erg s^{-1})	Range (keV)	Ref.[∥]
1978K	IIL	NGC1313	4.5	~1978 Jun 10[††]	~13	12.2 yr	ROSAT, Asuka, ASCA	5.3 ± 3.4	0.5–2	1,2
1979C	IIP	NGC4321=M100	17.1	~1978 May 25[††]	~14.5	16 yr	ROSAT	~3	0.5–2	3
1980K	IIL	NGC6946	5.1	1979 Apr 19	≤12	44 d	Einstein	1.0 ± 0.1	0.1–2.4	4
1986J	IIpec	NGC891	9.6	1980 Nov 05	11.7 ± 0.1	~9 yr	ROSAT, ASCA[‡‡]	~0.5	0.2–4	1,5
1987A	IIP	LMC	0.05	1983 Jan ??	?	154 d	Ginga, Röntgen, ROSAT	16–70	0.1–2.4	1,6,7
1987A	IIP	LMC	0.05	1987 May 09	3.5			~0.015	10–30	1,8,9
								~10^{-5}	0.5–2	10
1988Z	IIP	MCG+03-28-022	95	1988 Dec 12	16.5	6.4 yr	ROSAT	150 ± 100	0.1–2.4	11
1993J	IIpec	NGC3031	3.63	1993 Apr 18	11.4	6 d	ROSAT, Asuka, ASCA, GRO	2.9 ± 0.2	0.1–2.4	1,12
								15 ± 4	1–10	13,14
1994I	Ic	NGC5194=M51	7.7	1994 Apr 08	13.77 ± 0.02	79–85 d	ROSAT	0.16 ± 0.05	0.1–2.4	15
1995N	IIn	MCG-2-38-017	24	1995 May 6–8	V ~ 17.5	1.2 yr	ROSAT	20–30	0.1–2.4	16
1998bw	Ib/c	ESO184-G82	38	1998 May 10.3	14.30 ± 0.05	1 d[¶¶]	BeppoSAX	47 ± 6	2–10	17,18

[†] Time interval between supernova explosion and first X-ray detection.
[‡] Spacecraft which observed the source. First listed is the one which first detected it.
[¶] Measured at first detection by the spacecraft first listed in preceding column. The energy range is specified in the next column.
[∥] References for X-ray observations. **1:** Schlegel (1995); **2:** Schlegel et al. (1996); **3:** Schlegel et al. (1999); **4:** Immler et al. (1998b); **5:** Canizares et al. (1982); **6:** Bregman & Pildis (1992); **7:** Houck et al. (1998); **8:** Dotani et al. (1987); **9:** Sunyaev et al. (1987); **10:** Beuermann et al. (1994); **11:** Fabian & Terlevich (1996); **12:** Zimmermann et al. (1994); **13:** Kohmura et al. (1994); **14:** Leising et al. (1994); **15:** Immler et al. (1998a); **16:** Lewin et al. (1996); **17:** Pian et al. (1999a); **18:** Pian et al. (1999b).
[††] Depending on assumed type, date and magnitude of optical maximum are as listed (Ref. 1).
[‡‡] f L(2–10 keV) = 20.4×10^{39} erg s^{-1} (Ref. 7).
[¶¶] Computed with respect to GRB980425 initial time.

TABLE 1. Supernovae detected at x-ray energies

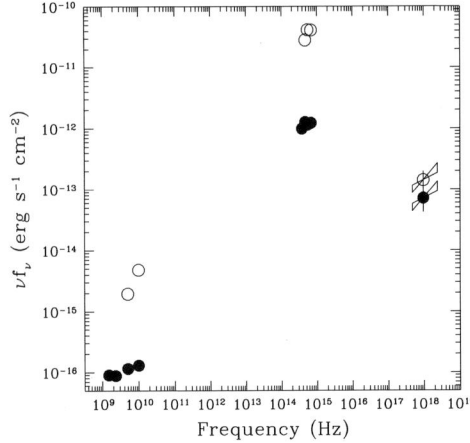

FIGURE 5. Quasi-simultaneous radio-to-X-ray spectral energy distributions of SN 1998bw in 3–5 May (open circles) and 10–12 November 1998 (filled circles). Power-law fits to the X-ray spectra are shown along with their 1-σ confidence ranges (Pian et al. 1999b). The optical magnitudes have been transformed to fluxes according to Fukugita et al. (1995) and corrected for Galactic absorption using $A_V = 0.2$ (Schlegel et al. 1998), although Patat et al. (1999) argue in favor of a lower value. For the first epoch, the optical and radio data have been taken from Galama et al. (1998) and Kulkarni et al. (1998a), respectively. For the second epoch the optical data have been either interpolated (bands V and I) between October 29 (McKenzie & Schaefer 1999) and November 26 (Vreeswijk et al. 1999c) measurements or extrapolated (bands B and R) using the late time exponential decay fitted to the light curves by Patat et al. (1999). The radio data are from Wieringa et al. (1999). Note that the *BeppoSAX* data represent the blend of the supernova and possible host galaxy emission and should then be considered upper limits on the X-ray emission of SN 1998bw. The optical supernova ejecta dominate the power output at both epochs. The radio and X-ray data could be consistent with a single radiation component.

photons of the thermal ejecta. The X-ray spectral index is consistent with that measured for the radio spectrum starting $\sim$ 15 days after the explosion (Kulkarni et al. 1998a; Wieringa et al. 1999; before that epoch the radio spectrum is significantly self-absorbed), and with the spectral slope connecting quasi-simultaneous radio and X-ray measurements ($\alpha \sim 0.8$). Therefore, in case the X-rays have a non-thermal origin, it is difficult to establish whether they are produced through the synchrotron or inverse Compton process (Fig. 5).

3. The GRB/Supernova Connection

Although the chance coincidence of GRB980425 and SN 1998bw has a very low probability (10^{-4}, Galama et al. 1998), the GRB community has not accepted unanimously the physical association of the GRB and the supernova. In fact, the faint source S2—possibly, but not clearly, fading—could be considered an afterglow candidate. The flux of S2 during the first *BeppoSAX* observation would be consistent with a power-law decay of index $p \simeq 1.3$ after the early X-ray emission observed by the WFC (see Fig. 4a). This is in the range of the power-law decay indices of "classical" X-ray afterglows (Costa et al. 1997; Nicastro et al. 1998; Dal Fiume et al. 1999; in 't Zand et al. 1998; Nicastro et al. 1999; Vreeswijk et al. 1999a; Heise et al. 1999). However, the second detection of S2 is not conclusive: it is marginally consistent with the first detection, but it is also marginally consistent with the power-law decay. The November 1998 upper limit is consistent with the detection level. Therefore, based on the present data, one cannot establish whether S2 is an afterglow exhibiting a small re-bursting (similar to GRB970508, although the time scale would be different, Piro et al. 1998b) or a permanent, perhaps modestly variable, X-ray emitter, like an active galactic nucleus or a Galactic binary (the chance probability of detecting a source of the level of S2 in the 8' radius WFC error box of GRB980425 is rather high, $\sim 12\%$). Optical observations have been equally inconclusive: no optical transient at a position consistent with S2 has been detected by early imaging of the GRB error box, and late epoch optical spectroscopy of sources brighter than $V \sim 18$ in the S2 error box failed to identify any active galaxy or binary stellar system having a compact object (Halpern 1998, and private communication).

At the time of GRB980425/SN 1998bw detection, five optical afterglows of GRBs had been detected, and for all of them, similarly to X-ray afterglows, a rapid power-law decay had been measured with index p in the range 1.1–2.1 (Van Paradijs et al. 1997; Fruchter et al. 1999a; Fruchter et al. 1999b, and references therein; Diercks et al. 1998; Halpern et al. 1998; Kulkarni et al. 1998b; Groot et al. 1998; Palazzi et al. 1998). The circumstance of detecting a supernova as the possible counterpart of a GRB was unprecedented. Therefore, it was proposed that this GRB might belong to a different class of events, with apparently indistinguishable high energy characteristics, but with different progenitors. Furthermore, assuming association with SN 1998bw, GRB980425 would be much closer than the GRBs for which a redshift measurement is available, which reinforced the idea that GRB980425 was physically dissimilar from GRBs exhibiting power-law decaying X-ray and optical remnants, predicted by the cosmological fireball model (Rees & Mészáros 1992; Piran 1999).

After the case of GRB980425/SN 1998bw, many authors have searched for statistical support of the possible association between GRBs and supernovae, and obtained different, and sometimes conflicting, results. The comparison of the BATSE catalog with supernovae compilations seems to suggest that some GRBs may be spatially (within an angular uncertainty of many degrees) and temporally (within ~ 20–30 days) consistent with Type Ib/c supernovae, while association with Type Ia is ruled out (Wang & Wheeler 1998. See however Kippen et al. 1998 and Graziani et al. 1998). Association has been specifically proposed for the cases of the Type II supernovae 1997cy and 1999E with GRB970514 and GRB980910, respectively, based on temporal and spatial proximity and on the outstanding optical properties of the two supernovae (Woosley et al. 1999; Germany et al. 1999; Thorsett & Hogg 1999; Turatto et al. 1999). However, limiting the GRB sample to the events with temporal profile similar to GRB980425 leads to no significant association (Bloom et al. 1998a). A negative result is also obtained by further restricting the subset to long, soft GRBs (Norris et al. 1998). This seems to suggest that the temporal and spectral characteristics of GRBs are not obvious tracers of possible association with supernovae.

More recent studies have shown that the optical afterglows of some GRBs exhibit deviations from a "pure" power-law decay, and these have been ascribed to the possible presence of a supernova underlying the afterglow (GRB970228, Reichart 1999; Galama et al. 1999b; GRB970508, Germany et al. 1999; GRB980326, Bloom et al. 1999; GRB990510, Fruchter et al. 1999c; Beuermann et al. 1999; GRB990712, Hjorth et al. 1999). This makes the association between GRB980425 and SN 1998bw more solid, and supports the speculation that all GRBs of long duration (> 1 s) are formed by extremely energetic supernova explosions ("failed" supernovae, hypernovae, or collapsars, Paczyński 1998; Woosley et al. 1999; MacFadyen & Woosley 1998). These observational hints and theoretical picture suggest that GRB980425 and some GRBs for which a counterpart has been detected at frequencies lower than the γ-rays belong to a same class and have similar progenitors, despite the different distance and behavior of the multiwavelength counterparts.

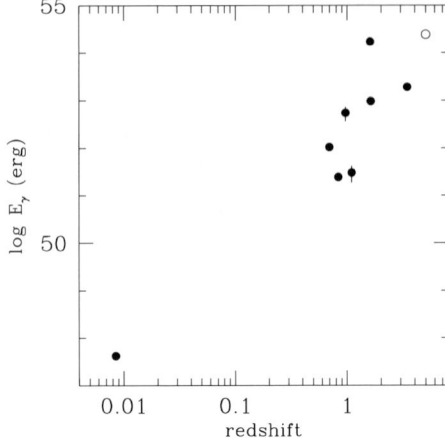

FIGURE 6. GRB energy output in the 40–700 keV range (assumed to be emitted isotropically) vs. redshift. The open circle corresponds to GRB980329, for which the redshift was only estimated with arguments based on the appearance of the optical spectrum (Fruchter 1999. This value is controversial: the Keck detection of the possible host galaxy of the GRB would point to a lower redshift, Fruchter 1999, private communication. See alternatively Draine 1999). Redshift measurements are from Djorgovski et al. (1999a); Bloom et al. (1998b); Kulkarni et al. (1998b); Tinney et al. (1998); Djorgovski et al. (1999b); Djorgovski et al. (1998); Kulkarni et al. (1999); Vreeswijk et al. (1999b). The references for γ-ray fluences measured by the *BeppoSAX* GRBM are Frontera et al. (1998); Piro et al. (1998b); Dal Fiume et al. (1999); in 't Zand et al. (1998); Pian et al. (1999a); Costa et al. (1999); Amati et al. (1998); Feroci et al. (1999); Amati et al. (1999). Adopted values for the Hubble constant and deceleration parameter are $H_0 = 70$ km s^{-1} Mpc^{-1} and $q_0 = 0.15$, respectively.

Indeed, the recent discovery of a GRB optical afterglow at the intermediate redshift $z = 0.43$ (GRB990712, Galama et al. 1999c) might support a continuity of properties between GRB980425 and the other precisely localized GRBs, perhaps based on the different amount of beaming, according to the degree of jet alignment (Eichler & Levinson 1999; Cen 1998; Postnov et al. 1999; Woosley et al. 1999). In this scenario, in highly beamed GRBs the non-thermal multiwavelength afterglow could overwhelm the underlying supernova emission. The latter should instead be detected more clearly in GRBs seen off-axis, like GRB980425, which also appear weaker. Assuming association with SN 1998bw and isotropic emission, the total energy of GRB980425 in the 40–700 keV range, $\sim 5 \times 10^{47}$ erg, is at least four order of magnitudes less than that of GRBs with known distance (see Figure 6).

4. X-ray Supernovae and Gamma-Ray Bursts

GRB980425 has become a milestone in the history of GRB research in that it provided a strong suggestion toward the determination of GRB progenitors. The *BeppoSAX* rapid turnaround allowed the most prompt detection ever of X-rays from a supernova, thus bringing to 10 the number of supernovae detected at these energies (barring supernova remnants). The complete list is reported in Table 1, which represents an update of Table 3 in the review by Schlegel (1995). In addition, X-ray luminosities at the discovery epoch are reported.

Since the epoch of Schlegel's review, the number of X-ray supernovae has doubled, and the detection of SN 1998bw has confirmed that medium energy X-rays are produced also from Type Ib/c supernovae. This result was predictable, given that these must have environments similar to Type II, namely a dense circumstellar medium produced by the slow wind of the progenitor, with which the supernova shock interacts producing both radio and X-ray emission.

Several issues have still to be clarified about GRB980425/SN 1998bw for what concerns X-ray emission. Particularly, an observation of the field with an instrument with good imaging angular resolution (like Chandra) is required to study in detail the host galaxy in X-rays and to disentangle it from the point-like supernova emission at the various epochs of *BeppoSAX* observation. A very sensitive instrument (like XMM) would instead allow a deep survey of the field of GRB980425 in medium energy X-rays, to detect the weak source S2 with a good signal-to-noise ratio, assuming it is broadly constant in the long term. (Its non detection would be perhaps more constraining toward its identification with the GRB afterglow.)

Based on the recent findings of possible supernova emission underlying the optical, power-law fading remnants of some GRBs, future research should exploit the X-ray observing facilities in search of analogous signatures in the X-ray afterglows. If GRBs are produced by supernovae, the same conditions which make detectable the afterglow, i.e. the presence of a sufficiently dense medium, should also favor the production of X-rays from the supernova. A possible past example might be the re-bursting of GRB970508 (Piro et al. 1998b), but it would imply a supernova X-ray luminosity four orders of magnitude larger than that of the most luminous X-ray supernova so far detected, 1988Z (Table 1). Therefore, the X-ray data available to date do not suggest any evidence of a supernova underlying a GRB afterglow. Clearly, more GRB localizations and observations of targets at redshifts no larger than ~ 0.1 are necessary to make a significant supernova detection affordable by the presently available X-ray instruments.

I am grateful to L. Amati, A. Antonelli, F. Boffi, R. Chevalier, E. Costa, J. Danziger, F. Frontera, A. Fruchter, T. Galama, P. Giommi, J. Halpern, J. Katz, S. Kulkarni, N. Masetti, E. Palazzi, N. Panagia, S. Perlmutter, L. Stanghellini, M. Turatto, P. Vreeswijk, C. Wheeler, T. Young for valuable comments and many technical and scientific inputs. I would like to thank Mario Livio and the other ST ScI May Symposium organizers for a pleasant and stimulating conference.

REFERENCES

AMATI, L., FRONTERA, F., COSTA, E., & FEROCI, M. 1998 GCN Circ. N. 146.
AMATI, L., FRONTERA, F., COSTA, E., & FEROCI, M. 1999 GCN Circ. N. 317.
BAND, D., ET AL. 1993 *ApJ*, **413**, 281.
BEUERMANN, K., BRANDT, S., & PIETSCH, W. 1994 *A&A*, **281**, L45.
BEUERMANN, K., ET AL. 1999, *A&A*, **352**, L26.
BLOOM, J. S., KULKARNI, S. R., HARRISON, F., PRINCE, T., PHINNEY, E. S., & FRAIL, D. A. 1998a *ApJ*, **506**, L105.
BLOOM, J. S., DJORGOVSKI, S. G., KULKARNI, & FRAIL, D. A. 1998b *ApJ*, **507**, L25.
BLOOM, J. S., ET AL. 1999 *Nature*, submitted (astro-ph/9905301).
BOELLA, G., ET AL. 1997 A&AS, **122**, 299.
BUTLER, C. & SCARSI, L. 1990 SPIE 1344, 46.
BREGMAN, J. N. & PILDIS, R. A. 1992 *ApJ*, **398**, L107.
CANIZARES, C., KRISS, G., FEIGELSON, E. D. 1982 *ApJ*, **253**, L17.
CEN, R. 1998 *ApJ*, **507**, L131.
CHEVALIER, R. A. & FRANSSON, C. 1994 *ApJ*, **420**, 268.

COSTA, E., ET AL. 1997 *Nature*, **387**, 783.

COSTA, E., ET AL. 1999, in preparation.

DAL FIUME, D., ET AL. 1999 *A&A* submitted.

DIERCKS, A. H., DEUTSCH, E. W., CASTANDER, F. J., CORSON, C., GILMORE, G., LAMB, D. Q., TURNER, E. L., & WYSE, R., 1998, *ApJ*, **503**, L105.

DJORGOVSKI, S. G., KULKARNI, S. R., BLOOM, J. S., GOODRICH, R., FRAIL, D. A., PIRO, L., & PALAZZI, E. 1998 *ApJ*, **508**, L17.

DJORGOVSKI, S. G., KULKARNI, S. R., BLOOM, J. S., & FRAIL, D. A. 1999a *GCN Circ.*, N. 289.

DJORGOVSKI, S. G., KULKARNI, S. R., BLOOM, J. S., FRAIL, D. A., CHAFFEE, F., & GOODRICH, R. 1999b *GCN Circ.*, N. 189.

DOTANI, T., ET AL. 1987 *Nature*, **330**, 230.

DRAINE, B. T. 1999 *ApJL*, submitted (astro-ph/9907232).

EICHLER, D. & LEVINSON, A. 1999 *ApJ*, **521**, L117.

FABBIANO, G. 1989 *Ann. Rev. Astr. & Ap.*, **27**, 87.

FABIAN, A. C. & TERLEVICH, R. 1996 *MNRAS*, **280**, L5.

FEROCI, F., PIRO, L., FRONTERA, F., TORRONI, V., SMITH, M., HEISE, J., & IN 'T ZAND, J. 1999 *IAU Circ.*, N. 7095.

FRONTERA, F., ET AL. 1998 *ApJ*, **493**, L67.

FRONTERA, F., ET AL. 1999 *ApJ*, submitted.

FRUCHTER, A. S. 1999 *ApJ*, **512**, L1.

FRUCHTER, A. S., ET AL. 1999a *ApJ*, **516**, 683.

FRUCHTER, A. S., ET AL. 1999b *ApJ*, in press (astro-ph/9903236).

FRUCHTER, A. S., ET AL. 1999c *GCN Circ.* N. 386.

FUKUGITA, M., SHIMASAKU, K., & ICHIKAWA, T., 1995, *PASP*, **107**, 945.

GALAMA, T. J., ET AL. 1998 *Nature*, **395**, 670.

GALAMA, T. J., ET AL., 1999a, Proc. of the Workshop GRBs in the Afterglow Era, *A&AS*, in press.

GALAMA, T. J., ET AL. 1999b *ApJ*, submitted (astro-ph/9907264).

GALAMA, T. J., ET AL. 1999c *GCN Circ.* N. 388.

GERMANY, L., REISS, D. J., SADLER, E. M., SCHMIDT, B. P., & STUBBS, C. W. 1999 *ApJ*, submitted (astro-ph/9906096).

GRAZIANI, C., LAMB, D. Q., & MARION, G. H. 1998 *ApJ*, submitted (astro-ph/9810374).

GROOT, P. J., ET AL. 1998 *ApJ*, **502**, L123.

HALPERN, J. P., THORSTENSEN, J. R., HELFAND, D. J., & COSTA, E. 1998 *Nature*, **393**, 41.

HALPERN, J. P. 1998 *GCN Circ.* N. 156.

HEISE, J., ET AL. 1999 *IAU Circ.* N. 7099.

HJORTH, J., FYNBO, J., DAR, A., COURBIN, F., & MOLLER, P. 1999 *GCN Circ.* N. 403.

HOUCK, J. C., BREGMAN, J. N., CHEVALIER, R. A., & TOMISAKA, K. 1998 *ApJ*, **493**, 431.

IMMLER, S., PIETSCH, W., & ASCHENBACH, B. 1998a *A&A*, **336**, L1.

IMMLER, S., PIETSCH, W., & ASCHENBACH, B. 1998b *A&A*, **331**, 601.

IWAMOTO, K., ET AL. 1998 *Nature*, **395**, 672.

KIPPEN, R. M. 1998 *GCN Circ.* N. 67.

KIPPEN, R. M., ET AL. 1998 *ApJ*, **506**, L27.

KOHMURA, Y., ET AL. 1994 *Pub. Astr. Soc. Japan*, **46**, L157.

KULKARNI, S. R., ET AL. 1998a *Nature*, **395**, 663.

KULKARNI, S. R., ET AL. 1998b *Nature*, **393**, 35.

KULKARNI, S. R., ET AL. 1999 *Nature*, **398**, 389.
LEISING, M. D., ET AL. 1994 *ApJ*, **431**, L95.
LEWIN, W. H. G., ZIMMERMANN, H.-U., & ASCHENBACH, B. 1996 *IAU Circ.* N. 6445.
LI, Z.-Y. & CHEVALIER, R. A. 1999 *ApJ*, in press (astro-ph/9903483).
MACFADYEN, A. & WOOSLEY, S. E. 1998 *ApJ*, submitted (astro-ph/9810274).
MCKENZIE, E. H. & SCHAEFER, B. E. 1999 *PASP*, **111**, 964.
NICASTRO, L., ET AL. 1998 *A&A*, **338**, L17.
NICASTRO, L., ET AL., 1999, Proc. of the Workshop GRBs in the Afterglow Era, *A&AS*, in press (astro-ph/9904169).
NORRIS, J. P., BONNELL, J. T., & WATANABE, K. 1998 *ApJ*, **518**, 901.
PACZYŃSKI, B. 1998 *ApJ*, **494**, L45.
PALAZZI, E., ET AL. 1998 *A&A*, **336**, L95.
PATAT, F. & PIEMONTE, A. 1998, *IAU Circ.* N. 6918.
PATAT, F., ET AL. 1999, in preparation.
PIAN, E., ANTONELLI, L. A., DANIELE, M. R., REBECCHI, S., TORRONI, V., GENNARO, G., FEROCI, M., & PIRO, L. 1998 *GCN Circ.* N. 61.
PIAN, E., ET AL., 1999a, Proc. of the Workshop GRBs in the Afterglow Era, *A&AS*, in press (astro-ph/9903113).
PIAN, E., ET AL. 1999b, in preparation.
PIRAN, T. 1999 *Phys. Rep.*, **314**, 575.
PIRO, L., BUTLER, C., FIORE, F., ANTONELLI, A., & PIAN, E. 1998a *GCN Circ.* N. 155.
PIRO, L., ET AL. 1998b, *A&A*, **331**, L41.
POSTNOV, K. A., PROKHOROV, M. E., LIPUNOV, V. M. 1998b *A&A*, in press (astro-ph/9908136).
REES, M. J. & MÉSZÁROS, P. 1992 *MNRAS*, **258**, 41.
REICHART, D. E. 1999 *ApJL*, in press (astro-ph/9906079).
SADLER, E. M., STATHAKIS, R., BOYLE, B. J., & EKERS, R. D., 1998, *IAU Circ.* N. 6901.
SCHLEGEL, D. J., FINKBEINER, D. P., & DAVIS, M. 1998 *ApJ*, **500**, 525.
SCHLEGEL, E. M. 1995 *Rep. Prog. Phys.*, **58**, 1375.
SCHLEGEL, E. M., PETRE, R., & COLBERT, E. J. M., 1996, *ApJ*, **456**, 187.
SCHLEGEL, E. M., RYDER, S., STAVELEY-SMITH, L., PETRE, R., COLBERT, E., DOPITA, M., & CAMPBELL-WILSON, D. 1999 *AJ*, in press (astro-ph/9908311).
SUNYAEV, R., ET AL. 1987 *Nature*, **330**, 227.
THORSETT, S. E. & HOGG, D. W. 1999 *GCN Circ.* N. 197.
TINNEY, C., STATHAKIS, R., CANNON, R., & GALAMA, T. J. 1998 *IAU Circ.* N. 6896.
TURATTO, M., ET AL. 1999, this conference.
VAN PARADIJS, J., ET AL. 1997 *Nature*, **386**, 686.
VREESWIJK, P. M., ET AL. 1999a, *ApJ*, in press (astro-ph/9904286).
VREESWIJK, P. M., ET AL. 1999b *GCN Circ.* N. 324.
VREESWIJK, P. M., ET AL. 1999c, in preparation.
WANG, L. & WHEELER, J. C. 1998 *ApJ*, **504**, L87.
WIERINGA, M. H., KULKARNI, S. R., & FRAIL, D. A. 1999 Proc. of the Workshop GRBs in the Afterglow Era, *A&AS*, in press (astro-ph/9906070).
WOOSLEY, S. E., EASTMAN, R. G., & SCHMIDT, B. P. 1999 *ApJ*, **516**, 788.
IN 'T ZAND, J. J. M., ET AL. 1998 *ApJ*, **505**, L119.
ZIMMERMANN, H.-U., ET AL. 1994 *Nature*, **367**, 621.

Direct analysis of spectra of Type Ic supernovae

By DAVID BRANCH

Department of Physics and Astronomy, University of Oklahoma, Norman, OK 73019

Synthetic spectra generated with the parameterized supernova synthetic-spectrum code SYNOW are compared with observed photospheric-phase optical spectra of the normal Type Ic SN 1994I and the peculiar Type Ic SNe 1997ef and 1998bw. The observed spectra can be matched fairly well with synthetic spectra that are based on spherical symmetry and that include lines of just a few ions that are expected to appear on the basis of LTE calculations. Spectroscopic estimates of the mass and kinetic energy of the line-forming layers of the ejected matter give conventional values for SN 1994I but high kinetic energy ($\sim 30 \times 10^{51}$ erg) for SN 1997ef and even higher ($\sim 60 \times 10^{51}$ erg) for SN 1998bw. It is likely that even if SNe 1997ef and 1998bw were non-spherical, they also were hyper-energetic.

1. Introduction

The photospheric–phase spectrum of a Type Ic supernova lacks the strong hydrogen lines of a Type II, the strong optical He I lines of a Type Ib, and the deep red Si II absorption of a Type Ia. A Type Ic (SN Ic) is thought to be the result of the core collapse of a massive star that either has lost its helium layer or ejects helium that remains insufficiently excited to produce conspicuous optical He I lines. For a review of observations of supernova spectra, see Filippenko (1997).

Since April, 1998, interest in SNe Ic has been very high because of the extraordinary SN 1998bw, which appears to have been associated with the gamma ray burst GRB 980425 (Galama et al. 1998; Kulkarni et al. 1998). Recently we (Millard et al. 1999) have used the parameterized supernova synthetic-spectrum code SYNOW to make a "direct analysis" of spectra of SN 1994I, a well observed, normal SN Ic. In this contribution, after summarizing our work on SN 1994I, I report some preliminary results of a similar analysis of the peculiar Type Ic SN 1997ef (Deaton et al. 1998) and then consider the related but even more peculiar SN 1998bw. The emphasis here is on establishing line identifications and making spectroscopic estimates of the mass and kinetic energy of the line-forming layers of the ejected matter.

2. SYNOW

In its simplest form the SYNOW code assumes spherical symmetry and that line formation takes place by resonant scattering outside a sharp photosphere that radiates a blackbody continuum. To a good approximation the simple explosion velocity law, $v = r/t$, holds. Consequently the velocity gradient is isotropic and homogeneous (unlike the case of a stellar wind, where even in the constant-velocity case the velocity gradient is neither isotropic nor homogeneous). From the point of view of an observer, the (non-relativistic) surfaces of constant radial velocity are planes perpendicular to the line of sight, and an unblended line formed by resonant scattering has an emission component that peaks at the line rest wavelength in the supernova frame and an absorption component whose minimum is blueshifted by an amount that corresponds approximately to the velocity at the photosphere (unless the line is strong).

SYNOW treats line formation in the Sobolev approximation, which is a good one for this purpose. The profile of an unblended line is determined by the adopted radial dependences of the line optical depth and the line source function. The line optical depth determines the strength of the line and is given by

$$\tau_l = \frac{\pi e^2}{m_e c} f \, \lambda \, t \, n_l(v) = 0.026 \, f \, \lambda_\mu \, t_d \, n_l(v),$$

where f is the oscillator strength, λ_μ is the line wavelength in microns, t_d is the time since explosion in days, and $n_l(v)$ is the population of the lower level of the transition in cm^{-3}. The correction for stimulated emission, although not written out here, is taken into account.

The line source function determines the extent to which the line is in emission or absorption. In the resonant-scattering approximation line photons are conserved, except for occultation effects, and the source function of an isolated line is just the product of the intensity of the photospheric continuum and the geometrical dilution factor. The source function of a line that interacts with one or more lines of shorter wavelength is altered by photons that are scattered by those lines. The essential role of SYNOW is to treat the multiple scattering, which in observers' language is line blending.

Various fitting parameters are available for a SYNOW synthetic-spectrum calculation. The parameter T_{bb} is the temperature of the blackbody continuum radiated by the photosphere. For each ion whose lines are introduced, the optical depth at the photosphere of a reference line is a parameter, and the optical depths of the other lines of the ion are calculated for Boltzmann excitation at excitation temperature T_{exc} (ordinarily is taken to be the same as T_{bb}). For the spectra shown here, the radial dependence of the line optical depths is a power law, $\tau \propto v^{-n}$. At each epoch the velocity at the photosphere, v_{phot}, is a parameter, and maximum and minimum velocities also can be imposed on an ion; when the minimum velocity exceeds v_{phot} the ion is said to be detached from the photosphere. The most interesting parameters are the velocity parameters and the "density" power-law index, n.

When deciding which ions to introduce, we are guided by experience and by the LTE calculations of line optical depths by Hatano et al. (1999), who considered six compositions that might be encountered in supernovae. The composition that is used as a guide for this work is the one in which hydrogen and helium have been burned to a mixture of carbon and oxygen, with the heavier elements present in their solar mass fractions. In this case, ions that are predicted to have lines of significant optical depth at the temperatures of interest include Ca II, Fe II, O I, Si II, C II, Mg II, O II, and Ti II.

3. The normal Type Ic SN 1994I

In Millard et al. (1999) we compare SYNOW synthetic spectra with observed spectra of SN 1994I obtained from 5 to 35 days after the assumed explosion date of March 30, 1994. A density power-law index of $n = 8$ is used for all epochs. Figure 1 compares a spectrum of SN 1994I obtained 16 days after explosion with a synthetic spectrum that has $v_{phot} = 10,000$ km s^{-1} and $T_{bb} = 8000$ K. Most of the observed features are well matched. Ions that certainly must be introduced to account for observed features are Ca II, O I, Na I, Fe II, and Ti II. (The Na I D–line feature is not predicted to be significant by Hatano et al. (1999), but as usual in supernova spectra the Na I feature is observed to be stronger than predicted. We are confident of the identification.) In this particular synthetic spectrum, lines of C II also are used, but detached at 16,000 km s^{-1} so that $\lambda 6580$ can account for most of the observed absorption near 6200 Å. Sometimes

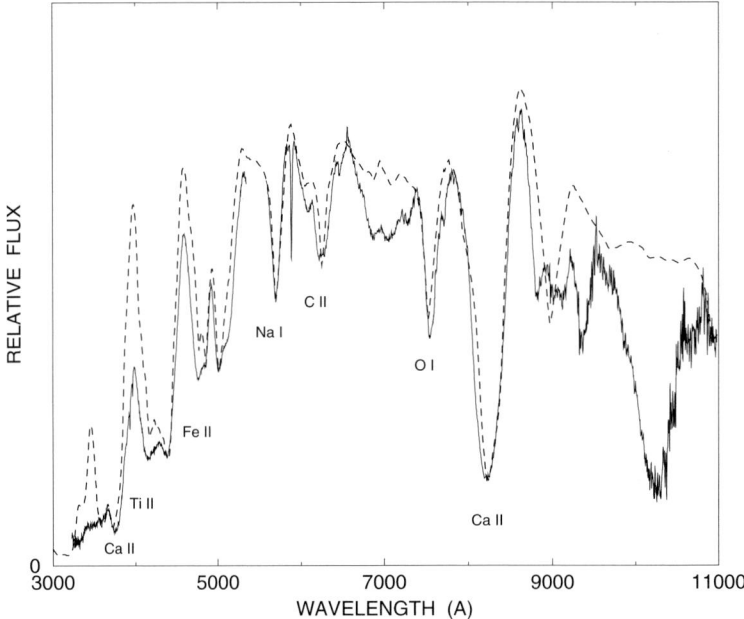

FIGURE 1. A spectrum of SN 1994I obtained 16 days after explosion is compared with a synthetic spectrum (dashed line). The flux is per unit frequency.

it is difficult to decide between detached C II λ6580 and undetached Si II λ6355, and the C II identification is not considered to be definite. In order to account for the observed absorption around 7000 Å we would have to introduce lines of O II (see below). The excessive height of the synthetic peaks in the blue part of the spectrum is not of great concern; the number of lines of singly ionized iron-peak elements rises rapidly toward short wavelengths, so SYNOW spectra often are underblanketed in the blue due to the absence of lines of iron-peak ions that are not introduced.

In Millard et al. (1999) we consider the identification of the observed absorption near 10,250 Å, which has been identified as He I λ10830 and taken as strong evidence that SN 1994I ejected some helium (Filippenko et al. 1995). We find that it is difficult to account for even just the core of the observed 10,250 Å absorption with He I λ10830 without compromising the fit in the optical (see also Baron et al. 1999). We suggest that the observed feature may be a blend of He I λ10830 and C I λ10695 (see also Woosley & Eastman 1997), or perhaps a blend of Si I lines. This is an important issue but it will not be discussed further here, since we have no evidence for or against the presence of the feature in SNe 1997ef and 1998bw.

Another comparison of observed and synthetic spectra, but for just five days after the assumed explosion date, is shown in Figure 2. (It can be helpful to consider photospheric-phase spectra in reverse chronological order, because at the earliest times line formation takes place in the highest-velocity layers and the blending is most severe.) The synthetic spectrum has $v_{\rm phot} = 17,500$ km s^{-1} and $T_{\rm bb} = 17,000$ K. Ions that certainly are needed are Ca II, O I, and Fe II. In this synthetic spectrum, lines of C II, Na I, Mg II, Si II, and O II (with only [O II] λ7320,7330 having a significant effect on the spectrum) also are introduced; they are considered probable but not definite.

The adopted values of $v_{\rm phot}$ can be used to estimate the mass and kinetic energy in the line-forming layers. For an r^{-n} density distribution, the mass (in M$_\odot$) and the kinetic

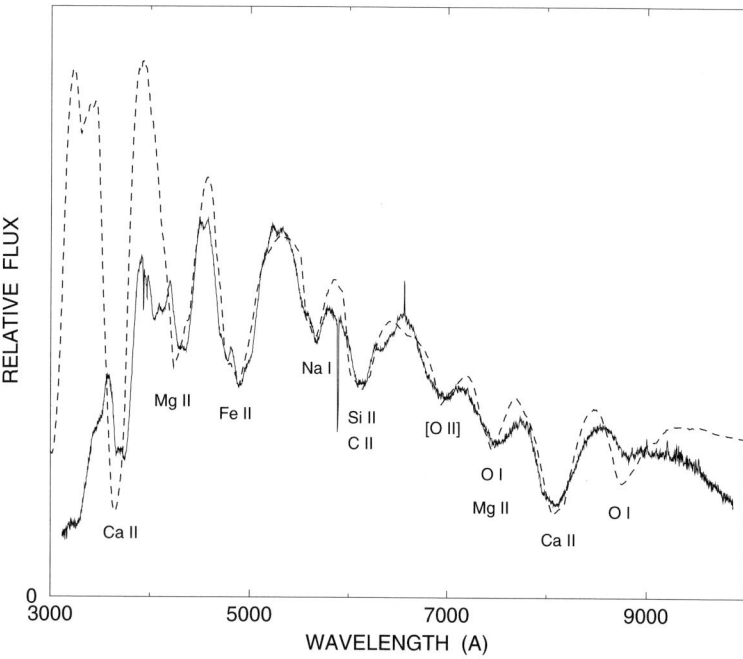

FIGURE 2. A spectrum of SN 1994I obtained 5 days after explosion is compared with a synthetic spectrum. The flux is per unit frequency.

energy (in foe, where 1 foe $\equiv 10^{51}$ erg) above the layer at which the electron-scattering optical depth is $\tau_{\rm es}$ can be expressed as

$$M = 1.2 \times 10^{-4} \, v_4^2 \, t_d^2 \, \mu_e \, \tau_{\rm es} \, f_M(n, v_{\rm max}),$$

$$E = 1.2 \times 10^{-4} \, v_4^4 \, t_d^2 \, \mu_e \, \tau_{\rm es} \, f_E(n, v_{\rm max}),$$

where v_4 is $v_{\rm phot}$ in units of 10^4 km s^{-1}, t_d is the time since explosion in days, μ_e ($\equiv Y_e^{-1}$) is the mean molecular weight per free electron, and the integration is carried out to velocity $v_{\rm max}$. The functions f_M and f_E always exceed unity.

For SN 1994I, we use $\mu_e = 14$ (a mixture of singly ionized carbon and oxygen), $\tau_{\rm es} = 2/3$ at the bottom of the line-forming layer, and integrate the steep density power-law to infinity ($f_M(8,\infty) = 1.4$, $f_E(8,\infty) = 2.3$). Figure 3 shows v_4, M, and E plotted against time. (E should increase monotonically with time. Its non-monotonic behavior just reflects the imprecision of our determinations of $v_{\rm phot}$; recall that $E \propto v_{\rm phot}^4$.) At 35 days after explosion, the mass moving faster than 7000 km s^{-1} is estimated to be about 1.4 M$_\odot$ and it carries a kinetic energy of about 1.2 foe. These numbers are reasonable, and similar to those that have been estimated for SN 1994I on the basis of light-curve studies (Nomoto et al. 1994; Iwamoto et al. 1994; Young, Baron, & Branch 1995; Woosley, Langer, & Weaver 1995).

Such spectroscopic estimates of mass and kinetic energy also come out to be reasonable for Type Ia supernovae (Branch 1980) and for SN 1987A (Jeffery & Branch 1990).

4. From SN 1994I to SNe 1997ef and 1998bw

Figure 4 compares spectra of SN 1994I at 16 days after explosion, SN 1997ef at 20 days after its assumed explosion date of November 15, 1997, and SN 1998bw at 16 days

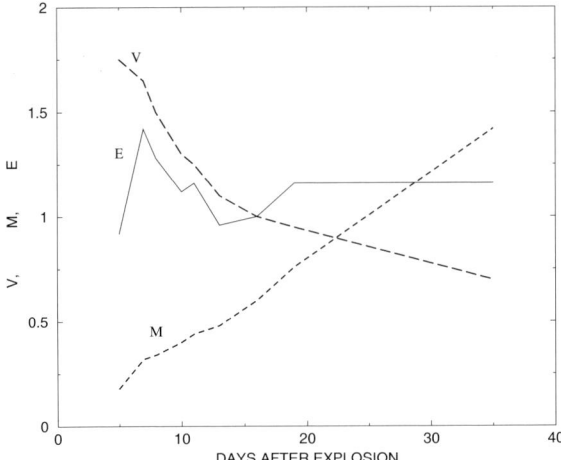

FIGURE 3. Velocity at the photosphere (in units of 10,000 km s^{-1}), and mass (in M$_\odot$) and kinetic energy (in foe) above the photosphere, are plotted against time for SN 1994I.

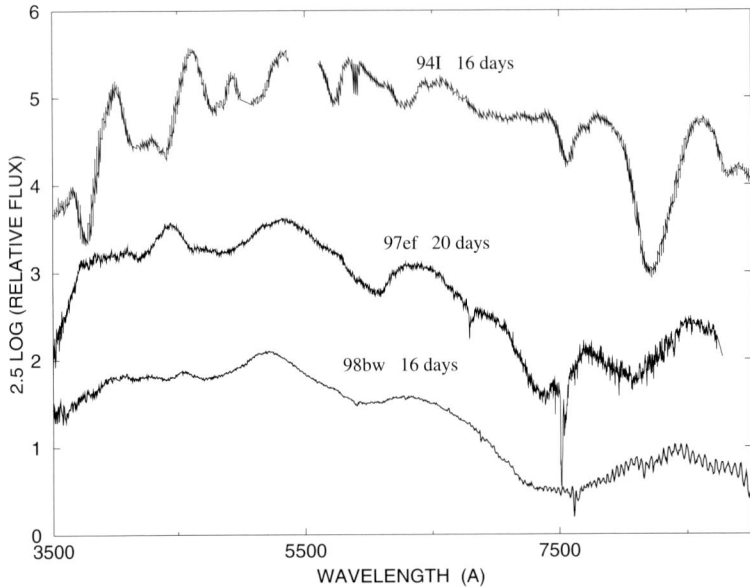

FIGURE 4. A spectrum of SN 1994I (Filippenko et al. 1995) is compared with spectra of SN 1997ef (P. Garnavich et al., in preparation) and SN 1998bw (F. Patat et al., in preparation). In this and subsequent figures the flux is per unit wavelength.

after its explosion date of April 25, 1998. It is clear that SNe 1997ef and 1998bw are spectroscopically related to each other and also, but less closely, to SN 1994I. Therefore it seems appropriate to refer to SNe 1997ef and 1998bw as "Type Ic peculiar." The observed absorption features are much broader and bluer in SNe 1997ef and 1998bw than in SN 1994I, which means that SNe 1997ef and 1998bw ejected more mass at high velocity. Figure 5 shows the effects, on the SYNOW synthetic spectrum of Figure 1 (for SN 1994I at 16 days), of raising $v_{\rm phot}$ from 10,000 to 30,000 km s^{-1}. Figure 6 shows the effects of dropping the density power-law index from $n = 8$ to $n = 2$. Raising $v_{\rm phot}$ and dropping n both cause the absorption features to become broader and bluer, and both

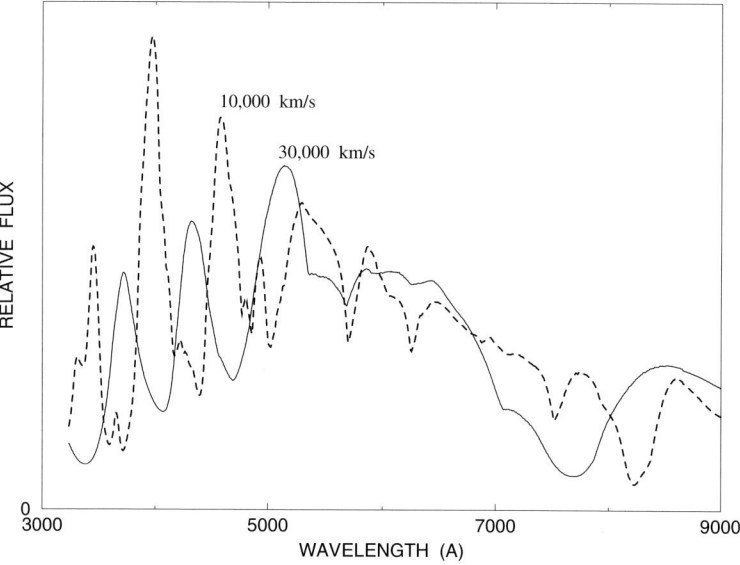

FIGURE 5. The dashed line is the synthetic spectrum of Figure 1. The solid line shows the effects of raising $v_{\rm phot}$ from 10,000 to 30,000 km s^{-1}.

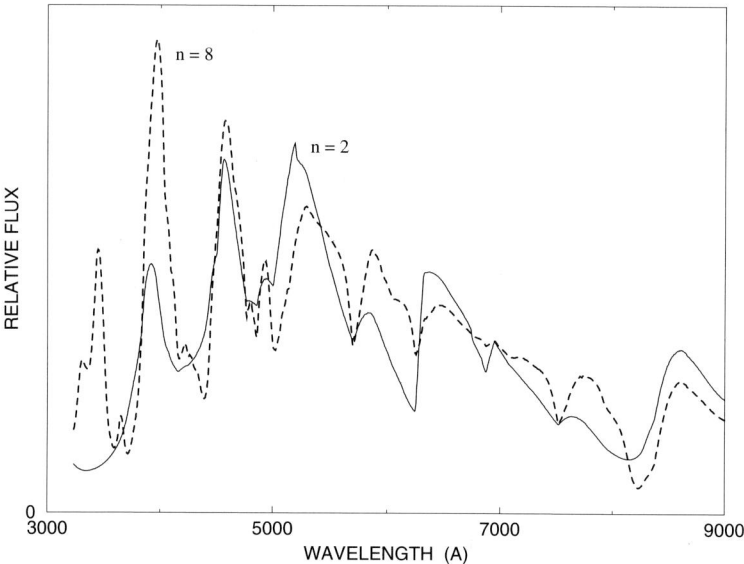

FIGURE 6. The dashed line is the synthetic spectrum of Figure 1. The solid line shows the effects of dropping n from 8 to 2.

appear to be necessary to obtain satisfactory SYNOW fits to spectra of SNe 1997ef and 1998bw. A value of $n = 2$ is used for all of the synthetic spectra shown below.

4.1. Fitting spectra of SN 1997ef

Figure 7 compares a spectrum of SN 1997ef obtained 34 days after explosion with a synthetic spectrum that has $v_{\rm phot} = 7000$ km s^{-1}, $T_{\rm bb} = 7000$ K, and uses only lines of Ca II, O I, Si II, Na I, and Fe II. This fit (and those to follow) could be improved by

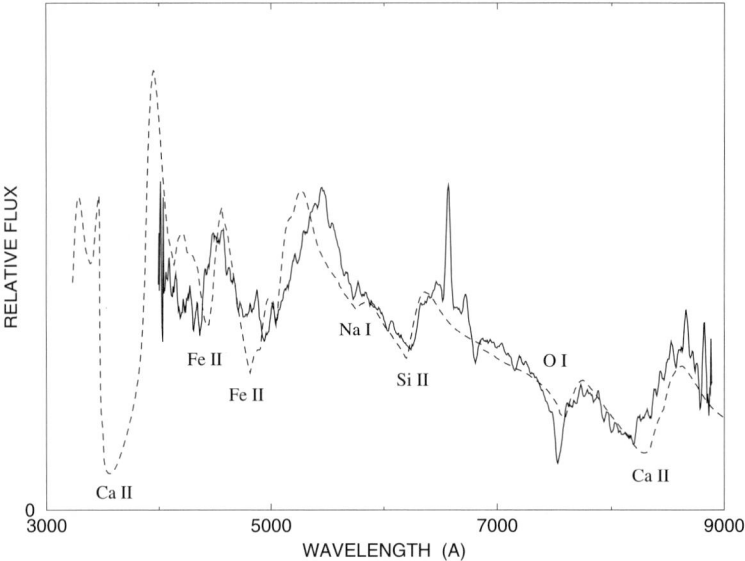

FIGURE 7. A spectrum of SN 1997ef (Y. Qiu et al., in preparation) obtained 34 days after explosion is compared with a synthetic spectrum.

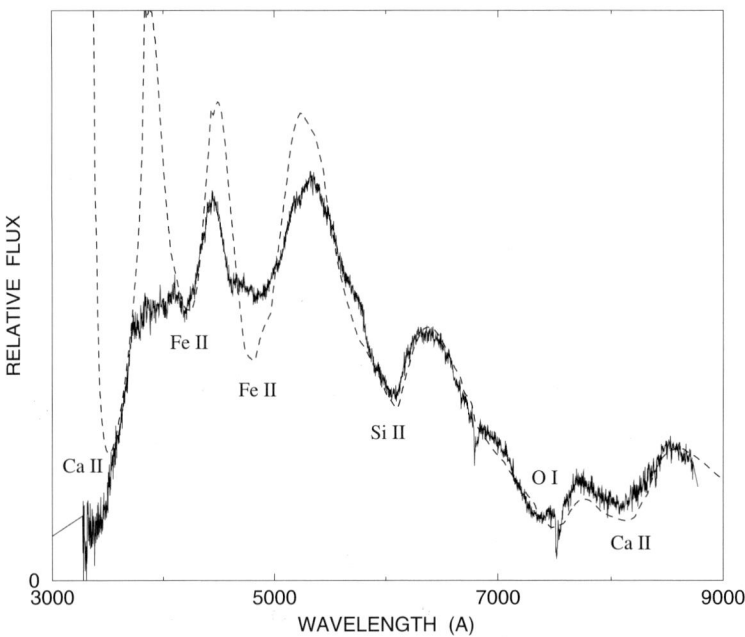

FIGURE 8. A spectrum of SN 1997ef (P. Garnavich et al., in preparation) obtained 20 days after explosion is compared with a synthetic spectrum.

introducing more ions (e.g. Ti II) and tuning the parameters, but as it stands it is good enough to indicate that we are on the right track.

Figure 8 compares a spectrum of SN 1997ef obtained 20 days after explosion with a synthetic spectrum that has $v_{\rm phot} = 12,000$ km s^{-1}, $T_{\rm bb} = 11,000$ K, and uses lines of Ca II, O I, Si II, Fe II, and Mg II. In the red part of the spectrum, the good fit indicates

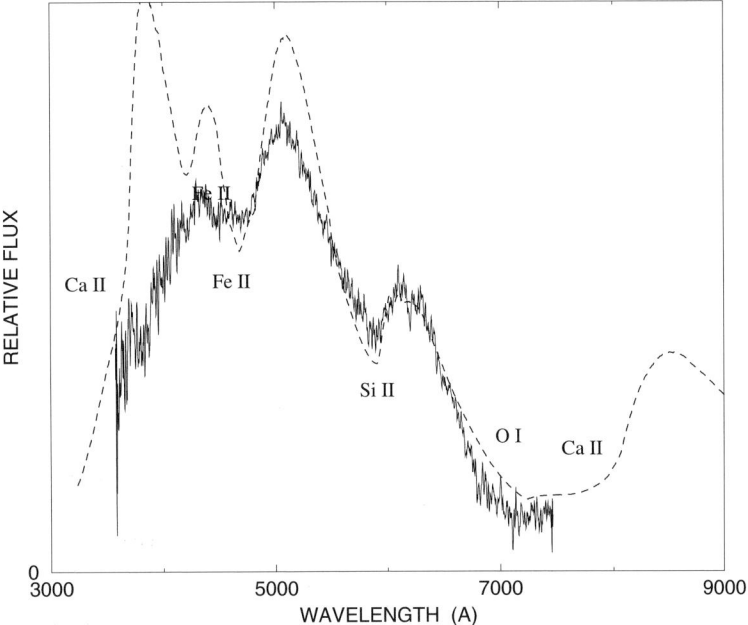

FIGURE 9. A spectrum of SN 1997ef (P. Garnavich et al., in preparation) obtained 10 days after explosion is compared with a synthetic spectrum.

that at this epoch just a few lines of Ca II, O I, and Si II are responsible for the features. In the blue, Fe II blends dominate. The synthetic spectrum is severely underblanketed in the blue, presumably due to missing lines of other singly ionized iron-peak elements.

Figure 9 compares a spectrum of SN 1997ef obtained 10 days after explosion with a synthetic spectrum that has $v_{\rm phot} = 22,000$ km s^{-1}, $T_{\rm bb} = 11,000$ K, and uses the same ions as in Figure 8. This is a good example of why it can be instructive to work backward in time; this interpretation of the spectrum might seem arbitrary if the later-epoch spectra had not already been discussed.

Figure 10 is exactly like Figure 9 except that the Fe II lines have been turned off. Comparison of Figures 9 and 10 shows how strongly the Fe II lines affect the blue, while having practically no effect in the red. The same is true for the 20 and 34 day spectra discussed above.

4.2. *Fitting spectra of SN 1998bw*

Figure 11 compares a spectrum of SN 1998bw obtained 28 days after explosion with a synthetic spectrum that has $v_{\rm phot} = 7000$ km s^{-1}, $T_{\rm bb} = 6000$ K, and uses lines of Ca II, O I, Si II, Na I, Ca II, and Fe II. The situation is much like that of SN 1997ef at 34 days.

Figure 12 compares a spectrum of SN 1998bw obtained 16 days after explosion with a synthetic spectrum that has $v_{\rm phot} = 17,000$ km s^{-1}, $T_{\rm bb} = 8000$ K, and uses only lines of Ca II, O I, Si II, and Fe II. The situation is like that of SN 1997ef at 20 days, but here the blending is more severe due to the higher $v_{\rm phot}$ of SN 1998bw.

Figure 13 compares a spectrum of SN 1998bw obtained 8 days after explosion with a synthetic spectrum that has $v_{\rm phot} = 30,000$ km s^{-1}, $T_{\rm bb} = 8000$ K, and uses only lines of O I, Si II, Ca II, and Fe II. The situation is like that of SN 1997ef at 10 days but with more blending due to the higher $v_{\rm phot}$.

Figure 14 is like Figure 13 except that the Fe II lines have been turned off. In SN 1998bw, as in SN 1997ef, Fe II affects only the blue part of the spectrum.

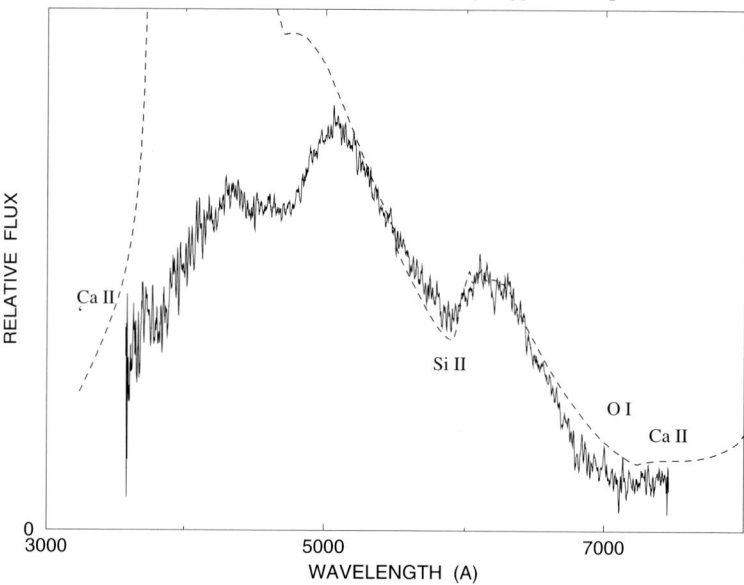

FIGURE 10. The synthetic spectrum is like that of Figure 9 except that the Fe II lines have been turned off.

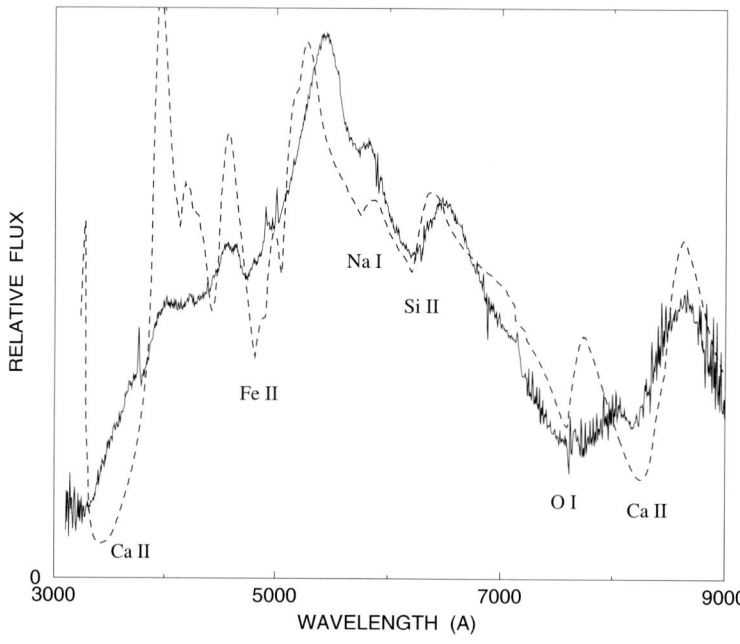

FIGURE 11. A spectrum of SN 1998bw (F. Patat et al., in preparation) obtained 28 days after explosion is compared with a synthetic spectrum.

4.3. Masses and kinetic energies of SNe 1997ef and 1998bw

Figure 15 compares the adopted values of $v_{\rm phot}$ versus time. Around 30 days after explosion the values converge to about 7000 km s^{-1}, but at earlier times the values for SN 1997ef are higher than those for SN 1994I, and the values for SN 1998bw are higher still.

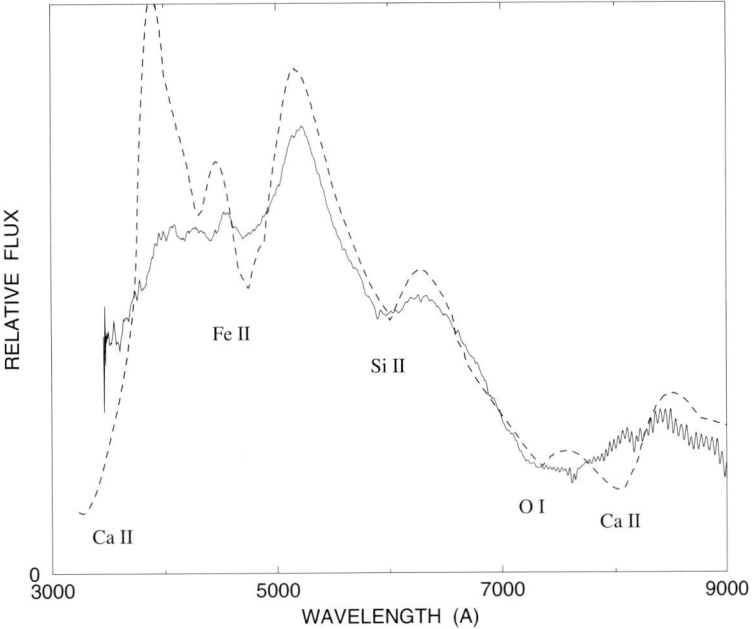

FIGURE 12. A spectrum of SN 1998bw (F. Patat et al., in preparation) obtained 16 days after explosion is compared with a synthetic spectrum.

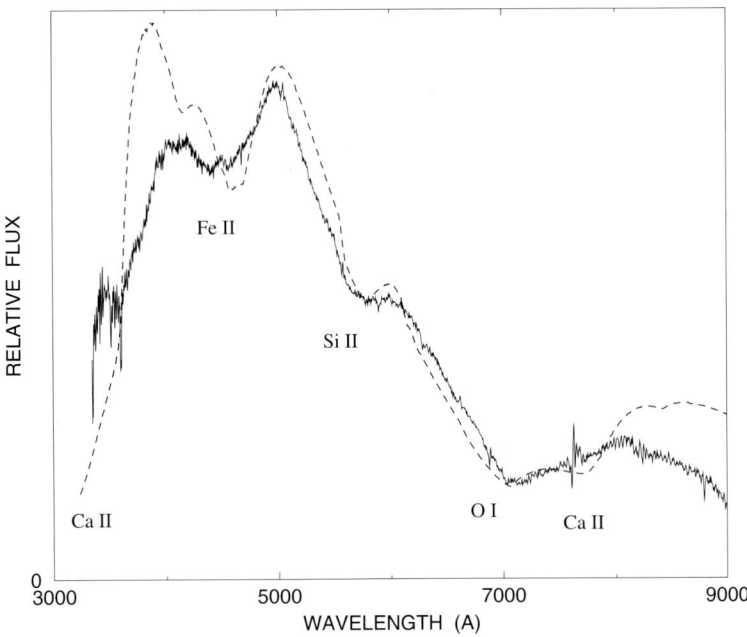

FIGURE 13. A spectrum of SN 1998bw (F. Patat et al., in preparation) obtained 8 days after explosion is compared with a synthetic spectrum.

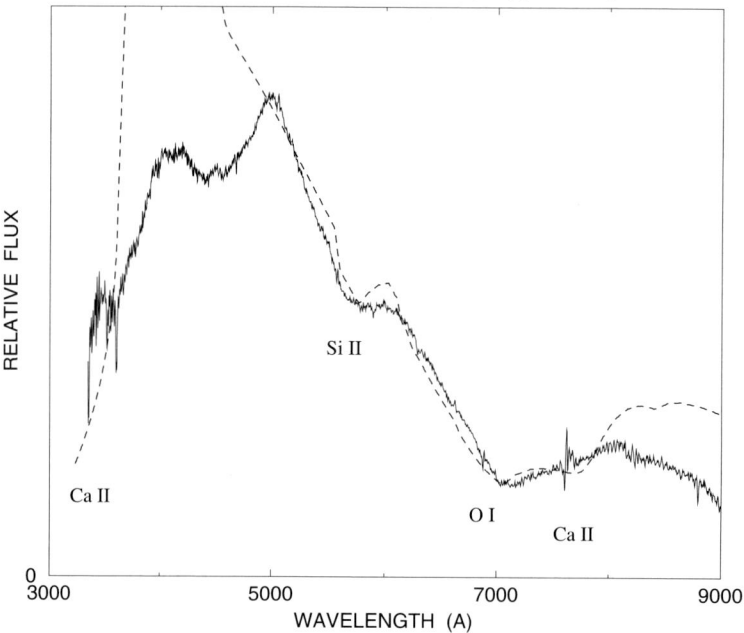

FIGURE 14. The synthetic spectrum is like that of Figure 13 except that the Fe II lines have been turned off.

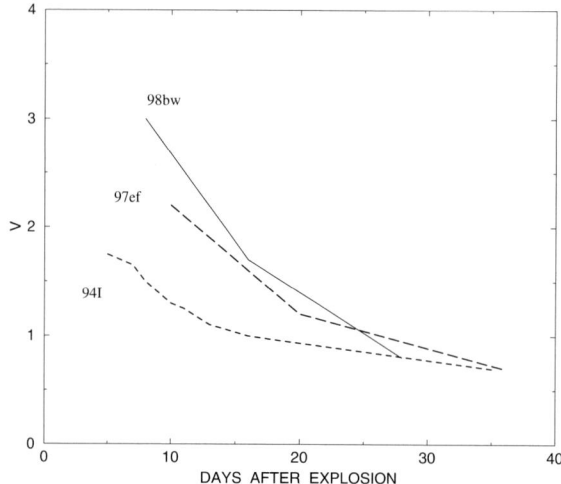

FIGURE 15. Velocity at the photosphere (in units of 10,000 km s^{-1}) is plotted against time.

When estimating the masses and kinetic energies of SNe 1997ef and 1998bw, the integration cannot be extended to infinity because $n = 2$ has been used for the synthetic spectra. Instead the integration is carried out only to $v_{\max} = 2v_{\text{phot}}$ for the earliest epoch considered, i.e. to 44,000 km s^{-1} for SN 1997ef and to 60,000 km s^{-1} for SN 1998bw. The results (with $f_M(2,2) = 2$, $f_E(2,2) = 4.7$) are shown in Figures 16 and 17. For both SNe 1997ef and 1998bw, the masses above 7000 km s^{-1} are estimated to be around 6 M$_\odot$. For SN 1997ef the kinetic energy above 7000 km s^{-1} comes out to be around 30 foe while that of SN 1998bw is around 60 foe. These kinetic energies are more likely to be too low than too high because most of the estimated kinetic energy comes from

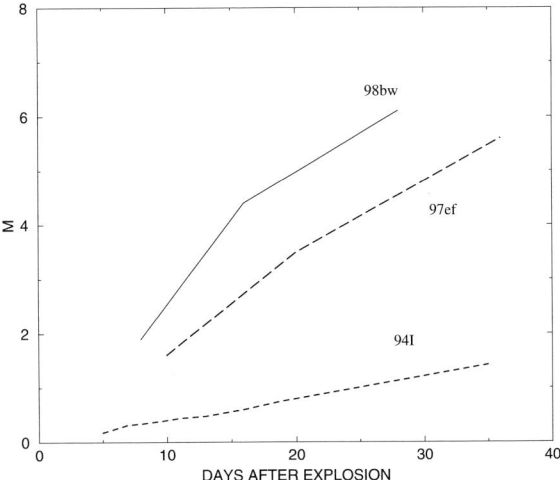

FIGURE 16. Spectroscopic estimates of mass (in $M_\odot$) above the photosphere are plotted against time.

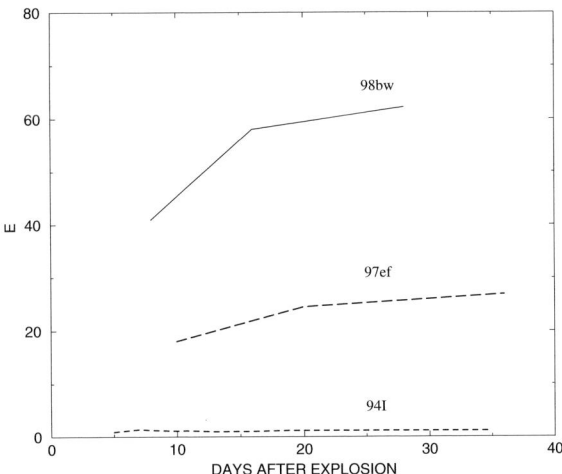

FIGURE 17. Spectroscopic estimates of kinetic energy (in foe) above the photosphere are plotted against time.

the earliest epochs considered, and the integrations are carried out to only $2v_{\rm phot}$ while the synthetic spectra actually go to higher velocities. Establishing lower limits on the velocities to which the integrations must be carried out will require further study. Of course, there also is more mass at velocities lower than 7000 km s^{-1}, but not much more kinetic energy.

In a preprint, Iwamoto et al. (1998) compare observed spectra of SN 1997ef with synthetic spectra calculated for a hydrodynamical model that has an ejected mass of about 4.6 $M_\odot$ and a kinetic energy of 1 foe. The prominent lines in their synthetic spectra are much the same as the ones that have been identified here but as they discuss, the lines in their synthetic spectra are too narrow and not sufficiently shifted to the blue. Synthetic spectra calculated for models having more mass and kinetic energy give much better fits to the SN 1997ef spectra (P. Mazzali and K. Nomoto, personal communication).

Iwamoto et al. (1999) compare observed spectra of SN 1998bw with synthetic spectra calculated for a hydrodynamical model that has an ejected mass of about 11 $M_\odot$ and a kinetic energy of 30 foe. Their synthetic spectra match the SN 1998bw spectra fairly well, and it appears that more mass at high velocity would lead to even better fits.

5. Conclusion

The spectroscopic mass and kinetic-energy estimates presented here for SNe 1997ef and 1998bw are preliminary and approximate. Nevertherless, it is clear that at least in the spherical approximation the kinetic energy of both events was much higher than the canonical one foe, as was reported by Deaton et al. (1998) for SN 1997ef and as in the models of Iwamoto et al. (1999) and Woosley, Eastman, & Schmidt (1999) for SN 1998bw.

Polarization spectra are much more sensitive than flux spectra to asymmetry. Core-collapse supernovae generally show detectable polarization, which indicates that they are significantly asymmetric (Wang et al. 1996). Höflich, Wheeler, & Wang (1999) calculate light curves of moderately asymmetric explosions and suggest that SN 1998bw was distinguished principally by having been viewed close to the symmetry axis, rather than by having a very high kinetic energy. It is true that to the extent that the ejecta of SNe 1997ef and 1998bw are "beamed," the kinetic energy estimates presented here could be too high. However, because the spectra of SNe 1997ef and 1998bw can be matched fairly well in a straightforward way with the spherical symmetry assumption, and the corresponding kinetic-energy estimates are so very high, and the Lorentz factor of the ejecta is not high enough to *be* a factor in the energy estimates (as it is for gamma-ray energy estimates), it is likely that even if SNe 1997ef and 1998bw were non-spherical, they also were hyper-energetic.

I am grateful to Peter Garnavich and Yulei Qiu for providing spectra of SN 1997ef, to Ferdinando Patat for providing spectra of SN 1998bw, and to Eddie Baron, Kazuhito Hatano, David Jeffery, Jennifer Millard, and Erica Snow for discussions and assistance.

REFERENCES

BARON, E., BRANCH, D., HAUSCHILDT, P. H., FILIPPENKO, A. V., & KIRSHNER, R. P. 1999 *ApJ*, in press.

BRANCH, D. 1980 in *Proceedings of the Texas Workshop on Type I Supernovae*, (ed. J. C. Wheeler), p. 66. University of Texas.

DEATON, J., BRANCH, D., BARON, E., FISHER, A., KIRSHNER, R. P., & GARNAVICH, P. 1998 *BAAS*, **30**, 824.

FILIPPENKO, A. V. 1997 *ARAA*, **35**, 309.

FILIPPENKO, A. V., ET AL. 1995 *ApJ*, **450**, L11.

GALAMA, T. J., ET AL. 1998 *Nature*, **395**, 670.

HATANO, K., BRANCH, D., FISHER, A., MILLARD, J., & BARON, E. 1999 *ApJS*, **121**, 233.

HÖFLICH, P., WHEELER, J. C., & WANG, L. 1999 *ApJ*, **521**, 179.

IWAMOTO, K., ET AL. 1994 *ApJ*, **437**, L115.

IWAMOTO, K., ET AL. 1998 astro-ph/9807060.

IWAMOTO, K., ET AL. 1999 *Nature*, **395**, 672.

JEFFERY, D. J. & BRANCH, D. 1990 in *Supernovae*, (eds. J. C. Wheeler, T. Piran, & S. Weinberg), p. 149. World Scientific.

KULKARNI, S. R., ET AL. 1998 *Nature*, **395**, 663.

MILLARD, J., ET AL. 1999 *ApJ*, in press.
NOMOTO, K., ET AL. 1994 *Nature*, **371**, 227.
WANG, L., WHEELER, J. C., LI, Z., & CLOCCHIATTI, A. 1996 *ApJ*, **467**, 435.
WOOSLEY, S. E. & EASTMAN, R. G. 1997 in *Thermonuclear Supernovae*, (eds. P. Ruiz-Lapuente, R. Canal, & J. Isern), p. 821. Kluwer.
WOOSLEY, S. E., EASTMAN, R. G., & SCHMIDT, B. P. 1999 *ApJ*, **516**, 788.
WOOSLEY, S. E., LANGER, N., & WEAVER, T. A. 1995 *ApJ*, **448**, 315.
YOUNG, T. R., BARON, E., & BRANCH, D. 1995 *ApJ*, **449**, L51

The interaction of supernovae and gamma-ray bursts with their surroundings

By ROGER A. CHEVALIER

Department of Astronomy, University of Virginia, P. O. Box 3818, Charlottesville, VA 22903

The emission from a supernova or GRB (gamma-ray burst) over a period of days to years yields clues about the nature of the surrounding medium, which can be a discriminant between possible progenitor objects. Observations of core collapse supernovae indicate interaction with a stellar wind, as expected around a massive star. Nonthermal radio emission shows that the shocked regions are capable of producing relativistic electrons and magnetic fields. For Type Ia supernovae, there are as yet no observations of an early interaction, but radio observations are close to a regime in which various progenitor models can be tested. The unusual Type Ic supernova SN 1998bw was likely associated with GRB 980425 and had bright radio emission that can be interpreted in a way similar to GRB afterglows. The emission was consistent with interaction with the wind from a Wolf-Rayet star, but not with a driving force from the outer supernova ejecta. Current observations of GRB afterglows suggest both wind and interstellar interactors. The former objects are likely to have massive star progenitors and the latter to have compact binary merger progenitors.

1. Introduction

Most of the light from a supernova is emitted at optical wavelengths from the photosphere. Beginning in the 1970s, analyses of optical light curves and photospheric spectra have led to our basic picture of the types of stars that are exploding. In the 1980s, multiwavelength studies of supernovae became possible, as well as following the explosions optically to late times ($\gtrsim$ few years). These observations have opened up the study of how supernovae interact with their immediate surroundings. The nature of the surroundings can be an important indicator of the supernova progenitor and can yield information on the evolution leading up to the explosion. The nearby event SN 1987A provides an excellent example of the complex circumstellar medium that can be created around a massive star.

In the case of GRBs (γ-ray bursts), a breakthrough occurred with the discovery of multiwavelength afterglow emission in 1997. This emission has been interpreted as synchrotron emission from the shock wave generated by the GRB moving into the surrounding medium. The properties and evolution of the emission are expected to depend on the nature of the surrounding medium, which in turn depends on the nature of the progenitor objects, as in the case of supernovae. The question of GRB progenitors is one of the most pressing in GRB astronomy.

Section 2 of this paper deals with circumstellar interaction around supernovae, starting with the basic physical process and continuing with applications to the various supernova types. SN 1998bw/GRB 980425, which provided the first clear link between supernovae and GRBs, is discussed in section 3. GRBs are treated in section 4 and the conclusions are in section 5.

2. Supernovae

2.1. Circumstellar interaction

The explosion of a supernova gives rise to freely expanding gas that interacts with the ambient medium. The initial acceleration of the supernova shock wave through the outer layers of the star gives rise to an approximately power law density profile in the outer layers (e.g. Matzner & McKee 1999 and references therein). Because the gas is freely expanding after several doubling times of the radius, the velocity of the gas is $v = r/t$, where t is the time since the explosion, and the density can be expressed as

$$\rho = B(r/t)^{-n}t^{-3}, \tag{2.1}$$

where B and n are constants. Typical values of n are in the range 8–12. The explosion of a relatively compact radiative star can be expected to yield $n \approx 9$–10, while the explosion of a more extended, convective star yields a somewhat steeper density profile in the outer layers (Matzner & McKee 1999). The outer layers of a supernova have an extended region with $n > 5$, which means that most of the energy in this region is at the lower velocities. The surrounding medium may be described by the density distribution

$$\rho = Ar^{-s}, \tag{2.2}$$

where A and s are constants. The cases of particular interest are $s = 0$ (constant density, interstellar medium) and $s = 2$ (steady circumstellar wind). In the wind case, $A = \dot{M}/4\pi v_w$, where $\dot{M}$ is the progenitor mass loss rate and v_w is the wind velocity.

The value of v_w is much smaller than the supernova shock velocity for reasonable parameters, so either the wind or the ISM (interstellar medium) can be regarded as stationary. The interaction problem is then determined by the two dimensional parameters A and B, and can be described by a self-similar solution (Chevalier 1982a). The outer shock radius is given by

$$R_{os} = a\left(\frac{B}{A}\right)^{1/(n-s)} t^{(n-3)/(n-s)}, \tag{2.3}$$

where a is a dimensionless constant. The $s = 2$ (wind) case is of most interest for supernovae because the progenitor stars can be commonly expected to have winds. The wind densities can be substantially larger than interstellar densities close to the star and such interactions will be the most observable. If the expansion parameter m is defined by $R_{os} \propto t^m$, then we expect m in the approximate range 0.83–0.90. The expansion is more rapid than that expected for a constant energy explosion ($m = 2/3$) because of the continual driving by the supernova ejecta. The freely expanding ejecta pass through a shock front (the reverse shock) and add energy to the shocked region.

The forward shock front has a velocity of $\sim 10^4$ km s^{-1} and can heat the shocked wind gas to a temperature $T \sim 10^9$ K. By contrast, the reverse shock front has a velocity $\sim 10^3$ km s^{-1}; the shocked supernova gas is cooler ($T \sim 10^7$ K) and denser than the shocked circumstellar gas, and is thus a stronger source of X-ray emission (Chevalier 1982b). The radiative losses may be sufficiently large that the reverse shock front is a cooling shock front. For expansion into the dense wind of a red supergiant progenitor star, the radiative phase can last for several years (Chevalier & Fransson 1994). The density declines with expansion so that eventually a transition is made to nonradiative evolution. This evolution is the opposite to that which occurs for expansion in the interstellar medium ($s = 0$), when radiative cooling is important for the hydrodynamics at late times.

Cooling at the reverse shock front creates a dense shell of shocked supernova ejecta which can absorb X-rays emitted near the shock transition. This energy is then reradiated at optical and UV (ultraviolet) wavelengths. In addition, the X-rays can be absorbed by the freely expanding ejecta that are interior to reverse shock wave. This gas is at a lower density that the shocked ejecta and the emission from close to the reverse shock is in lines of highly ionized ions (Chevalier & Fransson 1994).

In addition to the thermal emission from the shock interactions, the fast shocks can give rise to particle acceleration and synchrotron emission from relativistic electrons. We do not have a first principles theory for the production of relativistic particles and magnetic fields necessary for synchrotron emission, so the theory here has been guided by observations. Radio synchrotron emission has been detected from a number of supernovae (see Weiler et al. 1998 for references to the observations). SNae 1979C and 1980K were the first well-observed Type II supernovae and led to the circumstellar interaction model for the radio emission (Chevalier 1982b). In this model, the magnetic fields and relativistic electrons are present in the shocked interaction region and are assumed to have an energy density proportional to the total postshock energy density. The relativistic electrons are probably produced by shock acceleration. The observations showed evidence for early absorption and Chevalier (1982b) considered free-free absorption, synchrotron self-absorption, and the Razin effect as possible mechanisms for the absorption. For the Type II supernovae SNae 1979C and 1980K, free-free absorption appeared to be the most likely mechanism and the magnitude of the effect could be used to estimate the circumstellar density.

If the assumptions leading to eq. (2.3) remain valid and the particle acceleration occurs rapidly compared to the age, the radio evolution can be expected to follow a power law in time. Deviations from power law evolution can be expected when either the supernova density profile or the ambient density profile vary from their power law distributions. The supernova density profile is generally expected to become flatter in the inner regions. When the reverse shock reaches these regions, the driving power for the shocked region declines and the radio emission can be expected to decline. The circumstellar density can increase or decrease depending on changes in mass loss from the star, with corresponding increases or decreases in radio flux. A particularly interesting case occurs when there is strong change in mass loss characteristics and a fast wind creates a bubble and surrounding shell in a slower wind.

2.2. Type II supernovae

Most Type II supernovae are expected to have red supergiant progenitors and thus have relatively dense circumstellar surroundings. The best observed case of a Type II supernova colliding with its dense progenitor wind is SN 1993J in M81. The supernova was sufficiently close that it could be imaged with radio VLBI (Very Long Baseline Interferometry) techniques. Marcaide et al. (1997) found that the radio emission was from a somewhat irregular shell that expanded in a self-similar way. The time evolution of the shell radius over the first 3 years could be described by $R \propto t^m$ with $m = 0.86 \pm 0.02$, in good agreement with expectations for supernova interaction with a circumstellar wind. The radio flux evolution of SN 1993J has been well observed and the initial interpretation of the rising part of the light curve as due to free-free absorption led to the suggestion of a $\rho \propto r^{-1.5}$ circumstellar density distribution (Van Dyk et al. 1994; Fransson, Lundqvist, & Chevalier 1996). This distibution is implausible on physical grounds. However, the recognition that both synchrotron self-absorption and free-free absorption probably contribute to the rise in the light curve shows that the data are consistent with a $\rho \propto r^{-2}$ circumstellar medium (Fransson & Björnsson 1998; Chevalier 1998). The evolution is sufficiently well constrained in this case that the standard assumptions about the magnetic and relativistic particle energy densities can be checked and are verified (Fransson & Björnsson 1998).

Within a matter of days of the explosion, SN 1993J was observed by a number of X-ray satellites, including *CGRO* (Leising et al. 1994). The early spectrum could be fitted by bremsstrahlung emission with $T \approx 10^9$ K (Fransson et al. 1996). After 200 days, the spectrum implied a lower temperature, $T \approx 10^7$ K. The transition could be explained by a model in which the softer X-rays were initially absorbed by a cool radiative shell downstream from the reverse shock front (Fransson et al. 1996); the early cooling phase and gradual decrease of the column density through the cool shell are distinctive features of expansion in a $\rho \propto r^{-2}$ wind. The reradiation of the X-rays at lower frequencies has been clear in SN 1993J. Within a year of the explosion the Hα line showed evidence for a broad flat-topped profile in the spectrum, which is characteristic of circumstellar shell emission (Filippenko, Matheson, & Barth 1994). The combined HST and Keck spectrum of SN 1993J obtained in February 1996 showed a very strong Mg II line as expected from the cool reverse shock shell (Chevalier 1997; Fransson et al., in preparation). The far-UV part of the *HST* spectrum showed emission lines of N II, N III, and N IV, presumably from the freely expanding supernova. The Lyman α line was strong, as expected. An analysis of the relative line strengths can yield the element abundances in the supernova ejecta. As the various heavy element layers of the star move through the reverse shock front, we can learn about the abundance structure of the exploded star.

In contrast to SN 1993J, the optical spectrum of SN 1987A did not show evidence for shock interactions during the first few years. However, radio emission was discovered that showed similar characteristics to other radio supernovae, albeit with a much lower radio luminosity (Turtle et al. 1987; Ball et al. 1995). The low luminosity can be attributed to the low density wind from the blue supergiant progenitor. The 895 MHz flux showed a clear power law evolution of the form $t^{-1.2}$ over a factor 100 in flux (Ball et al. 1995). SN 1987A later showed the clearest case of late deviation from power law behavior for a radio supernova. The radio flux started to rise in 1990 and eventually returned to similar levels to its early peak (e.g. Ball et al. 1995). There is little doubt that the rise is connected to interaction with denser gas lost during an early, dense wind evolutionary phase, probably red supergiant mass loss photoionized by the blue progenitor (Chevalier & Dwarkadas 1995). The inner edge of the photoionized gas was at a radius of $\sim 4 \times 10^{17}$ cm. In the past few years, the interaction with dense gas has also been observed at optical wavelengths (Kirshner, these proceedings).

Two other Type II supernovae have shown weaker deviations from power law radio evolution. Observations of SN 1980K in 1994–96 showed the radio flux to be a factor of 2 below a power law extrapolation of the earlier evolution (Montes et al. 1998). Montes et al. interpret this as evidence for an earlier phase of lower density mass loss. Another possibility is that the reverse shock front has made its way back into supernova ejecta with a flatter density profile. In the case of SN 1979C, observations in 1996 showed the radio flux to be a factor of 2 above an extrapolation of the earlier evolution (Weiler et al. 1999); interaction with denser mass loss is likely. These were normal Type II supernovae interacting with red supergiant winds.

In addition to the normal Type II supernovae, there are some that are especially strong in radio and X-ray emission, indicating strong circumstellar interaction. These supernovae, examples of which are SNae 1978K, 1986J, 1988Z, and 1995N, have relatively narrow line emission (FWHM $\lesssim 2000$ km s^{-1}) and thus belong to the class of Type IIn supernovae. The narrow line emission is an indication that these supernovae are colliding with dense clumps in the wind (Chugai & Danziger 1994). The strong radio emission from these objects is probably due to a combination of a dense presupernova wind and clumps, which can enhance the turbulent motions in the shocked region (Jun, Jones, & Norman 1996).

2.3. Type Ib and Ic supernovae

The spectra and locations in galaxies of Type Ib and Ic supernovae suggest that they are the explosions of massive stars that have lost their hydrogen envelopes. These are observed as Wolf-Rayet stars, which are known to have fast, low density winds. The estimated mass loss rate is $\dot M \sim 10^{-5}$ $M_\odot$ yr^{-1} with a wind velocity $\sim 1,000$ km s^{-1} (Willis 1991). In view of the low density, it is not surprising that observations of circumstellar interaction are much less extensive than in the Type II supernova case. The reverse shock is not expected to be radiative and there have been no optical observations of circumstellar interaction. In the radio there is some light curve information for SNae 1983N and 1984L (Type Ib) and the Type Ic SNae 1990B and 1994I (for references, see Weiler et al. 1998 and Chevalier 1998), excluding SN 1998bw for the time being. These 4 radio supernovae have remarkably similar peak luminosities, within a factor of a few compared to the factor 10^6 spread for the luminosities of Type II radio supernovae. The high peak brightnesses at an early time for these supernovae implies that their initial rise in flux is due to synchrotron self-absorption and that the shell velocity at the time of the peak flux is 30,000–50,000 km s^{-1} (Slysh 1990; Chevalier 1998).

The radio flux of these supernovae can be explained by interaction with a Wolf-Rayet star wind if there is a high efficiency putting the postshock energy density into relativistic electrons and magnetic fields (Chevalier 1998). An alternative is that the interaction is with a denser wind from a companion star (Van Dyk et al 1993). It is possible that mass loss is driven by binary evolution leading up to the explosion (e.g. the model of Nomoto et al. 1994 for SN 1994I). Another handle on this question is the detection of emission at other wavelengths. SN 1994I was detected as an X-ray source with a 0.1–2.4 keV luminosity of 1.6×10^{38} erg s^{-1} (Immler, Pietsch, & Aschenbach 1998). The emission is likely to be from the reverse shock wave, which involves metal enriched ejecta. Immler et al. (1998) give some estimates of the amount of mass loss from the progenitor, but the emission estimates need to be examined for the composition expected in the outer parts of a Type Ic supernova.

In the same way as the progenitor of SN 1987A, the progenitors of Type Ib and Ic supernovae may go through a red supergiant phase and subsequently create a bubble in the dense wind. Bubble nebulae are sometimes observed around Wolf-Rayet stars (Chu 1991). The supernova remnant Cas A is a possible example of interaction with a bubble shell (Chevalier & Liang 1989; Borkowski et al. 1996). The shell in this case has a radius of about 1.6 pc.

2.4. Type Ia supernovae

Type Ia supernovae have not yet been observed to interact with their surroundings during the first few years of evolution. This is an important constraint on progenitor models for Type Ia supernovae (e.g. Branch et al. 1995). Most current models for these progenitors involve the explosion of a white dwarf in a binary system. One candidate for the progenitor of a Type Ia explosion is a double degenerate dwarf. In this case, the presupernova wind is likely to be weak and the supernova can interact directly with the ISM (interstellar medium). In other scenarios, there may be significant winds from the companion star or from the white dwarf progenitor (Hachisu et al. 1999a,b).

With current techniques, the most sensitive way to detect circumstellar interaction appears to be at radio wavelengths. The best limit at present is for SN 1986G in Centaurus A (Eck et al. 1995). Eck et al. have used this limit to infer a mass loss rate from the progenitor $\lesssim 10^{-7}$ $M_\odot$ yr^{-1} for a wind velocity of 10 km s^{-1}. Eck et al. assumed that the primary absorption mechanism would be due to free-free absorption, but synchrotron self-absorption may play a role, as is apparently the case in Type Ib and Ic supernovae

(Chevalier 1998). The limit on $\dot{M}$ then depends on the uncertain efficiency of production of the synchrotron emission. In any case, it is clear that radio observations are in an interesting regime and should be pursued.

Both SN 1006 and SN 1572 in our Galaxy are likely remnants of Type Ia supernovae. In these cases, the expansion rate of the remnant is close to the Sedov solution in a uniform medium ($R \propto t^{0.4}$) and the objects are probably interacting with the ISM. The swept up mass in both cases is $\sim$ 7–8 M$_\odot$, so these objects do not provide good tests for small amounts of progenitor mass loss.

3. SN 1998bw/GRB 980425

SN 1998bw was discovered in a search of the error box of GRB 980425; the estimated probability of a chance coincidence of a supernova and GRB occurring at the same time in the same small area was 10^{-4} (Galama et al. 1998b). The connection was strengthened by the unusual optical and radio properties of the supernova. The optical data imply a Type Ic supernova, involving the explosion of a Wolf-Rayet star with an energy of $(2-3) \times 10^{52}$ ergs (Iwamoto et al. 1998; Woosley, Eastman, & Schmidt 1999). The radio data show a peak luminosity that is a factor $\sim$ 100 larger than the luminosity of other Type Ib and Ic supernova, and indicate relativistic motion (Kulkarni et al. 1998).

Li & Chevalier (1999) analyzed the radio emission in terms of a thin shell model in which constant fractions of the postshock energy density went into relativistic electrons and magnetic field. They included relativistic effects, synchrotron self-absorption, and a lower limit to the (power law) electron distribution function. One result was that the ambient density was consistent with that expected for a Wolf-Rayet star, with an $s = 2$ density gradient. Another result was that the shell was not being driven by a steep power density distribution, as is expected and observed for normal supernovae. Over much of the observation period, the evolution was consistent with a constant energy blast wave. This is also true for GRB afterglows, in which the energy deposition in the surrounding medium is thought to be associated with the initial γ-ray burst (e.g. Mészáros & Rees 1997). However, the radio emission from SN 1998bw showed a marked rise over days 20–40, even at optically thin frequencies (Kulkarni et al. 1998). An interpretation of this event as an increase in the blast wave energy (by a factor of 2–3) is preferred over a model involving interaction with a denser medium (Li & Chevalier 1999). The additional energy might come from a shell of matter that was ejected at the time of the initial burst, but took some time to catch up with the blast wave.

The model of Li & Chevalier (1999) does not account for the radio evolution before day 10, but it does account for the subsequent evolution to day 250. The model is spherically symmetric, which, for the mildly relativistic velocities that are involved, implies that there was not a significant collimation of the flow. The nature of the radio source implies that it was powered by a burst from a central engine, not from high velocity supernova ejecta. The power from the central engine must form a collimated flow to escape through the star (e.g. MacFadyen & Woosley 1999). It appears that the collimation was lost upon, or shortly after, the escape of the energetic flow from the stellar surface.

4. Gamma-ray bursts

The fireball model for GRBs (gamma-ray bursts) led to predictions of the afterglow emission that might be expected when the energetic shock wave encountered the surrounding medium (Katz 1994; Mészáros & Rees 1997). The subsequent optical and

X-ray observations of the afterglow from GRB 970228 appeared to confirm the predictions of the simplest afterglow model (Wijers, Rees, & Mészáros 1997). This model involved synchrotron emission from electrons accelerated to a power law energy spectrum ($N(E) \propto E^{-p}$) in a relativistic blast wave expanding into a constant density, presumably interstellar medium (ISM). In particular, the expected relation between the flux spectral index and the power law rate of flux decay was in approximate accord with the observations. This model has become the "standard model" for the interpretation of GRB afterglow observations (e.g. Piran, these proceedings). It was used to make predictions of bright optical emission in the early phases when a reverse shock front is present (Sari & Piran 1999b). The observation of a bright flash from GRB 990123 (Akerlof et al. 1999) gave basic confirmation of this aspect of the model (Sari & Piran 1999a). The expectation of jets for the initial energy deposition led to predictions of the effects on the light curve as the jet slowed (Rhoads 1999; Sari, Piran, & Halpern 1999). The observations of the afterglow of GRB 990510 confirmed the basic features expected for jet deceleration (Harrison et al. 1999).

These successes of the standard model give confidence in many aspects of the model. However, there has been increasing observational evidence that at least some GRBs have massive star progenitors. Paczyński (1998) noted that the available evidence on the location of GRBs in their host galaxies indicated a link to star formation. A more direct link to massive stars is provided by the presence of a supernova, the light from the exploded star matter. GRB 980425, discussed above, was the first burst with a likely supernova association. Evidence for supernova type emission has now been found in GRB 980326 (Bloom et al. 1999) and GRB 970228 (Reichart 1999; Galama et al. 1999).

The importance of a massive star origin for the afterglow evolution is that the GRB blast wave should be expanding into the stellar wind of the progenitor star. Chevalier & Li (1999a,b) made estimates of afterglow evolution for the case of expansion in the wind of a Wolf-Rayet star. At an age of a few days, when many afterglow observations are made, the shock wave is expected to be in a region with a density comparable to the interstellar density. Thus a comparable luminosity can be expected at this time, unlike the supernova case where the shock velocity is lower. At early times, the shock wave does interact with a high density medium and at late times ($\gtrsim 1$ year) the interaction is with a low density medium, so that the transition to nonrelativistic expansion is delayed compared to the interstellar case.

Another aspect of these two cases is the difference in evolution of the spectrum related to the sychrotron processes of self-absorption and cooling (cf. Sari, Piran, & Narayan 1998 for the interstellar case). For adiabatic evolution, the optical luminosity is expected to drop more rapidly for wind interaction. Both GRB 980326 and GRB 980519 had relatively steep declines that were consistent with adiabatic evolution in a wind with an electron spectrum $p = 3$ (Groot et al. 1998; Halpern et al. 1999; Chevalier & Li 1999a). The problem is that this evolution is the same as that expected for a cooling jet with $p = 2$ in either a constant density ISM (Sari et al. 1999) or a wind. There is a possibility that radio observations can break the degeneracy. Chevalier & Li (1999a) found that the early rise that appears to be present in the radio evolution of GRB 980519 could be accounted for in a wind model; it is not clear whether an ISM model can do the same.

GRB 970228 is an interesting case because the initial finding that the rate of decline is compatible with the standard interstellar interaction model (Wijers et al. 1997) appears to be at odds with the later finding of supernova emission (Reichart 1999; Galama et al. 1999), which implies a massive star progenitor. However, when the supernova emission is subtracted off from the light curve, the time dependence of the nonthermal afterglow steepens and becomes compatible with an adiabatic wind interaction model (Chevalier

& Li 1999b). The non-detection of radio emission from this burst (Frail et al. 1998) can also be accomodated by a wind interaction model. A jet model in which the jet is in the laterally expanding phase and $F_\nu \propto t^{-p}$ seems unlikely for GRB 970228 because $p < 2$ would be required. However, jet evolution in some intermediate phase probably cannot be ruled out.

In a paper advocating an $s = 0$ blast wave model for the well-observed afterglow of GRB 970508, Galama et al. (1998a) end by describing three deficiencies of the model: (1) the peak flux should be constant but is observed to decrease with time; (2) the self-absorption frequency is predicted to be time-independent, but is observed to decrease with time, and the rise of the radio fluxes is slower than expected; and (3) the decay after maximum at mm wavelengths is perhaps somewhat faster than expected. All of these problems are addressed by the wind model, which provides approximate quantitative agreement with the observations (Chevalier & Li 1999b). In this case, it is the radio evolution that provides support the wind interaction picture. At optical wavelengths, the power law evolution over the period 2–100 days is consistent with cooling interstellar interaction, but cooling wind interaction predicts exactly the same evolution. The position of GRB 970508 within 70 pc of the center of the host galaxy (Fruchter et al. 1999) provides further support for a massive star progenitor. However, the power law optical light curve indicates that a supernova would have to be at least 1 magnitude fainter than SN 1998bw (Fruchter, these proceedings). Some diversity of supernova properties seems plausible.

Both GRB 990123 and GRB 990510 showed early afterglow evolution that was consistent with adiabatic interstellar interaction (Kulkarni et al. 1999; Harrison et al. 1999). In the case of GRB 990123, there are two additional pieces of support for interstellar interaction. One is the more rapid decrease of the X-ray emission, which is expected for interstellar interaction if X-ray wavelengths are in cooling regime. This is not expected for wind interaction, for which the relative optical and X-ray evolution is expected to be the opposite. Second, the evolution of the prompt optical emission (Akerlof et al. 1999) can be explained by reverse shock wave emission in an interstellar interaction model (Sari & Piran 1999a). For wind interaction, the strong initial cooling yields an evolution that is incompatible with the observations (Chevalier & Li 1999b). For GRB 990510, the flat initial evolution would be compatible with wind interaction only if there were a driving force for the blast wave. However, the evolution to the steep evolution expected for a laterally expanding jet suggests that such a driving force was not present.

For other GRB afterglows, there are uncertainties in the interpretation of the observational results. At optical wavelengths, the extinction can be high, so that the optical spectral index and flux level are uncertain; an example is GRB 980703 (Vreeswijk et al. 1999). At radio wavelengths, scintillation effects complicate the comparison with theoretical light curves (e.g. Waxman et al. 1998). In addition, current theoretical models suffer from a number of assumptions: constant efficiencies of production of magnetic field and relativistic particles and constant p in energy and time. Of these assumptions, the constancy of p is in doubt because it appears to vary among observed sources (Chevalier & Li 1999b). These problems can be best addressed by observing favorable afterglows over a wide span of wavelengths (radio to X-ray) and over a wide time span.

5. Conclusions

Our understanding of the early supernova and GRB interactions with their surroundings varies for the different types of objects. The optical light from supernovae is generally dominated by thermal emission from shock heated gas and by gas heated by radioactivity.

Type II supernovae interacting with the dense winds of their red supergiant progenitors show emission over a broad range of wavelengths and some details of the interaction, such as clumping, can be deduced. Nonthermal emission has been observed only at radio wavelengths. Interactions around typical Type Ib and Ic supernovae have been observed primarily at radio wavelengths; there is still some uncertainty whether the emission is compatible with interaction with a Wolf-Rayet star wind or whether a denser wind is necessary. Observations of Type Ia supernovae have not shown evidence for interaction. The existing limit on the radio emission appears to be close to a range in which some models can be ruled out, although there are still uncertainties in the interaction models for the radio emission.

The Type Ic supernova SN 1998bw was apparently associated with the weak GRB 980425. The afterglow emission was weak compared to that from GRBs at cosmological distances, but the nonthermal radio emission was a factor 100 more luminous than that from other Type Ib or Ic supernovae. The emission was consistent with interaction with the wind from a Wolf-Rayet star, but the explosion was not consistent with a driving force from the outer density profile of a supernova. Over most of the observation period, it was consistent with a constant energy explosion, as is also inferred in typical GRB afterglows. The energy inferred for this explosion, $\sim 10^{49} - 3 \times 10^{50}$ ergs is considerably less than the energy inferred for the supernova, $(2\text{--}3) \times 10^{52}$ ergs.

The nonthermal afterglow emission dominates the radiation from cosmological GRBs, although there is evidence for supernova-type light in some cases. The models developed for GRB afterglows are essentially relativistic versions of those previously developed for radio supernovae, although the GRBs can be observed over a much broader frequency range. In many cases there are ambiguities in the interpretation of afterglow observations: wind vs. interstellar interaction, jet vs. spherical expansion, relativistic vs. nonrelativistic expansion. Resolution of these ambiguities is needed to make progress on the problems of GRB progenitors and total energies. There is already intriguing evidence among the relatively small number of observed afterglows from long duration bursts for both wind and interstellar interactors. The implication is that the GRB mechanism may operate in a variety of progenitor situations. As GRBs come under more intense observational scrutiny, these fascinating issues should be clarified.

I am grateful to Claes Fransson and Zhi-Yun Li for enjoyable collaborations on supernova and gamma-ray burst problems. This work was supported in part by NASA grant NAGW-8232.

REFERENCES

AKERLOF, C. W., ET AL. 1999 *Nature*, **398**, 400.
BALL, L., CAMPBELL-WILSON, D., CRAWFORD, D. F., & TURTLE, A. J. 1995 *ApJ*, **453**, 864.
BLOOM, J. S., ET AL. 1999 *Nature*, in press.
BORKOWSKI, K., SZYMKOWIAK, A. E., BLONDIN, J. M., & SARAZIN, C. L. 1996 *ApJ*, **466**, 866.
BRANCH, D., LIVIO, M., YUNGELSON, L. R., BOFFI, F. R., & BARON, E. 1995 *PASP*, **107**, 1019.
CHEVALIER, R. A. 1982a *ApJ*, **258**, 790.
CHEVALIER, R. A. 1982b *ApJ*, **259**, 302.
CHEVALIER, R. A. 1997 *Science*, **276**, 1374.
CHEVALIER, R. A. 1998 *ApJ*, **499**, 810.
CHEVALIER, R. A. & DWARKADAS, V. V. 1995 *ApJ*, **452**, L45.

Chevalier, R. A. & Fransson, C. 1994 *ApJ*, **420**, 268.
Chevalier, R. A. & Li, Z.-Y. 1999a *ApJ*, **520**, L29.
Chevalier, R. A. & Li, Z.-Y. 1999b *ApJ*, submitted (astro-ph/9908272).
Chevalier, R. A. & Liang, E. P. 1989 *ApJ*, **344**, 332.
Chu, Y.-H. 1991 in *Wolf-Rayet Stars and Interrelations with Other Massive Stars in Galaxies* (ed. K. A. van der Hucht & B. Hidayat), p. 349. Kluwer.
Chugai, N. N. & Danziger, I. J. 1994 *MNRAS*, **268**, 173.
Eck, C. R., Cowan, J. J., Roberts, D. A., Boffi, F. R., & Branch, D. 1995 *ApJ*, **451**, L53.
Filippenko, A. V., Matheson, T., & Barth, A. J. 1994 *AJ*, **108**, 2220.
Frail, D. A., Kulkarni, S. R., Shepherd, D. S., & Waxman, E. 1998 *ApJ*, **502**, L119.
Fransson, C. & Björnsson, C.-I. 1998 *ApJ*, **509**, 861.
Fransson, C., Lundqvist, P., & Chevalier, R. A. 1996 *ApJ*, **461**, 993.
Fruchter, A. S., et al. 1999 preprint (astro-ph/9903236).
Galama, T. J., Wijers, R. A. M. J., Bremer, M., Groot, P. J., Strom, R. G., Kouveliotou, C., & van Paradijs, J. 1998a *ApJ*, **500**, L97.
Galama, T. J., et al. 1998b *Nature*, **395**, 670.
Galama, T. J., et al. 1999 *ApJ*, submitted (astro-ph/9907264).
Groot, P. J., et al. 1998 *ApJ*, **502**, L123.
Hachisu, I., Kato, M., Nomoto, K., & Umeda, H. 1999a *ApJ*, **519**, 314.
Hachisu, I., Kato, M., & Nomoto, K. 1999b *ApJ*, **522**, 487.
Halpern, J. P., Kemp, J., Piran, T., & Bershady, M. A. 1999 *ApJ*, **517**, L105.
Harrison, F. A., et al. 1999 *ApJ*, in press (astro-ph/9905306)
Immler, S., Pietsch, W., & Aschenbach, B. 1998 *A&A*, **336**, L1.
Iwamoto, K., et al. 1998 *Nature*, **395**, 672.
Jun, B.-I., Jones, T. W., & Norman, M. L. 1996 *ApJ*, **468**, L59.
Katz, J. I. 1994 *ApJ*, **422**, 248.
Kulkarni, S. R., Frail, D. A., Wieringa, M. H., Ekers, R. D., Sadler, E. M., Wark, R. M., Higdon, J. L., Phinney, E. S., & Bloom, J. S. 1998 *Nature*, **395**, 663.
Kulkarni, S. R., et al. 1999 *Nature*, **398**, 389.
Leising, M. D., et al. 1994 *ApJ*, **431**, L95.
Li, Z.-Y. & Chevalier, R. A. 1999 *ApJ*, in press (astro-ph/9903483).
MacFadyen, A. & Woosley, S. E. 1999 *ApJ*, in press (astro-ph/9810274).
Matzner, C. D. & McKee, C. F. 1999 *ApJ*, **510**, 379.
Marcaide, J. M., et al. 1997 *ApJ*, **486**, L31.
Mészáros, P. & Rees, M. J. 1997 *ApJ*, **476**, 232.
Montes, M. J., Van Dyk, S. D., Weiler, K. W., Sramek, R. A., & Panagia, N. 1998 *ApJ*, **506**, 874.
Nomoto, K., et al. 1994 *Nature*, **371**, 227.
Paczyński, B. 1998 *ApJ*, **494**, L45.
Reichart, D. E. 1999 *ApJ*, **521**, L111.
Rhoads, J. E. 1999 *ApJ*, in press (astro-ph/9903399).
Sari, R. & Piran, T. 1999a *ApJ*, **517**, L109.
Sari, R. & Piran, T. 1999b *ApJ*, **520**, 641.
Sari, R., Piran, T., & Halpern, J. P. 1999 *ApJ*, **519**, L17.
Sari, R., Piran, T., & Narayan, R. 1998 *ApJ*, **497**, L17.
Slysh, V. I. 1990 *Sov. Astr. Lett.*, **16**, 339.

TURTLE, A. J., ET AL. 1987 **Nature**, **327**, 38.
VAN DYK, S. D., SRAMEK, R. A., WEILER, K. W., & PANAGIA, N. 1993 *ApJ*, **409**, 162.
VAN DYK, S. D., WEILER, K. W., SRAMEK, R. A., RUPEN, M. P., & PANAGIA, N. 1994 *ApJ*, **432**, L115.
VREESWIJK, P. M., ET AL. 1999 *ApJ*, **523**, 171.
WAXMAN, E., KULKARNI, S. R., & FRAIL, D. A. 1998 *ApJ*, **497**, 288
WEILER, K. W., MONTES, M. J., VAN DYK, S. D., SRAMEK, R. A., & PANAGIA, N. 1999, in *SN 1987A: 10 Years Later*, (ed. M. Phillips & N. Suntzeff). ASP, in press.
WEILER, K. W., VAN DYK, S. D., MONTES, M. J., PANAGIA, N., & SRAMEK, R. A. 1998 *ApJ*, **500**, 51.
WIJERS, R. A. M. J., REES, M. J., & MÉSZÁROS, P. 1997 *MNRAS*, **288**, L51.
WILLIS, A. J. 1991 in *Wolf-Rayet Stars and Interrelations with Other Massive Stars in Galaxies* (ed. K. A. van der Hucht & B. Hidayat), p. 265. Kluwer.
WOOSLEY, S. E., EASTMAN, R. G., & SCHMIDT, B. P. 1999 *ApJ*, **516**, 788.

Magnetars, Soft Gamma-ray Repeaters and Gamma-ray Bursts

By ALICE K. HARDING

NASA Goddard Space Flight Center, Greenbelt, MD 20771

Recent observations of soft gamma-ray repeaters (SGRs) and anomalous X-ray pulsars are providing evidence for the existence of magnetars, a class of neutron stars with surface magnetic fields exceeding the critical field of 4.4×10^{13} Gauss. Magnetars are the first astrophysical sources believed to derive their primary radiation power from magnetic energy, which substantially exceeds their rotational energy. In addition, the physics of supercritical magnetic fields operates in a more exotic realm than even that of normal pulsars. The discovery of long periods and high period derivatives in soft gamma-ray repeaters (SGRs) over the past year strongly suggests that SGRs are magnetars. But do magnetars have any relation to classical gamma-ray bursts (GRBs)? I will review the evidence that SGRs are magnetars and discuss the question of whether any of the classical GRBs could be magnetars.

1. Introduction

Until a little over a year ago, neutron stars having supercritical fields in the range 10^{14}–10^{15} G lay in the realm of theoretical fantasy. They were first proposed to explain a number of properties of soft gamma-ray repeaters (SGRs), including the highly super-Eddington luminosities (Katz 1982, Paczynski 1992), the young age and long period of SGR 0526−66 (the March 5, 1979 source) and the tremendous energy of the SGR giant bursts (Thompson & Duncan 1995). Recently (see review by Murakami 1999), long ($P > 5$ s) periods and high period derivatives ($\dot{P} \sim 10^{-10}\,\mathrm{s\,s^{-1}}$), the major prediction of the magnetar model, have been detected in both SGR sources and in anomalous X-ray pulsars (AXPs). The derived surface magnetic fields of these sources, assuming dipole spin-down such that $B_0 = 6.4 \times 10^{19}\,(P\dot{P})^{1/2}$, are indeed in the range 10^{14}–1.6×10^{15} G. The rotational energy loss rate of these sources is too low to account for their observed radiation, but the magnetic energy released through field decay is more than adequate.

2. Theory of Magnetars

2.1. Formation

There have been several mechanisms proposed for generating the huge magnetar fields. Duncan & Thompson (1992) noted that the vigorous convection due to neutrino diffusion in core collapse supernovae, as has since been found in 3D numerical simulations (e.g. Herant et al. 1994), might drive an $\alpha - \Omega$ dynamo in a rapidly spinning core. They were originally studying the generation of fields in pulsars and other "normal" neutron stars. But they found that magnetic fields as high as 3×10^{17} G $(P_i/1\mathrm{ms})^{-1}$ could be generated by this mechanism, so that neutron stars that are initially spinning with periods P_i of a few ms could, in the seconds following their birth, acquire fields above 10^{15} G. Usov (1992) suggested that if a magnetic white dwarf having a field around 10^9 G, the very upper range of the observed WD magnetic fields, underwent accretion-induced collapse, then it would form a neutron star with a magnetic field of $\sim 10^{15}$ G by flux conservation. He postulated that the energy of 10^{51} erg, released over a few seconds during the initial rapid spin-down, could produce a classical gamma-ray burst at cosmological distances (more discussion about this follows in Section 4).

2.2. Spin down

Isolated neutron stars with magnetar-strength fields spin down much faster than do neutron stars with fields in the normal pulsar range of 10^{11}–10^{13} G. The spin-down timescale, or characteristic age,

$$\tau_{\mathrm{SD}} = \frac{P}{2\dot{P}} = 1600 \left(\frac{P}{B_{14}}\right)^2 \mathrm{yr}, \qquad (2.1)$$

from integrating the rotational energy loss rate for magnetic dipole radiation, $\dot{E}_D = I\Omega\dot{\Omega} \propto \Omega^4$, where B_{14} is the surface field in units of 10^{14} G, I is the neutron star moment of inertia and P and Ω are the present period and rotation frequency, predicts that magnetars will reach relatively long periods approaching 10 s in an age of several thousand years. As will be discussed in Section 3.2.2, particle wind flows may also contribute to the rotational energy loss of some magnetars, affecting both the characteristic age and the derived surface magnetic field.

The magnetic energy of a magnetar dominates over rotational energy at an early age,

$$\tau_B \simeq 1.5 \times 10^5 \, \mathrm{yr} \left(\frac{B_0}{10^{14}\,\mathrm{G}}\right)^{-4}, \qquad (2.2)$$

and it becomes a plentiful supply of free energy to power both the bursting and quiescent emission observed in the SGRs.

2.3. Field decay

Magnetar fields can decay on relatively short timescales and this decay could be the mechanism by which the field energy is released to power the stellar activity (Thompson & Duncan 1995 [TD95]). Ohmic diffusion, in which the magnetic flux moves through the charged particle component, produces field decay on timescales too long to be useful for energy generation. However, ambipolar diffusion, in which the magnetic flux carries the charged particles with it as it moves out of the neutron star core, can operate on much shorter timescales in magnetar fields. The interactions of the charged particles produce a drag force that sets the diffusion timescale, which depends on the core temperature. Goldreich & Reisnegger (1992) noted that the charged particle flux can be separated into solenoidal and irrotational modes, having different growth rates. The ambipolar diffusion itself heats the core by magnetic energy release. The dominate mode of cooling of the core at the resulting high temperatures is neutrino emission via the modified URCA reactions. One therefore must first find the equilibrium core temperature by balancing the rates of heating and cooling, and the equilibrium temperature sets the diffusion timescale. Thompson & Duncan (1996), arguing that only the irrotational mode of ambipolar diffusion is important in magnetar fields, find a diffusion timescale of

$$\tau_{\mathrm{decay}} \simeq 10^4 \, \mathrm{yr} \left(\frac{B_{\mathrm{core}}}{5 \times 10^{15}\,\mathrm{G}}\right)^{-14}, \qquad (2.3)$$

where B_{core} is the core field strength, which can be larger than the surface field strength. This expression has a very strong dependence on core magnetic field, and would give a very small range of magnetar fields having interesting decay timescales. However, Goldreich & Reisnegger (1992) derived the timescales for both modes, and found that the growth timescale of the irrotational mode is limited by pressure gradients, such that the minimum decay timescale is

$$\tau_{\mathrm{decay}}^{\mathrm{min}} \simeq 10^5 \, \mathrm{yr} \left(\frac{B_{\mathrm{core}}}{10^{15}\,\mathrm{G}}\right)^{-2}. \qquad (2.4)$$

The estimated decay timescales for ambipolar diffusion in magnetar-strength fields seem to fall in the range of the derived characteristic ages of the SGRs and AXPs. Field decay is thus a promising source of energy for these objects.

2.4. Strong field radiation physics

The extremely strong fields of magnetars also push the radiation physics into a different regime from that of ordinary neutron stars. In particular, there are a number of new effects in the photon transport to be considered. The total magnetic Compton scattering cross section, in the non-relativistic limit, has the behavior

$$\sigma_\perp = \left(\frac{\omega}{\omega_B}\right)^2 \sigma_T \tag{2.5}$$

$$\sigma_\parallel = \sin^2(\theta_{kB}) \sigma_T, \tag{2.6}$$

where $\perp$ and $\parallel$ refer to the photon polarization modes having electric vector perpendicular and parallel to the plane defined by the photon wavevector and the magnetic field, $\omega_B = mc^2 (B/B_{\rm cr})$ is the cyclotron energy, θ_{kB} is the angle between the photon wavevector and the field, and σ_T is the Thompson cross section. Paczynski (1992) pointed out that the $\perp$ mode cross section is greatly suppressed in fields $B \gg B_{\rm cr}$, because the typical photon energies of 10–100 keV are orders of magnitude below the cyclotron energy of 10 to 100 MeV. This might lead to an effective Eddington luminosity,

$$L_{\rm Edd}^B \simeq 2 \times 10^{40} \,\text{erg s}^{-1} \left(\frac{B}{B_{\rm cr}}\right)^{4/3} \left(\frac{R}{R_0}\right)^{2/3}, \tag{2.7}$$

where R_0 is the neutron star radius and R is the radius of emission. If SGRs were neutron stars with fields above 10^{14} G, then this effect could explain the observed burst luminosities that are several orders of magnitude above the Eddington luminosity, $L_{\rm Edd}^B \simeq 10^{38}$ erg s^{-1}, for a neutron star with a conventional field. However, Miller (1995) argues that although most of the photons may escape in the $\perp$ mode, the dominant scattering into the $\parallel$ mode will keep the radiation pressure high in hydrostatic atmospheres and limit the effective Eddington luminosity to more modest values. Magnetic confinement therefore may play a larger role in allowing the radiation of highly super-Eddington luminosities.

In fields above 10^{13} G, photon splitting, a QED process in which a single photon splits into two lower energy photons, becomes an important attenuation mechanism. It has no threshold and can therefore operate at photon energies well below the threshold for one-photon pair production, $2mc^2/\sin\theta_{\rm kB}$. Although photon splitting attenuation in neutron star magnetospheres becomes more effective than one-photon pair production above $B \sim 10^{13}$ G (Harding et al. 1997), the full implications of photon splitting for SGR models is not yet clear, due to an incomplete understanding of the behavior of the process in ultra-strong fields. According to the kinematic selection rules, proven by Adler (1971) to operate in weakly dispersive regimes ($B \lesssim B_{\rm cr}$), energy-momentum conservation in the magnetized vacuum permit splitting of only one polarization mode, $\perp \to \parallel\parallel$. These rules thus apply in normal pulsars fields, and prevent photon splitting from completely suppressing pair production. However, the kinematic selection rules have never been investigated in magnetar-strength fields, where high vacuum dispersion is present. Baring & Harding (1998) suggest that if all three modes of photon splitting allowed by CP invariance in QED operate at higher field strengths, so that both photon polarization states are allowed to split, then pair production would be completely suppressed. This would have several important consequences for SGRs. First, photon splitting cascades

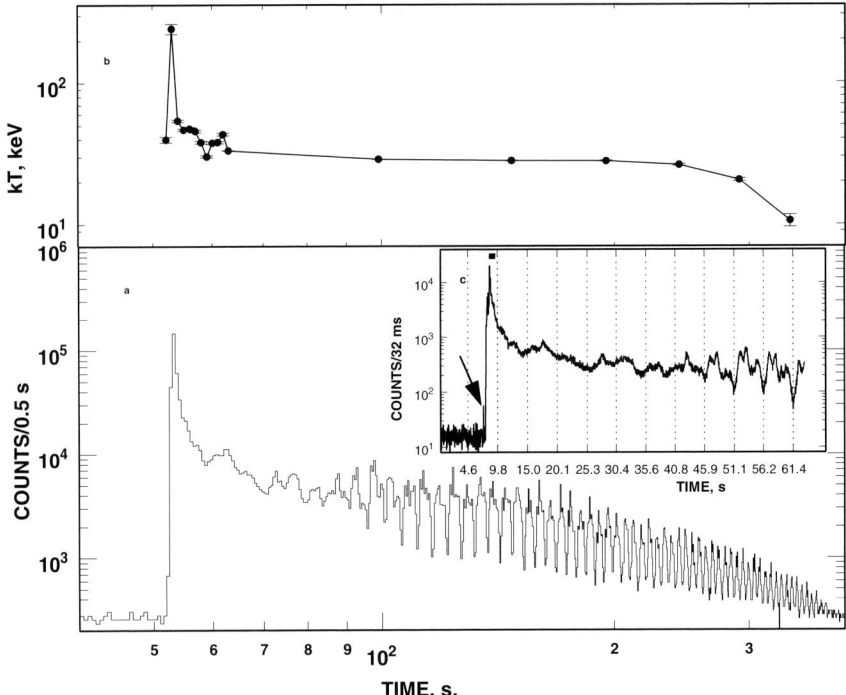

FIGURE 1. Time history of the August 27, 1998 giant burst from SGR1900+14 at 25–150 keV, showing 5.16 s periodicity in the decay phase (from Hurley et al. 1999)

would efficiently degrade energies of radiated photons down to around 30 keV (these models will be discussed in more detail in Section 3.2.1 below). Second, the complete suppression of pair production would cause these sources to be radio quiet, according to standard pulsar models which require electron-positron pairs for the production of coherent radio emission.

3. Magnetars and Soft Gamma-ray Repeaters

3.1. *Observational evidence*

3.1.1. *Periodicities*

Observed characteristics of SGRs, the class of γ-ray burst sources that undergo repeated outbursts, furnish strong evidence for the existence of magnetars. There are four known SGR sources (one newly discovered; Woods et al. 1999), all of which have been associated with young ($\tau < 10^5$ yr) supernova remnants. The first detection of a periodicity in an SGR source occurred during the decay phase of the giant outburst of SGR 0526−66 on Mar. 5 1979 (Mazets et al. 1981). The relatively long period of 8 s (for an isolated neutron star) seemed at first unlikely to be a rotation period, since a normal pulsar would not have spun down so far in the age of the associated supernova remnant, N49. Furthermore, a period derivative has never been established for this source, because the quiescent emission needed for an additional period measurement was not detected. The breakthrough came with the detection of 7 s pulsations in the quiescent X-ray emission of SGR 1806−20 (Kouveliotou et al. 1998) at several different epochs, and the resulting measurement of a period derivative, $\dot{P} = 8.3 \times 10^{-11}$ s s^{-1}, several orders of magnitude higher than that of typical radio pulsars. Late last year SGR 1900+14

Source	P (s)	$\dot{P}$ ($s\,s^{-1}$)	Age (yr)	B (Gauss)	Data from
SGR 1806−20	7.47	8.3×10^{-11}	1.4×10^3	1.6×10^{15}	Kouveliotou et al. 1998
SGR 1900+14	5.16	1.1×10^{-10}	7.8×10^2	1.5×10^{15}	Hurley et al. 1999
SGR 0526−66	8				Mazets et al. 1981
SGR 1627−41	6.41				Woods et al. 1999
1E 1841−045	11.77	4.2×10^{-11}	4.0×10^3	1.5×10^{15}	Vasisht & Gotthelf 1997
1E 1048−5937	6.45	2.2×10^{-11}	4.6×10^3	4.6×10^{14}	Oosterbroek et al. 1998
4U 0142+615	8.69	2.3×10^{-12}	6.0×10^4	2.9×10^{14}	Wilson et al. 1998
1E 2259+586	6.98	7.3×10^{-13}	1.5×10^5	1.4×10^{14}	Mereghetti & Stella 1995
1RXS J170849.0−400910	11	2.3×10^{-11}	8.0×10^3	1.1×10^{15}	Israel et al. 1999

TABLE 1. SGRs and Anomalous X-Ray Pulsars

became very active, undergoing a giant burst having energy $\sim 10^{45}$ erg, followed by a number of smaller bursts. The giant burst decay phase (see Fig. 1) revealed many cycles of a 5.16 s periodicity (Hurley et al. 1999), the same period as had been detected earlier in the quiescent emission (Kouveliotou et al. 1999).

Another group of sources having similar P and $\dot{P}$ are the anomalous X-ray pulsars (AXPs), a group of six or seven pulsating X-ray sources with periods in the range 6–12 s and period derivatives in the range 10^{-12}–$10^{-11}\,\mathrm{s\,s^{-1}}$ (Gotthelf & Vasisht 1998). These sources have shown only quiescent emission with no bursting behavior, but which are anomalous in comparison with average characteristics of known accreting X-ray pulsars. They are bright, steady X-ray sources having luminosities $L_X \sim 10^{35}$ erg s^{-1}, show no sign of any companion, are steadily spinning down and have ages $\tau \leq 10^5$ years (e.g. Vasisht & Gotthelf 1997). Their derived surface magnetic fields, assuming that they are spinning down due to dipole radiation torques, are between 10^{14} and 10^{15} G. It is notable that the observed X-ray luminosities of AXPs (as well the quiescent luminosities of SGRs) exceed their spin-down luminosity, so that an additional power source is necessary. Table 1 shows the observed spin parameters for all of the detected SGR sources and those AXPs with measured $\dot{P}$.

3.1.2. *Bursts*

As well as the giant bursts observed in two SGR sources, the SGRs emit many smaller bursts which repeat over a wide range of intervals and with a range of luminosities. Intervals between the small bursts from SGR 1806−20 have been shown to follow a log-normal distribution (Hurley et al. 1994). The size spectrum (Number vs. energy) of SGR 1806−20 small bursts have a power-law spectrum with index −1.6, similar to the power law in the size spectrum of Earthquakes (Cheng et al. 1995). The spectrum of the small bursts from SGRs seems to be very similar in all bursts (Fenimore et al. 1994, Mazets et al. 1999), regardless of their energy, showing a relatively soft, quasi-Maxwellian peaking around 30 keV.

3.2. *Models*

3.2.1. *Burst radiation*

The large magnetic flux diffusing out of the neutron star core, and eventually through the crust, creates imbalances and stresses. TD95 postulate that a large-scale rearrangement of the field, either through an interchange instability of both the internal and

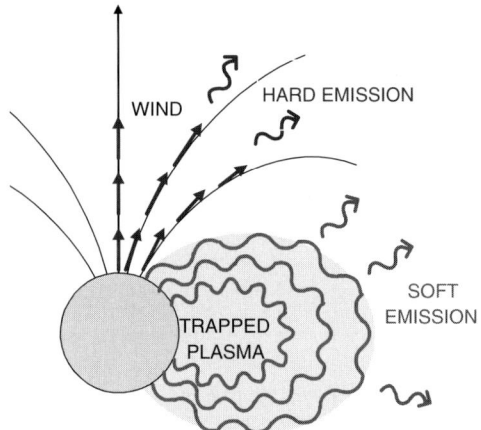

FIGURE 2. Schematic illustration of a burst from a soft gamma-ray repeater (based on model of Thompson & Duncan 1995).

external field or shearing of the external field, can cause a massive reconnection event. These events could release a significant portion of the magnetic field energy,

$$\frac{B_{\rm core}^2}{8\pi} R^3 \simeq 4 \times 10^{46}\, {\rm erg} \left(\frac{B_{\rm core}}{10^{15}\,{\rm G}}\right)^2 \qquad (3.8)$$

and may be responsible for the giant SGR flares observed from SGR 0526−66 and SGR 1900+14 having energies $\sim 10^{45}$ erg.

The small SGR bursts, having average energies of 10^{41} ergs, may be due to cracking of the neutron star crust. Ambipolar diffusion of the magnetic flux through the crust will built up stresses which can cause it to crack, producing small horizontal displacements of the magnetic footpoints, and producing an energy release,

$$E_{\rm SGR} \simeq 10^{41}\, {\rm erg\, s}^{-1} \left(\frac{B_{\rm core}}{10^{15}\,{\rm G}}\right)^{-2} \left(\frac{l}{1{\rm km}}\right)^2 \left(\frac{\theta_{\rm max}}{10^3}\right)^2, \qquad (3.9)$$

where l is the scale length of the cracking. The sudden motion of the field can drive transient Alfven waves into the magnetosphere (TD95) or induce parallel electric fields (Miller et al. 1992), either of which can accelerate particles.

The radiation from particles trapped on closed field lines is expected to fall in the soft γ-ray range (see Figure 2). TD95 suggested that the radiation would form an optically thick pair plasma at temperatures around 20–30 keV, the peak energies of the observed SGR bursts. However, Baring (1995) pointed out that photon splitting (see Section 2.4) dominates over one-photon pair production in magnetar fields. If enough splitting modes operate at these high fields to allow photons of both polarizations to split rather than produce pairs (which is not certain), then a photon splitting cascade will produce a quasi-thermal spectrum (see Figure 3) peaking around 30 keV (Harding & Baring 1996). In addition, the model spectra from photon splitting cascades when $B \geq 4B_{\rm cr}$ do not vary with emission location in the magnetosphere, if the emission occurs in equatorial regions. This characteristic seems to be born out in the data (see Section 3.1.2) and would favor burst emission in the closed-field region of the neutron star magnetosphere. The particles accelerated on open field lines may produce the harder emission seen in the giant bursts, and would drive a powerful wind.

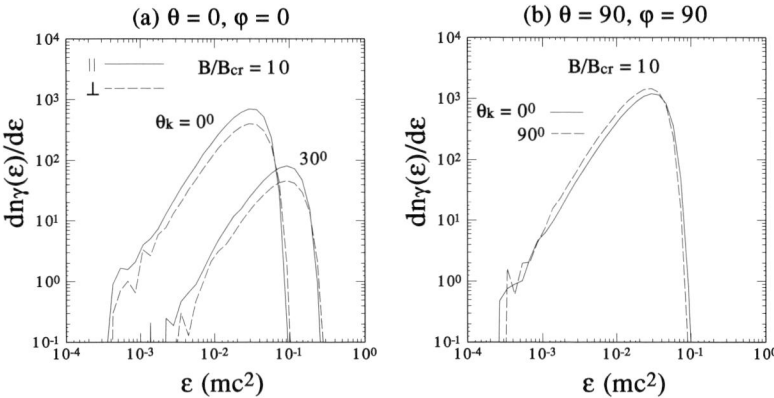

FIGURE 3. Photon splitting cascade spectra in curved space for photons emitted at the (a) pole and (b) equator of a dipole field and for different emission directions θ_k to the dipole axis (from Harding & Baring 1996).

3.2.2. *Spin dynamics*

If one derives surface fields of SGRs on the basis of an assumption of pure magnetic dipole spin-down (see Section 2.2), then magnetar-strength fields result. However, the characteristic ages are much smaller than the estimated ages (Eqn 2.1) of the surrounding supernova remnants. For example, the characteristic age derived for SGR 1806−20 assuming pure dipole spin down is 1500 yr, compared to the SNR age, thought to be about 10^4 yr. Accretion torques have been proposed as the spin-down mechanism for AXPs (Mereghetti et al. 1995, van Paradijs et al. 1995), but there are problems with this hypothesis, especially since the connection to SGRs has been proposed (Thompson & Duncan 1996). The accretion scenario requires a narrow range of periods, not borne out for SGRs, and the derived ages are much longer than the SNR ages. But the powerful particle winds which probably accompany the bursts, and may in fact be necessary to power the surrounding plerionic nebulae (Harding 1995, Frail et al. 1997), will dominate the spin-down torque. The wind will cause the field lines to open up at radius $r_{\rm open}$, inside the light cylinder, $R_{\rm LC}$, and increase the usual dipole torque by a factor $(R_{\rm LC}/r_{\rm open})^2$ (Thompson & Blaes 1998), which gives a rotational energy-loss rate $\dot E_W \propto B_0 \Omega^2 L_p^{1/2}$, where L_p is the wind luminosity, rather than the pure dipole energy-loss rate $\dot E_D \propto B_0^2 \Omega^4$. But a continuous wind of luminosity $L_p = 10^{37}$ erg s^{-1}, required to match the age of the SNR G10.0−0.3 surrounding SGR 1806−20 will decrease the derived neutron star surface magnetic field to $B_0 \simeq 3 \times 10^{13}$ G, well below the magnetar range.

So do we need to give up the magnetar models in order to accommodate a particle wind in SGR sources? Harding et al. (1999) point out that if the wind outflows are not continuous but episodic, following the SGR bursts, then the average power output depends on the wind duty cycle, D_p, as $\langle L_p \rangle = L_p D_p$, whereas the rotational energy loss $\dot E_W \propto L_p^{1/2} D_p$. The surface magnetic field, derived assuming both dipole and episodic wind spin-down torques, is

$$B_{\rm wind} = -\sqrt{6c^3} L_p^{1/2} \frac{D_p P^2}{8\pi^2 R_0^3} F(P, \dot P) \qquad (3.10)$$

where,

$$F(P,\dot{P}) = \left\{1 - \left(1 + \frac{4\dot{E}}{L_p D_p^2}\right)^{1/2}\right\}, \qquad (3.11)$$

and

$$\dot{E} = \frac{4\pi^2 I \langle \dot{P} \rangle}{P^3}. \qquad (3.12)$$

The associated characteristic age becomes

$$\tau_{\text{wind}} \simeq -\frac{4\pi^2 I}{L_p D_p^2 P^2} \frac{\ln\left[1 - 2/F(P,\dot{P})\right]}{F(P,\dot{P})}. \qquad (3.13)$$

where I and R_0 are the moment of inertia and radius of the neutron star. It is then possible, for $D_p < 1$, to have a magnetar-strength field that is consistent with a neutron star age within a factor of two of the SNR age. In this case, one should observe sudden increases in the $\dot{P}$ when SGR bursts occur, followed by a decrease back to the dipole $\dot{P}$ after some time. In fact, Marsden et al. (1999) have observed evidence for a spin-up of SGR 1900+14 following the giant burst of August, 1998. But only frequent monitoring of the SGR periods and period derivatives will be able to answer this question.

4. Magnetars and Gamma-ray Bursts

Finally, in this section I address the question that is perhaps most relevant to this Symposium: could some classical GRBs be magnetars? I will discuss both the issue of the energetics and of the magnetar formation rate.

4.1. Energetics

Although the magnetic energy dominates over rotational energy (see Eqn 2.2) in the SGR sources known in our galaxy and in the LMC, the rotational energy at birth

$$E_R = \frac{1}{2}I\Omega^2 = 2 \times 10^{52} \text{ erg } P_{\text{ms}}^{-2}, \qquad (4.14)$$

where P_{ms} is the period in ms, was much higher and dominated over the magnetic energy. Much of this energy is released in the seconds after the birth of the magnetar. The initial magnetic dipole energy loss rate,

$$\dot{E}_D^{\text{init}} = \frac{B_0^2 R^6 \Omega^4}{6c^3} = 10^{49} \text{ erg s}^{-1} \left(\frac{B_0}{10^{15} \text{ G}}\right)^2 P_{\text{ms}}^{-4} \qquad (4.15)$$

is high enough to release 10^{51} erg in the first 100 s (Usov 1992). However, this assumes that gravitational radiation energy loss is not important in the newborn neutron star. The initial rotational energy given in Eqn (4.14) if much less than the neutron star binding energy

$$E_B \sim 2.6 \times 10^{53} \text{ erg } (M/M_\odot)^2 \qquad (4.16)$$

necessary (and possibly not even sufficient) for the most energetic cosmological GRBs. However, the rotational energy released following the birth of a magnetar could certainly be seen at cosmological distances out to ~ 100 Mpc and might be a less luminous class of GRBs.

The impulsive phase of giant SGR bursts, having energies 10^{45} erg, could be seen outside our local group of galaxies, but only out to distances of only ~ 4 Mpc.

4.2. Rate

Aside from the issue of the energetics, we need to know the rate at which magnetars form to assess the contribution of magnetars to the GRB population. The rate of magnetar formation may be estimated from the number of known magnetars in our galaxy, which is about 10 SGRs and AXPs, and their ages, which are $\leq 10^4$ yr. This gives a rate of about 10^{-3}/yr/galaxy, which is about $sim 10^3$ times the rate of neutron star-neutron star mergers! So, we have a population of objects that may produce emission at cosmological distances, which could contribute to the overall population at some level. It is now well established that classical GRBs have a bimodal duration distribution (Kouvetiotou et al. 1993), with short duration bursts peaking at around 0.2 s and the long duration bursts peaking at around 20–30 s. There is a correlation between duration and hardness, such that the short bursts tend to be harder than the long bursts. Studies of different burst classes based on hardness and duration have found that there are significant differences in their count spectra (Pizzichini 1998, Tavani 1998). Tavani (1998) divides bursts into three classes, long-short, long-hard and short-hard, and finds that the long-soft and the short-hard bursts have count spectra that are consistent with a Euclidean distribution, whereas the long-hard bursts are the only class showing a non-Euclidean distribution. This would imply that the long-hard bursts are the only class of bursts seen at very large distances of $z > 1$ (≥ 1 Gpc) and that the others are more local. It is not at all clear whether magnetar formation makes up any part of these GRB classes, but they could plausibly be contributing to the long-soft class.

REFERENCES

ADLER, S. 1971 *Ann. Phy.*, **67**, 559.

BARING, M. G. 1995 *ApJ*, **440**, L69.

BARING, M. G. & HARDING, A. K. 1998 *ApJ*, **507**, L55.

CHENG, B., ET AL. 1995 *Nature*, **382**, 518.

DUNCAN, R. C. & THOMPSON, C. 1992 *ApJ*, **392**, L9.

FENIMORE, E. E., LAROS, J. G., & ULMER, A. 1994 *ApJ* **432**, 742.

FRAIL, D. A., VASISHT, G. & KULKARNI, S. R. 1997 *ApJ*, **480**, 1129.

GOLDREICH, P. & REISNEGGER, A. 1992 *ApJ*, **395**, 250.

GOTTHELF, E. V. & VASISHT, G. 1998 *New Astronomy*, **3**, 293.

HARDING, A. K. 1995, in High Velocity Neutron Stars and Gamma-Ray Bursts, (ed. R. E. Rothschild & R. E. Lingenfelter), p. 118. AIP Conf. Proc. 366.

HARDING, A. K. & BARING, M. G. 1996 in Proc. 3rd Huntsville Workshop on Gamma-Ray Bursts (ed. C. Kouveliotou, M. Briggs & G. Fishman), p. 941. AIP, New York.

HARDING, A. K., BARING, M. G., & GONTHIER, P. L. 1997 *ApJ*, **476**, 246.

HARDING, A. K., CONTONPOULOS, I., & KAZANAS, D. 1999, *ApJL*, submitted.

HERANT, M., ET AL. 1994 *ApJ*, **435**, 339.

HURLEY, K., ET AL. 1999 *ApJ*, **510**, L110.

HURLEY, K. J., MCBREEN, B., RABBETTE, M., & STEEL, S. 1994 *A&A*, **288**, L49.

ISRAEL, G. L., ET AL. 1999 *ApJ*, **518**, 107.

KATZ, J. 1982, *ApJ*, **260**, 371.

KOUVELIOTOU, C., ET AL. 1993 *ApJ*, **413**, L101.

KOUVELIOTOU, C., ET AL. 1998 *Nature*, **393**, 235.

KOUVELIOTOU, C., ET AL. 1999 *ApJ*, **510**, L115.

MARSDEN, D., ROTHSCHILD, R. E., & LINGENFELTER, R. E. 1999 *ApJ*, **520**, L107.

MAZETS, E. P., ET AL. 1981 *Nature*, **290**, 378.
MAZETS, E. P., ET AL. 1999 *Astronomy Letters*, in press.
MELIA, F. & FATUZZO, M. 1991 *ApJ*, **376**, 673.
MEREGHETTI, S. & STELLA, L. 1995 *ApJ*, **442**, L17.
MILLER, J. C. 1995 *ApJ*, **448**, L29.
MILLER, G. & EPSTEIN, R. I. 1992 in Gamma-Ray Bursts (ed. C. Ho, R. Epstein & E. Fenimore), p. 24. Cambridge Univ. Press.
MURAKAMI, T. 1999, in Proc. of the 4th ASCA Symposium.
OOSTERBROEK, T., PARMAR, A. N., MEREGHETTI, S. & ISRAEL, G. L. 1998 *A&A*, **334**, 925.
PACZYNSKI, B. 1992 *Acta Astron.*, **42**, 145.
PIZZICHINI, G., 1998 *AAS*, **192**, 8505.
TAVANI, M. 1998 **497**, *ApJ*, L21.
THOMPSON, C. & BLAES, O. 1998 *Phys. Rev. D*, **57**, 3219.
THOMPSON, C. & DUNCAN, R. C. 1995 *MNRAS*, **275**, 255 [TD95].
THOMPSON, C. & DUNCAN, R. C. 1996 *ApJ*, **473**, 322.
VASISHT, G. & GOTTHELF, E. V. 1997 *ApJ*, **486**, L129.
WILSON, C. A., ET AL. 1999 *ApJ*, **513**, 464.
WOODS, P. M., ET AL. 1999 *ApJ*, **519**, 139.
VAN PARADIJS, J., TAAM, R. E., & VAN DEN HEUVEL, E. P. J 1995 *A&A*, **299**, L41.
USOV, V. V. 1992 *Nature*, **357**, 472.

Super-luminous supernova remnants

By YOU-HUA CHU, C.-H. ROSIE CHEN
AND SHIH-PING LAI

Astronomy Department, University of Illinois, 1002 W. Green St., Urbana, IL 61801

Some extragalactic SNRs are more than two orders of magnitude more luminous than the young Galactic SNR Cas A. These SNRs are called super-luminous or ultra-luminous SNRs. Their high luminosities can be caused by chance superpositions of multiple objects, interactions with a very dense environment, or unusually powerful supernova explosions. Four super-luminous SNRs are known: one in NGC 4449, one in NGC 6946, and two in M101. The two remnants in M101, NGC 5471B and MF83, are recently suggested to be "hypernova remnants" possibly connected to the GRBs. We have obtained new or archival *HST WFPC2* images and new high-dispersion echelle spectra of these super-luminous SNRs, in order to examine their stellar and interstellar environments and to analyze their energetics. We discuss the physical nature of these four SNRs, with a special emphasis on the two "hypernova remnants" in M101.

1. Introduction

Supernova remnants (SNRs) usually exhibit three distinguishing characteristics: nonthermal radio emission, bright X-ray emission, and large optical [S II]/Hα line ratios. These characteristics have been used as identification criteria for SNRs. In the Galaxy, SNRs are most effectively surveyed at radio wavelengths because heavy extinction in the Galactic plane hampers optical and X-ray observations (e.g. Green 1988). In the Magellanic Clouds, where interstellar extinction is small, SNRs can be surveyed at radio, X-ray, or optical wavelengths (Mathewson et al. 1983, 1984, 1985; Smith 1999). In distant galaxies, the sensitivity of available radio and X-ray instruments becomes a limiting factor; thus, surveys of SNRs are carried out in the optical (e.g. D'Odorico, Dopita, & Benvenuti 1980; Blair & Long 1997; Matonick & Fesen 1997).

Only a small number of the optically identified extragalactic SNR candidates can be confirmed at radio or X-ray wavelengths. These confirmed SNRs are among the most luminous remnants. In fact, some of them are so luminous that they were first identified in radio or X-rays and subsequently confirmed by optical spectroscopic observations. These luminous SNRs are often more than two orders of magnitude more luminous than the young Galactic SNR Cas A, and hence they have been called super-luminous or ultra-luminous SNRs.

Several factors may contribute to the brightness of extragalactic super-luminous SNRs. First, they might be composite objects, with X-ray binaries contributing to the X-ray emission, multiple SNRs contributing to the radio emission, or H II regions contributing to the Hα emission. Second, these SNRs might be interacting with a very dense circumstellar or interstellar environment. As optical and X-ray emissions are dependent on (density)$^2\times$(volume) and the energy is dependent on (density)$\times$(volume), for the same amount of energy, a greater amount of emission can be generated if the density is high. Finally, unusually powerful supernova explosions produce more energetic SNRs which may generate more emission and become super-luminous SNRs. For example, two super-luminous SNRs in M101 have been suggested to be "hypernova remnants" that require explosion energies as large as 10^{53}–10^{54} ergs (Wang 1999).

In this paper, we report our analysis of four super-luminous SNRs. Two of these four remnants are the "hypernova remnants" in M101, and they are the main emphasis of

SNRs	Distance (Mpc)	L(X) (10^{36} erg/s)	L(408 MHz) (Cas A)	L(Hα) (10^{37} erg/s)	Size (pc)
Cas A	0.003	3	1	—	4
NGC 4449	5	800	25	<0.15	<0.6
NGC 6946	5.1	2800	1	19	20×30
NGC 5471B	7.2	170	3	160	60
MF83	7.2	100	—	17	267

TABLE 1. Super-luminous supernova remnants

this paper. We have obtained high-resolution images to examine the nebular morphology and stellar content, and high-dispersion spectra to determine the expansion velocity and kinetic energy. These new observations allow us to investigate the emission mechanisms and whether unusually energetic supernova explosions are needed for these remnants.

2. Super-luminous supernova remnants

Four super-luminous SNRs have been reported: one in NGC 4449 (Kirshner and Blair 1980), one in NGC 6946 (Schlegel 1994; Blair & Fesen 1994; Van Dyk et al. 1994), and two in M101 (Matonick & Fesen 1998; Wang 1999). The X-ray, radio, and optical Hα luminosities of these four remnants and Cas A (for comparison) are listed in Table 1. The numbers in Table 1 are taken from the above cited references and this paper. The radio luminosity at 408 MHz has been normalized to Cas A for easy comparisons.

It is clear from Table 1 that these super-luminous SNRs are most remarkable in the X-ray. The SNR in NGC 4449 is the smallest, the SNR in NGC 6946 has the highest X-ray luminosity, NGC 5471B has the highest Hα luminosity, and MF83 is the largest.

3. Observations and datasets

We intend to determine the physical nature of these super-luminous SNRs and the causes of their high luminosities. Useful optical observations for this study include high-resolution images and high-dispersion spectra. High-resolution images in the continuum bands can be used to study the stellar environment of a remnant, while flux-calibrated Hα images can be used to determine the Hα luminosity and rms electron density of the remnant. The [S II]/Hα ratio map can be used to examine the spatial extent of the SNR shocks. High-dispersion spectra are useful in separating shocked and unshocked components, determining expansion and shock velocities, and measuring diagnostic nebular line ratios.

We have used the following datasets for the analysis presented in this paper:

(1) *HST FOC* [O III] image of the SNR in NGC 4449 (archival);

(2) *HST WFPC2* Hα, [S II], and continuum images of the SNR in NGC 6946 (archival), NGC 5471B (proprietary);

(3) MDM 2.4m B, V, I, and Hα images of MF83 (courtesy of Eva Grebel); and

(4) KPNO 4m echelle spectra of the SNR in NGC 6946, NGC 5471B, and MF83 (proprietary).

We have adopted the analysis of the super-luminous SNR in NGC 4449 by Blair & Fesen (1998), but carried out our own analysis for the other three super-luminous SNRs. The detailed analysis and final results will be reported later in refereed journals.

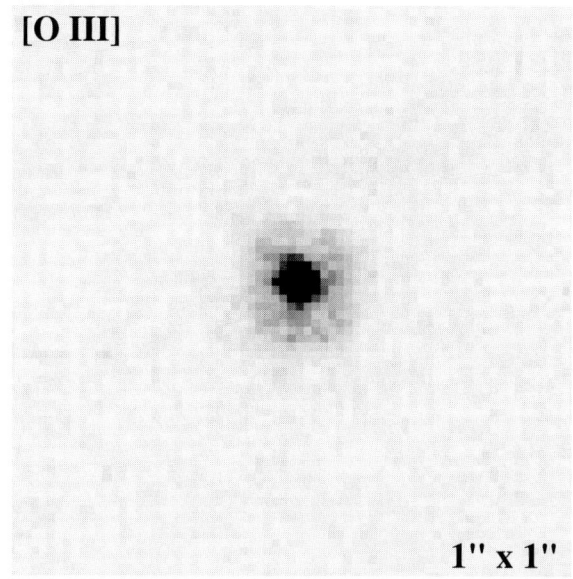

FIGURE 1. *HST FOC* [O III]λ5007 image of the super-luminous SNR in NGC 4449. The image is unresolved and shows basically the point spread function of the *HST*. The field of view is $1'' \times 1''$ ($1'' = 24$ pc).

4. The Super-luminous SNR in NGC 4449

The super-luminous SNR in NGC 4449 ($\alpha_{2000} = 12^{\rm h}28^{\rm m}10\overset{\rm s}{.}9$, $\delta_{2000} = +44°06'47\overset{''}{.}7$) was first diagnosed by its nonthermal radio emission (Seaquist & Bignell 1978). Subsequent spectroscopic observations show narrow emission lines belonging to an H II region and broad emission lines of oxygen that are associated with dense supernova ejecta (Kirshner & Blair 1980). The widths of these oxygen emission lines suggest an expansion velocity of 3,500 km s^{-1}. The absence of broad [O II]λ3727 emission indicates an electron density greater than 10^5 cm^{-3}, and the [O III]λ4363 and λ5007 line ratio implies an electron temperature of $\sim 40,000$ K. These properties strongly point to a case of young SNR dominated by oxygen-rich ejecta of a supernova from a progenitor of mass > 25 M$_\odot$. The expansion of the high-density supernova ejecta into an H II region is responsible for the $\sim 10^{39}$ erg s^{-1} X-ray luminosity (Blair, Kirshner, & Winkler 1983).

The super-luminous SNR in NGC 4449 has been recently studied by Blair & Fesen (1998) with high-resolution images taken with the *HST Faint Object Camera (FOC)*, and with high-S/N spectra taken with the *HST Faint Object Spectrograph* and the 2.4m telescope at the *MDM Observatory*. The *HST FOC* images show that the SNR is unresolved by the point spread function of the *HST* (Figure 1), placing an upper limit of $0\overset{''}{.}028$ (or 0.6 pc) on its diameter. The new spectra show broad wings of many emission lines indicating expansion velocities greater than 6,000 km s^{-1}. Blair and Fesen conclude that the super-luminous SNR in NGC 4449 is less than 100 years old, and that its high luminosity is owed to its youth and expansion into a dense circumstellar environment.

5. The Super-luminous SNR in NGC 6946

The super-luminous SNR in NGC 6946 ($\alpha_{2000} = 20^{\rm h}35^{\rm m}00\overset{\rm s}{.}75$, $\delta_{2000} = +60°11'30\overset{''}{.}6$) was initially identified in an optical survey (Blair & Fesen 1994). It was immediately obvious that this SNR is associated with an extremely luminous X-ray source. With an

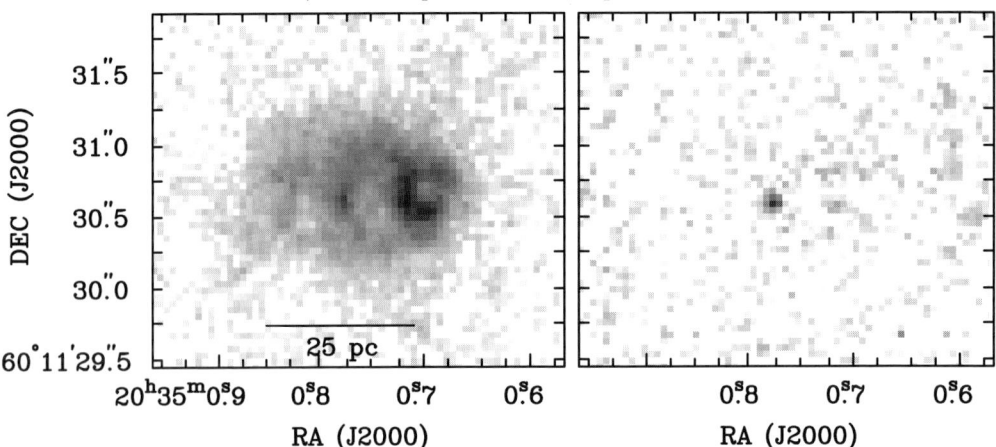

FIGURE 2. *HST WFPC2* images of the super-luminous SNR in NGC 6946 in Hα (left) and F439W blue continuum (right). These images are downloaded from the *HST* archive and plotted by Bryan Dunne.

X-ray luminosity of 2.8×10^{39} ergs s^{-1}, this SNR in NGC 6946 has the highest X-ray luminosity among all known SNRs (Schlegel 1994). Radio observations show that this SNR is three times as luminous as Cas A (Van Dyk et al. 1994). Such high luminosities are often associated with young SNRs; however, no high-velocity gas with expansion velocity greater than 600 km s^{-1} is detected (Blair & Fesen 1994).

HST Wide Field Planetary Camera 2 (*WFPC2*) images of the super-luminous SNR in NGC 6946 were recently reported by Blair, Fesen, & Schlegel (1997). The WFPC2 Hα (F656N filter) image shows that this SNR is in an isolated environment and that the remnant has a multiple-loop morphology with an oblong outermost shell 20×30 pc in size (Figure 2). Based on this morphology, Blair et al. suggested that the high X-ray luminosity is caused by colliding SNRs, with the small, bright loop (~ 8 pc in diameter) on the western part being a young SNR and the outermost shell being an older SNR.

To investigate the kinematic properties of the super-luminous SNR in NGC 6946, we have obtained high-dispersion spectra using the echelle spectrograph on the 4m telescope at Kitt Peak National Observatory (Dunne, Gruendl, & Chu 1999). As shown in Figure 3, the nebular emission lines are resolved into a narrow component superposed on a broad component. The velocity width and [N II]$\lambda 6583$/Hα line ratio can be measured separately for the broad component and the narrow component. These measurements are listed in Table 2.

The broad component must originate in SNR shocked material, while the narrow component consists of unshocked gas. The base of the broad component, the full width at zero intensity (FWZI), of the Hα line extends over 450 km s^{-1}, indicating an expansion velocity of at least 225 km s^{-1}. This expansion velocity is not particularly high compared to those of SNRs in the Large Magellanic Cloud (Chu & Kennicutt 1988), and is usually associated with SNRs that are $\sim 10^4$ yrs old. The lack of material expanding at higher velocities is inconsistent with the identification of the bright, small (8-pc across) loop as a young SNR. The simple stellar environment, shown by the *WFPC2* blue continuum image in Figure 2, also does not support a high supernova occurrence rate. It is thus not likely that the super-luminous SNR in NGC 6946 consists of colliding SNRs.

An effective clue to the nature of the super-luminous SNR in NGC 6946 is provided by the [N II]/Hα ratios measured in the echelle spectrum. The [N II]/Hα ratio is 0.8 in the narrow component and 1.0 in the broad component. An [N II]/Hα ratio of 0.8

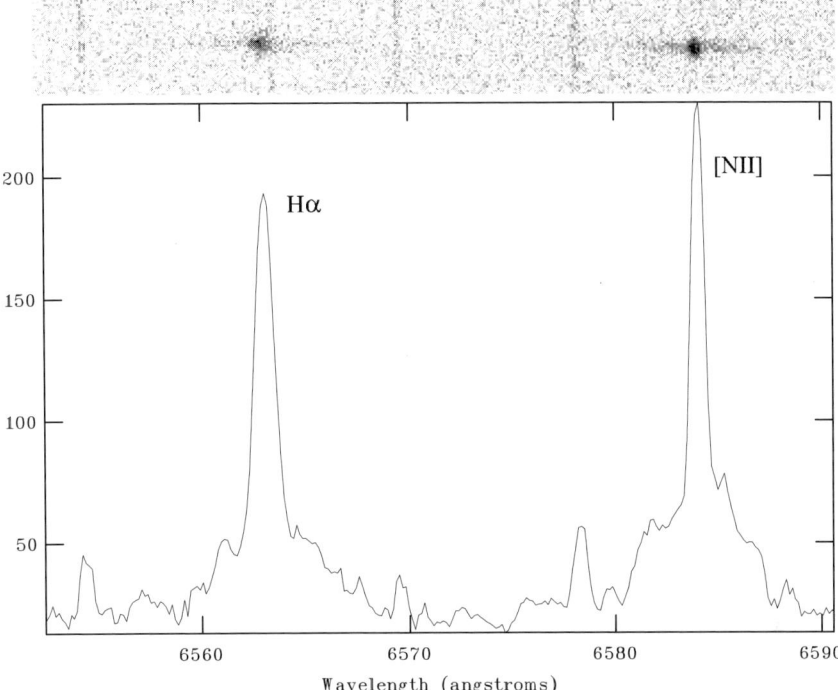

FIGURE 3. KPNO 4m echelle spectra of the super-luminous SNR in NGC 6946 in the region of Hα and [N II] lines. Both lines show a narrow component superposed on a broad component. The weak narrow lines are the geocoronal Hα and telluric OH lines.

is hardly ever seen in an interstellar H II region, but is frequently seen in ring nebulae around Wolf-Rayet (WR) stars or luminous blue variables (LBVs). The [N II] line is strong because the nebulae contain N-rich material ejected by the central stars. It is possible that the progenitor of the supernova in NGC 6946's super-luminous SNR was a WR or LBV star, and the supernova ejecta interact with the dense circumstellar nebula and produce the high luminosity (Dunne et al. 1999).

The Hα emission from the SNR shocked material can be determined from the flux contained in the broad component of the velocity profile. Assuming that the SNR shocked material is distributed in a shell whose thickness is 1% of its radius, Dunne et al. (1999) have derived a rms electron density of ~ 185 cm^{-3}, a mass of $\sim 1,300$ M$_\odot$, and a kinetic energy of $\sim 7 \times 10^{50}$ ergs for the SNR shell. This kinetic energy is somewhat high, but does not need an explosion energy as high as those provided by GRBs.

6. The super-luminous SNR NGC 5471B in M101

NGC 5471B is the B component of the giant H II region NGC 5471 in M101. The super-luminous SNR of NGC 5471B was initially discovered by its nonthermal radio emission and confirmed by its high [S II]/Hα ratio (Skillman 1985). High-dispersion echelle spectroscopic observations clearly show a broad emission line component at NGC 5471B (Figure 4). Using the high-dispersion echelle spectra, Chu & Kennicutt (1986) decomposed the Hα velocity profile into a broad SNR component and a narrow H II region component, and derived a mass of $6,500 \pm 3,000$ M$_\odot$ and a kinetic energy of $(2.5 \pm 1) \times 10^{50}$ ergs for the SNR. ROSAT observations of NGC 5471 show an X-ray source centered at the B component with a luminosity of 3×10^{38} ergs s^{-1} (Williams & Chu 1995; Wang

			Broad Component	Narrow Component
Hα	FWHM	(km s^{-1})	285	42
[N II]	FWHM	(km s^{-1})	250	25
Hα	FWZI	(km s^{-1})	450	...
[N II]	FWZI	(km s^{-1})	400	...
[N II]/Hα flux ratio			1.0	0.8
Hα, broad/narrow flux ratio			1.5	...
[N II], broad/narrow flux ratio			1.8	...

TABLE 2. Hα and [N II] lines of the super-luminous SNR in NGC 6946

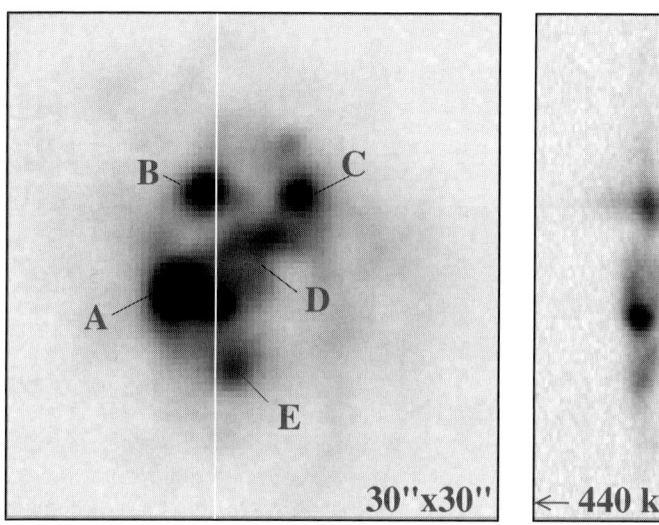

FIGURE 4. *KPNO* 2.1m Hα CCD image of NGC 5471 (left) and *KPNO* 4m echelle Hα line image (right), plotted with the same image scale. The A–E components of NGC 5471 and the echelle slit position are marked on the CCD image. The field of view of this image is 30″ × 30″ (1″ = 35 pc). The echelle Hα line image covers 440 km s^{-1} along the X-axis. The broad emission component is clearly seen at the B component of NGC 5471.

1999). From this X-ray luminosity, Wang (1999) derived a supernova explosion energy of $\sim 3 \times 10^{52}$ ergs, and suggested that NGC 5471B is a "hypernova remnant."

To study the physical nature of NGC 5471B, we have obtained *HST WFPC2* images in Hα (F656N), [S II] (F673N), and continuum bands (F547M and F675W). Comparisons between the Hα and [S II] images of NGC 5471 show three shells with enhanced [S II]/Hα ratios (Figure 5). The brightest of these [S II]-bright shells is the super-luminous SNR in the B component.

A closer examination of NGC 5471B (Figure 6) shows that the super-luminous SNR is embedded in a very complex environment. The Hα image shows roughly a shell structure, with a bright compact H II region at the southeast rim. The blue continuum (F547M) image shows individual supergiants, OB associations, and clusters. The compact H II region on the shell rim is coincident with an OB association/cluster. The [S II] image shows the brightest [S II] emission on the western side of the shell, where the underlying

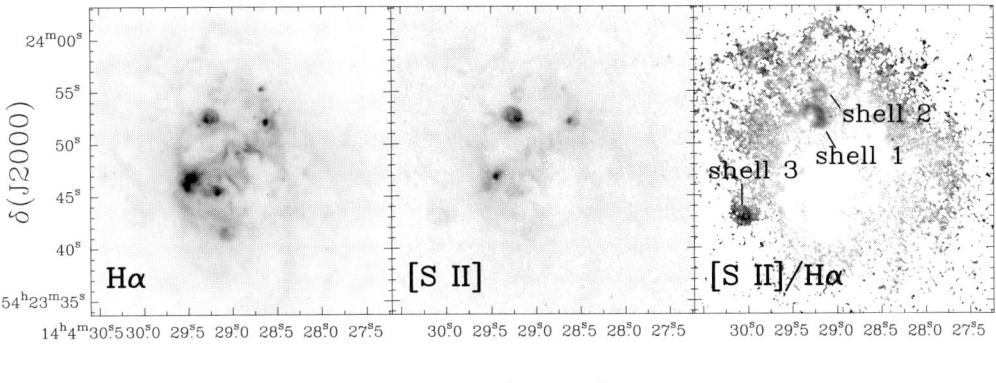

FIGURE 5. *HST WFPC2* images of NGC 5471. The left two images are in Hα and [S II] lines, respectively. The right panel shows the [S II]/Hα ratio map. The three [S II]-enhanced shells are labeled. Shell 1 is the super-luminous SNR in NGC 5471B.

stellar emission is the lowest. The areas of enhanced [S II]/Hα ratio extend beyond the southwest rim of the shell.

To study the physical properties of the super-luminous SNR in NGC 5471B, it is necessary to separate the SNR emission from the background H II region emission. This can be achieved kinematically, using high-dispersion spectra. We have obtained new observations of NGC 5471 with the echelle spectrograph on the *KPNO* 4m telescope in 1999. These new spectra are deeper and have higher S/N that those presented by Chu & Kennicutt (1986). The new echelle data are analyzed similarly. As shown in Figure 7, the Hα velocity profile can be fitted by two Gaussian components, with the broad component corresponding to SNR emission and the narrow component the background H II region emission. The broad component contributes to ∼ 70% of the total flux. Its FWHM is 135 ± 5 km s^{-1}, but the faint wings extends over ± 210 km s^{-1}.

The Hα luminosity of the SNR in NGC 5471B can be determined using the Hα flux derived from the *HST WFPC2* Hα image, the broad/total flux ratio derived from the Hα velocity profile, and the extinction derived spectroscopically by Kennicutt & Garnett (1996). The luminosity of the SNR so derived is 1.6×10^{39} ergs s^{-1}. Assuming that the SNR shell has a shell thickness 1% of the shell radius, we derive a mass of 1.8×10^4 M$_\odot$. Adopting an expansion velocity of 210 km s^{-1} (one half of the full velocity extent), we derive a shell kinetic energy of 8.2×10^{51} ergs. This kinetic energy is more than one order of magnitude higher than those seen in SNRs in the Magellanic Clouds (Williams et al. 1997, 1999). If the SNR in NGC 5471B is indeed a SNR caused by one single supernova explosion, the shell kinetic energy implies that the supernova explosion energy must be 3–30 times higher, depending on whether Sedov's solution or Chevalier's (1974) model is adopted. An explosion energy of 2×10^{52}–2×10^{53} ergs is 1–2 orders of magnitude higher than the canonical explosion energy of 10^{51} ergs (Jones et al. 1998).

A crucial question to ask is whether the energetic, expanding shell in NGC 5471B is indeed energized by a single supernova with an extraordinary explosion energy or a large number of stars over a long period of time. As the shell interior does not have a high concentration of stars and the [S II]-bright part of the shell is particularly devoid of stars, it is quite likely that the shell structure is a SNR, instead of a well-developed superbubble recently energized by an interior supernova. Further discussion is given in Section 8.

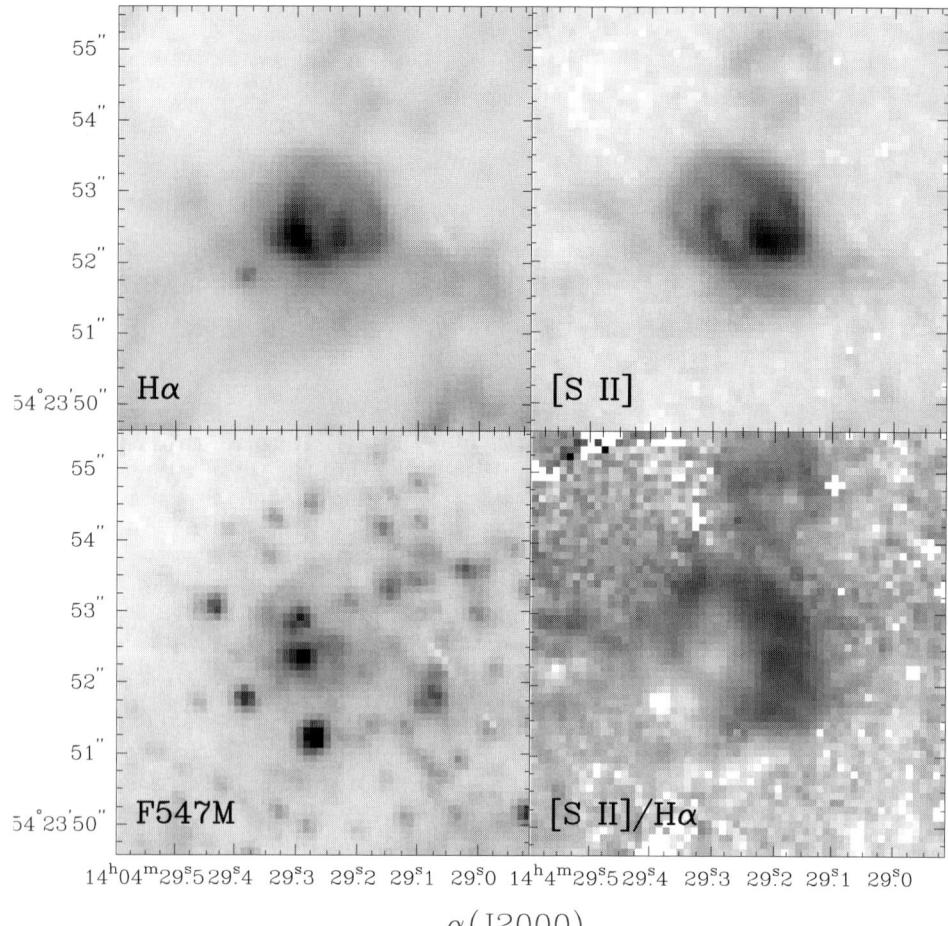

FIGURE 6. *HST WFPC2* images of NGC 5471B in Hα (F656N) and [S I] (F673N) lines and a blue continuum (F547M) band. The lower right panel is the [S I]/Hα ratio map. The Hα image shows the ionized gas, [S I] image the shocked gas, and the F547M image the stars. The brightest concentrations of stars are OB associations or clusters.

7. The Super-luminous SNR MF83 in M101

The super-luminous SNR MF83 was cataloged by Matonick & Fesen (1997). Hα and V-band images of MF83 and surrounding regions have been obtained with the *MDM* 2.4m telescope by Eva Grebel, who has kindly made these images available to us. Figure 8 shows that MF83 is located between two spiral arms on the eastern part of M101. The Hα image in Figure 9 shows that this remnant does not have a clear shell morphology. The bright spots on the south rim of the remnant corresponds to stars, which are better seen in the V-band image in Figure 9.

We have obtained high-dispersion echelle spectra of MF83 with the *KPNO* 4m telescope in 1999. The echelle observations detected MF83 in Hα, [N II], and [S II] lines. The Hα line, having the highest S/N ratio, is displayed in Figure 10. The line image shows clearly an expanding shell structure, with the extreme velocities extending over ~ 145 km s^{-1}. The integrated Hα velocity profile of MF83, after subtracting the telluric OH line, can be fitted with a Gaussian component with a FWHM of 82 ± 14 km s^{-1} (see

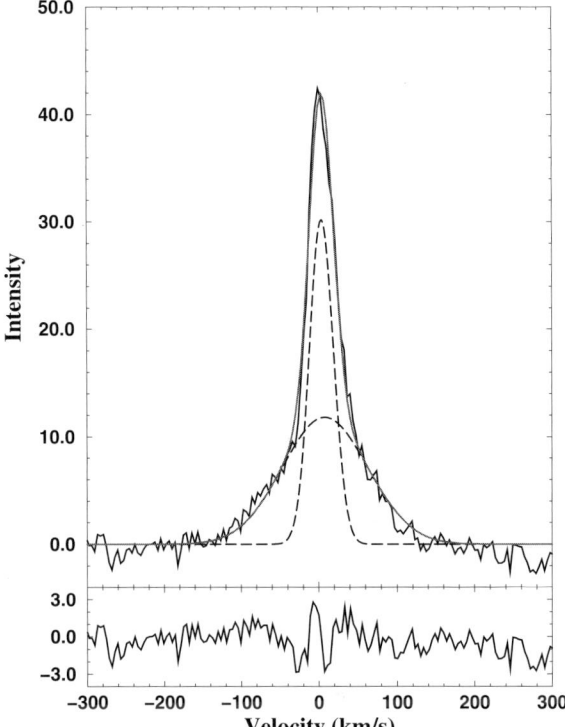

FIGURE 7. Hα line profile of NGC 5471B. The profile is fitted by two Gaussian components. The dashed curves are the two components. The bottom panel plots the residuals.

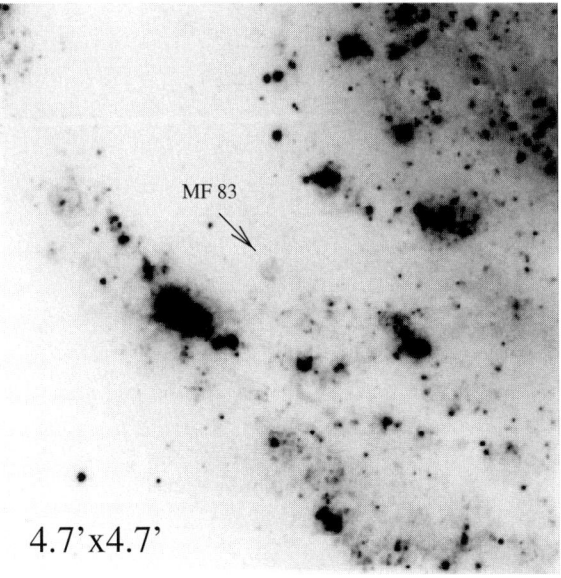

FIGURE 8. *MDM* 2.4m Hα image of MF83 and the eastern portion of M101. (Photo credit: Eva Grebel)

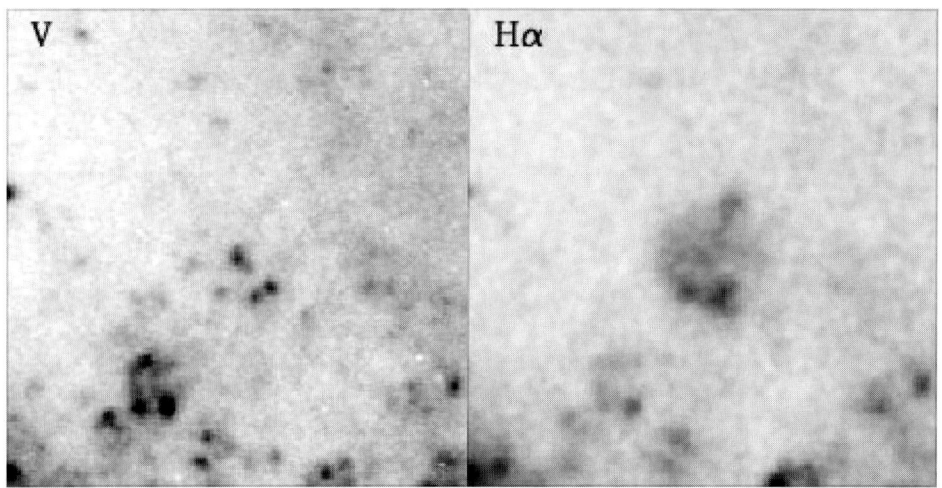

FIGURE 9. *MDM* 2.4m images of MF83 in the V-band and the Hα line. The field of view is 45″ × 45″. (Photo credit: Eva Grebel)

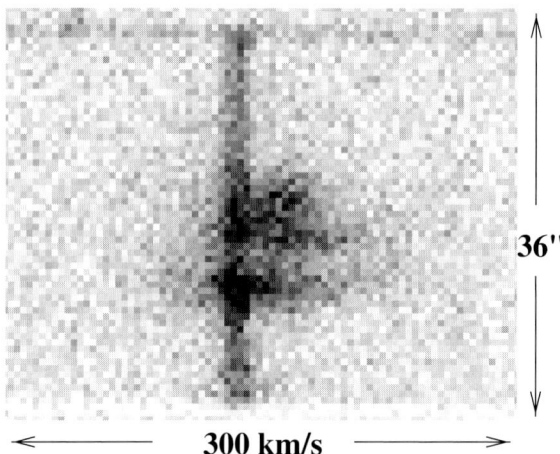

FIGURE 10. *KPNO* 4m echelle image of the Hα line. The emission from MF83 shows a position-velocity ellipse indicating an expanding-shell structure. The narrow, constant component is a telluric OH line.

Figure 11). We adopt the full extent of the Hα profile as twice the expansion velocity. The expansion velocity is thus ~ 70 km s^{-1}.

We adopt the Hα luminosity 1.7×10^{38} ergs s^{-1} from Matonick & Fesen (1997) and a diameter of 267 pc for MF83. Assuming a shell geometry with a shell thickness of 1% the shell radius, we derive a shell mass of 5.5×10^4 M$_\odot$ and a kinetic energy of 3×10^{51} ergs. If MF83 is formed by one single supernova explosion, the explosion energy would need to be 10^{52}–10^{53} ergs. This is 1–2 orders of magnitude higher than the canonical explosion energy for a normal supernova.

Is MF83 formed by one single supernova? The answer is likely to be negative because (1) MF83 has a cluster at its center, (2) MF83's shell size is comparable to those of superbubbles around OB associations or clusters, and (3) MF83's shell expansion velocity is too low to produce X-ray emission. This will be discussed further in the next section.

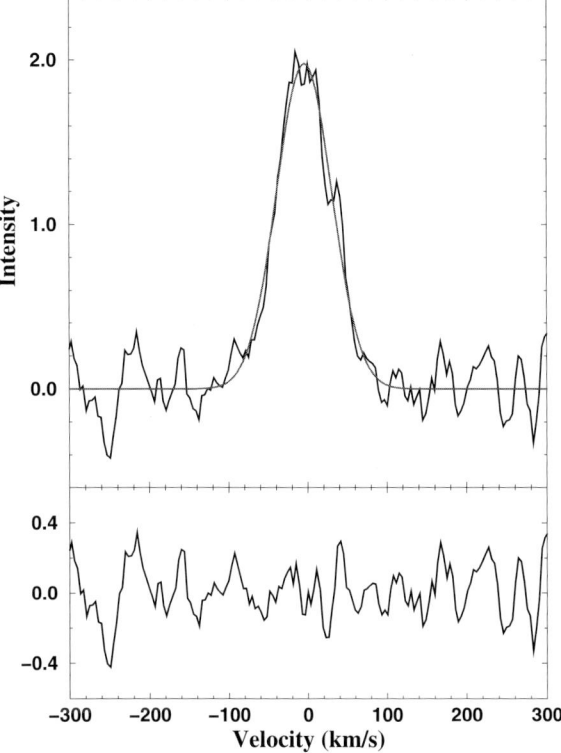

FIGURE 11. Integrated Hα velocity profile of MF83. The superposed telluric OH line has been removed. The smooth curve is the Gaussian fit. The residuals are plotted in the bottom panel.

Parameter	NGC 5471B	MF83	Units
Size	60	270	pc
V$_{exp}$	210	70	km s^{-1}
Mass	1.8×10^4	5.5×10^4	M$_\odot$
L(Hα)	1.6×10^{39}	1.7×10^{38}	ergs s^{-1}
L(X)	1.7×10^{37}	1×10^{37}	ergs s^{-1}
Kinetic Energy	8×10^{51}	3×10^{51}	ergs
Explosion Energy	2×10^{52}–2×10^{53}	10^{52}–10^{53}	ergs
Radio spectrum	nonthermal	?	
X-ray spectrum	soft	?	
OB association	off-center	at center	
Powerful Explosion	very likely	less likely	

TABLE 3. Physical properties of NGC 5471B and MF83

8. Are NGC 5471B and MF83 remnants related to GRBs?

It has been suggested that NGC 5471B and MF83 are hypernova remnants that require explosion energies comparable to the energies frequently associated with GRBs (Wang 1999). We have analyzed these two remnants using *HST WFPC2* images, *MDM* 2.4m CCD images, and *KPNO* 4m echelle spectra. The results of our analysis are summarized in Table 3. While we find that the kinetic energies of these two remnants do require explosion energies of 10^{52}–10^{53} ergs, we also find that they are in complex environments which make it difficult to assess whether they were formed by single supernova explosions.

NGC 5471B has nonthermal radio emission and soft X-ray emission that are characteristic of SNRs. It has an OB association/cluster at its rim, but its shell size is smaller than most known superbubbles. It is unlikely that the NGC 5471B shell is a superbubble produced by this off-center OB association/cluster. There is no obvious counter-evidence for the super-luminous SNR NGC 5471B to be produced by a single explosion. We consider that a single explosion is at least as likely as multiple explosions.

MF83, on the other hand, is not a known nonthermal radio source, and its ROSAT X-ray observations were too noisy to provide spectral information. It has an OB association/cluster at its center and its shell size is comparable to those commonly seen in superbubbles. MF83 was initially identified as a SNR based on its high [S II]/Hα ratio, but many superbubbles are known to have high [S II]/Hα ratios as well, e.g. N185 and N186E in the Large Magellanic Cloud (Lasker 1977). Its expansion velocity is on the high side of the known expansion velocities of superbubbles, but is similar to that of the aforementioned [S II]-bright superbubble N185 (70 ± 10 km s^{-1}, Rosado et al. 1982) and is too low to generate much X-ray emission. We therefore conclude that there is no compelling evidence for MF83 to be produced by one single powerful hypernova explosion.

Finally, are NGC 5471B and MF83 remnants of GRBs? The explosion energies we have derived have large uncertainties; however, the lower end of the energy range, a few times 10^{52} ergs, is compatible with some supernova energies determined recently by Branch (1999, this volume). We conclude that the association of these two super-luminous SNRs with GRBs is at best remote.

REFERENCES

Blair, W. P. & Fesen, R. A. 1994 *ApJ*, **424**, L103.

Blair, W. P. & Fesen, R. A. 1998 *AAS*, **193**, 74.04.

Blair, W. P., Fesen, R. A., & Schlegel, E. M. 1997 *AAS*, **190**, 27.06.

Blair, W. P., Kirshner, R. C., & Winkler, P. F. 1983 *ApJ* **272**, 84.

Blair, W. P. & Long, K. S. 1997 *ApJS*, **108**, 261.

Chevalier, R. A. 1974 *ApJ*, **188**, 501.

Chu, Y.-H. & Kennicutt, R. C., Jr. 1986 *ApJ*, **311**, 85.

Chu, Y.-H. & Kennicutt, R. C., Jr. 1988 *AJ*, **95**, 1111.

D'Odorico, S., Dopita, M. A., & Benvenuti, P. 1980 *A&AS*, **40**, 67.

Dunne, B. C., Gruendl, R. A., & Chu, Y.-H. 1999 *AAS*, **194**, 85.05.

Green, D. A. 1988 *Ap&SS*, **148**, 3.

Jones, T. W., et al. 1998 *PASP* **110**, 125.

Kennicutt, R. C., Jr. & Garnett, D. R. 1996 *ApJ*, **456**, 504.

Kirshner, R. C. & Blair, W. P. 1980 *ApJ*, **236**, 135.

Lasker, B. M. 1977 *ApJ*, **212**, 390.

Mathewson, D. S., Ford, V. L., Dopita, M. A., Tuohy, I. R., Long, K. S., & Helfand, D. J. 1983 *ApJS*, **51**, 345.

Mathewson, D. S., Ford, V. L., Dopita, M. A., Tuohy, I. R., Mills, B. Y., & Turtle, A. J. 1984 *ApJS*, **55**, 189.

Mathewson, D. S., Ford, V. L., Tuohy, I. R., Mills, B. Y., & Turtle, A. J. 1985 *ApJS*, **58**, 197.

Matonick, D. M. & Fesen, R. 1997 *ApJS*, **112** 49.

ROSADO, M., GEORGELIN, Y. M., GEORGELIN, Y. P., LAVAL, A., & MONNET, G. 1982 *A&A*, **115**, 61.

SCHLEGEL, E. M. 1994 *ApJ*, **424**, L99.

SEAQUIST, E. R. & BIGNELL, R. C. 1978 *ApJ*, **226**, L5.

SKILLMAN, E. D. 1985 *ApJ*, **290**, 449.

SMITH, R. C., ET AL. 1999 In *New Views of the Magellanic Clouds* (ed. Y.-H. Chu et al.) IAU Symposium No. 190.

VAN DYK, S. D., SRAMEK, R. A., WEILER, K. W., HYMAN, S. D., & VIRDEN, R. E. 1994 *ApJ*, **425**, L77.

WANG, Q. D. 1999 *ApJ*, **517**, L27.

WILLIAMS, R. M. & CHU, Y.-H. 1995 *ApJ*, **439**, 132.

WILLIAMS, R. M., CHU, Y.-H., DICKEL, J. R., BEYER, R., PETRE, R., SMITH, R. C., & MILNE, D. K. 1997 *ApJ*, **480**, 618.

WILLIAMS, R. M., CHU, Y.-H., DICKEL, J. R., SMITH, R. C., MILNE, D. K., & WINKLER, P. F. 1999 *ApJ*, **514**, 798.

The properties of hypernovae: SNe Ic 1998bw, 1997ef, and SN IIn 1997cy

By K. NOMOTO,[1,2] P. A. MAZZALI,[2,3] T. NAKAMURA,[1]
K. IWAMOTO,[4] K. MAEDA,[1] T. SUZUKI,[1,2]
M. TURATTO,[5] I. J. DANZIGER,[3] AND F. PATAT[6]

[1]Department of Astronomy, School of Science, University of Tokyo, Tokyo, Japan

[2]Research Center for the Early Universe, School of Science, University of Tokyo, Tokyo, Japan

[3]Osservatorio Astronomico di Trieste, via G. B. Tiepolo, Trieste, Italy

[4]Department of Physics, College of Science and Technology, Nihon University, Tokyo, Japan

[5]Osservatorio Astronomico di Padova, vicolo dell'Osservatorio, Padova, Italy

[6]European Southern Observatory, Garching, Germany

We discuss the properties of the hyper-energetic Type Ic supernovae (SNe Ic) 1998bw and 1997ef and Type IIn supernova (SN IIn) 1997cy. SNe Ic 1998bw and 1997ef are characterized by their large luminosity and the very broad spectral features. Their observed properties can be explained if they are very energetic SN explosions with the kinetic energy of $E_K \gtrsim 1 \times 10^{52}$ erg, originating probably from the core collapse of the bare C+O cores of massive stars ($\sim$ 30–40 M$_\odot$). At late times, both the light curves and the spectra suggest that the explosions may have been asymmetric; this may help us understand the claimed connection with GRBs. The Type IIn SN 1997cy is even more luminous than SN 1998bw and the light curve declines more slowly than ^{56}Co decay. We model such a light curve with circumstellar interaction, which requires the explosion energy of $\sim 5 \times 10^{52}$ erg. Because these kinetic energies of explosion are much larger than in normal core-collapse SNe, we call objects like these SNe "hypernovae." The mass of ^{56}Ni in SN 1998bw is estimated to be as large as 0.5–0.7 M$_\odot$ from both the maximum brightness and late time emission spectra, which suggests that the asymmetry may not be extreme.

1. Introduction

Recently, there have been an increasing number of candidates for the gamma-ray burst (GRB)/supernova (SN) connection (Woosley 1993; Paczyński 1998). The first example of such a candidate was provided by SN 1998bw. SN 1998bw was discovered in the error box of GRB980425 (Kulkarni et al. 1998), only 0.9 days after the date of the gamma-ray burst and was very possibly linked to it (Galama et al. 1998).

Early spectra were rather blue and featureless, showing some similarities with the spectra of Type Ic SNe (SNe Ic), but with one major difference (Figs. 1, 2): the absorption lines were so broad in SN 1998bw that they blended together, giving rise to broad absorption trough separated by apparent 'emission peaks' (Iwamoto et al. 1998; Patat et al. 2000; Stathakis et al. 2000). This supernova was immediately recognized to be very powerful and bright (Figs. 3, 4).

Velocities in the Si II 6355 Å line are as high as 30,000 km s^{-1}. Also, the SN was very bright for a SN Ic: the observed peak luminosity, $L \sim 1.4 \times 10^{43}$ erg s^{-1}, is almost ten times higher than that of previously known SNe Ib/Ic. Models which described the SN as the energetic explosion of a C+O core of an initially massive star could successfully fit the first 60 days of the light curve (Iwamoto et al. 1998; hereafter IMN98).

The very broad spectral features and the light curve shape have led to the conclusion that SN 1998bw had an extremely large *kinetic* energy of explosion, $E_K \sim 3 \times 10^{52}$ ergs (IMN98; Woosley, Eastman, & Schmidt 1999). This is more than one order of magnitude

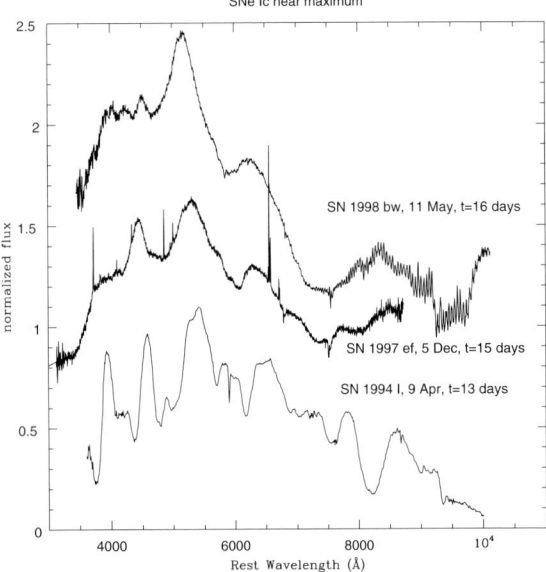

FIGURE 1. Observed spectra of Type Ic supernovae 1998bw, 1997ef, and 1994I.

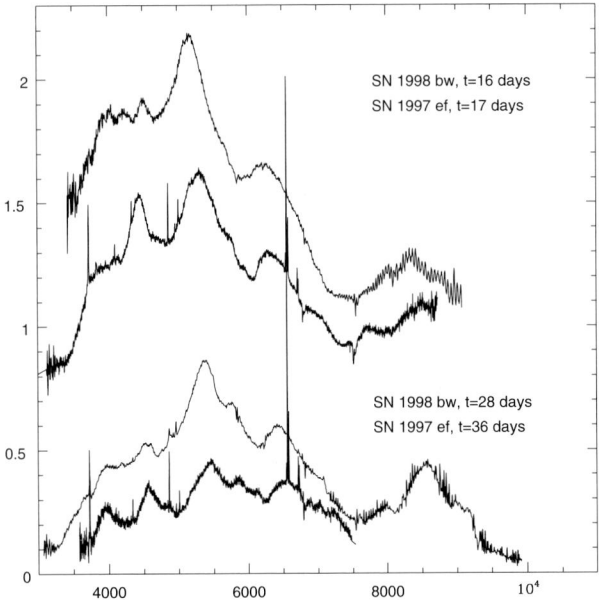

FIGURE 2. Observed spectra of Type Ic supernovae 1998bw and 1997ef.

larger than the energy of typical supernovae, thus SN 1998bw was termed a "hypernova" (IMN98). "Hypernova" is a term we use to describe the events of $E_K \gtrsim 10^{52}$ erg without specifying whether the central engine is a collapsar or magnetar or pair-instability.

SN 1997ef was also noticed for its unique light curve and spectra (Figs. 1–4). At early times, the spectra were dominated by broad oxygen and iron absorption lines, but did not show any clear feature of hydrogen or helium (Garnavich et al. 1997a; Hu et al. 1997), which led us to classify SN 1997ef as a SN Ic. The most striking and peculiar

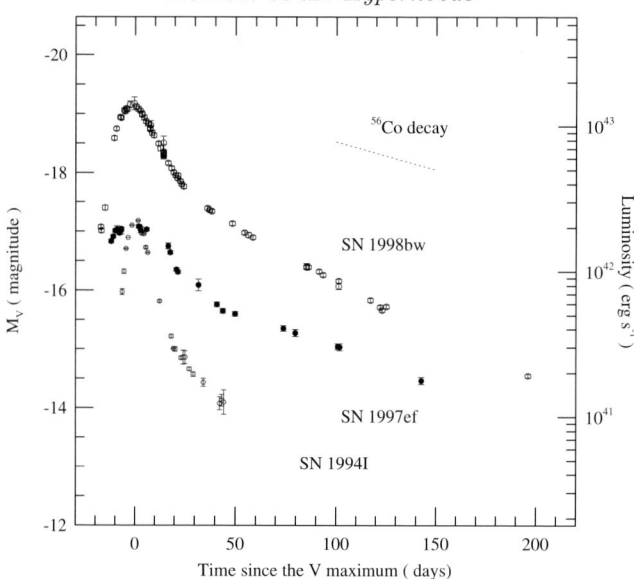

FIGURE 3. Absolute magnitudes of Type Ic supernovae: the ordinary SN Ic 1994I (Richmond et al. 1996a, b), and the hypernovae SN 1998bw (Galama et al. 1998) and SN 1997ef (Iwamoto et al. 2000). The dashed line indicates the ^{56}Co decay rate.

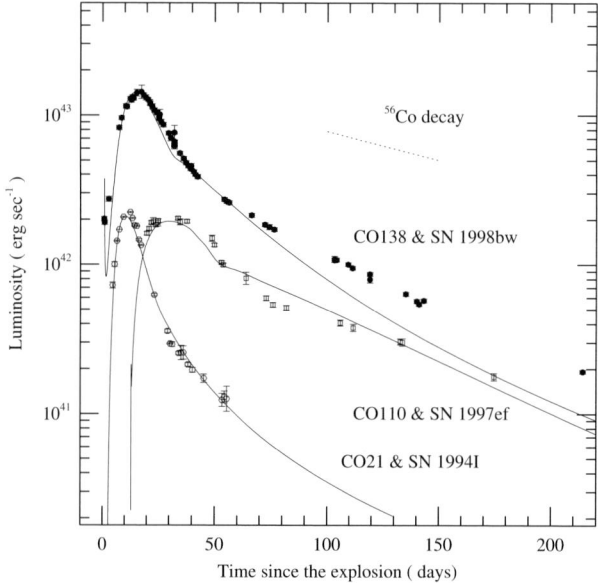

FIGURE 4. Same as Figure 3 but with theoretical models.

characteristic of SN 1997ef is the width of its line features. Such broad spectral features were later recognized to be a distinguishing property of the spectra in SN 1998bw (Figs. 1, 2). The spectral similarities between SN 1997ef and SN 1998bw suggest that SN 1997ef may also be a hypernova (Iwamoto et al. 2000; Nomoto et al. 1999).

In Figure 3 the visual light curve of SNe 1998bw (Galama et al. 1998) and 1997ef (Garnavich et al. 1997b, c) are compared with the ordinary SN Ic 1994I (Richmond et al. 1996a, b). Despite the spectral similarity, the light curve of SN 1997ef is quite different from those of SN 1998bw and SN 1994I. It has a flat peak, much broader than those of the other SNe Ic. Besides, the tail of the light curve of SN 1997ef starts late and the rate of its decline is much slower than in other SNe Ic. Since the light curves are rather diverse, even in this limited number of samples, a range of energies and/or progenitor masses of SN Ic explosions may be implied.

Among the SNe with a possible GRB counterpart, we show here that the Type IIn SN 1997cy (Germany et al. 1999; Turatto et al. 1999ab) is also characterized by an extremely large kinetic explosion energy, $E_K \gtrsim 10^{52}$ erg, thus belonging to "hypernovae." Also, SN IIn 1999E has a spectrum very similar to that of SN 1997cy (Cappellaro et al. 1999), and so it is probably a similar object.

Despite the success of our 'hypernova' model in reproducing the early light curves of SN 1998bw and SN 1997ef, the model light curve tail declines more rapidly than the observations for both SNe. Another unexpected feature appeared in the late phase spectra of SN1998bw. Measuring the velocity of each element (Patat et al. 2000), we note that iron expands faster than oxygen, which is contrary to expectations. This may indicate asymmetry in the ejecta.

Here we summarize the photometric and spectroscopic properties of these hypernovae and the estimated explosion energies and ejecta mass using the hydrodynamical models. We also study nucleosynthesis in hypernovae in 1D and 2D.

2. Explosion models for supernovae and hypernovae

We construct hydrodynamical models of an ordinary SN Ic and a hypernova as follows. Since the light curve of SN Ic 1994I was successfully reproduced by the collapse-induced explosion of C+O stars (Nomoto et al. 1994; Iwamoto et al. 1994), we adopted C+O stars as progenitor models for SNe 1997ef and 1998bw as well. We calculate the light curves and spectra for various C+O star models with different values of E_K and M_{ej}. These parameters can be constrained by comparing the calculated light curves, the synthetic spectra, and the photospheric velocities with the data of SNe 1998bw and 1997ef.

(1) In the ordinary SN Ic model (model CO60), a C+O star with a mass $M_{CO} = 6.0$ $M_\odot$ (which is the core of a 25 $M_\odot$ main-sequence star; Nomoto & Hashimoto 1988) explodes with kinetic energy of explosion $E_K = 1.0 \times 10^{51}$ ergs and ejecta mass $M_{ej} = M_{CO} - M_{cut} = 4.6$ $M_\odot$. Here M_{cut} (= 1.4 $M_\odot$) denotes the mass of the compact star remnant (either a neutron star or a black hole).

(2) In the hypernova model (CO100), a C+O star of $M_{CO} = 10$ $M_\odot$ is constructed from a 10 $M_\odot$ He star (which has a 8 $M_\odot$ C+O core) by removing the outermost 2 $M_\odot$ of He layer and extending the C+O layer up to 10 $M_\odot$. This model corresponds to 30–35 $M_\odot$ on the main-sequence. The star explodes with $E_K = 8.0 \times 10^{51}$ ergs, ejecting $M_{ej} = 7.6$ $M_\odot$, i.e., $M_{cut} = 2.4$ $M_\odot$.

(3) For the hypernova models CO138H and CO138L, C+O stars of $M_{CO} = 13.8$ $M_\odot$ are constructed from a 16 $M_\odot$ He star. This model corresponds to ~ 40 $M_\odot$ on the main-sequence. These models are exploded with $E_K = 6 \times 10^{52}$ erg (CO138H) and $E_K = 3 \times 10^{52}$ erg (CO138L) and $M_{ej} \simeq 11 M_\odot$, i.e., $M_{cut} \simeq 3$–4 $M_\odot$.

These model parameters are summarized in Table 1, together with model CO21 for SN 1994I. The position of the mass cut is chosen so that the ejected mass of ^{56}Ni is the value required to explain the observed peak brightness of SN 1997ef and SN 1998bw by radioactive decay heating. The compact remnant in CO60 is probably a neutron star because $M_{cut} = 1.4$ $M_\odot$, while it may be a black hole in CO100 and CO138 because M_{cut} may well exceed the maximum mass of a stable neutron star.

model	$M_{\rm ms}(M_\odot)$	$M_{\rm C+O}$	$M_{\rm ej}$	^{56}Ni mass	$M_{\rm cut}$	$E_{\rm K}$ (10^{51} erg)	SN
CO21	~ 15	2.1	0.9	0.07	1.2	1	1994I
CO60	~ 25	6.0	4.4	0.15	1.4	1	
CO100	$\sim 30\text{--}35$	10.0	7.6	0.15	2.4	8	1997ef
CO138H	~ 40	13.8	10	0.5	4	60	1998bw
CO138L	~ 40	13.8	11	0.5	3	30	1998bw

TABLE 1. Parameters of the C+O star models

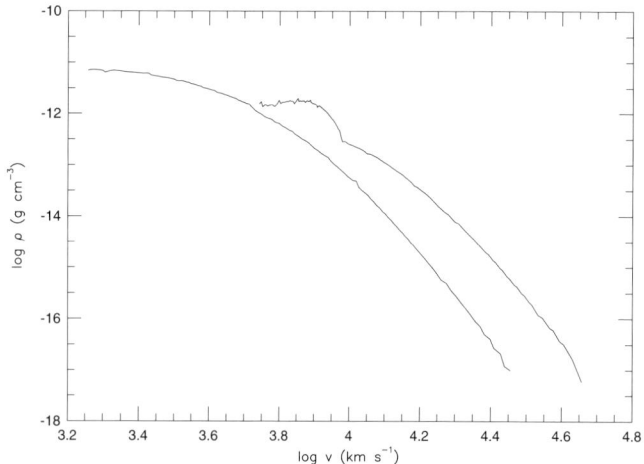

FIGURE 5. Density distributions against the velocity of homologously expanding ejecta for CO60 and CO100.

The hydrodynamics at early phases was calculated by using a Lagrangian PPM code (Colella & Woodward 1984). The explosion is triggered by depositing thermal energy in a couple of zones just below the mass cut so that the final kinetic energy has the required value. The explosive nucleosynthesis are discussed in §7.

The expansion soon becomes homologous so that $v \propto r$. The solid lines in Figure 5 show the density distributions in velocity space for CO60 and CO100 at $t = 16$ days. The expansion velocities are clearly higher in CO100 than in CO60. We performed detailed radiation transfer calculations to obtain light curves and spectra for the explosion models. The results were compared with observations of SNe 1998bw and 1997ef, to derive explosion energies and the ejecta masses, and thus to determine whether the SNe were ordinary SNe Ic or hypernovae.

3. SN 1997ef

3.1. Light Curve Models

In Figure 6 we compare the calculated V light curves for models CO60 and CO100 with the observed V light curve of SN 1997ef. We adopt a distance of 52.3 Mpc (a distance modulus of $\mu = 33.6$ mag) as estimated from the recession velocity, 3,400 km s^{-1}

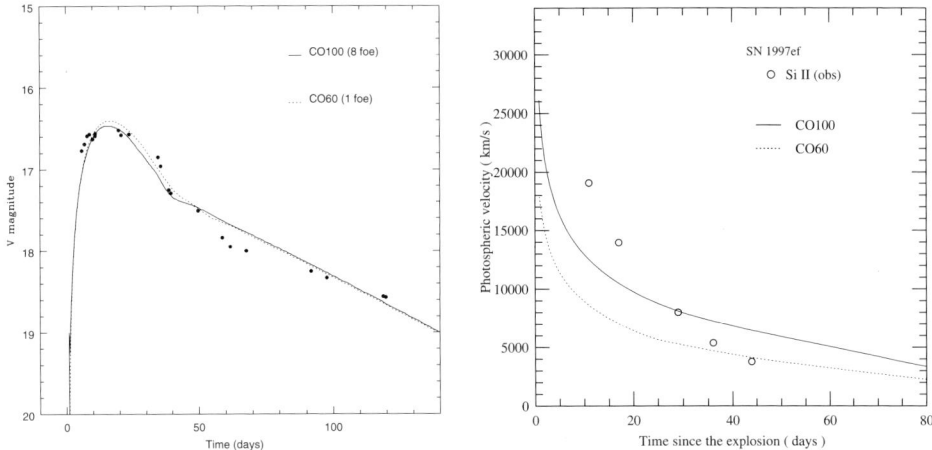

FIGURE 6. Left panel: Calculated Visual light curves of CO60 and CO100 compared with that of SN 1997ef. Right panel: Evolution of the calculated photospheric velocities of CO60 and CO100 (solid lines) compared with the observed velocities of the Si II 634.7, 637.1 nm line measured in the spectra at the absorption core.

(Garnavich et al. 1997a) and a Hubble constant $H_0 = 65$ km s^{-1} Mpc^{-1}. We assume no color excess, $E(B-V) = 0.00$; this is justified by the fact that no signature of a narrow Na I D interstellar absorption line is visible in the spectra of SN 1997ef at any epochs (Garnavich et al. 1997a). The light curve of SN 1997ef has a very broad maximum, which lasts for ~ 25 days. The light curve tail starts only ~ 40 days after maximum, much later than in other SNe Ic.

The light curve of SN 1997ef can be reproduced basically with various explosion models with different energies and masses. In general, the properties of the light curve are characterized by the decline rate in the tail and the peak width, τ_{peak}. The peak width scales approximately as

$$\tau_{\text{peak}} \propto \kappa^{1/2} M_{\text{ej}}^{3/4} E_{\text{K}}^{-1/4}, \qquad (3.1)$$

where κ denotes the optical opacity (Arnett 1996). This is the time-scale on which photon diffusion and hydrodynamical expansion become comparable. Since the model parameters of CO100 and CO60 give similar τ_{peak}, the light curves of the two models look similar: both have quite a broad peak and reproduce the light curve of SN 1997ef reasonably well (Figure 6).

The light curve shape depends also on the distribution of ^{56}Ni, which is produced in the deepest layers of the ejecta. More extensive mixing of ^{56}Ni leads to an earlier rise of the light curve. For SN 1997ef, the best fit is obtained when ^{56}Ni is mixed almost uniformly to the surface for both models. Without such extensive mixing, the rise time to V = 16.5 mag would be ~ 30 d for CO100, which is clearly too long to be compatible with the spectroscopic dating.

Model CO60 has the same kinetic energy ($E_{\text{K}} = 1 \times 10^{51}$ erg) as model CO21, which was used for SN Ic 1994I (see Table 1 for the model parameters). Since the light curve of SN 1997ef is much slower than that of SN 1994I, the ejecta mass of CO60 is ~ 5 times larger than that of CO21.

The ejecta mass of CO100 is a factor of ~ 2 larger than that of CO60, and it is only $\sim 20\%$ smaller than that of model CO138, which was used for SN 1998bw (Table 1). Thus the explosion energy of CO100 should be ~ 8 times larger than that of CO60 to reproduce the light curve of SN 1997ef. This explosion is very energetic, but still much weaker than the one in CO138. The smaller E_K for a comparable mass allows CO100 to reproduce the light curve of SN 1997ef, which has a much broader peak than that of SN 1998bw.

The light curve of SN 1997ef enters the tail around day 40. Since then, the observed V magnitude declines linearly with time at a rate of $\sim 1.1 \times 10^{-2}$ mag day^{-1}, which is slower than in other SNe Ic and is close to the ^{56}Co decay rate 9.6×10^{-3} mag day^{-1}. Such a slow decline implies much more efficient γ-ray trapping in the ejecta of SN 1997ef than in SN 1994I. The ejecta of both CO100 and CO60 are fairly massive and are able to trap a large fraction of the γ-rays, so that the calculated light curves have slower tails compared with CO21.

However, the light curves for both models decline somewhat faster in the tail than the observations. A similar discrepancy has been noted for the Type Ib supernovae (SNe Ib) 1984L and 1985F (Swartz & Wheeler 1991; Baron et al. 1993). The late time light curve decline of these SNe Ib is as slow as the ^{56}Co decay rate, so that the inferred value of M is significantly larger (and/or E_K is smaller) than those obtained by fitting the early light curve shape. Baron et al. (1993) suggested that the ejecta of these SNe Ib must be highly energetic and as massive as ~ 50 M$_\odot$. In §4.1, we will suggest that such a discrepancy between the early- and late-time light curves might be an indication of asphericity in the ejecta of SN 1997ef and that it might be the case in those SNe Ib as well.

3.2. Photospheric velocities

As we have shown, light curve modeling provides direct constraints on M_{CO} and E_K. However, it is difficult to distinguish between the ordinary SN Ic and the hypernova model from the light curve shape alone, since models with different values of M_{ej} and E_K can reproduce similar light curves. However, these models are expected to show different evolutions of the photospheric velocity and the spectrum as will be discussed in the following sections.

The photospheric velocity scales roughly as $v_{ph} \propto M_{ej}^{-1/2} E_K^{1/2}$, so that M_{ej} and E_K can be constrained by v_{ph} in a different way from by means of the light curve width. Figure 6 (right) shows the evolution of the observed velocities of the Si II line measured in the spectra at the absorption core, and the velocities at the grey photosphere computed by the light curve code for models CO60 and CO100. The velocities of the Si II line are somewhat higher than that of the photosphere, reaching $\sim 20,000$ km s^{-1} at the earliest times.

In model CO60 the photosphere forms at velocities much smaller than those of the observed lines, while CO100 gives photospheric velocities as high as the observed ones. It is clear, from this comparison, that the hyper-energetic model CO100 is preferable to the ordinary model CO60. The apparent discrepancy that still exists between the model CO100 and observations might be related to the morphology of the ejecta, i.e., a deviation from spherical symmetry, as was also suggested in the case of SN 1998bw (Höflich et al. 1999; IMN98).

3.3. Synthetic Spectra

To strengthen the arguments in §3.2, we compare the emergent spectra for the two explosion models. Using detailed spectrum synthesis, we can distinguish between different models more clearly, because the spectrum contains much more information than a single-band light curve.

Around maximum light, the spectra of SN 1997ef show just a few very broad features, and are quite different from those of ordinary SNe Ib/c, but similar to SN 1998bw. However, at later epochs the spectra develop features that are easy to identify, such as

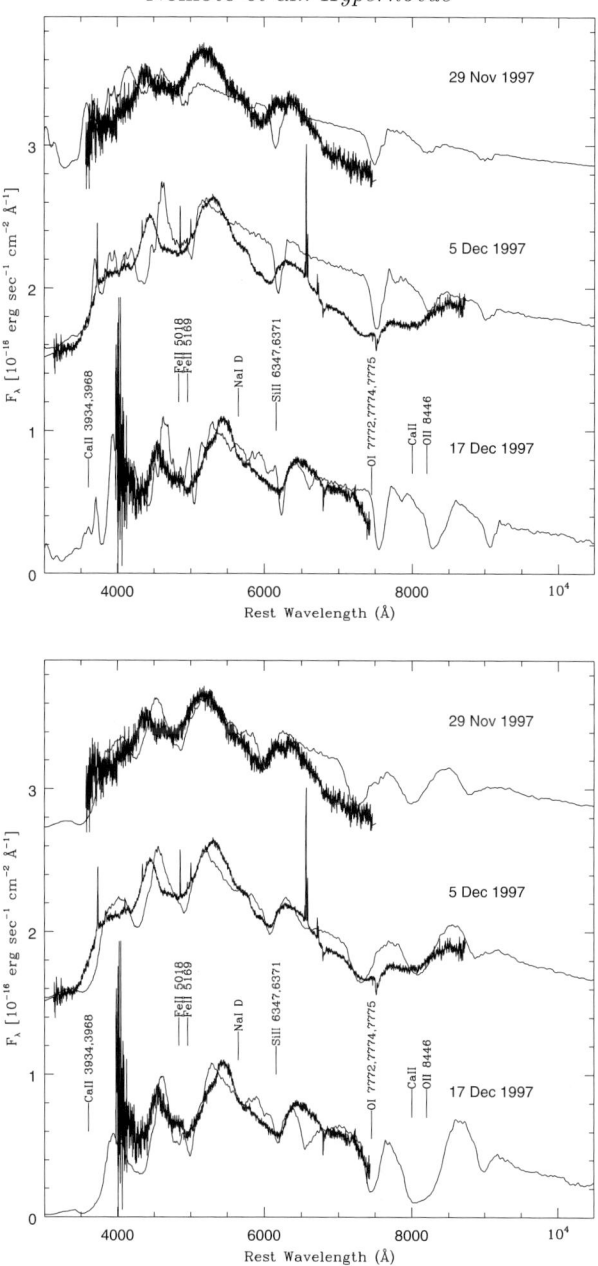

FIGURE 7. Upper panel: Observed spectra of SN 1997ef (bold lines) and synthetic spectra computed using model CO60. The lines in the synthetic spectra are much too narrow. Lower panel: Observed spectra of SN 1997ef (bold lines) and synthetic spectra computed using model CO100 (fully drawn lines).

the Ca II IR triplet at ~ 8200 Å, the O I absorption at 7500 Å, several Fe II features in the blue, and they look very similar to the spectrum of the ordinary SN Ic 1994I.

We computed synthetic spectra with a Monte Carlo spectrum synthesis code using the density structure and composition of the hydrodynamic models CO60 and CO100. The code is based on the pure scattering code described by Mazzali & Lucy (1993), but has been improved to include photon branching, so that the reprocessing of the radiation from the blue to the red is followed more accurately and efficiently (Lucy 1999; Mazzali 2000).

We produced synthetic spectra for three epochs near maximum, of SN 1997ef: Nov 29, Dec 5, and Dec 17. These are early enough that the spectra are very sensitive to changes in the kinetic energy. As in the light curve comparison, we adopted a distance modulus of $\mu = 33.6$ mag, and $E(B-V) = 0.0$.

In Figure 7 (above) we show the synthetic spectra computed with the ordinary SN Ic model CO60. The lines in the spectra computed with this model are always much narrower than the observations. This clearly indicates a lack of material at high velocity in model CO60, and suggests that the kinetic energy of this model is much too small.

Synthetic spectra obtained with the hypernova model CO100 for the same 3 epochs are shown in Figure 7 (below). The spectra show much broader lines, and are in good agreement with the observations. In particular, the blending of the Fe lines in the blue, giving rise to broad absorption troughs, is well reproduced, and so is the very broad Ca-O feature in the red. The two 'emission peaks' observed at ~ 4400 and 5200 Å correspond to the only two regions in the blue that are relatively line-free.

The spectra are characterized by a low temperature, even near maximum, because the rapid expansion combined with the relatively low luminosity (from the tail of the light curve we deduce that SN 1997ef produced about 0.15 $M_\odot$ of ^{56}Ni, compared to about 0.6 $M_\odot$ in a typical SN Ia and 0.5 $M_\odot$ in SN 1998bw) leads to rapid cooling. Thus the Si II 6355 Å line is not very strong.

Although model CO100 yields rather good synthetic spectra, it still fails to reproduce the observed large width of the O I–Ca II feature in the only near-maximum spectrum that extends sufficiently far to the red (5 Dec 1997). An improvement can be obtained by introducing an arbitrary flattening of the density profile at the highest velocities.

3.4. Possible Aspherical Effects

We have shown that the light curve, the photospheric velocities, and the spectra of SN 1997ef are better reproduced with the hyper-energetic model CO100 than with the ordinary SN Ic model CO60. However, there remain several features that are still difficult to explain with model CO100.

(1) The observed velocity of Si II decreases much more rapidly than models predict. It is as high as $\sim 30,000$ km s^{-1} at the earliest phase, but it gets as low as $\sim 3,000$ km s^{-1} around day 50 (Figure 6, right). We find that it is difficult to get such a rapid drop of the photospheric velocity not only in models CO100 and CO60, but also in other models that can reproduce the light-curve shape reasonably well. Models with higher energies and/or smaller masses would be able to reproduce the fast evolution of the photospheric velocity, but such models would inevitably produce light curves with a narrower peak and a faster tail.

(2) Obviously, the observed light curve decline is slower than model CO100 in the tail part, and it is also a bit flatter than the model around maximum (Figure 6). Models with lower energies and/or larger masses give better fits to both the peak and the tail of the light curve, but then it gets very difficult to reproduce the large photospheric velocities observed at early times in SN 1997ef.

This dilemma might be overcome if we introduce multiple components of the light curve from different parts of ejecta moving at different velocities. In fact, the discrepancies may be interpreted as a possible sign of asphericity in the ejecta: A part of the ejecta moves faster than average to form the lines at high-velocities at early phases, while the other part of ejecta expands with a lower velocity so that the low-velocity Si II line comes up at later epochs. Having a low-velocity component would also make it easier to reproduce the slow tail.

(3) Extensive mixing of ^{56}Ni is required to reproduce the short rise time of the light curve. According to hydrodynamical simulations of the Rayleigh-Taylor instability in the ejecta of envelope-stripped supernovae (Hachisu et al. 1991; Iwamoto et al. 1997), large scale mixing is not expected to occur in massive progenitors, because in the core of such massive stars the density gradient is not steep enough around the composition interfaces. One possibility to induce such mixing in the velocity space is an asymmetric explosion. Higher velocity ^{56}Ni could reach the ejecta surface so that the effect of radioactive heating comes up as early as is required from light curve modeling.

In order to realize higher densities at low velocity regions without increasing the mass of the ejecta significantly, it may be necessary that the explosion is somewhat aspherical. If the explosion is aspherical, the shock would be stronger and the material would expand at a larger velocity in a certain direction, while in the perpendicular direction the shock would be weaker, ejecting lower velocity material (e.g., Höflich et al. 1999). The density of the central region could be high enough for γ-rays to be trapped even at advanced phases, thus giving rise to a slowly declining tail. In the extremely asymmetric cases, material ejection may take place in a jet-like form. A jet could easily bring some ^{56}Ni from the deepest layers out to the high velocity surface. Detailed spectral analysis of observed spectra for different epochs is necessary to investigate this issue further.

4. SN 1998bw

In this section models are presented for SN 1998bw which reproduce most of the early data. The very bright and relatively broad light curve of SN 1998bw can be reproduced by a family of models with various values of the fundamental parameters (M_{ej}, E_K), but in all the models these parameters are much larger than in the case of a typical SN Ic. All models require $M(^{56}\text{Ni}) \sim 0.5$ M$_\odot$ to power the bright light curve peak. This is about an order of magnitude larger than in typical core-collapse SNe. Models with different E_K yield different synthetic spectra, and by comparing with the observed early-time spectra of SN 1998bw and trying to fit the very broad absorption features, we selected a model CO138H with M_{ej}= 10 M$_\odot$, E_K= 6×10^{52} erg (Nakamura et al. 1999b). The large value of E_K easily qualifies SN 1998bw as 'the' Type Ic Hypernova. The mass of the progenitor C+O star is 13.8 M$_\odot$, which implies a main sequence mass of ~ 40 M$_\odot$. We also find that imposing a flatter density structure at high velocities ($v > 30,000$ km s^{-1}) results in more high-velocity absorption and in an even better-looking spectrum: significant absorption at $v \sim 60,000$ km s^{-1} is actually necessary to reproduce the observations. We then discuss the discrepancies with the later data, both light curve and spectra, and suggest that asymmetry may be playing a key role.

4.1. Model light curves

IMN98 modeled the SN 1998bw as the energetic explosion of a massive C+O core, and obtained a good fit to the SN light curve in the first 60 days. The model was selected from a set of degenerate models because it gave the best fit to the observed velocity of the Si II line and yielded sufficiently broad-lined spectra.

Nevertheless, the lines in the synthetic spectra were still noticeably narrower than the observed features (IMN98, Fig. 2). Also, photometry after day ~ 60 show that SN 1998bw declined significantly more slowly than the rate predicted by the model. Therefore, we recompute the light curve of SN 1998bw using progenitors of different masses and explosions of different energies. All models are spherically symmetric. The

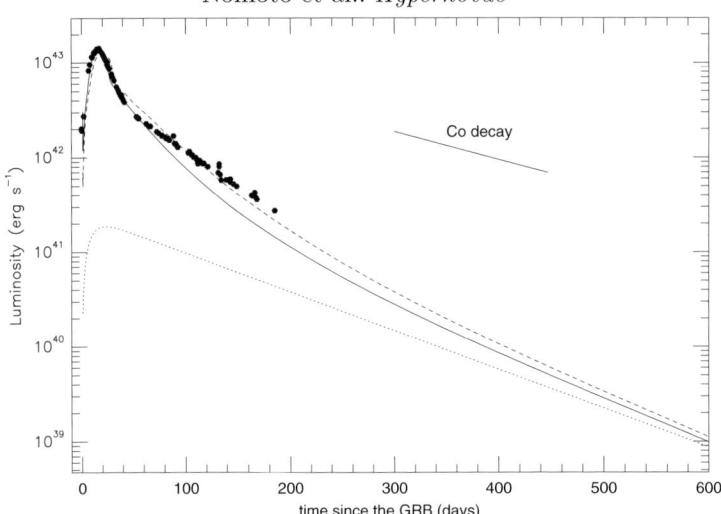

FIGURE 8. The light curves of models CO138H ($E_K = 6 \times 10^{52}$ erg; solid) and CO138L ($E_K = 3 \times 10^{52}$ erg; dashed) compared with the observations of SN1998bw (Galama et al. 1999; McKenzie & Schaefer 1999). A distance modulus of $\mu = 32.89$ mag and $A_V = 0.0$ are adopted. The dotted line indicates the energy deposited by positrons for CO138H.

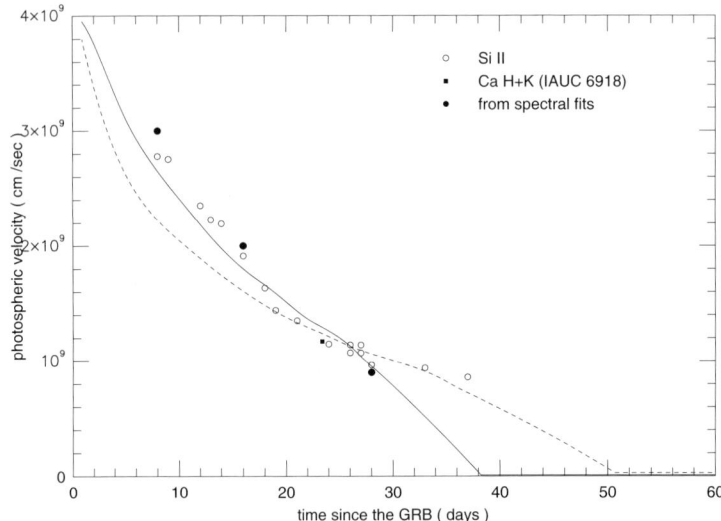

FIGURE 9. Photospheric velocities of models CO138H and CO138L compared with the observations of SN1998bw.

ejected ^{56}Ni is assumed to be rather centrally distributed, and its mass was determined by fitting the light curve around maximum. We assume that the progenitor was a C+O star, and searched a wide range of parameters.

We find that the model that give the best agreement to both the light curve and the spectra is that of the explosion of a 13.8 M$_\odot$ C+O star, ejecting 10 M$_\odot$ of material with $E_K = 6 \times 10^{52}$ erg, including 0.5 M$_\odot$ of ^{56}Ni (CO138H). In Figures 8 and 9 we compare the bolometric light curves and the photospheric velocities of model CO138H (solid) with the V photometry of SN 1998bw. We use $\mu = 32.97$ mag, $A_V = 0.0$ mag, and assume that $BC = 0.0$. This model has the same mass as that published in IMN98, but it has a larger E_K, which is necessary to improve the fit to the spectra (§4.2).

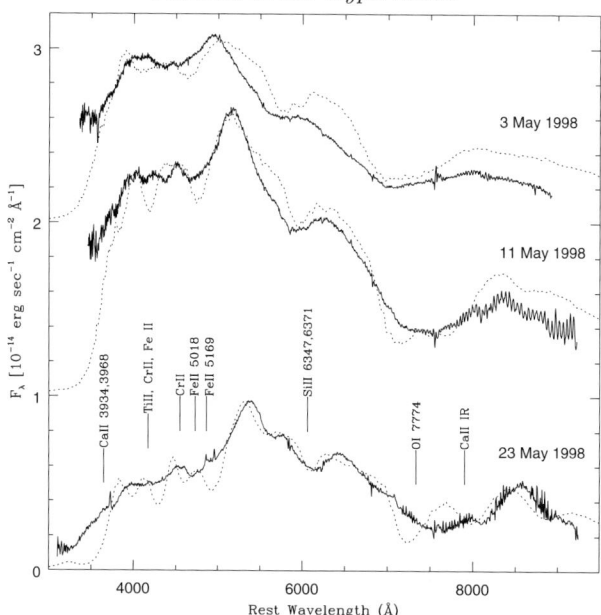

FIGURE 10. Observed spectra of SN1998bw (full lines) and synthetic spectra calculated using model CO138H ($E_K = 6 \times 10^{52}$ erg; dashed lines).

However, CO138H has difficulties reproducing the apparently exponential decline after day 60. On the other hand, CO138L ($E_K = 3 \times 10^{52}$; dashed) is in better agreement after day 90, although the early light curve and photospheric velocities do not fit well.

After day ~ 200 the decline of the model light curve becomes slower, and it approaches the half-life of ^{56}Co decay around day 400. At $t \gtrsim 400$ days most γ-rays escape from the ejecta, while positrons emitted from the ^{56}Co decay are mostly trapped and their energies are thermalized. Therefore, positron deposition determines the light curve at $t \gtrsim 400$ days (dotted line in Fig. 8). If the observed tail should follow the positron-powered light curve, the ^{56}Co mass could be determined directly.

The comparison between SN 1998bw and the model light curve of CO138H (which fits better at early phases) and CO138L (which is better for late phases) in Figure 8 suggests that there is a significant amount of mass expanding at very low velocity, containing some ^{56}Ni. In this case the trapping time for the γ-rays would be quite long, and the light curve might be explained. Indications for a low-velocity, high-density region are found for SN 1997ef, a lower-energy analogue of SN 1998bw.

This might indicate a departure from spherical symmetry. We will discuss this further in the next section in combination with the spectra. In any case, it is clear that the hydrodynamical model obtained from the explosion has to be altered.

4.2. Early Time Spectra

We have used model CO138H as a basis to compute synthetic spectra for the near-maximum phase of SN 1998bw. In particular, the model density structure and composition were used as input into the Monte Carlo code described above.

In Figure 10 we show the synthetic spectra obtained for the same 3 epochs fitted in IMN98. The observed spectra used here are the 'definitive', fully reduced version of the same ESO spectra shown in IMN98, and are calibrated with respect to the V photometry. The residual correction factors for the other bands are usually very close to 1, but they are ∼ 1.1 for B in the May 11 and 23 spectra. Therefore, the new models have somewhat different parameters than those of IMN98. Both the luminosity and the photospheric velocity have increased somewhat. The photospheric velocity now is in even better agreement with the measured velocity of the Si II line (IMN98, Fig. 3). All spectra are computed assuming $\mu = 32.97$ mag and $A_V = 0.0$. The assumption of zero reddening is supported by the upper limit of 0.1 Å in the equivalent width (EW) of the Na I D line obtained from high-resolution spectra (Patat et al. 2000).

The synthetic spectra clearly improve over those of IMN98. In particular, those absorptions not due to broad blends, i.e. the Si II feature near 6000 Å, and the O I+Ca II feature between 7000 and 8000 Å are now much broader, in significantly better agreement with the data. Nevertheless, the blue sides of those absorptions are still too narrow, indicating that even the new model may not contain enough mass at the highest velocities. Therefore we introduced an arbitrary change to the original CO138 density structure. Several possibilities were tested, and satisfactory results were found when the density slope was reduced from $\rho \propto r^{-8}$ to $\rho \propto r^{-6}$ at $v > 30,000$ km s^{-1}. This does not introduce a significant change in $M_{\rm ej}$, and increases E_K by only about 10%, but it does increase the density at high velocities, leading to significant absorption at $v \sim 60,000$ km s^{-1} in the strongest lines, especially the Ca II IR triplet, extending the absorption troughs to the blue. The corresponding synthetic spectra are shown as the dotted lines in Figure 10. The effect of the change is largest at the earliest epochs. The overall agreement with the observed spectra is better, although several problems remain, the most severe of which is clearly the excessive strength of the O I line at 7200 Å on May 11 and 23. The composition is dominated by O, and it is difficult to make that line become weaker. On 23 May, the synthetic Ca II IR triplet matches the weak feature at 8000 Å, which is first seen on 11 May and which continues to grow until it finally causes the wavelength of the absorption minimum of the entire broad feature to shift to ∼ 8200 Å (Patat et al. 2000). This is rather a peculiar behavior, because on 3 May the O I and Ca II lines had to blend much more to give rise to the observed broad feature, which then had a minimum at 7000 Å.

A very flat ($\rho \propto r^{-2}$) density distribution was also used by Branch (2000) to fit the spectrum of SN 1998bw. This dependence is however too flat when we use our MC model, because the ionization of e.g. Ca II does not fall as steeply as assumed by Branch (2000). On the other hand, Branch's E_K (5×10^{52} erg) is similar to ours, but he quotes a mass of 6 $M_\odot$ above 7000 km s^{-1}, while in our case the mass above that velocity is as large as ∼ 10 $M_\odot$.

Clearly, a definitive solution has not been found yet. It is quite possible that only by taking into account departures from spherical symmetry it will be possible to get a really accurate fit to the spectra. Nevertheless, considering the complexity of the problem, our fits at least demonstrate that a large E_K is necessary, and that a Type Ic SN O-dominated composition yields quite a reasonable reproduction of the observations.

4.3. Late time evolution

Thus far we have presented the results of an analysis of SN 1998bw based on the spherically symmetric model CO138H, which explain the observations around the maximum. However, peculiarities in the spectrum began to appear soon after maximum, when the broad feature in the red appeared to show a 'double' Ca II absorption, or possibly an O I absorption at much lower velocity than predicted by the spherically symmetric models, which place O at the highest velocity.

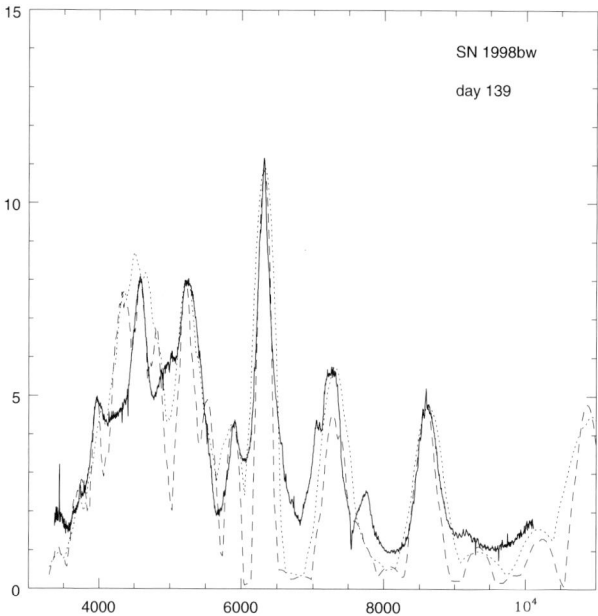

FIGURE 11. A nebular spectrum of SN 1998bw on 12 Sept 1998 (rest frame epoch 139 days) is compared to synthetic spectra obtained with a NLTE nebular model based on the deposition of gamma-rays from ^{56}Co decay in a nebula of uniform density. Two models were computed. In one model (dotted line) we tried to reproduce the broad FeII] lines near 5300 Å. The ^{56}Ni mass is 0.65 M$_\odot$, and the outer nebular velocity is 11,000 km s^{-1}, and the O mass is 3.5 M$_\odot$. The average electron density in the nebula is log $n_e = 7.47$ cm^{-3}. In the other model (dashed line), we tried to reproduce only the narrow [OI] 6300 Å emission line. These model has smaller ^{56}Ni mass (0.35 M$_\odot$) and O mass (2.1 M$_\odot$), and an outer velocity of 7500 km s^{-1}. The density is similar to that of the 'broad-lined' model. The filling factor used is 0.1 for both models.

Following the evolution of the spectra as they become nebular, we see that the SN showed a 'composite' spectrum (Fig. 11; Danziger et al. 1999; Patat et al. 2000): Fe II] lines, typical of SNe Ia, were strong, and so were lines of O I] and Mg I], which are typical of SNe Ib/c. At the same time, Fe III] lines, also typical of SNe Ia, were absent. The O I] and Mg I] lines grew stronger with time relative to the Fe II] lines, but had a narrower profile, maybe the composite of a broad and a narrow profile, which dominates more and more with time. The emergence of the narrow profiles occurs at about the same time as the light curve deviates from the model prediction of CO138H. In Fig. 11 we show how one nebular spectrum can be reproduced by two alternative models, one trying to fit the broad lines, and the other aimed at fitting the narrower O I] line. The ^{56}Ni mass estimated from the broad-line fit is comparable to the value obtained from the light curve calculations.

We suggest that a rather large mass of O-dominated material is also present at low velocity. Spherical explosions do not allow that: they are very effective at 'emptying' the central region, and always place the unburned elements at the top of the ejecta. So we suggest that the explosion was highly asymmetric, leaving large quantities of unburned material expanding at low velocity in directions away from the axis along which most of the energy was released and ^{56}Ni synthesized. Our vantage point must have been very close to that axis, because we also detected the GRB. At early times, the fast-expanding lobes were much brighter than the rest, and so we observed the broad-lined spectra and the bright light curve. The ^{56}Ni mass estimate of ~ 0.6 M$_\odot$ should not change much if the explosion was not spherical. The fast-moving regions rapidly became thin, though, and soon emission lines appeared. Initially those were broad, dominated by the hyper-energetic lobes. Later, though, the γ-rays from the fast-moving ^{56}Co could escape that region more and more easily, and a significant fraction of them could penetrate down into the low-velocity region and excite the O and Mg there. Some ^{56}Ni may also be present in the low-velocity region, but that could only be determined with better S/N observations of the nebular spectrum, which require 8m-class telescopes at this point. Additional γ-ray deposition in the low-velocity region may in turn increase the deposition function above what our spherically symmetric model estimates, and thus explain the slowly declining tail of the light curve.

Both the need for a high density region and the velocity inversion as well as polarization measurements (Patat et al. 2000) might indicate that the explosion is aspherical. If the outburst in SN 1998bw took the form of a prolate spheroid, for example, the explosive shock along the long axis was probably strong, ejecting material with large velocities and producing abundant ^{56}Ni. In directions away from the long axis, on the other hand, oxygen is not much burned and the density is high enough for γ-rays to be trapped even at advanced phases, thus giving rise to the slowly declining tail.

That the SN 1998bw explosion was asymmetric is not a new suggestion: first the polarization measurements indicated an axial ratio of 2–3:1 (Höflich et al. 1999) and the calculation of the explosion of a rotating core (MacFadyen & Woosley 1999) also gave similar results, in an effort to explain the connection between SN 1998bw and GRB980425. More detailed results have to await detailed numerical models in two dimension. For the time being we can comment that if the explosion was asymmetric, most likely our results for E_K are overestimated, because those would only refer to the fast-moving part of the ejecta. As for the value of M_{ej}, this can only be determined via 3D hydrodynamical models of the explosion, but we hope that careful analysis of the spectra, especially at late times, when both the fast and the slow components are observable, can yield at least some preliminary results.

We note that our estimate of the ^{56}Ni mass of ~ 0.6 M$_\odot$ from the nebula spectra in Figure 11 does not much depend on the asphericity. This is in good agreement with the spherical models CO138. Since Höflich et al. (1999) suggested that the ^{56}Ni mass can be as small as 0.2 M$_\odot$ if aspherical effects are large, our results suggest that the aspherical effects might be modest in SN 1998bw.

5. Type IIn SN 1997cy

SN 1997cy displayed narrow Hα emission on top of broad wings, which lead to its classification as a Type IIn (Germany et al. 1999; Turatto et al. 1999ab). Assuming $A_V = 0.00$ for the galactic extinction (NED) we get an absolute magnitude at maximum $M_v \leq -20.1$. It is the brightest SN II discovered so far. The light curve of SN 1997cy does not conform to the classical templates of SN II, namely Plateau and Linear, but resembles the slow evolution of the Type IIn SN 1988Z. As seen from the *uvoir* bolometric light curve in Figure 12, the SN light curve decline is slower than the ^{56}Co decay rate between day 120 to 250, suggesting circumstellar interaction for the energy source. (Here the outburst is taken to be coincident with GRB970514.)

In the interaction model, collision of the SN ejecta with the slowly moving circumstellar matter (CSM) converts the kinetic energy of the ejecta into light, thus producing the observed intense light display of the SN. Our exploratory model considers the explosion of a massive star of $M = 25$ M$_\odot$ with a parameterized kinetic energy E_K. We assume that the collision starts near the stellar radius at a distance r_1, where the density of the

CSM is ρ_1, and adopt for the CSM a power-law density profile $\rho \propto r^n$. The parameters E_K, ρ_1, and n, are constrained from comparison with the observations.

The regions excited by the forward and reverse shock emit mostly X-rays. The density in the shocked ejecta is so high that the reverse shock is radiative and a dense cooling shell is formed (e.g., Suzuki & Nomoto 1995; Terlevich et al. 1992). The X-rays are absorbed by the outer layers and the core of the ejecta, and re-emitted as UV-optical photons.

Narrow lines are emitted from the slowly expanding unshocked CSM photoionized by the SN UV outburst or by the radiation from the shocks; intermediate width lines come from the shock-heated CSM; broad lines come from either the cooler region at the interface between ejecta and CSM.

Figure 12 shows the model light curve which best fits the observations. The model parameters are: $E_K = 5 \times 10^{52}$ erg, $\rho_1 = 4 \times 10^{-14}$ g cm^{-3} at $r_1 = 2 \times 10^{14}$ cm (which corresponds to a mass-loss rate of $\dot{M} = 4 \times 10^{-4}$ M$_\odot$ yr^{-1} for a wind velocity of 10 km s^{-1}), and $n = -1.6$. The large mass-loss episode giving rise to the dense CSM is supposed to occur after the progenitor makes a loop in the HR diagram from BSG to RSG. In this model, the mass of the low-velocity CSM is ~ 5 M$_\odot$, which implies that the transition from BSG to RSG took place about 10^4 yr before the SN event.

The large CSM mass and density are necessary to have large shocked masses and thus to reproduce the observed high luminosity, and so is the very large explosion energy. For models with low E_K and high ρ_1, the reverse shock speed is too low to produce a sufficiently high luminosity. For example, a model with $E_K = 10^{52}$ erg and ρ_1 as above yields a value of $L_{\rm UVOIR}$ lower than the observed luminosity by a factor of ~ 5. For high E_K or low ρ_1, the expansion of the SN ejecta is too fast for the cooling shell to absorb enough X-rays to sustain the luminosity. Thus in this model E_K and $\dot{M}$ are constrained within a factor of ~ 3 of the reported values.

The shape of the light curve constrains the circumstellar density structure. For $n = -2$, the case of a steady wind, $L_{\rm UVOIR}$ decreases too rapidly around day 200. To reproduce the observed decrease after day ~ 300, the CSM density is assumed to drop sharply at the radius the forward shock reaches at day 300, so that the collision becomes weaker afterwards. (Such a change of the CSM density corresponds to the transition from BSG to RSG of the progenitor $\sim 10^4$ yr before the SN explosion.) This is consistent with the simultaneous decrease in the Hα luminosity.

The observed light curve drops sharply after day 550. We reproduce such a light curve behavior (Figure 12) assuming that when the reverse shock propagates through ~ 5 M$_\odot$, it encounters exceedingly low density region and thus it dies. In other words, the model for the progenitor of SN 1997cy assumes that most of the core material has fallen into a massive black hole of, say, ~ 10 M$_\odot$, while the extended H/He envelope of ~ 5 M$_\odot$ has not collapsed. Then material is ejected from the massive black hole possibly in a jet-like form, and the envelope is hit by the "jet" and ejected at high velocity.

In this model, the ejecta are basically the H/He layers and thus contain the original (solar abundance) heavy elements plus some heavy elements mixed from the core (before fall back) or jet materials. This might explain the lack of oxygen and magnesium lines in the spectra particularly at nebular phases (Turatto et al. 1999ab).

6. Possible evolutionary scenarios to Hypernovae

Here we classify possible evolutionary paths leading to C+O star progenitors. In particular, we explore the paths to the progenitors that have rapidly rotating cores with

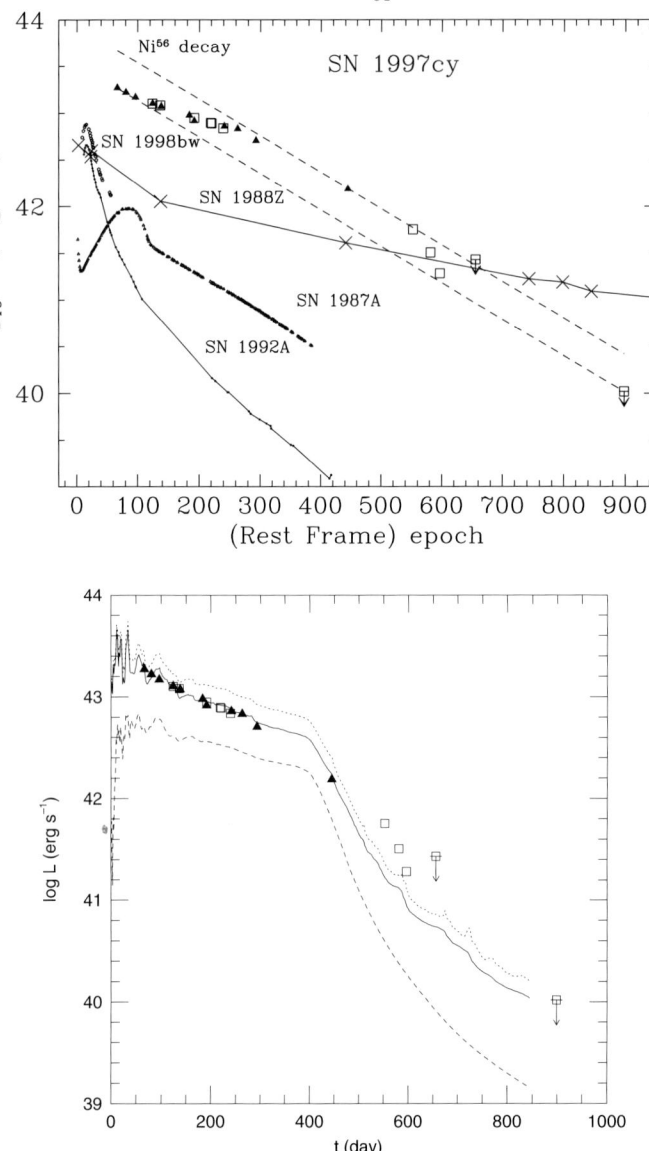

FIGURE 12. Upper panel: Observed light curves for SNe 1997cy, 1988Z, 1987A, 1992A, & 1998bw. Lower panel: The *uvoir* bolometric light curve of SN 1997cy compared with the synthetic light curve obtained with the CSM interaction model.

a special emphasis, because the explosion energy of hypernovae may be extracted from rapidly rotating black holes (Blandford & Znajek 1977).

(1) Case of a single star: If the star is as massive as $M_{\rm ms} \gtrsim 40$ M$_\odot$, it could lose its H and He envelopes in a strong stellar wind (e.g., Schaller et al. 1992). This would be a Wolf-Rayet star.

(2) Case of a close binary system: Suppose we have a close binary system with a large mass ratio. In this case, the mass transfer from star 1 to star 2 inevitably takes place in a non-conservative way, and the system experiences a common envelope phase where star 2 is spiraling into the envelope of star 1. If the spiral-in releases enough energy to remove

the common envelope, we are left with a bare He star (star 1) and a main-sequence star (star 2), with a reduced separation. If the orbital energy is too small to eject the common envelope, the two stars merge to form a single star (e.g., van den Heuvel 1994).

(2-1) For the non-merging case, possible channels from the He stars to the C+O stars are as follows (Nomoto, Iwamoto, & Suzuki 1995).

(a) Small-mass He stars tend to have large radii, so that they can fill their Roche lobes more easily and lose most of their He envelope via Roche lobe overflow.

(b) On the other hand, larger-mass He stars have radii too small to fill their Roche lobes. However, such stars have large enough luminosities to drive strong winds to remove most of the He layer (e.g., Woosley, Langer, & Weaver 1995). Such a mass-losing He star would corresponds to a Wolf-Rayet star.

Thus, from the non-merging scenario, we expect two different kinds of SNe Ic, fast and slow, depending on the mass of the progenitor. SNe Ic from smaller mass progenitors (channel 2-1-a) show faster light-curve and spectral evolutions, because the ejecta become more quickly transparent to both gamma-ray and optical photons. The slow SNe Ic originate from the Wolf-Rayet progenitors (channels 1 and 2-1-b). The presence of both slow and fast SNe Ib/Ic has been noted by Clocchiatti & Wheeler (1997).

(2-2) For the merging case, the merged star has a large angular momentum, so that its collapsing core must be rotating rapidly. This would lead to the formation of a rapidly rotating black hole from which possibly a hyper-energetic jet could emerge. If the merging process is slow enough to eject the H/He envelope, the star would become a rapidly rotating C+O star. Such stars are the candidates for the progenitors of Type Ic hypernovae like SNe 1997ef and 1998bw. If a significant amount of H-rich envelope remains after merging, the rapidly rotating core would lead to a hypernova of Type IIn possibly like SN 1997cy (or Type Ib).

7. Nucleosynthesis in hypernovae

Since hypernovae explode with much higher explosion energies than usual supernovae, explosive nucleosynthesis could have some special features. Also hypernovae have shown some aspherical signatures. Here we investigate how the explosive nucleosynthesis results depend on the explosion energy and asphericity (Nomoto et al. 1998; Nakamura et al. 1999c, 2000; Maeda et al. 2000).

7.1. *Nucleosynthesis in spherical explosions*

We calculate explosive nucleosynthesis in hypernovae in the same way as has been done for normal supernovae; we use a detailed nuclear reaction network including 211 isotopes up to ^{71}Ge (Thielemann, Nomoto, & Hashimoto 1996; Hix & Thielemann 1996; Nakamura et al. 1999a) (Figure 13: left). Nucleosynthesis in normal supernovae ($E_{\rm K} = 1 \times 10^{51}$ erg) is also shown in Figure 13 (right) for comparison.

A similar comparison is made in Figure 14, which shows the composition structure of models CO60 and CO100 against the expansion velocity and the Lagrangian mass coordinate of the progenitor. In CO100, the Fe and Si-rich layers expand much faster than in CO60. The total amount of nucleosynthesis products are summarized in Table 2.

From this figure, we can see the following characteristics of nucleosynthesis with the very large explosion energy.

(1) The complete Si-burning region is extended to the outer, lower density region. Whether this region is ejected or not depends on the mass cut. The large amount of ^{56}Ni observed in hypernovae (e.g., ~ 0.5 M$_\odot$ for SN1998bw and 0.15 M$_\odot$ for SN 1997ef) implies

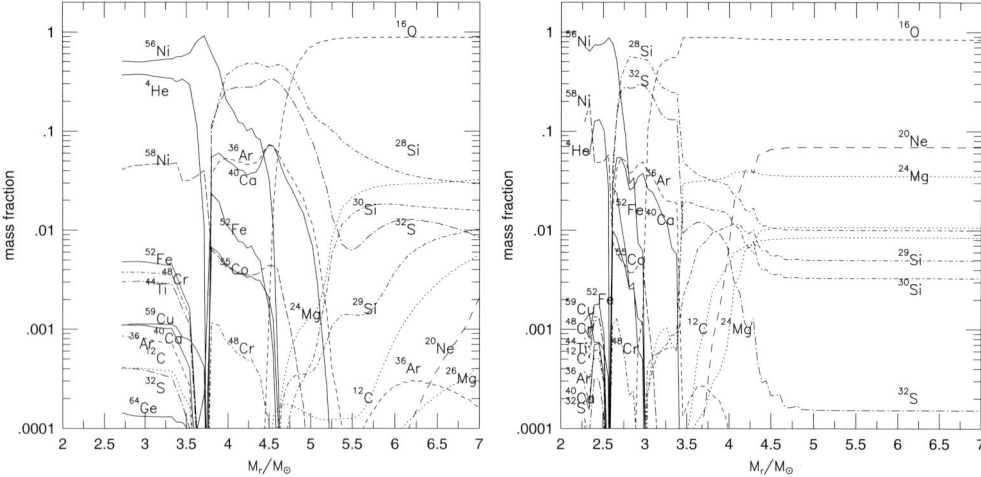

FIGURE 13. The isotopic composition of ejecta of the hypernova ($E_K = 3 \times 10^{52}$ erg; left) and the normal supernova ($E_K = 1 \times 10^{51}$ erg; right) for a 16 $M_\odot$ He star. Only the dominant species are plotted. The explosive nucleosynthesis is calculated using a detailed nuclear reaction network including a total of 211 isotopes up to ^{71}Ge.

model	C	O	Ne	Mg	Si	S	Ca	Ti	Fe	Ni
CO60	0.082	3.0	0.62	0.24	0.10	0.037	0.006	0.0003	0.16	0.017
CO100	0.58	5.6	0.38	0.22	0.42	0.19	0.025	0.0003	0.19	0.021
CO138H	0.11	6.6	0.35	0.29	0.95	0.52	0.088	0.0011	0.50	0.028

model	^{44}Ti	^{56}Ni	^{57}Ni
CO60	2.1×10^{-4}	0.15	5.7×10^{-3}
CO100	4.5×10^{-5}	0.15	5.7×10^{-3}
CO138H	2.2×10^{-4}	0.50	1.5×10^{-2}

TABLE 2. Yields of hypernova and supernova models ($M_\odot$)

that the mass cut is rather deep, so that the elements synthesized in this region such as ^{59}Cu, ^{63}Zn, and ^{64}Ge (which decay into ^{59}Co, ^{63}Cu, and ^{64}Zn, respectively) are likely to be ejected more abundantly. In the complete Si-burning region of the hypernova, elements produced by α-rich freezeout are enhanced because nucleosynthesis proceeds under lower densities than in usual supernovae (Fig. 15). Figure 13 clearly shows a trend that a larger amount of ^{4}He is left in more energetic explosion. Hence, elements synthesized through α-captures such as ^{40}Ca (stable), ^{44}Ti and ^{48}Cr (decaying into ^{44}Ca and ^{48}Ti, respectively) become more abundant.

(2) The more energetic explosion produces a broader incomplete Si-burning region. The elements produced mainly in this region such as ^{52}Fe, ^{55}Co, and ^{51}Mn (decaying into ^{52}Cr, ^{55}Mn, and ^{51}V, respectively) are synthesized more abundantly with the larger explosion energy.

(3) Oxygen burning takes place in more extended, lower density region for the larger explosion energy, so that the abundances of elements like O, C, Al are smaller. On the other hand, a larger amount of ash products such as Si, S, Ar are synthesized by oxygen burning.

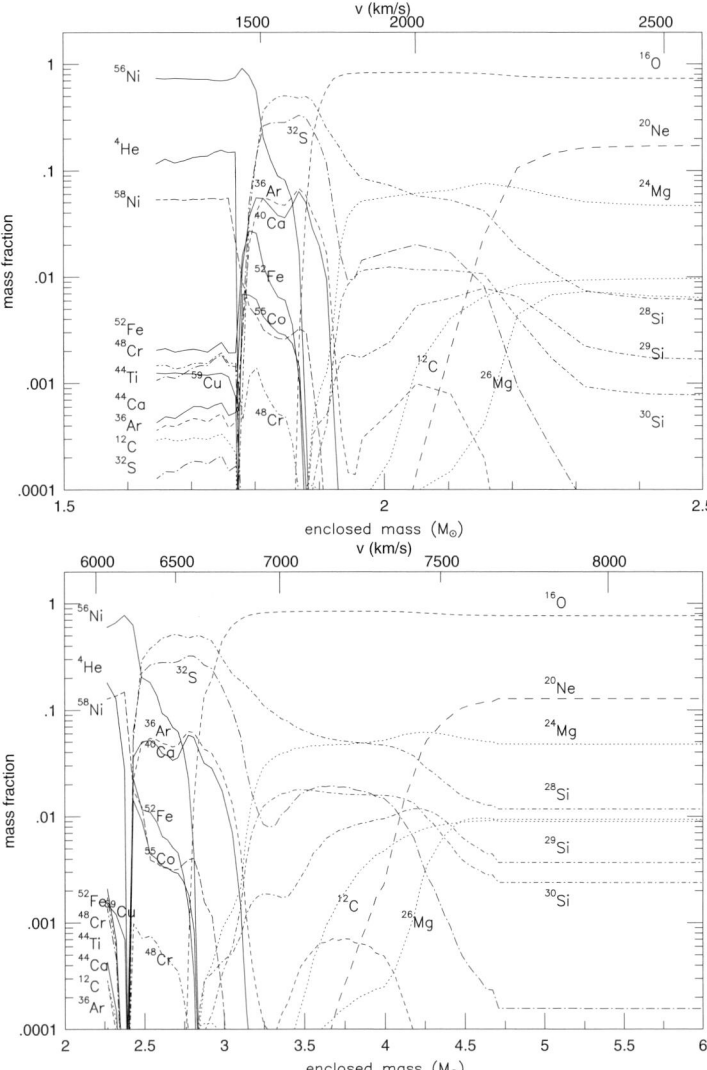

FIGURE 14. The composition structure of models CO60 (left) and CO100 (right).

Figure 16 shows the abundances of stable isotopes relative to the solar values for 3×10^{52} erg and 1×10^{51} erg. The progenitor is the 16 $M_\odot$ He star and products from H-rich envelope are not included. The isotopic ratios relative to ^{16}O with respect to the solar values are shown. As a whole, intermediate mass nuclei and heavy nuclei are more abundant for the more energetic explosion, except for the elements being consumed in oxygen burning like O, C, Al. Especially, the amounts of ^{44}Ca and ^{48}Ti are increased significantly because of the enhanced α-rich freezeout.

7.2. Nucleosynthesis in asymmetric explosions

In §4, we speculate that the expansion velocity of Fe and O in SN 1998bw betrays the effect of the asymmetry in the explosion. To confirm this we calculate the explosive nucleosynthesis in an axisymmetric explosion. Figure 17 shows the isotopic compositions of the ejecta of the axisymmetric explosion in the direction of the jet (left) and perpen-

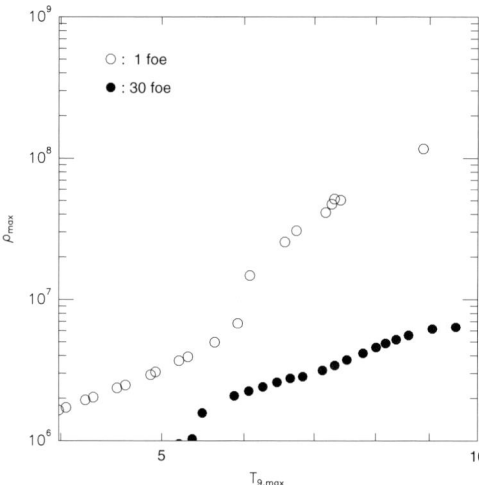

FIGURE 15. The maximum ρ–T conditions of individual mass zones in the normal supernova ($E_K = 1$ foe $= 1 \times 10^{51}$ erg) and the hypernova ($E_K = 30$ foe).

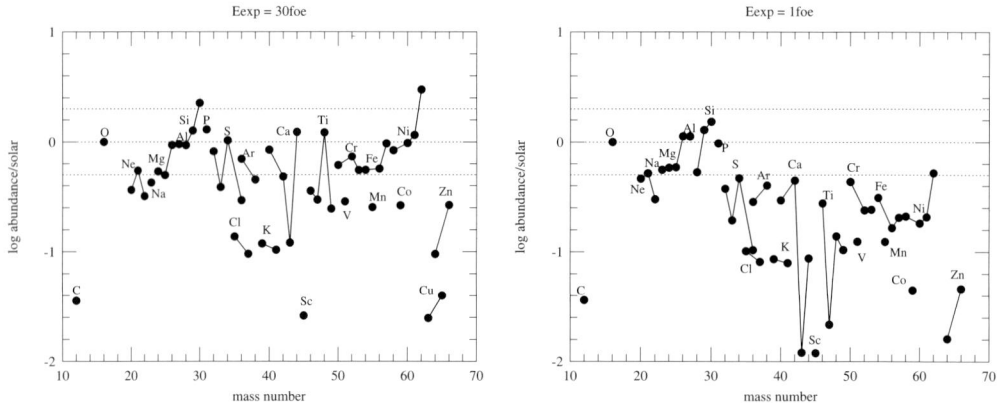

FIGURE 16. Abundances of stable isotopes relative to the solar values for 3×10^{52} ergs and 1×10^{51} ergs (Figure 13). The progenitor is a 16 $M_\odot$ He star (H-rich envelope is not included).

dicular to the jet (right). The progenitor model is CO138. The the explosion energy is $E_K = 1 \times 10^{51}$ erg. Starting the hydrodynamical simulation, we deposit the energy as 50% thermal energy and 50% kinetic energy toward the jet (z) below the mass cut that divides the ejecta and the collapsing core.

The shock is stronger and the post-shock temperatures are higher along the jet direction (z), so that explosive nucleosynthesis takes place in a more extended, lower density region compared with the perpendicular direction (r). A larger amount of ^{56}Ni is produced in the jet direction. In addition, elements produced by α-rich freezeout are enhanced because nucleosynthesis proceeds at higher entropies than in the region away from the jet. Figure 13 clearly shows that in the jet direction a larger amount of ^{4}He is left after the shock decomposition. Hence, elements synthesized through capturing α-particles such as ^{44}Ti and ^{48}Cr (decaying into ^{44}Ca and ^{48}Ti, respectively) become more abundant (see also Nagataki et al. 1997). In contrast, little ^{56}Ni is produced in the r-direction. Also the expansion velocities are lower than those in the z-direction. Therefore, the Fe velocities (mostly z-direction) can exceed the O velocities (in the r-direction),

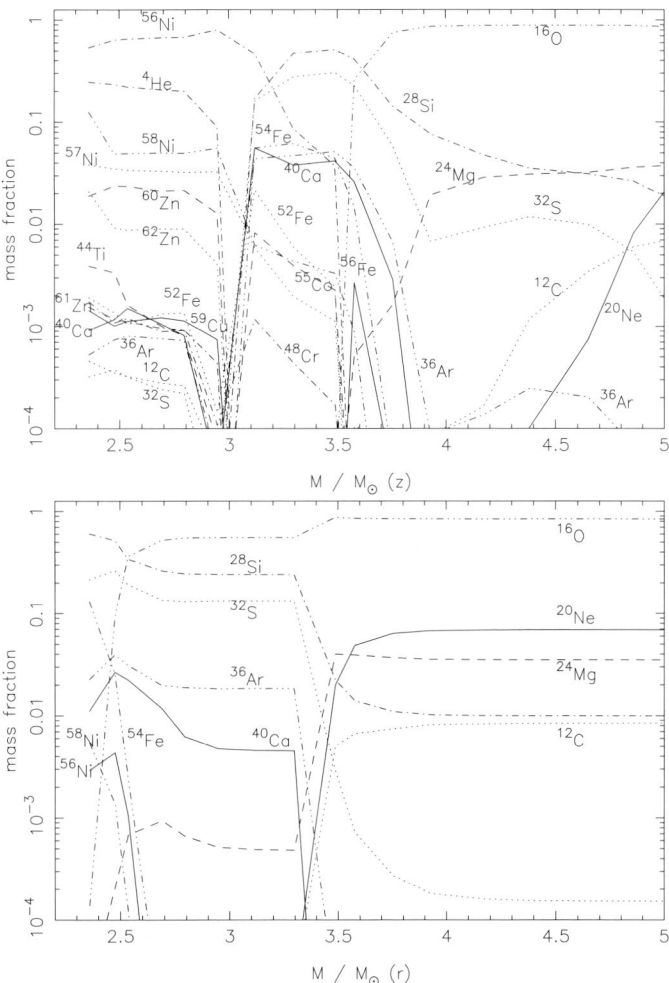

FIGURE 17. The isotopic compositions of the ejecta of the axisymmetric explosion in the direction of the jet (top) and the perpendicular to the jet (bottom) with $E_\mathrm{K} = 1 \times 10^{51}$ erg (Maeda et al. 2000).

as observed in SN 1998bw. Oxygen in the z-direction has the highest velocities but the densities may become too low to be excited by gamma-rays.

Such an asymmetric ejection of nucleosynthesis products may explain the abundance features observed in X-ray Nova Sco (GRO J1655-40), which consists of a massive black hole and a low mass companion (e.g., Brandt et al. 1995; Nelemans et al. 2000). The companion star is enriched with Ti, S, Si, Mg, and O but not much Fe (Israelian et al. 1999). This is compatible with heavy element ejection from a black hole progenitor. In order to eject large amount of Ti, S, and Si and to have at least ~ 4 M$_\odot$ below mass cut and thus form a massive black hole, the explosion should be highly energetic (Fig. 13; Israelian et al. 1999; Brown et al. 2000a; Podsiadlowski et al. 2000). Suppose that an asymmetric explosion occurred when the black hole formed in Nova Sco. Then it is likely that the companion star captured material ejected in the r-direction (i.e., on the orbital plane) which contains relatively little Fe compared with the z-direction, where burning is more effective (Podsiadlowski et al. 2000; Brown et al. 2000b). Quantitatively, nucleosynthesis in the r-direction for $E_\mathrm{K} = 1 \times 10^{52}$ erg is in good agreement with Nova Sco (Maeda et al. 2000).

7.3. The mass of ejected ^{56}Ni

For the study of the chemical evolution of galaxies, it is important to know the mass of ^{56}Ni, $M(^{56}\mathrm{Ni})$, synthesized in core-collapse supernovae as a function of the main-sequence mass M_ms of the progenitor star (e.g., Nakamura et al. 1999a). From our analysis of SNe 1998bw and 1997ef, we can add new points on this diagram.

We evaluate the uncertainty in our estimates of $M(^{56}\mathrm{Ni})$ and M_ms. We need 0.15 $M_\odot$ of ^{56}Ni to get a reasonable fit to the light curve of SN 1997ef at a distance $D = 52.3$ Mpc. The expected 10% uncertainty in the distance leads to a 20% uncertainty in the ^{56}Ni mass, i.e., $M(^{56}\mathrm{Ni}) = 0.15 \pm 0.03$ $M_\odot$. The distribution of ^{56}Ni affects the peak luminosity somewhat, but the effect is found to be much smaller than that of the uncertainty in the distance. A 10 $M_\odot$ C+O star corresponds to a $M_\mathrm{ms} = 30$–35 $M_\odot$, but the uncertainty involved in the conversion of the core mass to M_ms may involve a larger uncertainty if the progenitor undergoes close binary evolution.

Figure 18 shows $M(^{56}\mathrm{Ni})$ against M_ms obtained from fitting the optical light curves of SNe 1987A, 1993J, and 1994I (e.g., Shigeyama & Nomoto 1990; Nomoto et al. 1993, 1994; Shigeyama et al. 1994; Iwamoto et al. 1994; Woosley et al. 1994; Young, Baron, & Branch 1995). The amount of ^{56}Ni appears to increase with increasing M_ms of the progenitor, except for SN II 1997D (Turatto et al. 1998).

This trend might be explained as follows. Stars with $M_\mathrm{ms} \lesssim 25$ $M_\odot$ form a neutron star, producing $\sim 0.08 \pm 0.03$ $M_\odot$ ^{56}Ni as in SN IIb 1993J, SN Ic 1994I, and SN 1987A (although SN 1987A may be a borderline case between neutron star and black hole formation). Stars with $M_\mathrm{ms} \gtrsim 25$ $M_\odot$ form a black hole (e.g., Ergma & van den Heuvel 1998); whether they become hypernovae or ordinary SNe may depend on the angular momentum in the collapsing core. For SN 1997D, because of the large gravitational potential, the explosion energy was so small that most of ^{56}Ni fell back onto a compact star remnant; the fall-back might cause the collapse of the neutron star into a black hole. The core of SN II 1997D might not have a large angular momentum, because the progenitor had a massive H-rich envelope so that the angular momentum of the core might have been transported to the envelope possibly via a magnetic-field effect. Hypernovae such as SNe 1998bw, 1997ef, and 1997cy might have rapidly rotating cores owing possibly to the spiraling-in of a companion star in a binary system. The outcome certainly depends also on mass-loss rate and binarity.

8. Gamma-Ray Bursts/Supernovae connection

Candidates for the GRB/SN connection include GRB980425/SN Ic 1998bw (Galama et al. 1998; IMN98), GRB971115/SN Ic 1997ef (Wang & Wheeler 1998), GRB970514/SN IIn 1997cy (Germany et al. 1999; Turatto et al. 1999ab), GRB980910/SN IIn 1999E (Thorsett & Hogg 1999), and GRB991002/SN IIn 1999eb (Terlevich et al. 1999).

Two other GRBs may also be associated with a SN: GRB980326 (Bloom et al. 1999) and GRB970228 (Reichart 1999; Galama et al. 1999). The optical afterglows of these GRBs showed that the decline of the light curve is slowed down at late phases, and this can be reproduced if a red-shifted SN 1998bw-like light curve is superposed on the power-law light component. A question arising from these two examples is whether the supernovae associated with GRBs have a uniform maximum luminosity, i.e., whether ~ 0.5 $M_\odot$ ^{56}Ni production as in SN 1998bw is rather common or not. However, the

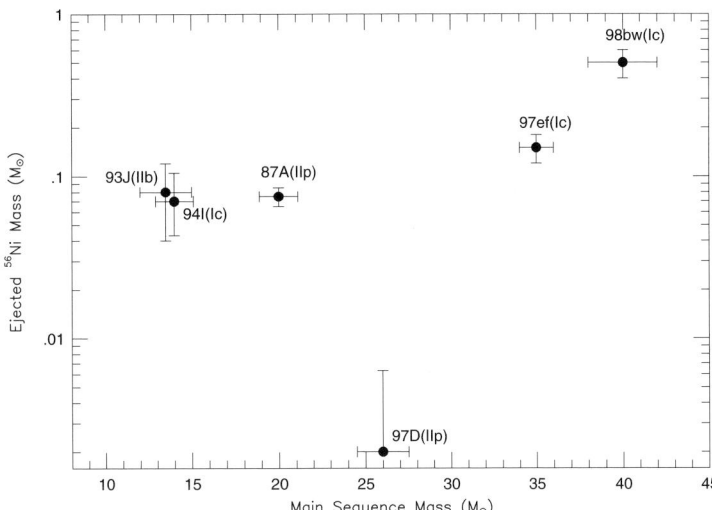

FIGURE 18. Ejected ^{56}Ni mass versus the main sequence mass of the progenitors of several bright supernovae obtained from light curve models.

present study of SN 1997ef shows that the ^{56}Ni mass and thus intrinsic maximum brightness of SN 1997ef is smaller than in SN 1998bw by a factor of 4–5. We certainly need more examples for defining the luminosity function and the actual distribution of masses of ^{56}Ni produced in supernovae/hypernovae.

Among the possible connections suggested above, the statistical significance for the case of SN 1997ef and GRB971115 is much weaker than for the case of SN 1998bw and GRB980425. Recently another SN Ic, 1998ey, showed a spectrum with very broad features, very similar to that of SN 1997ef on Dec 17 (Garnavich et al. 1998); but no GRB counterpart has been proposed for SN 1998ey. Although this may cast some doubt on the general association between hypernovae and GRBs, it must be noted that both SNe 1997ef and 1998ey were less energetic events than SN 1998bw. It is possible that a weaker explosion is less efficient in collimating the γ-rays to give rise to a detectable GRB (GRB980425 was already quite weak compared to the average GRBs), or that some degree of inclination of the beam axis to the line-of-sight results in a seemingly weaker supernova and in the non-detection of a GRB. Only the accumulation of more data will allow us to address these questions.

9. Concluding remarks

We have calculated the light curves and spectra for various C+O star models with different values of E_K and M_{ej} and reached several striking conclusions.

We have shown that the spectra of SNe 1998bw and 1997ef are much better reproduced with the hypernova models than with the ordinary SN Ic model. Since SN 1998bw was connected to a highly non-spherical event like a GRB, departure from spherical symmetry could be expected. Early polarization measurements confirmed this: polarization of $\sim 1\%$, decreasing with time, was detected.

The evidence for asphericity in SN 1998bw becomes even stronger with the extended time coverage: the light curve decline is slower than predicted by our spherically symmetric model; the composite nebular spectra have different velocities in lines of different elements, with iron expanding more rapidly than oxygen; the O I] nebular line declines more slowly than the Fe II ones, signaling deposition of γ-rays in a slowly-moving O-dominated region.

The smaller line velocities at advanced phases and the flat light curve tail of SN 1997ef may also suggest the presence of a low-velocity, relatively dense core, while the high line velocities at early phases imply the presence of an even higher-velocity component of the ejecta. This discrepancy between models and observations, as well as the extensive mixing of ^{56}Ni required to explain the early rise of the light curve, seems to indicate that the explosion of SN 1997ef was at least somewhat aspherical.

Therefore, we suggest that SNe 1997ef, 1998ey, and 1998bw form a new class of hyperenergetic Type Ic supernovae, which we may call "Type Ic" hypernovae. SN 1998bw produced ~ 0.5–0.7 M$_\odot$ of ^{56}Ni, as much as a SN Ia, while SN 1997ef produced less, only ~ 0.15 M$_\odot$, but still more than in ordinary SNe Ic. SN 1997ef also appeared to be less energetic than SN 1998bw. This may be a real difference, but it may also result from different inclination or beaming properties, since no GRB counterpart was positively observed for SN 1997ef.

SNe 1997cy, 1999E, and 1999eb may form a class of "Type IIn" hypernovae. They are also distinguished by their large kinetic energies, 8–60 times larger than in ordinary supernovae, but it is not easy to determine how much ^{56}Ni they produced since their light curves and spectra are dominated by interaction with a massive CSM. Simulations of the interaction can reproduce the observed light curve of SN 1997cy, indicating that the progenitor must have been a massive star, which possibly underwent spiral-in of the companion star in a close binary system.

Continuing observations and theoretical modeling of this interesting class of objects are certainly necessary.

This work has been supported in part by the grant-in-Aid for COE Scientific Research (07CE2002) of the Ministry of Education, Science, Culture and Sports in Japan.

REFERENCES

ARNETT, W. D. 1996 Supernovae and Nucleosynthesis. Princeton University Press.

BARON, E., YOUNG, T. R., & BRANCH, D. 1993 ApJ, **409**, 417.

BRANCH, D. 2000 this volume.

BLANDFORD, R. D. & ZNAJEK, R. L. 1977 MNRAS, **179**, 433.

BLOOM, J. S., ET AL. 1999 Nature, **401**, 453.

BRANDT, W. N., PODSIADLOWSKI, PH., & SIGURDSSEN, S. 1995 MNRAS, **277**, L35.

BROWN, G. E., LEE, C.-H., WIJERS, R. A. M. J., LEE, H. K., ISRAELIAN, G., & BETHE, H. A. 2000a, preprint.

BROWN, G. E., LEE, C.-H., LEE, H. K., & BETHE, H. A. 2000b, in Cosmic Explosions (ed. S. S. Holt & W. W. Zhang). AIP, in press.

CAPPELLARO, E., TURATTO, M., & MAZZALI, P. 1999, IAU Circ. No. 7091.

CLOCCHIATTI, A. & WHEELER, J. C. 1997 ApJ, **491**, 375.

COLELLA, P. & WOODWARD, P. R. 1984 J. Comput. Phys., **54**, 174.

Danziger, I. J., et al. 1999, this volume.

ERGMA, E. & VAN DEN HEUVEL, E. P. J. 1998 A&A, **331**, L29.

GALAMA, T. J., ET AL. 1998 Nature, **395**, 670.

GALAMA, T. J., ET AL. 1999 ApJ, submitted.

GARNAVICH, P., JHA, S., KIRSHNER, R., & CHALLIS, P. 1997a IAU Circ. No. 6778.

GARNAVICH, P., JHA, S., KIRSHNER, R., CHALLIS, P., & BALAM, D. 1997b IAU Circ. No. 6786

GARNAVICH, P., JHA, S., KIRSHNER, R., & CHALLIS, P. 1997c IAU Circ. No. 6798.

GARNAVICH, P., JHA, S., & KIRSHNER, R. 1998 IAU Circ. No. 7066.

GERMANY, L. M., REISS, D. J., SCHMIDT, B. P., STUBBS, C. W., SADLER, E. M. 1999 *ApJ*, submitted (astro-ph/9906096).

HACHISU, I., MATSUDA, T., NOMOTO, K., & SHIGEYAMA T. 1991 *ApJ*, **368**, 27.

HIX, W. R. & THIELEMANN, F.-K. 1996 *ApJ*, **460**, 869.

HÖFLICH, P., WHEELER, J. C., & WANG, L. 1999 *ApJ*, **521**, 179.

HU, J. Y., ET AL. 1997 IAU Circ. No. 6783.

ISRAELIAN, G., REBOLO, R., BASRI, G., CASARES, J., & MARTIN, E. L. 1999 *Nature*, **401**, 142.

IWAMOTO, K., NOMOTO, K., HÖFLICH, P., YAMAOKA, H., KUMAGAI, S., & SHIGEYAMA, T. 1994 *ApJ*, **437**, L115.

IWAMOTO, K., YOUNG, T. R., NAKASATO, N., SHIGEYAMA, T., NOMOTO, K., HACHISU, I., & SAIO, H. 1997 *ApJ*, **477**, 865.

IWAMOTO, K., MAZZALI, P. A., NOMOTO, K., ET AL. 1998 *Nature*, **395**, 672 (IMN98).

IWAMOTO, K., NAKAMURA, T., NOMOTO, K., MAZZALI, P. A., DANZIGER, I. J., GARNAVICH, P., KIRSHNER, R., JHA, S., BALAM, D., & THORSTENSEN, J. 2000 *ApJ*, **534**, in press (astro-ph/9807060).

KULKARNI, S. R., ET AL. 1998 *Nature*, **395**, 663.

LUCY, L. B. 1999 *A&A*, in press.

MACFADYEN, A. I. & WOOSLEY, S. E. 1999, *ApJ*, **524**, 262.

MAEDA, K., NAKAMURA, T., NOMOTO, K., & HACHISU, I. 2000 in Origin of Matter and Evolution of Galaxies (ed. S. Kubono, et al.). World Scientific, in press.

MAZZALI, A. P. & LUCY, L. B. 1993 *A&A*, **279**, 447.

MAZZALI, A. P. 2000 *A&A*, submitted.

MCKENZIE, E. H. & SCHAEFER, B. E. 1999 *PASP*, **111**, 964.

NAGATAKI, S., HASHIMOTO, M., SATO, K., & YAMADA, S. 1997 *ApJ*, **486**, 1026.

NAKAMURA, T., UMEDA, H., NOMOTO, K., THIELEMANN, F.-K., & BURROWS, A. 1999a *ApJ*, **517**, 193.

NAKAMURA, T., MAZZALI, P. A., NOMOTO, K., IWAMOTO, K., & UMEDA, H. 1999b *Astron. Nachrichten*, **320**, 363.

NAKAMURA, T., NOMOTO, K., IWAMOTO, K., UMEDA, H., MAZZALI, P.A., & DANZIGER, I. J. 1999c *Memorie della Società Astronomica Italiana*, in press.

NAKAMURA, T., MAEDA, K., IWAMOTO, K., SUZUKI, T., NOMOTO, K., MAZZALI, P. A., TURATTO, M., DANZIGER, I. J., & PATAT, N. 2000 in IAU Symposium 195, *Highly energetic physical processes and mechanisms for emission from astrophysical plasmas* (ed. S. Tsuruta). ASP.

NELEMANS, G., TAURIS, T. M., & VAN DEN HEUVEL, E. P. J. 2000 *A&A*, in press (astro-ph/9911054).

NOMOTO, K. & HASHIMOTO, M. 1988 *Phys. Rep.*, **163**, 13.

NOMOTO, K., SUZUKI, T., SHIGEYAMA, T., KUMAGAI, S., YAMAOKA, H., & SAIO, H. 1993 *Nature*, **364**, 507.

NOMOTO, K., YAMAOKA, H., POLS, O. R., VAN DEN HEUVEL, E. P. J., IWAMOTO, K., KUMAGAI, S., & SHIGEYAMA, T. 1994 *Nature*, **371**, 227.

NOMOTO, K., IWAMOTO, K., & SUZUKI, T. 1995 *Phys. Rep.*, **256**, 173.

NOMOTO, K., NAKAMURA, T., IWAMOTO, K., UMEDA, H., & MAZZALI, P. A. 1998 in Nuclei in the Cosmos V (eds. N. Prantzos). Editions Frontieres. p. 252.

NOMOTO, K., IWAMOTO, K., MAZZALI, P.A., & NAKAMURA, T. 1999 *Astron. Nachrichten*, **320**, 265.

PACZYŃSKI, B. 1998 *ApJ*, **494**, L45.

Patat, F., et al. 2000 *ApJ*, submitted.
Podsiadlowski, Ph., Nomoto, K., Mazzali, P. A., & Schmidt, B. 2000 preprint.
Reichart, D. E. 1999 *ApJ*, **521**, L111.
Richmond, M. W., et al. 1996a, *AJ*, **111**, 327.
Richmond, M. W., Treffers, R. R., Filippenko, A. V., & Paik, Y. 1996b *AJ*, **112**, 732.
Schaller, G., Schaerer, D., Meynet, G., & Maeder, A. 1992 *A&AS*, **96**, 269.
Shigeyama, T. & Nomoto, K. 1990 *ApJ*, **360**, 242.
Shigeyama, T., Suzuki, T., Kumagai, S., Nomoto, K., Saio, H., Yamaoka, H. 1994 *ApJ*, **420**, 341.
Stathakis, R. A., et al. 2000 *MNRAS*, submitted (astro-ph/0001497).
Suzuki, T. & Nomoto, K. 1995 *ApJ*, **455**, 658.
Swartz, D. A. & Wheeler, J. C. 1991 *ApJ*, **379**, L13.
Terlevich, R., Tenorio-Tagle, G., Franco, J., & Melnick, J. 1992 *MNRAS*, **255**, 713.
Terlevich, R., Fabian, A., & Turatto, M. 1999 IAU Circ. No. 7269.
Thielemann, F.-K., Nomoto, K., & Hashimoto, M. 1996 *ApJ*, **460**, 408.
Thorsett, S. E. & Hogg, D. W. 1999 GCN Cir. No. 197.
Turatto, M., Mazzali, P. A., Young, T. R., Nomoto, K., Iwamoto, K., Benetti, S., Cappellaro, E., Danziger, I. J., de Mello, D. F., Phillips, M. M., Suntzeff, N. B., Clocchiatti, A., Piemonte, A., Leibundgut, B., Covarrubias, R., Maza, J., Sollerman, J. 1998 *ApJ*, **498**, L129.
Turatto, M., Mazzali, P., Suzuki, T., Young, T., Nomoto, K., Benetti, S., Cappellaro, E., Danziger, I. J., Patat, R. 1999a, this volume.
Turatto, M., Suzuki, T., Mazzali, P.A., Benetti, S., Cappellaro, E., Nomoto, K., Nakamura, T., Young, T. R., Patat, F. 1999b *ApJ*, submitted (astro-ph/9910324).
van den Heuvel, E. P. J. 1994 in Interacting Binaries (eds. H. Nussbaumer & A. Orr). Springer Verlag, p. 263.
Wang, L. & Wheeler, J. C. 1998 *ApJ*, **504**, L87.
Woosley, S. E. 1993 *ApJ*, **405**, 273.
Woosley, S. E., Eastman, R. G., & Schmidt, B. P. 1999 *ApJ*, **516**, 788.
Woosley, S. E., Eastman, R. G., Weaver, T. A., & Pinto, P. A. 1994 *ApJ*, **429**, 300.
Woosley, S. E., Langer, N., & Weaver, T. A. 1995 *ApJ*, **448**, 315.
Young, T., Baron, E., & Branch, D. 1995 *ApJ*, **449**, L51.

Collapsars, Gamma-Ray Bursts, and Supernovae

By S. E. WOOSLEY, A. I. MACFADYEN
AND ALEXANDER HEGER

Department of Astronomy and Astrophysics, UCSC, Santa Cruz, CA 95064

A diverse range of phenomena is possible when a black hole experiences very rapid accretion from a disk due to the incomplete explosion of a massive presupernova star endowed with rotation. In the most extreme case, the outgoing shock fails promptly in a rotating helium star, a black hole and an accretion disk form, and a strong gamma-ray burst (GRB) results. However, there may also be more frequently realized cases where the black hole forms after a delay of from several tens of seconds to several hours as ~ 0.1 to 5 $M_\odot$ falls back into the collapsed remnant following a mildly successful supernova explosion. There, the same MHD mechanisms frequently invoked to produce GRBs would also produce jets in stars already in the process of exploding. The presupernova star could be a Wolf-Rayet star or a red or blue supergiant. Depending upon its initial pressure, the collimation of the jet may also vary since "hot" jets will tend to diverge and share their energy with the rest of the star. From these situations, one expects diverse outcomes ranging from GRBs with a large range of energies and durations, to asymmetric, energetic supernovae with weak GRBs. SN 1998bw may have been the explosion of a star in which fall back produced a black hole and a less collimated jet than in the case of prompt black hole formation.

1. Introduction

In recent years, our theoretical understanding of common GRBs has moved out of the "dark ages" of the 1980s into a BATSE and Beppo-Sax inspired "rennaisance." The burst and its afterglow in various wavelengths have been successfully modeled as the interaction of a highly relativistic jet ($\gamma \gtrsim 100$) with itself (internal shocks) and with circumstellar or interstellar material—the so called "relativistic fireball model" (e.g. Piran 1999; Meszaros 1999). The origin of the jet is still widely debated, but is generally believed to involve the formation of a stellar mass (approximately 2 to 5 $M_\odot$) black hole and the rapid accretion of matter into that hole from a disk. Modes of forming the black hole vary (e.g Fryer, Woosley, & Hartmann 1999), as do assumptions regarding the accretion rate, duration, and means of extracting disk binding energy and converting it into the relativistic motion of the jet. For accretion rates in excess of ~ 0.05 $M_\odot$ s^{-1} neutrino energy transport may be efficient (Popham, Woosley, & Fryer 1999). For lower accretion rates, and perhaps also for the higher ones, MHD processes—magnetic field reconnection in the disk, extraction of black hole spin energy, Alfven waves, magneto-centrifugal winds, etc.—are invoked.

As our understanding of GRBs has improved, an interesting "paradigm shift" has also been going on in the modeling of supernovae. For the last 30 years, most researchers have assumed a Type II (or Ib) supernova to be a consequence of neutrinos extracting a portion of the binding energy of a newly formed neutron star. The neutrinos then deposit a portion of their energy in a low density region just outside the neutron star and the resulting "bubble" of pairs and radiation explodes the rest of the star. There have been interesting exceptions along the way (e.g. LeBlanc & Wilson 1970; Bodenheimer & Woosley 1983), but, for the most part, researchers have preferred their supernovae round and without magnetic fields.

Three things have happened lately to make us suspect that this is not always the way supernovae work (though, admittedly, the exceptions may be rare). First, we have observed supernovae, notably SN 1997cy and SN 1998bw, that do not fit the traditional mold (Germany et al. 1999; Galama et al. 1998), supernovae that seem to require an order of magnitude more energy than the traditional mechanism provides and which may be associated with GRBs. Second, models for GRBs have converged on a massive presupernova star—and its explosion as a "hypernova"—as one leading candidate. Finally, supernovae may have been observed as the counterparts to two or more GRBs (990425, Galama et al. 1998; 980326, Bloom et al. 1999; 970228, Reichart 1999). Rather suddenly, the supernova community and GRB community have awakened to realize just how much they have in common.

In this paper, we explore this interface between GRBs and supernovae. We find that massive stars can produce a variety of energetic explosions ranging from traditional supernovae (by far the most frequent occurrence) to energetic GRBs, and seemingly all points in between. Ordinary supernovae still come from neutron star formation in the approximately spherically symmetric explosion of a massive ($M \gtrsim 8\ M_\odot$) star with little or no fall back, but failed or weak explosions in rotating stars give hyper-accreting black holes whose jets can both explode the star in a grossly asymmetric way and produce a variety of high energy phenomena.

2. GRB models

One leading model for a GRB involves a neutron star merging with another neutron star or with a black hole. Either way, after the merger, a black hole ends up accreting $\sim 0.01\ M_\odot$ (neutron star companion) to, at most, $\sim 0.5\ M_\odot$ (black hole companion) from a Keplerian disk. Even for this relatively simple model, assumptions and results vary widely. If the disk viscosity is high, say $\alpha \gtrsim 0.01$, the disk becomes very hot and emits its binding energy as neutrinos. Neutrino annihilation along the axis may then energize the jet (Ruffert & Janka 1999; Janka, Ruffert, & Eberl 1999; Janka et al. 1999; Rosswog et al. 1999). Since the efficiency for neutrino annihilation is small, typically $\lesssim 1\%$, and the viscous time scale, short ($\lesssim 100$ ms), this variety of model produces relatively weak, brief jets, perhaps appropriate for the class of short, hard GRBs, but unlikely to explain long energetic events like those localized by BeppoSax. The MHD variety of this model (e.g. Meszaros 1999) assumes a much lower viscosity and thus a longer time scale for the accretion, up to tens of seconds. The merit of this sort of model is that one can assume (within a large error bar) a high efficiency for extracting energy from the disk or rotating hole. This greater energy and longer time scale are both necessary and sufficient to explain the most energetic bursts observed so far.

Another leading model, and the main subject of this paper, is the *collapsar*. A collapsar is a black hole formed by the incomplete explosion of a rapidly rotating massive star (Woosley 1993; MacFadyen & Woosley 1999, henceforth MW99). It sets up the same sorts of circumstances as the merging neutron star model for GRBs, but with a number of important distinctions: 1) the event occurs only in the most massive stars and thus tracks star formation directly; 2) a supernova is produced by every GRB because the jet not only makes a GRB, but explodes the star; 3) the amount of matter available for accretion (and thus the maximum energy available for the GRB) is one to three orders of magnitude greater than for merging compact objects; 4) the duration of the jet is set by the collapse time scale of the star, not by the disk viscous time scale; no very short bursts are possible; 5) the accretion rate is likely to be lower than for the neutrino version of merging neutron stars, but faster than some MHD versions; 6) the engine is deeply embedded in a star that the jet must penetrate in order to make the GRB; 7) the star is surrounded by an extended presupernova "wind zone" in which the mass density is proportional to r^{-2}; and 8) compared to merging neutron stars, the gravitational radiation accompanying the burst is very weak. The angular momentum one invokes in the collapsar model is also much less certain than for compact objects merging by gravitational radiation. Once the disk is set up, however, the same physics that makes jets in merging neutron star models, be it neutrinos or MHD, should work equally well for collapsars. The interaction of this jet with the rest of the star and with the stellar wind is a challenging problem in radiation-hydrodynamics, but one that is tractable.

One often sees allusions to both a "hypernova" (Paczynski 1998) model and a collapsar model. We make no distinction here. We avoid the term hypernova as applied to GRBs because one of us previously used the same word to mean a super-bright pair-instability supernova (Woosley & Weaver 1982). However, to the extent that the term hypernova is used by the GRB community, it is an observational phenomenon caused by a collapsar.

3. Supernova fallback

The simplest way, conceptually, to form a black hole in a massive star, and thus set up the conditions for the collapsar model, is for the traditional neutrino powered explosion to fail. The iron core collapses and within a second or so has made a black hole into which the rest of the star proceeds to accrete. This may be the common case for stars above about 35–40 $M_\odot$ (Fryer 1999), although uncertainties in convection, mass loss, rotationally induced mixing, and the explosion mechanism itself make this an uncertain number—and one that may vary with redshift and metallicity. If the star loses its hydrogen envelope along the way, and if the jet produced by the accretion maintains its energy and focus for a longer time than it takes the jet to tunnel through the star, about 5–10 s, a common GRB is produced (MW99). Otherwise a weaker, less collimated GRB results (helium star case; MacFadyen & Woosley 1998), or an energetic asymmetric supernova (§6).

However, there should also be a range of stellar masses for which a black hole is not made promptly, but after a "successful" shock has already been launched. The binding energies of stellar helium cores outside the collapsing iron core increases with their mass. The energy of the neutrino engine seems, if anything, to decrease with mass. Thus there is a range of masses, estimated by Fryer to be roughly 20 to 40 $M_\odot$, where a supernova occurs, but so much matter fails to achieve escape and falls back onto the neutron star that it turns into a black hole. This delayed production of a black hole is probably a more frequent occurence than prompt black hole formation.

As a representative case, consider a 25 $M_\odot$ main sequence star evolved with mass loss, rotationally induced mixing, and angular momentum transport (Heger, Woosley, & Langer 1999). This star ends its life as a red supergiant with an iron core of 1.90 $M_\odot$, a helium core of 8.06 $M_\odot$, and a low density envelope of 6.57 $M_\odot$ (total mass 14.6 $M_\odot$). The presupernova stellar radius is 8.1×10^{13} cm. The model has sufficient angular momentum in the equator ($j \sim 10^{17}$ cm^2 s^{-1}) to form an accretion disk outside the black hole.

Explosions were simulated in this star using a piston at the edge of the iron core (MacFadyen, Woosley, & Heger 1999). The motion of this piston was varied so as to produce a kinetic energy at infinity for the ejecta ranging from 0.255×10^{51} erg (Model 25A1) to 2.09×10^{51} erg (Model 25A16). The subsequent evolution was followed using two different one-dimensional hydrodynamics codes, KEPLER (an implicit Lagrangian hydrodynamics code) and PROMETHEUS (an explicit Eulerian code). For similar assumptions regarding the launch of the shock and the inner boundary condition, the

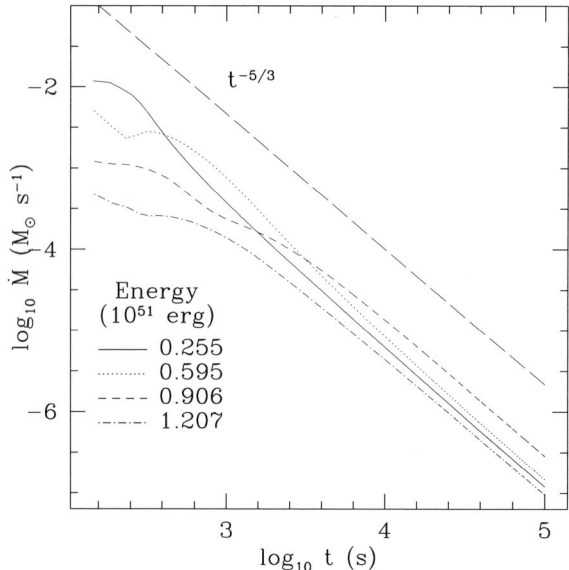

FIGURE 1. Accretion rates for fall back in five different explosions of a 25 M$_\odot$ presupernova star (see text). These five explosions gave kinetic energies at infinity for their ejecta of 0.255, 0.595, 0.906, and 1.207 × 10^{51} erg. The integrated fall back masses for these spherically symmetric calculations were 3.71, 2.85, 1.39 and 0.48 M$_\odot$ respectively. Characteristic time scales for the fallback are 100, 450, 1140, and 1060 s. Calculations were carried out using a one-dimensional version of the PROMETHEUS hydrodynamics code.

results of the two codes agreed. For energies above 1.5×10^{51} erg, all matter external to the piston was ejected, but for lower energies an increasing amount of mass fell back to the origin (Fig. 1). At late times the accretion rate followed the $t^{-5/3}$ scaling predicted by Chevalier (1989).

It is noteworthy that the accretion rate during the time most of the mass falls back, about 0.001 to 0.01 M$_\odot$ s^{-1}, is very similar to that frequently invoked in the MHD version of the merging neutron star model (§2), especially for the lower explosion energies. If jets are to form in one place, surely they should form in the other. However, for these relatively low accretion rates, the disk temperature will be too cool to emit neutrinos efficiently. Any jet that forms must be powered by MHD processes. If we make a simple *ansatz* that the jet energy, at any point in time, is an efficiency factor, ϵ, times $\dot{M}c^2$, with $\epsilon \sim 0.001$–0.01 (certainly modest compared to many assumptions in the literature), then the energy potentially available for making a jet in Model A01 is $\sim 10^{52}$–10^{53} erg. This is large compared both to the energy of the initial shock in Model 25A1 and the energy of a typical supernova.

4. Some general considerations

Unfortunately, while a compelling case can be made, both on observational grounds (e.g. Pringle 1993; Livio 1999) and from theory (MacFadyen, Woosley, & Heger 1999) for linking the jet energy to the accretion rate, the energy alone does not define the model. One still needs to know the initial partition between internal and kinetic energy and the beaming angle. In "thermal" models, such as the neutrino version of merging neutron stars or collapsars, the initial energy is overwhelmingly in the form of radiation and pairs. In fact, the plasma starts at rest with $aT^4/\rho c^2 \sim \gamma_f \sim 100$. Expansion of the radiation converts internal energy into kinetic energy very far from the source. For Poynting flux models, on the other hand, the jet may be born relatively cold. The initial collimation of the jet may be either by pressure and density gradients, as in the collapsar of MW99, or by magnetic fields, or both. Lacking details of the jet formation process, ambiguity in the collimation angle and mass to energy ratio makes predictions difficult, but hot, poorly collimated jets will clearly have a harder time penetrating the star.

One also expects some systematic differences between cases in which the black hole forms promptly (Case A) or by fall back (Case B) that may bear on this issue of collimation. The lower accretion rate in Case B suggests a smaller disk mass in steady state (Popham et al. 1999) and the confining pressure of the medium through which the jet initially propagates will also be less in Case B, because the star has already partly exploded. In both Case A and B there will still be an inner disk that will help to collimate the initial outflow, but, depending upon how much mass falls back and its angular momentum, that disk may not extend to such large radii in Case B. All in all, one expects that the geometrical focusing of the jet at least, may not be so great in Case B, especially for thermal models. The extent of MHD collimation is, however, unknown.

Given an initially well collimated jet, one still faces a formidable computational task following its propagation out to, say, 1000 Schwarzschild radii. The jet is an inherently relativistic and can only be described accurately by a special relativistic (SR) calculation. To do less gives, at best, a qualitative description of the jet propagation while possibly generating unrealistic artifacts such as superluminal speeds (MW99). Special relativistic codes are available (e.g. Aloy et al. 1999) and can be adapted to the problem, but, unfortunately, results are not yet available.

There are several SR effects worth keeping in mind though. First, a jet of radiation and matter has quite different properties, in SR, from one composed only of matter. In particular, the equivalent "dynamical" density, which must be regarded as a vector, is related to the rest mass density, n, by (Rosen et al. 1999)

$$\rho = 2n\gamma^2 \left(\frac{\gamma}{\gamma+1} + \frac{\Gamma p}{(\Gamma-1)nc^2} \right) \quad (4.1)$$

which clearly shows the increase of the effective ρ with γ and p. Here $\gamma = (1-(v/c)^2)^{-1/2}$, n is the rest mass density, Γ, the adiabatic index, and p, the pressure. As noted earlier, for a thermal model, p/nc^2 is initially about 100. As p turns into γ by expansion, the relativistic correction to the momentum becomes anisotropic and greatest along the jet. As a result, SR jets of radiation and matter have much more penetrating power than Newtonian jets with rest mass density, n.

Time dilation also plays an important role. In the frame of the jet, the star is crossed in a shorter time than in the lab frame. Yet perpendicular to the jet, motions remain sub-relativistic and clocks run at similar rates. Thus a SR jet loaded with radiation will diverge, in the laboratory frame, less than a similar Newtonian jet loaded with radiation. Indeed, in a Newtonian code, the sound speed and the jet speed would both be $\sim c$.

Together these effects help to explain why a jet, initially focused by the geometry of the accretion disk or by the magnetic field near the hole, but loaded with radiation, might maintain its collimation while its internal energy is converted into kinetic energy. Eventually, if the star is not too big, the jet escapes, reaches its asymptotic γ, and produces a GRB by running into circumstellar material.

However, we shall be particularly interested here in another case—jets that lose their energy before breaking out, share that energy with the star, and thus become only mildly relativistic. Our present calculations are, of necessity, carried out using a Newtonian version of PROMETHEUS, but we have attempted to capture the flavor of mildly relativistic jets as they propagate through the helium core and red giant envelope of an exploding star. To do so, we picked an inner boundary radius, 10^9 cm, which is computationally expedient (i.e. not too small), but still well within the helium core, and at about the radius where radiation and rest mass might start to become comparable (see, e.g. Fig. 26 of MW99), especially for MHD models in which the initial thermal loading of the jet is not so large. Besides the supernova structure when the jet starts to propagate, there are three key ingredients to the model, all specified at 10^9 cm: 1) the kinetic energy of the jet as a function of time, given by $\epsilon \dot{M} c^2$; 2) the opening angle of the jet, assumed to have a 10 degree half-angle, and 3) the ratio of internal pressure to kinetic energy, f_P. This last parameter turns out to be quite important. If the jet pressure is large compared to the stellar surroundings in which it propagates, the jet will diverge. If it is less, the jet may, under some circumstances, be hydrodynamically focused to a still smaller opening angle. For the calculations we shall consider, the pressure in the jet is dominantly due to radiation—though not by a large margin for the smaller values of f_P.

The principal effect of f_P is to increase the tendency of the jet to diverge. This divergence may, in fact have already occurred inside 10^9 cm. For a relativistic jet, the effective value of f_P would actually be much larger owing to the previously mentioned modification of the dynamical density and time dilation. In order to keep our jet velocities on our Newtonian grid below c however, we are compelled to study only $f_P \lesssim 1$.

5. Some representative calculations

To illustrate the possible characteristics of supernovae exploded by jets, we calculated the two-dimensional evolution of Model 25A1 incorporating parameterized jets. Details of these and other similar calculations will be presented in a forthcoming paper (MacFadyen, Woosley, & Heger 1999). The spherically symmetric explosion, followed until 100 s after the launch of a weak shock in the KEPLER code, was remapped onto the Eulerian grid of a two-dimensional version of PROMETHEUS. This grid used 150 radial zones spaced logarithmically between an inner boundary at 10^9 cm and the outer boundary at 8.1×10^{13} cm. Forty angular zones concentrated near the pole were used to simulate one quadrant of the stellar volume, assuming axial and reflection symmetry across the equatorial plane. The angular resolution varied from $1.25°$ at the pole to $3.5°$ at the equator. At 100 s, the inner 1.99 $M_\odot$ of the star was removed and replaced by an open (zero radial gradient of all variables) boundary condition at 10^4 km. The 1.99 $M_\odot$ continued to contribute to the gravitational potential as a central point mass and mass accreting through the inner boundary was added to the point mass during the calculation. At this time the weak initial shock was already at 1.1×10^5 km when the jet was turned on at the inner boundary.

We gave the jet a constant velocity at this inner boundary, 10^{10} cm s^{-1}, a compromise between what the code could realistically calculate (v less than c) and the true relativistic nature of the initial jet. This velocity, the radius of the inner boundary, and the (Newtonian) kinetic energy of the jet implied a jet density, 1.9×10^3 g cm^{-3} ($\dot{M}/0.01\ M_\odot$ s^{-1}) ($\epsilon/0.01$). This assumed that any internal energy deposited in the jet near the black hole had been decompressed by adiabatic expansion to the point where, at 10^9 cm, it was small compared to ρv^2.

We considered four cases, $\epsilon = 0.001$, $f_P = 0.001$, 0.01, and 0.1 and $\epsilon = 0.01$, $f_P = 0.01$. The results are summarized in Table 1 and Figs. 2–4. Here the name follows the convention "AMN" where "A" indicates the model was based upon the weakest explosion considered of a 25 $M_\odot$ (main sequence mass) supernova (0.255×10^{51} erg; Fig. 1), "M" is the exponent of the efficiency factor, $\epsilon = 10^{-M}$, and "N" is the exponent of the pressure

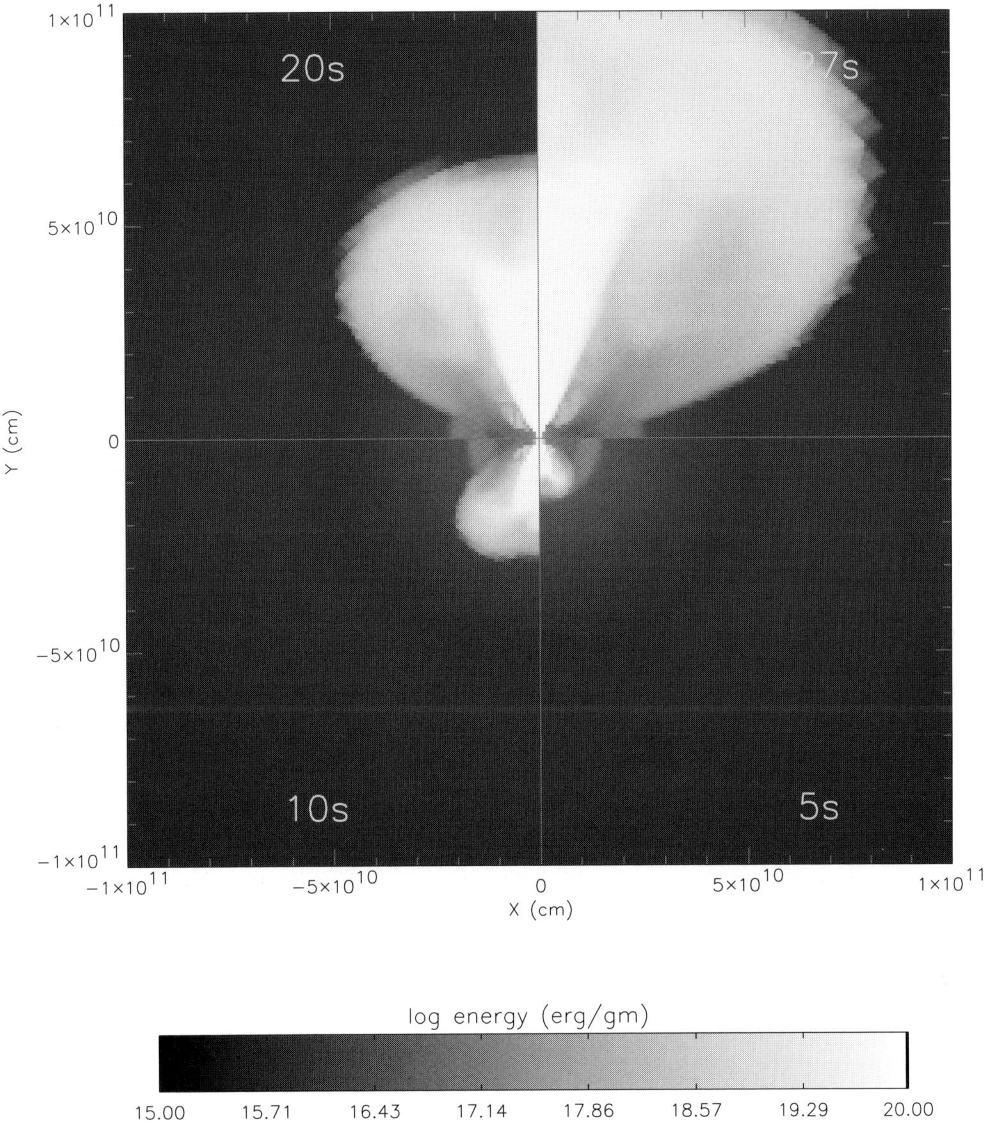

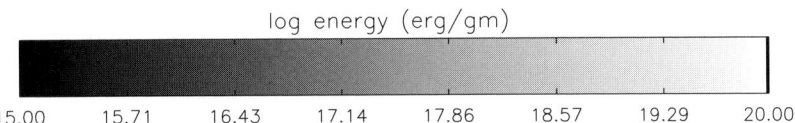

FIGURE 2. The total energy density of the jet and explosion is shown at times of 5, 10, 20 and 27 s after initiation for jet Model A22 (Table 1). The passage of the jet initiates a shock that propagates to lower latitudes, eventually exploding the entire star. The supernova shock can be seen at a radius of about 2×10^{10} cm.

factor, $f_P = 10^{-N}$. The mass accreted, ΔM in Table 1, is smaller than the 3.71 $M_\odot$ computed without a jet (Fig. 1) for Model 25A1, because the jet impeded the accretion at high latitude and because the accretion was not quite over at after 500 seconds (Fig. 1). The total energy input by the jet was still $\epsilon \Delta M c^2$, but the number in Table 1 was reduced by the work done up to 500 s in unbinding the star and by the internal and kinetic energy which passed inside the inner boundary. The 2.55×10^{50} erg due to the initial shock has been subtracted in Table 1 so that E_{tot} reflects only the energy input by the jet.

The angular factor $R(\theta > 10°)$ is the ratio of the integral of the kinetic energy due to the jet outside 10 degrees polar angle (98.5% of the sky) to the total kinetic energy in

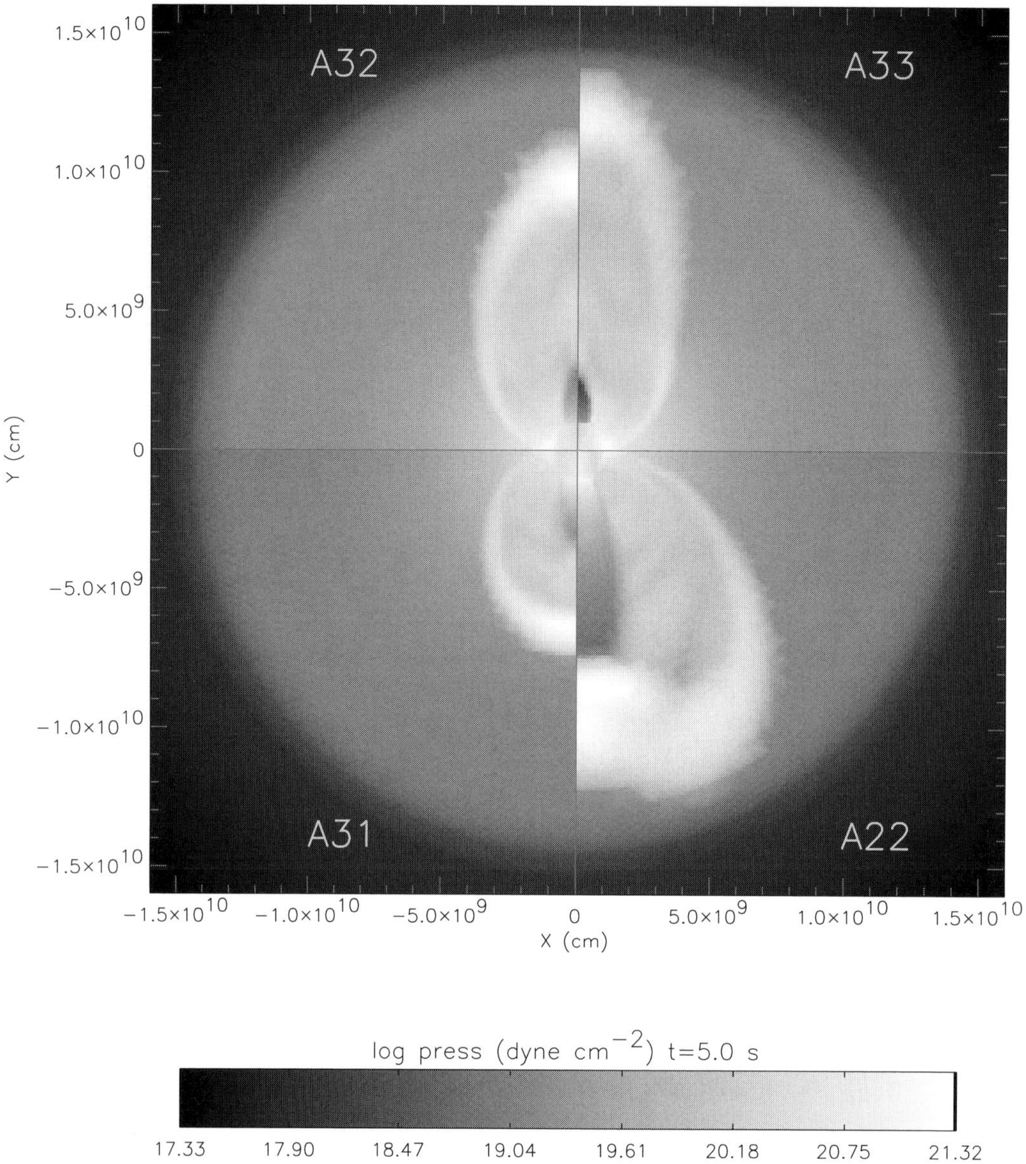

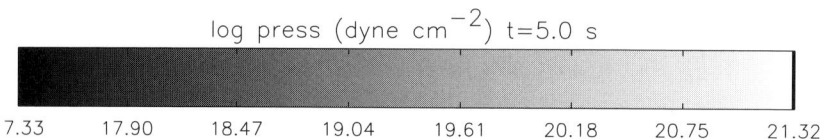

FIGURE 3. Pressure in the jet and surrounding star at 5.0 s after the initiation of the jet in four different models. Higher presure leads to greater jet divergence, more mass swept up, and slower propagation. Model A22 had a higher jet energy than the other models (Table 1).

the star due to the jet (see Fig. 4). These energies were computed by taking the total kinetic energy at 400 s after jet initiation in both regions and subtracting the kinetic energy of the initial supernova shock. $R(\theta > 10°)$ measures the extent to which the jet spread laterally and shared its energy with the rest of the star. The limiting case R=0 would correspond to a jet that shared none of its energy with the supernova outside an initial 10° polar angle. This sort of behavior is expected for "cold" jets with internal pressure small compared to the exploding helium core. The other extreme, where the jet shared its energy evenly with the entire star and produced a spherical explosion, would correspond to $R = \cos\theta = 0.985$. Our "hot" jets lie somewhere between these two limits.

Name	ϵ	f_P	ΔM ($M_\odot$)	E_{tot} (10^{51} erg)	$R(\theta > 10°)$	$R(\theta > 20°)$
A33	0.001	0.001	2.76	3.38	0.075	0.037
A32	0.001	0.01	2.69	3.23	0.102	0.047
A31	0.001	0.1	2.51	3.00	0.425	0.256
A22	0.01	0.01	1.72	19.91	0.429	0.230

TABLE 1. Explosion characteristics at t = 400 s after jet initiation

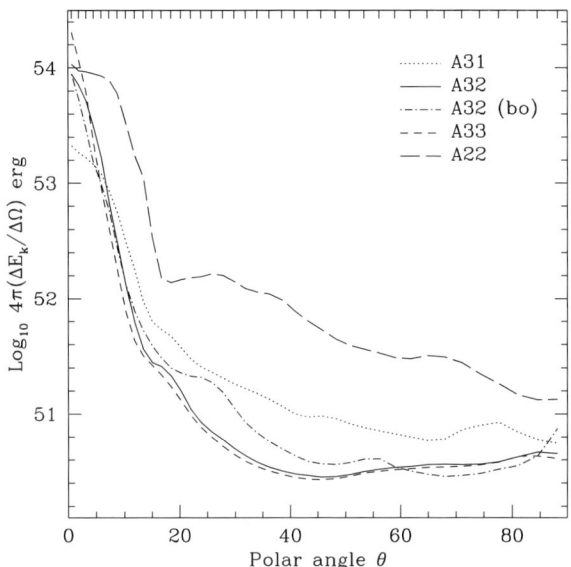

FIGURE 4. The "equivalent isotropic kinetic energy" as a function of polar angle for four models having variable energy efficiency factors and internal pressures (Table 1 and text). Model A32 is shown at two times, once at 400 s after the initiation of the jet and later, at 7716 s, as the jet penetrated the surface of the star at 8×10^{13} cm; dash-dot line. Other models are also shown for comparison at 400 s. Note that the degree of collimation is strongly dependent upon f_P. Equivalent isotropic kinetic energy is defined as the integral from the center to surface of the star of its kinetic energy in the solid angle subtended by θ and $\theta + \Delta\theta$ divided by the solid angle, $2\pi(\cos\theta - \cos(\theta + \Delta\theta))$ and multiplied by 4π. The injected energy at the base of the jet would be a flat line out to ten degrees with a value equal to $66\,\epsilon\Delta M c^2$ with ΔM in Table 1 and $66 = (1 - \cos(10°))^{-1}$. Tick marks along the top axis give the angular zoning of the two dimensional code.

The quantity $R(\theta > 20°)$ was similarly computed for a polar angle of 20°. The isotropic limit there would be 0.940.

In all cases a very energetic asymmetric supernova resulted. Since the integrated mass of the (Newtonian) jet in our code was comparable to that of the stellar material within 10 degrees, the time for jet break out was approximately the stellar radius divided by the jet input speed. In reality, that would be $\sim R/c$, or for a red supergiant several thousand seconds. Since the energy of the jet engine had declined greatly by that time, due to the declining accretion rate (Fig. 1), and the jet had swept up far more than γ^{-1} of its rest mass, the jet that broke out was only mildly relativistic. Both the long time scale and the low energy input are inconsistent with what is seen in common GRBs. However, if the hydrogen envelope had been lost, a longer than typical GRB could have resulted.

Figs. 3 and 4 illustrate how the pressure balance between the jet and the star through which it propagates affected its collimation properties. The interaction at late times with the hydrogen envelope had relatively little effect on the angular energy distribution which was set chiefly by f_P and the interaction with the helium core. Model A33 had the lowest internal pressure (note that the actual value of the initial pressure depends upon the product of ϵ and f_P). The final jet was collimated even more tightly than given by its initial injection. That is, a jet initially of 10 degrees half width will exit the star with a FWHM of less than two degrees, about 0.06% of the sky (though the angular resolution of the code is questionable for such small angles). Meanwhile the energy at larger angles was not much greater than that given by the initial, weak spherically symmetric explosion, $10^{50.4}$ erg. There was little sharing of the jet energy with the star and, except for the jet, the supernova energy remained low.

This is to be contrasted with Models A22 and A31 where the jet collimation was much weaker and much more energy was shared with the star. Note that though Model A22 had about 6 times the total energy of A31 owing to its larger ϵ the fraction of energy at large angles in both these models was significantly greater than in Models A32 and A33. Model A22 would be an especially powerful supernova as well as one accompanied by a jet.

6. Supernovae and GRB diversity

Provided the necessary conditions for the collapsar model can be met—black hole formation in a massive star with sufficient angular momentum to make a disk—the discussion and results of the previous two sections suggest a wide variety of possible outcomes, including, besides ordinary GRBs:

"Smothered" and broadly beamed gamma-ray bursts; GRB 980425

These can occur in helium stars in which the jet either fails to maintain sufficient focus (e.g. is too "hot" compared to the star through which it propagates), or loses its energy input before breaking out of the star ($\lesssim$ 10 s; MW99). An energetic supernova still occurs (SN 1998bw, in this case) and a weak GRB is produced, not by the jet itself, but by a strong, mildly relativistic shock from break out interacting with the stellar wind. (Woosley, Eastman, & Schmidt 1999). Because these events are so low in gamma-ray energy, many could go undetected by BATSE. Indeed these could be the most common form of GRB in the universe. Because the initial jet may be less effectively collimated in GRBs made by supernova fall back, it is tempting to associate these phenomena with delayed black hole formation and the stronger GRBs with prompt black hole formation. More study is needed.

Long gamma-ray bursts; $\tau_{burst} \gtrsim$ 100 s

Though typical "long, complex bursts" observed by BATSE last about 20 seconds, there are occasionally much longer bursts. For example, GRB 950509, GRB 960621, GRB 961029, GRB 971207, and GRB 980703 all lasted over 300 s. These long durations may simply reflect the light crossing time of the region where the jet dissipates its energy (modulo γ^{-2}), especially in the "exterior shock model" for GRBs. However, if the event is due to internal shocks, the duration depends on the time the engine operates. Such long bursts would imply enduring accretion on a much longer time scale than one expects in the simplest collapsar model where the black hole forms promptly. The fallback powered models discussed in this paper could maintain a GRB for these long time scales (Fig. 1).

Very energetic supernovae—SN 1997cy

Germany et al. (1999) have called attention to this extremely bright supernova with an unusual spectrum. The supernova was Type IIn and its late-time light curve, which approximately followed the decay rate of ^{56}Co, would require $\gtrsim 2$ M$_\odot$ of ^{56}Ni to explain its brightness. Perhaps this was a pair-instability supernova (Woosley & Weaver 1982; Heger, Woosley, & Waters 1999). On the other hand, circumstellar interaction could be the source of the energy and the agreement with $\tau_{1/2}$ (^{56}Co) merely fortuitous. This would require both a very high explosion energy and a lot of mass loss just prior to the supernova. The sort of model described in §5, especially Model A22, could provide the large energy in a massive star that would be naturally losing mass at a high rate when it died. But the radius is too large and the jet would share its energy with too great a mass to make a common GRB. Therefore we regard the detection of a short, hard GRB from the location of SN 1997cy as spurious.

Nucleosynthesis—^{56}Ni and the r-process

An explosion of 10^{52} erg focused into 1% of the star (or 10^{53} erg into 10%) will have approximately the same shock temperature as a function of radius as an isotropic explosion of 10^{54} erg. From the simple expression $\frac{4}{3}\pi r^3 aT^4 \sim 10^{54}$ erg (Woosley & Weaver 1995), we estimate that a shock temperature in excess of 5 billion K will be reached for radii inside 4×10^9 cm. The mass inside that radius external to the black hole (assumed mass initially 2 M$_\odot$) depends on how much expansion (or collapse) the star has already experienced when the jet arrives. Provided the star has not expanded much before the jet arrives, an approximate number comes from the presupernova model, 3 M$_\odot$ times the solid angle of the explosion divided by 4π, or ~ 0.03 M$_\odot$. Additional ^{56}Ni is probably synthesized by the wind blowing off the accretion disk (MW99; Stone, Pringle, & Begelman 1999) and this may be the dominant source in supernovae like SN 1998bw.

The composition of the jet itself depends upon details of its acceleration that are hard to calculate. However it should originate from a region of high density and temperature (Popham, Woosley, & Fryer 1999). The high density will promote electron capture and lower Y_e. The high entropy, low Y_e, and rapid expansion rate are what is needed for the r-process (Hoffman, Woosley, & Qian 1997). The mass of the jet, $\sim 10^{-4}$ M$_\odot$ (corrected for relativity) is enough to contribute significantly to the r-process in the Galaxy even if the event rate was $\lesssim 1\%$ that of supernovae and the jet carried only a fraction of its mass as r-process.

Soft x-ray transients from shock breakout

Focusing a jet of order 10^{52} ergs into 1–10% of the solid angle of a supernova results in a shock wave of extraordinary energy (Fig. 4). As it nears the surface of the star, this shock is further accelerated by the declining density gradient. MacFadyen, Woosley, & Heger (1999) estimate, for a 10^{54} erg (isotropic equivalent) shock, a break out transient of 10^{49} erg s^{-1} (times $(1-\cos\theta_j)$, the solid angle of the jet at break out divided by 4π, where θ_j is the half opening angle of the jet at breakout) for ~ 10 s. The color temperature at peak would be approximately 2×10^6 K (see also Matzner & McKee 1999). A 10^{53} erg shock gave a transient about half as hot and ten times longer and fainter. The impact of the mildly relativistic matter could give an enduring x-ray transient like the afterglows associated with some GRBs, even though the time scale is too long for the x-ray burst to be a common GRB itself.

Mixing in supernovae—SN 1987A

It is generally agreed (Arnett et al. 1989) that the explosion that gave rise to SN 1987A initially produced a neutron star of approximately 1.4 M$_\odot$. There may have been ~ 0.1 M$_\odot$ of fallback onto that neutron star (Woosley 1988) and a black hole may or may not have formed. Again invoking our *ansatz* that L$_{\rm jet} = \epsilon \dot{M} c^2$, even for $\epsilon \sim 0.003$, we have a total jet energy of 6×10^{50} erg. This is about half of the total kinetic energy inferred for SN 1987A. Thus very appreciable mixing and asymmetry would be introduced by such a jet—*provided the material that fell back had sufficient angular momentum to accumulate in a disk outside the compact object*. However this would not be enough energy to make a powerful gamma-ray burst as proposed by Cen (1999).

Still to be discovered

It may be that, especially with common GRBs, we have just seen the "tip of the iceberg" of a large range of high energy phenomena powered by hyper-accreting, stellar mass black holes. We already mentioned the possibility of a large population of faint, soft bursts like GRB 980425. Other possibilities include very long GRBs below the threshold of BATSE, "orphan" x-ray afterglows from jet powered Type II supernovae, supernova remnants having toroidal structure, GRBs from the first explosions of massive stars after recombination, and more. It is an exciting time.

7. Does it all work?

Exciting that is, if it all works as described. That a hyper-accreting black hole (M$_{\rm hole}$ = 2 to 10 M$_\odot$, accreting 10^{-1}–10^1 M$_\odot$ s^{-1}) gives rise to an energetic jet with dramatic observational consequences seems to us unavoidable. True the physics of jet formation is poorly understood, but the ubiquity of jets in all sorts of systems where disk accretion is going on, the success of the basic idea of AGNs as accreting massive black holes, and the identity of "microquasars" as accreting black holes all argue that this is an assumption worth exploring. That supernovae sometimes form black holes, both promptly and in a delayed manner, also seems unavoidable. Our calculatiions show that if a jet forms in a massive collapsing star, and if that jet has only a fraction of a per cent of the energy potentially available from the accretion process, that energetic supernovae and GRBs are a likely outcome.

The weakest assumption in all the models discussed here is that the requisite amount of angular momentum is present to form a disk. The best available stellar evolution models suggest it is there (Heger, Langer, & Woosley 1999), but these calculations have left out magnetic field effects that might lead to the dramatic slowing of the rotation of the helium core, especially in red supergiants (Spruit & Phinney 1998). These models also imply that neutron stars may be born rotating near break up. Whether either of these concerns will ultimately prove fatal to the model remains to be seen. Since GRBs are much rarer in the universe than supernovae, it is of course possible that the production of GRBs demands some very special circumstances, e.g. the merging of two stripped helium stars already in a late stage of evolution (Fryer, Woosley, & Hartmann 1999).

This work has been supported by NASA (NAG5-8128 and MIT SC A292701), the NSF (AST 97-31569), and the Department of Energy ASCI Program (W-7405-ENG-48), and by the A. V. Humboldt-Stiftung (1065004).

REFERENCES

Aloy, M. A., Ibanez, J. M., Marti, J. M., & Müller, E. 1999 *ApJS*, **122**, 151, astro-ph/9903352.
Arnett, W. D., Bahcall, J. N., Kirshner, R. P., & Woosley, S. E. 1989 *ARAA*, **27**, 629.
Bloom, J. S. 1999 *Nature*, submitted, astro-ph/9905301.
Bodenheimer, P. & Woosley, S. E. 1983 *ApJ*, **269**, 281.
Cen, R. 1999 *ApJ*, submitted, astro-ph/9904147.
Chevalier, R. A. 1989 *ApJ*, **346**, 847.
Fryer, C. L. 1999 *ApJ*, in press, astro-ph/9902315.
Fryer, C. L., Woosley, S. E., & Hartmann, D. H. 1999 *ApJ*, in press, astro-ph/9904122.
Galama, T., et al. 1998 *Nature*, **395**, 670.
Germany, L., Reiss, D. J., Sadler, E. M., Schmidt, B., & Stubbs, C. W. 1999 *ApJ*, submitted, astro-ph/9906096.
Heger, A., Woosley, S. E., & Waters, R. 1999 *ApJ*, in preparation.
Heger, A., Langer, N., & Woosley, S. E. 1999 *ApJ*, in press, astro-ph/9904132.
Heger, A., Woosley, S. E., & Langer, N. 1999 *ApJ*, in preparation.
Hoffman, R. D., Woosley, S. E., & Qian, Y. 1997 *ApJ*, **482**, 951.
Janka, H.-Th., Ruffert, M., & Eberl, T. 1999 in *Nuclei in the Cosmos, V* (eds. N. Prantzos & S. Harissopulos), p. 325. Editions Frontieres. astro-ph/9810057.
Janka, H.-Th., Eberl, T., Ruffert, M., & Fryer, C. 1999 *ApJ*, submitted, astro-ph/9908290.
LeBlanc, J. M. & Wilson, J. R. 1970 *ApJ*, **161**, 541.
Livio, M. 1999 *Phys. Rept.*, **311**, 225.
MacFadyen, A. & Woosley, S. E. 1998 *BAAS*, **30**, No. 4, 1311.
MacFadyen, A. & Woosley, S. E. 1999 *ApJ*, in press, astro-ph/9810274.
MacFadyen, A. I., Woosley, S. E., & Heger, A. 1999 *ApJ*, in preparation.
Matzner, C. D. & McKee, C. F. 1999 *ApJ*, **510**, 379, astro-ph/9807046.
Meszaros, P. 1999 *Proc. 19th Texas Symposium*, astro-ph/9904038, *Nuc. Phys. B*, in press.
Paczynski, B. 1998 *ApJL*, **494**, 45.
Piran, T. 1999 *D. N. Schramm Memorial Volume*, to be published, astro-ph/9907392. See also *Physics Reports*, **314**, 575.
Popham, R., Woosley, S. E., & Fryer, C. 1999 *ApJ*, in press, astro-ph/9807028.
Pringle, J. E. 1993, in *Astrophysical Jets* (eds. D. Burgarella, M. Livio, & C. P. O'Dea), p. 1. Cambridge Univ. Press.
Reichart, D. E. 1999 *ApJL*, in press, astro-ph/9906079.
Rosen, A., Hughes, P. A., Duncan, G. C., & Hardee, P. E. 1999 *ApJ*, **516**, 729, astro-ph/9906491.
Rosswog, S., Liebendörfer, M., Thielemann, F.-K., Davies, M., Benz, W., & Piran, T. 1999 *A&A*, **341**, 499.
Ruffert, M. & Janka, H.-Th. 1999 *A&A*, **344**, 573.
Spruit, H. C. & Phinney, E. S. 1998 *Nature*, **393**, 139.
Stone, J. M., Pringle, J. E., & Begelman, M. C. 1999 *MNRAS*, in press, astro-ph/9908185.
Woosley, S. E. 1988 *ApJ*, **330**, 218.
Woosley, S. E. 1993 *ApJ*, **405**, 273.
Woosley, S. E. & Weaver, T. A. 1982, in *Supernovae: A Survey of Current Research* (eds. M. J. Rees & R. J. Stoneham), p. 79. Reidel.
Woosley, S. E. & Weaver, T. A. 1995 *ApJS*, **101**, 181.
Woosley, S. E., Eastman, R. G., Schmidt, B. P. 1999 *ApJ*, **516**, 788, astro-ph/9806299.

Pre-Supernova evolution of massive stars

By NINO PANAGIA[1,2] AND GIUSEPPE BONO[3]

[1]Space Telescope Science Institute, 3700 San Martin Drive, Baltimore, MD 21218

[2]On assignment from the Space Science Department of ESA

[3]Osservatorio Astronomico di Roma, Via Frascati 33, Monte Porzio Catone, Italy

We present the preliminary results of a detailed theoretical investigation on the hydrodynamical properties of Red Supergiant (RSG) stars at solar chemical composition and for stellar masses ranging from 10 to 20 $M_\odot$. We find that the main parameter governing their hydrodynamical behavior is the effective temperature, and indeed when moving from higher to lower effective temperatures the models show an increase in the dynamical perturbations. Also, we find that RSGs are pulsationally unstable for a substantial portion of their lifetimes. These dynamical instabilities play a key role in driving mass loss, thus inducing high mass loss rates (up to almost 10^{-3} $M_\odot$ yr^{-1}) and considerable variations of the mass loss activity over timescales of the order of 10^4 years. Our results are able to account for the variable mass loss rates as implied by radio observations of type II supernovae, and we anticipate that comparisons of model predictions with observed circumstellar phenomena around SNII will provide valuable diagnostics about their progenitors and their evolutionary histories.

1. Introduction

More than 20 years of radio observations of supernovae (SNe) have provided a wealth of evidence for the presence of substantial amounts of circumstellar material (CSM) surrounding the progenitors of SNe of type II and Ib/c (see Weiler et al., this Conference, and references therein). Also, the radio measurements indicate that (*a*) the CSM density falls off like r^{-2}, suggesting a constant velocity, steady wind, and that (*b*) the density is so high as to require a ratio of the mass loss rate, $\dot{M}$, to the wind velocity, w, to be higher than $\dot{M}/w \sim 10^{-7}$ $M_\odot$ yr^{-1}/km s^{-1}. These requirements are best satisfied by red supergiants (RSG), with original masses in the range 8–30 $M_\odot$, that indeed are the putative progenitors of SNII. Note that in the case of SNe Ib/c, the stellar progenitor cannot provide such a dense CSM directly and that a wind from a binary companion must be invoked to explain the observations (Panagia & Laidler 1988, Boffi & Panagia 1996, 2000).

This scenario is able to account for the basic properties of all radio SNe. However, the evolution of SN 1993J indicated that the progenitor mass loss rate had declined by almost a factor of 10 in the last few thousand years before explosion (Van Dyk et al. 1994). In addition, there are SNe, such as SN 1979C (Montes et al. 2000), SN 1980K (Montes et al. 1998), and SN 1988Z (Lacey et al. 2000), that have displayed relatively sudden changes in their radio emission evolution about 10 years after explosion, which also cannot be explained in terms of a constant mass loss rate. Since a SN shock front, where the radio emission originates, is moving at about 10,000 km s^{-1} and a RSG wind is typically expanding at 10 km s^{-1}, a sudden change in the CSM density about ten years after explosion implies a relatively quick change of the RSG mass loss rate about 10,000 years before it underwent the SN explosion. These findings are summarized in Figure 1 that, for several well studied RSNe, displays the mass loss rate implied by radio observations as a function of the look-back time, calculated simply as the actual time since explosion multiplied by a factor of 1000, which is the ratio of the SN shock velocity to the RSG wind velocity.

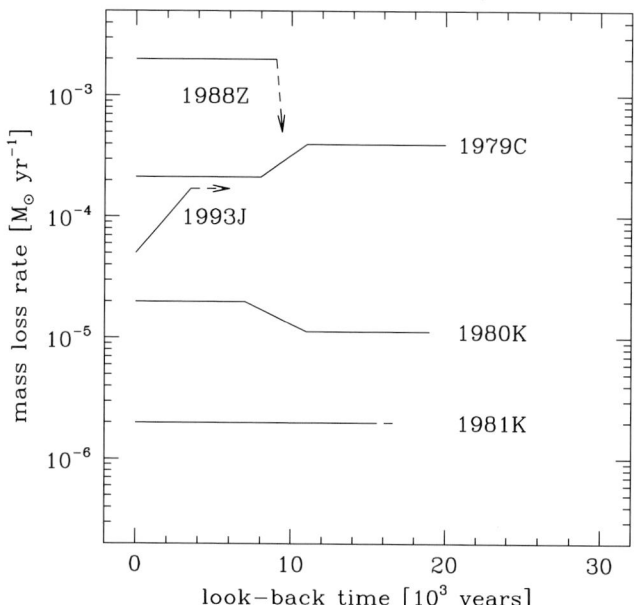

FIGURE 1. Mass loss rates as a function of look-back time as measured for a number of radio supernovae (adapted from Weiler et al. 2000; schematic)

Additional evidence for enhanced mass loss from SNII progenitors over time intervals of several thousand years is provided also by the detection of relatively narrow emission lines with typical widths of several 100 km s^{-1} in the spectra of a number of SNII (e.g. SN 1978K: Ryder et al. 1993, Chugai, Danziger & Della Valle 1995, Chu et al. 1999; SN 19997ab: Salamanca et al. 1998; SN 1996L: Benetti et al. 1999), that indicate the presence of dense circumstellar shells ejected by the SN progenitors in addition to a more diffuse, steady wind activity.

We note that a time of about 10,000 years is a sizeable fraction of the time spent by a massive star in the RSG phase and implies a kind of variability which is not predicted by standard stellar evolution. In particular, a timescale of $\sim 10^4$ years is considerably shorter than the H and He burning phases but is much longer than any of the successive nuclear burning phases that a massive star goes through before core collapse (e.g. Chieffi et al. 1999). Therefore, some other phenomenon is to be sought to properly account for the observations.

Another problem which needs to be addressed is the actual rate of mass loss for red supergiants. The observational evidence is that mass loss rates in the range 10^{-6}–10^{-4} $M_\odot$ yr^{-1} are commonly found in RSG, with a relatively steep increase in mass loss activity for the coolest stars (e.g. Reid, Tinney & Mould 1990, Feast 1991). On the other hand, there is no satisfactory theory to predict mass loss rates in these phases of stellar evolution, and current parametrizations fall short of describing the phenomenon in detail. For example, let us consider the classical formula by Reimers (1975),

$$\log(\dot{M}) = -12.6 + \log\left(\frac{LR}{GM}\right) + \log(\eta)$$

which can be rewritten as:

$$\dot{M} \propto \eta \frac{L^{1.5}}{M\, T_{\text{eff}}^2}$$

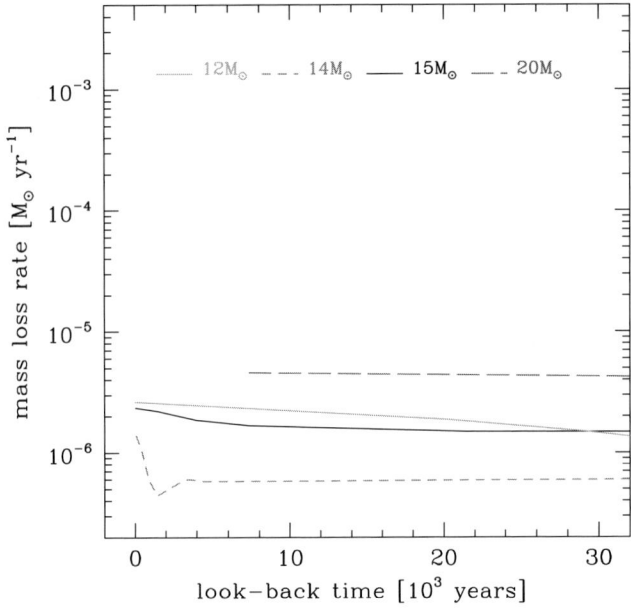

FIGURE 2. Mass loss rates predicted from canonical stellar evolution theory by using Reimers' formula.

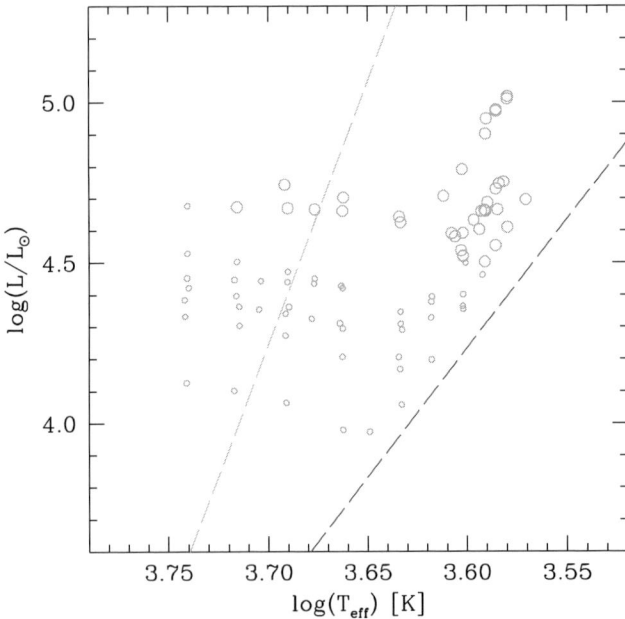

FIGURE 3. Predicted mass loss rates as a function of the position in the HR diagram adopting Reimers' formula. The sizes of the circles are proportional to the logarithm of the mass loss rate: the highest values are several 10^{-6} $M_\odot$ yr^{-1} and the lowest ones about 10^{-7} $M_\odot$ yr^{-1}. The dashed lines represent the analytical relations for the fundamental blue and red edges of the Cepheid instability strip defined by Bono et al. (2000a).

This formula was devised to dimensionally account for the mass loss from low-mass red giants, but has also been widely adopted for evolutionary track calculations. We see that the predicted mass loss rate varies rather slowly when a star is moving from the blue to the red region (i.e. during H-shell burning and/or He-core burning) of the HR diagram, the main functional dependence being a 1.5 power of the luminosity. The corresponding mass loss rates, computed using the evolutionary tracks by Bono et al. (2000b) for stars in the mass range 10–20 $M_\odot$, are shown in Figures 2 and 3. It is apparent that not only the rates are not as high as suggested by spectroscopic observations of RSGs (this aspect alone could easily be "fixed" by increasing the efficiency factor η) but, more importantly, are very slowly varying with time and, therefore, cannot account for radio observations of SNe, either.

Other parametrizations of the mass loss rate in the HR diagram have been proposed by different authors (e.g. De Jager, Nieuwenhuijzen & van der Hucht 1988, Salasnich, Bressan & Chiosi 1999), but insofar for RSGs the main dependence of $\dot{M}$ is a power of ~ 2 of the luminosity, they all are unable to reproduce appreciable mass loss variations over a timescale of roughly 10^4 years.

Actually, one notices that for masses above 10 $M_\odot$, the last phases of the RSG evolution fall within the extrapolation of the Cepheid instability strip (see Figure 4), as calculated by Bono et al. (1996), and therefore, one may expect that pulsational instabilities could represent the additional mechanism needed to trigger high mass loss rates. Indeed, the pioneering work of Heber et al. (1997), based on both linear and nonlinear pulsation models, demonstrated that RSG stars are pulsationally unstable. In particular, they found that, for periods approaching the Kelvin-Helmotz timescale, these stars display large luminosity amplitudes, which could trigger a strong enhancement in their mass loss rate before they explode as supernovae. According to these authors this pulsation behavior should take place during the last few 10^4 yrs before the core collapse, due to the large increase in the luminosity to mass ratio experienced by RSG stars during these evolutionary phases.

However, the nonlinear calculations performed by Heber et al. (1997) were hampered by the fact that their hydrodynamic code could not properly handle pulsation destabilizations characterized both by small growth rates due to numerical damping, and by large pulsation amplitudes due to the formation and propagation of strong shock waves during the approach to limit cycle stability. Also, as Heber et al. (1997) pointed out, their main theoretical difficulty in dealing with the dynamical instabilities of RSG variables resided in the coupling between convection and pulsation. In fact, they constructed the linear models by assuming that the convective flux is frozen in, and the nonlinear ones by assuming that the convective flux is instantaneously adjusted. However, this treatment does not account for the driving and/or quenching effects caused by the interaction between pulsation and convection: this shortcoming may explain why their nonlinear models could not approach a stable limit cycle.

It is clear that a more general approach must be adopted to solve the problem. This motivated us to start a systematic study of the pulsational properties of massive stars. In the following we shall illustrate briefly the procedures adopted and the first results obtained (Section 2), and will present and discuss our findings on the mass loss rates in the late phases of the evolution of massive stars (Sections 3 and 4).

2. Theoretical framework

The procedures employed to construct both linear and nonlinear models of high-mass radial variables have been described in detail in a number of papers (Bono & Stellingwerf

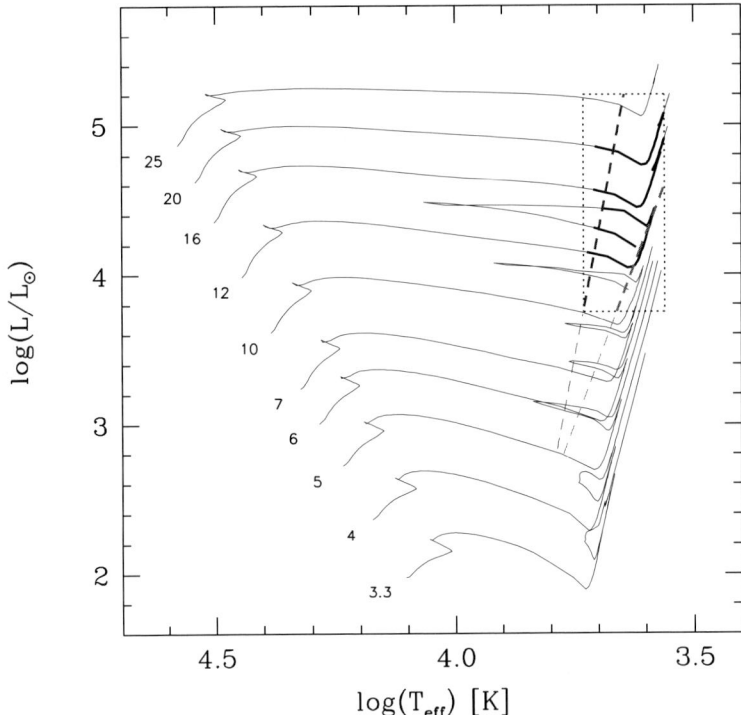

FIGURE 4. Evolutionary tracks for stars in the range 3.3–25 $M_\odot$. The marked area is the portion of the HR diagram shown in more detail in Figure 5. The dashed lines are the extrapolation of the fundamental boundaries of the Cepheid instability strip according to the analytical relations provided by Bono et al. (2000a), which are based on a detailed investigation of Cepheid models at solar chemical composition and stellar masses ranging from 5 to 11 $M_\odot$. Note that a substantial portion of both H-shell and He burning phases for stars with masses higher than about 10 $M_\odot$ occur within the instability strip.

1994; Bono, Caputo, & Marconi 1998; Bono, Marconi & Stellingwerf 1999), so that a brief outline of the methods adopted will suffice here. In particular:

• We constructed a set of limiting amplitude, nonlinear, convective models of Red Super Giant (RSG) variables during both hydrogen and helium burning evolutionary phases.

• A Lagrangian one-dimensional hydrocode was used in which local conservation equations are simultaneously solved with a nonlocal, time-dependent convective transport equation (Stellingwerf 1982; Bono & Stellingwerf 1994; Bono et al. 1998).

• Nonlinear effects such as the coupling between convection and pulsation, the convective overshooting and the superadiabatic gradients are fully taken into account.

• As for the opacity, which is a key ingredient for constructing stellar envelope models, we adopted OPAL opacities (Iglesias & Rogers 1996) for $T > 10,000$ K and molecular opacities (Alexander & Ferguson 1994) $T < 10,000$ K. The method adopted for handling opacity tables was discussed in Bono et al. (1996).

• Artificial viscosity was included following Stellingwerf (1975) prescriptions.

• To provide accurate predictions on the limit cycle behavior of these objects, the governing equations were integrated in time for a number of periods ranging from 1000 to 5000.

In order to make a detailed analysis of RSG pulsation properties during both H and He burning phases, to be compared with actual properties of real RSG stars, we con-

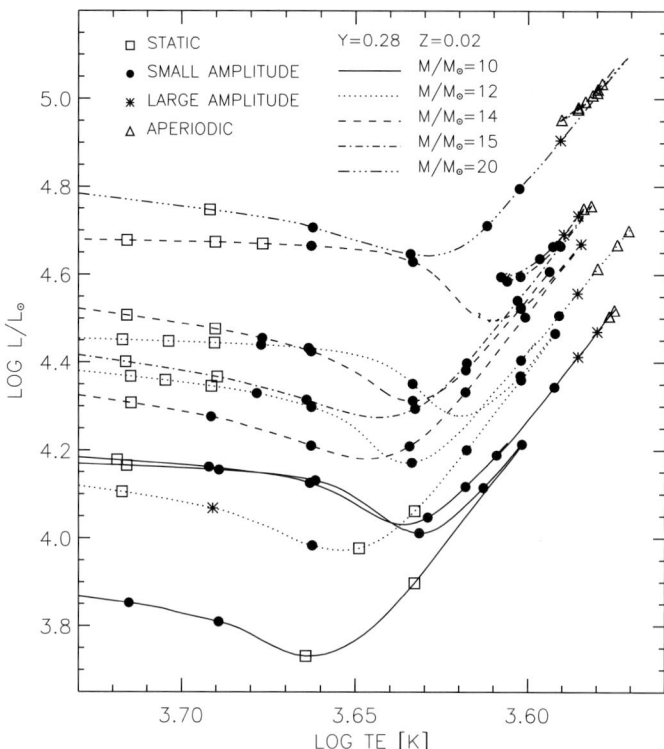

FIGURE 5. Evolutionary tracks and pulsation models at solar chemical composition in the HR diagram. Evolutionary models constructed by adopting different stellar masses are plotted by using different line styles, while pulsation models characterized by different limiting amplitude behaviors are plotted using different symbols. See text for further details.

structed several sequences of models at fixed chemical composition (Y = 0.28, Z = 0.02) which cover a wide range of stellar masses $10 \leq M_\odot \leq 20$. Moreover, since we are interested in mapping the properties of RSG stars from H-shell burning up to the central He exhaustion, both the luminosities and the effective temperature values were selected directly along the evolutionary tracks (see Figure 4). The evolutionary calculations were performed at fixed mass—i.e. no mass-loss—and neglecting the effects of both convective core overshooting and rotation. Since the pulsational properties of a star depend mostly on the physical structure of the envelope regions in which H and He undergo partial ionization, i.e. well above the layers in which the nuclear burning takes place, the use of constant mass evolutionary tracks for our study does not limit the qualitative value of our conclusions. However, self-consistent evolutionary models which properly include a parametrization of mass loss will eventually be needed for a full, quantitative description of the phenomenon (see Section 4).

The input physics and physical assumptions adopted for constructing evolutionary and pulsational models will be described in detail in a forthcoming paper (Bono & Panagia 2000). Figure 5 shows the location in the HR diagram of both evolutionary tracks and pulsation models we constructed. The pulsation models characterized by different limiting amplitude behavior are plotted with different symbols. The squares refer to models which are pulsationally stable, i.e. those in which, after the initial perturbation, the radial motions decay and the structure approaches once again the static configuration. Filled circles and asterisks refer to models which show small ($\Delta \log L < 0.4$) and large

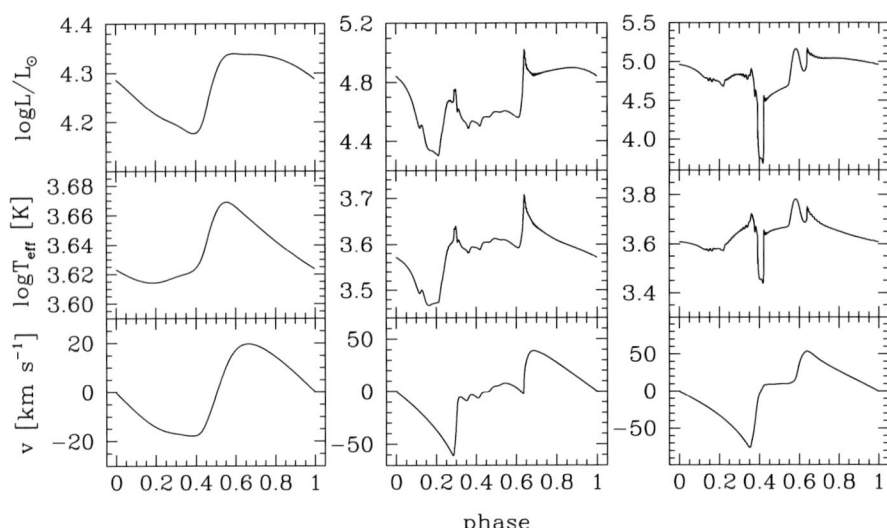

FIGURE 6. The cycle variation of luminosity, temperature and velocity for a 15 $M_\odot$ red supergiant for three distinct regimes, small amplitude (left panel), large amplitude (central panel), and aperiodic extreme pulsations (right panel).

($\Delta \log L > 0.4$) pulsation amplitudes together with a periodic behavior (stable limit cycle). Triangles denote the models which not only present large pulsation amplitudes but also aperiodic radial displacements (unstable limit cycle). The behavior of pulsation properties discloses several interesting features:

1) For effective temperatures lower than approximately 5100 K high-mass models are, with few exceptions, pulsationally unstable in the fundamental mode both during H shell and He burning phases.

2) The pulsational behavior is mainly governed by the effective temperature and to a less extent by the luminosity. In fact, the transition from small to large pulsation amplitudes ($T_e \approx 3900$ K) and from periodic to aperiodic behavior ($T_e \leq 3800$ K) take place roughly at constant temperature.

3) Current models support the evidence suggested by Li & Gong (1994) on the basis of linear, nonadiabatic models that RSG variables are pulsating in the fundamental mode. In fact we find that throughout this region of the HR diagram the fundamental mode is pulsationally unstable, whereas the first overtone is stable.

4) The region in which the models attain small amplitudes is the natural extension of the classical Cepheid instability strip. This confirms the empirical evidence originally brought out by Eichendorf & Reipurth (1979) and more recently by Kienzle et al. (1998), as well as the theoretical prediction by Soukup & Cox (1996).

5) Interestingly enough, theoretical light curves of periodic large amplitude models show the characteristic RV Tauri behavior, i.e. alternating deep and shallow minima, observed in some RSG variables (Eichendorf & Reipurth 1979).

3. Pulsationally induced mass loss

As discussed above, in the reddest part of their RSG evolution, massive stars are found to pulsate with large amplitudes in both luminosity and velocity. In these phases, the

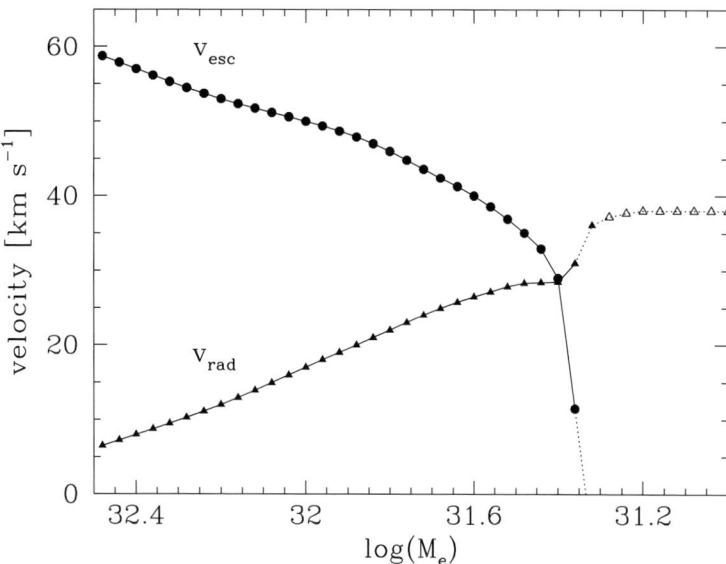

FIGURE 7. Radial velocity and escape velocity as a function of the mass as measured from the stellar surface for a 15 $M_\odot$ star at $T_{\rm eff} = 3836$ K and $\log(L/L_\odot) = 4.75$.

radial velocity near the stellar surface may reach values of 50 km s^{-1} or higher, which may become higher than the effective escape velocity, i.e. the one computed by including both inward gravitational force and outward radiation acceleration. Therefore, the outer layers may become unbound and be lost from the system, thus producing a high mass loss rate. As an illustration, Figure 6 shows the variation of luminosity, temperature and velocity for a 15 $M_\odot$ red supergiant for three distinct regimes, small amplitude (left panel), large amplitude (central panel), and aperiodic extreme pulsations (right panel).

In order to calculate pulsation induced mass loss, for each model we identify the outer layers for which

$$v_{\rm rad} > v_{\rm esc} + c_s$$

where $v_{\rm rad}$ is the radial velocity of a given layer, $v_{\rm esc}$ is the effective escape velocity that includes the effects of radiation forces, and c_s is the sound speed (Hill & Willson 1979). Those layers are effectively unbound and, therefore, are lost from the star. An example is shown in Figure 7 where we plot the actual velocity of the stellar envelope layers as a function of the external mass, $M_e = M^* - M(r)$, and compare them with the effective escape velocity. The mass present above the radius where the actual velocity exceeds the escape velocity represents the amount of mass which is lost in a pseudo-impulsive event.

The characteristic time between successive pseudo-impulsive events is the Kelvin-Helmotz time, i.e.

$$\tau_{KH} \simeq \frac{GM^2}{RL} = 63 \left[\frac{M}{10\ M_\odot}\right]^2 \left[\frac{R}{500\ R_\odot}\right]^{-1} \left[\frac{L}{10^5\ L_\odot}\right]^{-1} \text{ yrs}$$

Thus, the mass loss rate is given by

$$\dot{M} = \frac{M(v > v_{\rm esc} + c_s)}{\tau_{KH}}$$

An inspection of Figure 7 shows that layers as massive as 10^{-2} $M_\odot$ may become unbound and be lost from the stellar surface within time intervals of several tens of

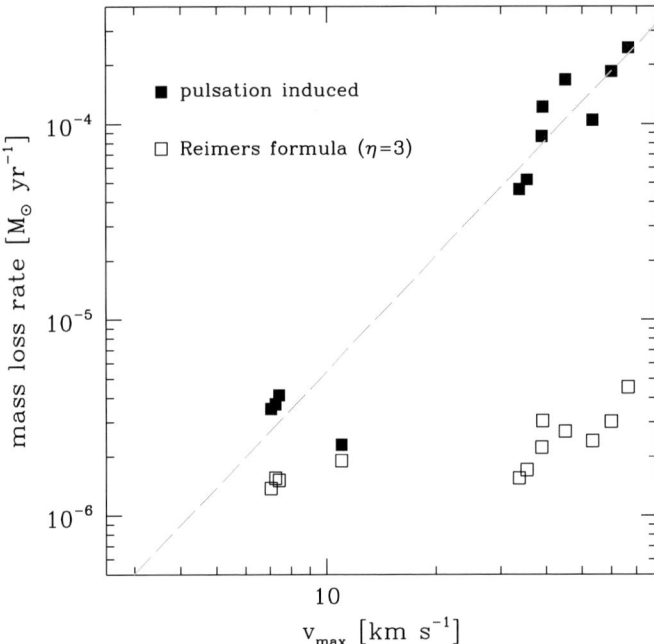

FIGURE 8. Mass loss rates as a function of the maximum velocity (see text). Filled symbols represent the pulsation-induced mass loss rates, and the open symbols the rates computed with Reimers' formula. The dashed line is a linear best-fit to the pulsation induced mass loss rates.

years, thus producing mass loss rates of the order of several 10^{-4} $M_\odot$ yr^{-1} or even higher.

Following this recipe, we have determined the mass loss rates for several values of the stellar mass and for a variety of locations in the HR diagram. As shown in Figure 8, we find that the pulsation induced mass loss can satisfactorily be represented as a power law function of the maximum expansion pulsational velocity (i.e. the maximum velocity that is attained by the outermost layers in a pulsational cycle), i.e.

$$\log(\dot{M}) = -7.24 + 1.97 \times \log(v_{\mathrm{max}})$$

Since pulsation-induced mass loss is an additional mass loss mechanism that comes on top of more conventional radiation-pressure induced mass loss, the total mass loss rate is assumed to be the straight sum of the two rates.

The main result is that including the effects of pulsations, the predicted mass loss rate in the RSG phase is a strong function of both the luminosity and the temperature, in the sense that the mass loss process is strongly enhanced by pulsations when a star is moving toward cooler effective temperatures. Figure 9 illustrates this result and shows that now it should only take small adjustment to fully reproduce the observations. Moreover, since a red supergiant may cross the instability strip rather quickly, its mass loss can change considerably on relatively short timescales, thus accounting for another observational fact. It is worth mentioning that Feast (1991) found, on the basis of IRAS data for 16 RSG variables in the LMC, a period-mass loss rate relation. This relation supports a similar behavior i.e. a steady increase in the mass-loss rate when moving from short to long period variables.

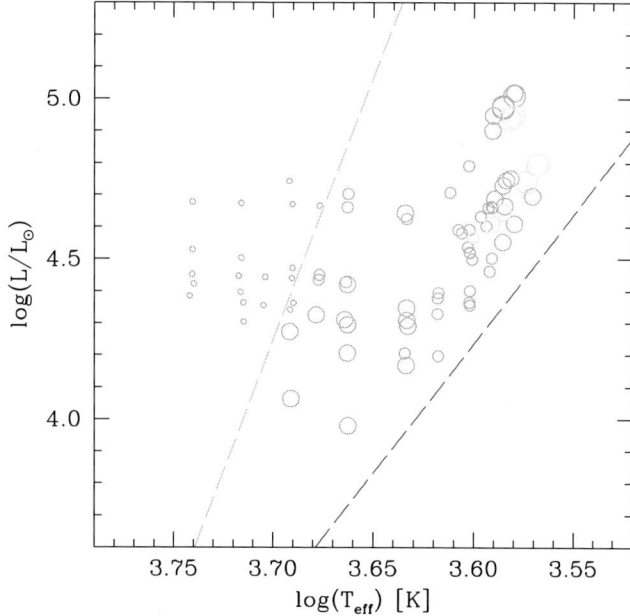

FIGURE 9. Pulsationally enhanced mass loss rates as a function of the position in the HR diagram. The size of the circles are proportional to the logarithm of the mass loss rate up to a maximum value of almost 10^{-3} $M_\odot$ yr^{-1}. Note the strong enhancement of the mass loss rate within the Cepheid instability strip (area within the dashed lines; Bono et al. 2000a).

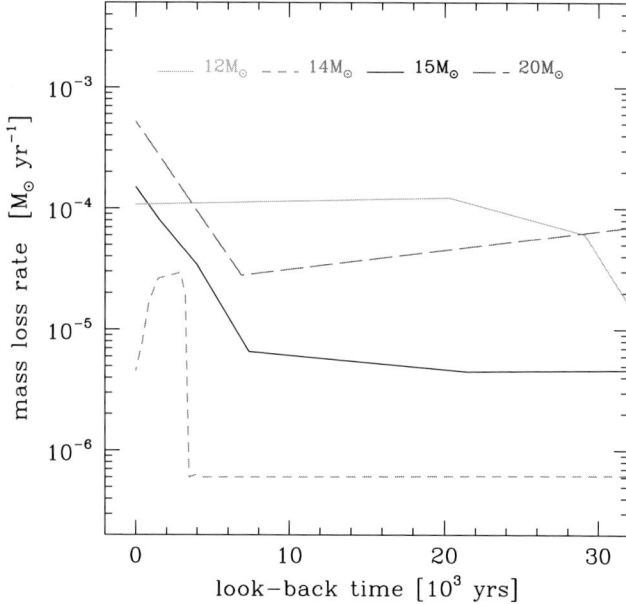

FIGURE 10. Pulsationally enhanced mass loss rates as a function of look-back time (short time-scale variation).

4. Discussion and conclusions

The computed mass loss rates, for stars in the range 12–20 $M_\odot$, as a function of look-back time are displayed in Figures 10–12 for short (0–30,000 years), medium (0–150,000 years) and long time-scales (0–1 Myrs), respectively. We see that the mass loss rates may be as high as almost 10^{-3} $M_\odot$ yr^{-1}, i.e. similar to what is measured for extreme red supergiants, and may vary by an order of magnitude over relatively short times, say, 10,000 years or less. In other words the predicted mass loss rates are able to account, at least qualitatively, for all of the features observed in radio supernovae. Moreover, since the predicted mass loss history is a critical function of how a massive star evolves within the pulsational instability strip, a comparison between observations and theory should lead to an accurate determination of the stellar progenitor mass. For example, the mass loss rate decline of a 20 $M_\odot$ star may be used to represent the apparent drop of emission of SN 1988Z about 9 years after explosion (cf. Figure 1). Similarly, the quick increase found for our 14 $M_\odot$ model closely resembles the behavior observed for SN 1993J. Of course, detailed comparisons will be meaningful only when we will have a fully self-consistent set of evolutionary tracks (see below).

The mid- and long-term behavior of the mass loss rate as a function of look-back time is also interesting because it allows one to make predictions about the radio emission, as well as on *any* other phenomenon linked to a SN shock front and/or ejecta interaction with a dense circumstellar medium, such as relatively narrow optical emission lines and X-ray emission. As we can see in Figures 11 and 12, massive stars are expected to display rather sudden variations of their mass loss rates of all timescales, both because of pulsational instabilities which arise with crossing the instability strip (e.g. the 12 $M_\odot$ star in the time range 20–60 × 10^3 years) and because of the so-called blue loops (an effect clearly apparent at look-back times around 0.4–1 Myrs) that are determined by a combination of core He-burning and shell H-burning (e.g. Brocato & Castellani 1993, Langer & Maeder 1995). Because of these effects, one may expect that in some cases, a SN may drop below detection limit for a while but still may have a renaissance, in the X-ray, optical and radio domains, several tens or hundreds of years later.

Also, we note in passing that our findings support the empirical evidence recently brought out by van Loon et al. (1999) on the basis of ISO data on RSG stars in the LMC. In fact, they found that the mass loss rates increase with increasing luminosities and decreasing effective temperatures and range from 10^{-6} up to 10^{-3} $M_\odot$ yr^{-1}. A strong dependence of the mass loss rate on the effective temperature in the stars at the tip of the AGB branch was recently suggested by Schröder, Winters and Sedlmayer (1999) on the basis of theoretical evolutionary models which account for carbon-rich wind driven by radiation pressure on dust.

Another interesting consequence of our results is that a more efficient mass loss in the RSG phase implies a lower mass cutoff to produce Wolf-Rayet stars and, therefore, one has to expect a more efficient mass return into the ISM than commonly adopted in Galactic evolution calculations.

Still there are improvements and refinements to apply to our models, because the calculations we presented here are not fully self-consistent in that we adopted evolutionary tracks computed either with no mass loss whatsoever, or with modest mass loss rates, and on them we performed our pulsational stability analysis and, thus, determined our new mass loss rates. Moreover, our models were constructed by adopting the diffusion approximation even in optically thin layers and therefore we neglected the dust formation processes (Arndt et al. 1997). A macroscopic example of the shortcomings of our current approach is that if we integrate the mass loss rates over time, in many cases we find that

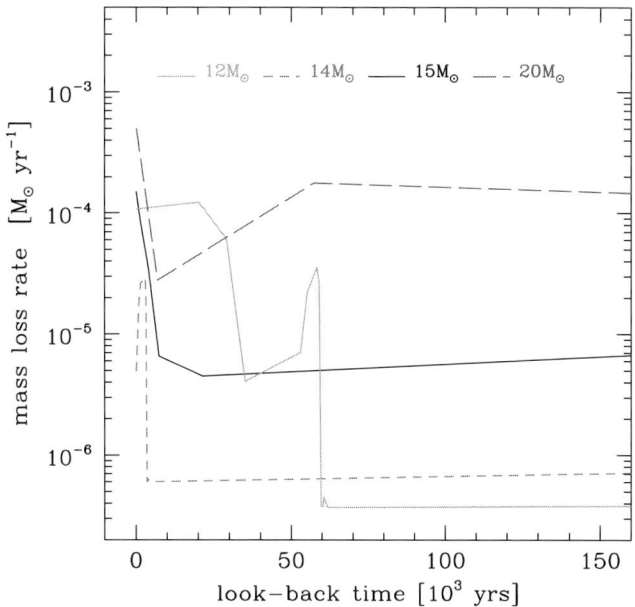

FIGURE 11. Pulsationally enhanced mass loss rates as a function of look-back time (medium timescale variation).

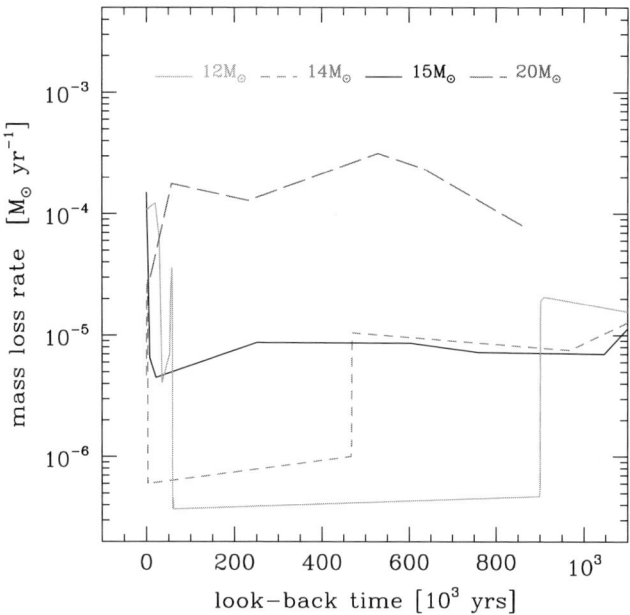

FIGURE 12. Pulsationally enhanced mass loss rates as a function of look-back time (long timescale variation).

the star looses a substantial fraction of its mass before reaching its evolutionary end. Although this is close to what one should expect on the basis of observations, definitely it is at variance with the assumptions that went into the adopted evolutionary model calculations.

It is clear that what we need to do now is to follow an iterative procedure in which we first use our present prescriptions to compute new evolutionary tracks, then we repeat our pulsational stability analysis, then we compute new mass loss rates, and we iterate the procedure until adequate convergence is achieved. This work is in progress and will be presented in future papers. For the time being, our conclusions can be summarized as follows:
- We have defined a new theoretical scenario for pulsation induced mass loss in RSGs.
- RSGs are pulsationally unstable for a substantial portion of their lifetimes.
- Dynamical instabilities play a key role in driving mass loss.
- Bright, cool RSGs undergo mass loss at considerably higher rates than commonly adopted in stellar evolution.
- Comparisons of model predictions with observed CSM phenomena around SNII will provide valuable diagnostics about their progenitors and their evolutionary history.
- More efficient mass loss in the RSG phase implies a lower mass cutoff to produce Wolf-Rayet stars and a more efficient return of polluted material into the ISM, thus affecting the expected chemical evolution of galaxies.

REFERENCES

ALEXANDER, D. R. & FERGUSON, J. W. 1994 *ApJ*, **437**, 879.

ARNDT, T. U., FLEISCHER, A. J., & SEDLMAYER, E. 1997 *A&A*, **327**, 614.

BENETTI, S., TURATTO, M., CAPPELLARO, E., DANZIGER, I. J., & MAZZALI, P. 1999 *MNRAS*, **305**, 811.

BOFFI, F. R. & PANAGIA, N. 1996, in *Radio emission from the stars and the sun*. ASP Conference Series, Volume 93, pp153–156. ASP.

BOFFI, F. R. & PANAGIA, N. 2000, in preparation.

BONO, G., CAPUTO, F., CASSISI, S., MARCONI, M., & TORNAMBÈ, A. 2000b *ApJ*, submitted.

BONO, G., CAPUTO, F., & MARCONI, M. 1998 *ApJ*, **497**, L43.

BONO, G., CASTELLANI, V., & MARCONI, M. 2000a *ApJ*, **529**, 293.

BONO, G., INCERPI, R., & MARCONI, M. 1996 *ApJ*, **467**, L97.

BONO, G., MARCONI, M., & STELLINGWERF, R. F. 1999 *ApJS*, **122**, 167.

BONO, G. & PANAGIA, N. 2000, in preparation.

BONO, G. & STELLINGWERF, R. F. 1994 *ApJS*, **93**, 233.

BROCATO, E. & CASTELLANI, V. 1993 *ApJ*, **410**, 99.

CHIEFFI, A., LIMONGI, M., & STRANIERO, O. 1998 *ApJ*, **502**, 737.

CHU, Y-H., CAULET, A., MONTES, M. J., PANAGIA, N., VAN DYK, S. D., & WEILER, K. W. 1999 *ApJ*, **512**, L51.

CHUGAI, N. N., DANZIGER, I. J., & DELLA VALLE, M. 1995 *MNRAS*, **276**, 530.

DE JAGER, C., NIEUWENHUIJZEN, H., & VAN DER HUCHT, K. A. 1988 *A&AS*, **72**, 295.

EICHENDORF, W. & REIPURTH, B. 1979 *A&A*, **77**, 227.

FEAST, M. W. 1991, in *Instabilities in Evolved Super- and Hyper-Giants* (eds. C. de Jager & H. Nieuwenhuijzen). p. 18–21. Amsterdam.

FOX, M. W. & WOOD, P. R. 1982 *ApJ*, **259**, 198.

HEBER, A., JEANNIN, L., LANGER, N., & BARAFFE, I. 1997 *A&A*, **327**, 224.

HILL, S. J. & WILLSON, L. A. 1979 *ApJ*, **229**, 1029.

IGLESIAS, C. A. & ROGERS, F. J. 1996 *ApJ*, **464**, 943.

KIENZLE, F., BURKI, G., BURNET, M., & MEYNET, G. 1998 *A&A*, **337**, 779.

LACEY, C., WEILER, K. W., SRAMEK, R. A., PANAGIA, N., & VAN DYK, S. D. 2000, in preparation.

LANGER, N. & MAEDER, A. 1995 å**295**, 685.

LI, Y. & GONG, Z. G. 1994 *A&A*, 289, 449.

MONTES, M. J., VAN DYK, S. D., WEILER, K. W., SRAMEK, R. A., & PANAGIA, N. 1998 *ApJ*, **506**, 874.

MONTES, M. J., VAN DYK, S. D., WEILER, K. W., SRAMEK, R. A., & PANAGIA, N. 2000 *ApJ*, , in press.

PANAGIA, N. & LAIDLER, V. G. 1988, in *Supernova Shells and Their Birth Events* (ed. W. Kundt). p. 187–191. Springer-Verlag.

REID, N., TINNEY, C., & MOULD, J. 1990 *ApJ*, **348**, 98.

REIMERS, D. 1975 Mém. Soc. Roy. Sci. Liége 6e Ser., **8**, 369.

RYDER, S., STAVELEY-SMITH, L., DOPITA, M., PETRE, R., COLBERT, E., MALIN, D., & SCHLEGEL, E. 1993 *ApJ*, **416**, 167.

SALAMANCA, I., CID-FERNANDES, R., TENORIO-TAGLE, G., TELLES, E., TERLEVICH, R. J., & MUNOZ-TUNON, C. 1998 *MNRAS*, **300**, L17.

SALASNICH, B., BRESSAN, A., & CHIOSI, C. 1999 *A&A*, **342**, 131.

SCHRÖDER, K. -P., WINTERS, J. M., & SEDLMAYER, E. 1999 *A&A*, **349**, 898.

SOUKUP, M. S. & COX, A. N. 1996 *A&AS*, **342**, 131.

STELLINGWERF, R. F. 1975 *ApJ*, **195**, 441.

STELLINGWERF, R. F. 1982 *ApJ*, **262**, 339.

STELLINGWERF, R. F. 1984 *ApJ*, **284**, 712.

STOTHERS, R. & LEUNG, K. C. 1971 *A&A*, **10**, 290.

VAN DYK, S. D., WEILER, K. W., SRAMEK, R. A., RUPEN, M.,& PANAGIA, N. 1994 *ApJ*, **432**, L115.

VAN LOON, J. TH., et al. 1999 *A&A*, **351**, 559.

WOOD, P. R., et al. 1999, in IAU Symposium No. 191 *AGB Stars*. ASP, in press.

Radio supernovae and GRB 980425

By KURT W. WEILER,[1] NINO PANAGIA,[2,3]
RICHARD A. SRAMEK,[4] SCHUYLER D. VAN DYK,[5]
MARCOS J. MONTES,[6] AND CHRISTINA K. LACEY[7,8]

[1]NRL, Code 7213, Washington, DC 20375-5320; weiler@rsd.nrl.navy.mil

[2]Space Telescope Science Institute, 3700 San Martin Drive, Baltimore, MD 21218; panagia@stsci.edu

[3]On assignment from the Astrophysics Division, Space Science Department of ESA.

[4]P.O. Box 0, NRAO, Socorro, NM 87801; dsramek@nrao.edu

[5]IPAC/Caltech, Mail Code 100-22, Pasadena, CA 91725; vandyk@ipac.caltech.edu

[6]NRL, Code 7212, Washington, DC 20375-5320; montes@rsd.nrl.navy.mil

[7]NRL, Code 7213, Washington, DC 20375-5320; lacey@rsd.nrl.navy.mil

[8]NRC Postdoctoral Fellow

Study of radio supernovae (RSNe) over the past 20 years includes two dozen detected objects and more than 100 upper limits. From this work we are able to identify classes of radio properties, demonstrate conformance to and deviations from existing models, estimate the density and structure of the circumstellar material and, by inference, the evolution of the presupernova stellar wind, and reveal the last stages of stellar evolution before explosion. It is also possible to detect ionized hydrogen along the line of sight, to demonstrate binary properties of the stellar system, and to show clumpiness of the circumstellar material. More speculatively, it may be possible to provide distance estimates to radio supernovae.

The interesting and unusual radio supernova SN 1998bw, which is thought to be related to the γ-ray burst GRB 980425, is discussed in particular detail. Its radio properties are compared and contrasted with those of other known RSNe.

1. Introduction

A series of papers published over the past 20 years on radio supernovae (RSNe) has established the radio detection and/or radio evolution for approximately two dozen objects: 2 Type Ib supernovae (SNe), 5 Type Ic SNe, and the rest Type II SNe. A much larger list of more than 100 additional SNe have low radio upper limits (Table 1).

In this extensive study of the radio emission from SNe, several effects have been noted: 1) Type Ia SNe are not radio emitters to the detection limit of the VLA;† 2) Type Ib/c SNe are radio luminous with steep spectral indices (generally $\alpha < -1$; $S \propto \nu^{+\alpha}$) and a fast turn-on/turn-off, usually peaking at 6 cm near or before optical maximum; and 3) Type II SNe show a range of radio luminosities with flat spectral indices (generally $\alpha > -1$) and a relatively slow turn-on/turn-off, usually peaking at 6 cm significantly after optical maximum. Type Ib/c may be fairly homogeneous in their radio properties while Type II, as in the optical, are quite diverse.

There are a large number of physical properties of SNe which we can determine from radio observations. VLBI imaging shows the symmetry of the explosion and the local CSM, estimates the speed and deceleration of the SN shock propagating outward from

† The VLA is operated by the NRAO of the AUI under a cooperative agreement with the NSF.

SN	Type	Radio	SN	Type	Radio	SN	Type	Radio	SN	Type	Radio
1895B	I		1901B	I		1909A	II		1914A	?	
1917A	I?		1921B	II		1921C	I		1923A	IIP	DT
1937C	Ia		1937F	II		1939C	II		1940A	IIL	
1945B	?		1948B	II?		1950B	II?	DT	1954A	I	
1954J	II		1957D	II?	DT	1959D	II		1959E	I	
1963J	I		1966B	II		1968D	II	DT	1968L	IIP	
1969L	IIP		1970A	II?		1970G	IIL	LC	1970L	I?	
1970O	?		1971G	I?		1971I	Ia		1971L	Ia	
1972E	Ia		1973R	IIP		1974E	?		1974G	Ia	
1975N	Ia		1977B	?		1978B	II		1978G	II	
1978K	II	LC	1979B	Ia		1979C	IIL	LC	1980D	IIP	
1980I	Ia		1980K	IIL	LC	1980L	?		1980N	Ia	
1980O	II		1981A	II		1981B	Ia		1981K	II?	LC
1982E	Ia?		1982R	Ib?		1983G	I		1983K	IIP/n	
1983N	Ib	LC	1984A	I		1984E	IIL/n		1984L	Ib	LC
1984R	?		1985A	Ia		1985B	Ia		1985F	Ib	
1985G	IIP		1985H	II		1985L	IIL	DT	1986A	Ia	
1986E	IIL	LC	1986G	Ia		1986I	IIP		1986J	IIn	LC
1986O	Ia		1987A	IIpec	LC	1987B	IIn		1987D	Ia	
1987F	IIpec		1987K	IIb		1987M	Ic		1987N	Ia	
1988I	IIn		1988Z	IIn	LC	1989B	Ia		1989C	IIn	
1989L	IIL		1989M	Ia		1989R	IIn		1990B	Ic	LC
1990K	IIL		1990M	Ia		1991T	Ia		1991ae	IIn	
1991ar	Ic		1991av	IIn		1991bg	Ia		1992A	Ia	
1992H	IIP		1992ad	IIP?	DT	1992bd	II		1993G	II	
1993J	IIb	LC	1993N	IIn		1993X	II		1994D	Ia	
1994I	Ic	LC	1994P	II		1994W	IIn		1994Y	IIn	
1994ai	Ic		1994ak	IIn		1995G	IIn		1995N	IIn	DT
1995X	IIP		1995ad	IIP		1995al	Ia		1996N	Ic	DT
1996W	IIpec		1996X	Ia		1996ae	IIn		1996an	II	
1996aq	Ic		1996bu	IIn		1996cb	IIb?	DT	1997X	Ic	DT
1998bw	Ic	LC									

TABLE 1. Observed Supernovae (DT = Detection; LC = Light Curve Available)

the explosion and, with assumptions of symmetry and optical line/radio sphere velocities, allows independent distance estimates to be made (see, e.g. Marcaide et al. 1997, Bartel et al. 1985).

Measurements of the multi-frequency radio light curves and their evolution with time show the density and structure of the CSM, evidence for possible binary companions, clumpiness or filamentation in the presupernova wind, mass-loss rates and changes therein for the presupernova stellar system and, through stellar evolution models, estimates of the ZAMS presupernova stellar mass and the stages through which the star passed on its way to explosion. It has also been proposed by Weiler et al. (1998) that the time from explosion to 6 cm radio maximum may be an indicator of the radio luminosity and thus an independent distance indicator.

A summary of the radio information on SNe can be found at *http://rsd-www.nrl.navy.mil/7214/weiler/sne-home.html*.

2. Models

All known RSNe appear to share common properties of: 1) nonthermal synchrotron emission with high brightness temperature; 2) a decrease in absorption with time, resulting in a smooth, rapid turn-on first at shorter wavelengths and later at longer wavelengths; 3) a power-law decline of the flux density with time at each wavelength after maximum flux density (optical depth ≈ 1) is reached at that wavelength; and 4) a final, asymptotic approach of spectral index α to an optically thin, nonthermal, constant negative value (Weiler et al. 1986, Weiler, Panagia, & Sramek 1990). Chevalier (1982a,b) has proposed that the relativistic electrons and enhanced magnetic field necessary for synchrotron emission arise from the SN shock interacting with a relatively high density circumstellar medium (CSM) which has been ionized and heated by the initial UV/X-ray flash. This CSM is presumed to have been established by a constant mass-loss ($\dot{M}$) rate, constant velocity (w) wind (i.e. $\rho \propto r^{-2}$) from a red supergiant (RSG) progenitor or companion. This ionized CSM is also the source of the initial absorption. A rapid rise in the observed radio flux density results from the shock overtaking more and more of the wind material, leaving progressively less of it along the line of sight to the observer to absorb the more slowly decreasing synchrotron emission from the shock region.

2.1. Parameterized radio light curves

The parameterized model of Weiler et al. (1986), Weiler, Panagia, & Sramek (1990), and Montes, Weiler, & Panagia (1997) may be written as:

$$S(\text{mJy}) = K_1 \left(\frac{\nu}{5\text{ GHz}}\right)^\alpha \left(\frac{t-t_0}{1\text{ day}}\right)^\beta e^{-(\tau+\tau'')} \left(\frac{1-e^{-\tau'}}{\tau'}\right), \quad (2.1)$$

where

$$\tau = K_2 \left(\frac{\nu}{5\text{ GHz}}\right)^{-2.1} \left(\frac{t-t_0}{1\text{ day}}\right)^\delta, \quad (2.2)$$

$$\tau' = K_3 \left(\frac{\nu}{5\text{ GHz}}\right)^{-2.1} \left(\frac{t-t_0}{1\text{ day}}\right)^{\delta'}, \quad (2.3)$$

and

$$\tau'' = K_4 \left(\frac{\nu}{5\text{ GHz}}\right)^{-2.1}. \quad (2.4)$$

K_1, K_2, and K_3 correspond, formally, to the unabsorbed flux density (K_1), and the uniform (K_2) and non-uniform (K_3) optical depths in the surrounding CSM at 5 GHz one day after the explosion date t_0. K_4 represents a non-time dependent HII absorption (EM $= 8.93 \times 10^7\, K_4\, [T_e/10^4 K]^{1.35}$ pc cm^{-6}, where T_e is the electron temperature of the ionized absorbing region; Eq. 1–223 of Lang 1986) along the line-of-sight to the radio emitting region. The term $e^{-\tau}$ describes the attenuation of a local medium with optical depth τ and time dependence δ that uniformly covers the emitting source ("uniform external absorption"); the term $(1-e^{-\tau'})\,\tau'^{-1}$ describes the attenuation produced by an inhomogeneous medium with optical depths distributed between 0 and τ' ("clumpy absorption") and time dependence δ'; and the term $e^{-\tau''}$ describes an absorption along the line-of-sight which is sufficiently far removed from the radio generating region to be constant with time. All absorbing media are assumed to be purely thermal, ionized hydrogen with opacity $\propto \nu^{-2.1}$.

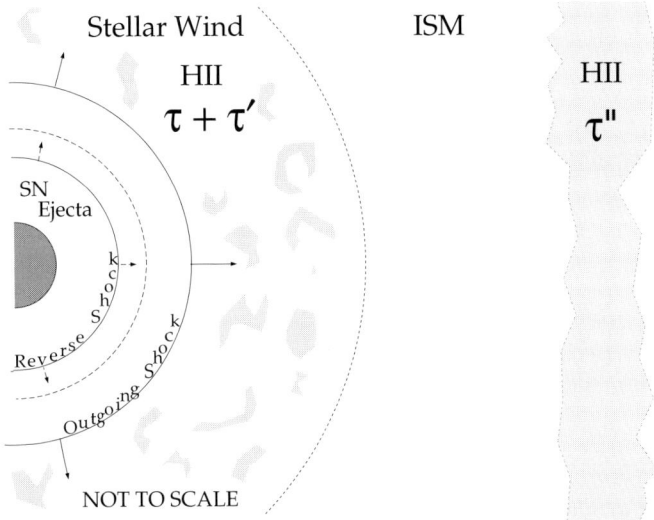

FIGURE 1. Cartoon, not to scale, of the SN and its shocks along with the stellar wind established CSM and more distant ionized material. The radio emission is thought to arise near the outgoing shock with the expected location of the several absorbing terms in Eqs. 2.2–2.4 illustrated.

A cartoon of the expected structure of the SN and its surrounding media is presented in Fig. 1. The radio emission is expected to arise near the outgoing shock (Chevalier & Fransson 1994).

3. Results

The success of the basic parameterization and model description can be seen in the relatively good correspondence between the model fits and the data for all three subtypes of RSNe, e.g. Type Ib SN 1983N (Fig. 2), Type Ic SN 1990B (Fig. 3), and Type II SN 1979C (Fig. 4a) and SN 1980K (Fig. 4b). (Note that after day ~ 4000, the evolution of the radio emission from both SN 1979C and SN 1980K deviates from the expected model evolution; see §4 for discussion of these changes.)

3.1. Mass-loss rate & change in mass-loss rate

From the Chevalier (1982a,b) model, the turn-on of the radio emission for RSNe provides a measure of the presupernova mass-loss-rate to stellar-wind-velocity ratio. Using the formulation of Weiler et al. (1986) Eq. 16, we can write

$$\frac{\dot{M}(\mathrm{M_\odot\ yr^{-1}})}{(w/10\mathrm{\ km\ s^{-1}})} = 3 \times 10^{-6}\ K_2^{0.5}\ m^{-1.5} \left(\frac{v_i}{10^4\mathrm{\ km\ s^{-1}}}\right)^{1.5} \left(\frac{1}{45\mathrm{\ days}}\right)^{1.5m} \left(\frac{T}{10^4\mathrm{\ K}}\right)^{0.68}$$
(3.5)

where $\dot{M}$ is the presupernova mass-loss rate, w is the presupernova wind velocity, K_2 is the same as in Eq. 2.2, m is the SN shock deceleration index (shock radius $\propto t^m$), v_i is the initial SN shock velocity ($\sim 13{,}000\mathrm{\ km\ s^{-1}}$), and T is the temperature of the circumstellar material ($\sim 30{,}000\mathrm{\ K}$).

From Eq. 3.5 the mass-loss rates from SN progenitors are generally estimated to be $\sim 10^{-6}\mathrm{\ M_\odot\ yr^{-1}}$ for Type Ib/c SNe and $\sim 10^{-4}\mathrm{\ M_\odot\ yr^{-1}}$ for Type II SNe. For the specific case of SN 1993J, where detailed radio observations are available starting just

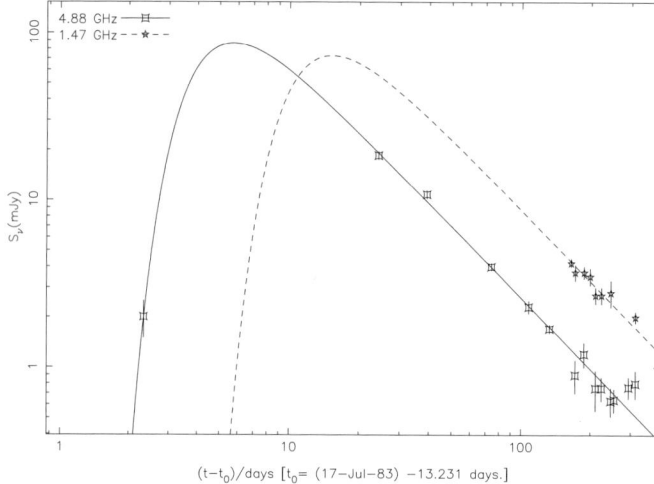

FIGURE 2. Type Ib SN 1983N at 6 cm (4.9 GHz; *squares, solid line*) and 20 cm (1.5 GHz; *stars, dashed line*).

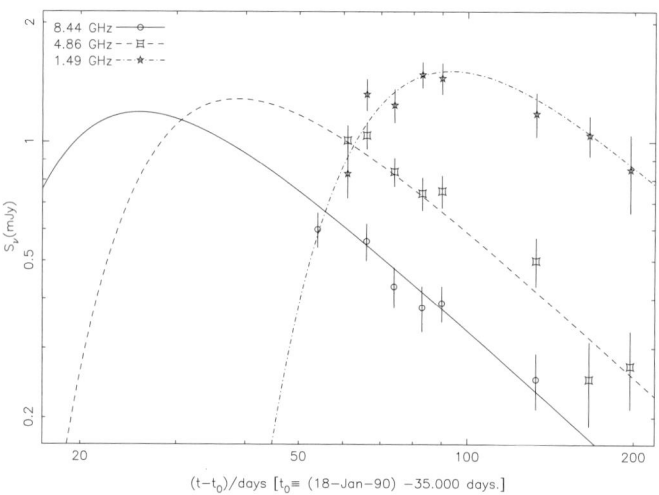

FIGURE 3. Type Ic SN 1990B at 3.4 cm (8.4 GHz; *circles, solid line*), 6 cm (4.9 GHz; *squares, dashed line*), and 20 cm (1.5 GHz; *stars, dash-dot line*).

a few days after explosion, Van Dyk et al. (1994) find evidence for a changing mass-loss rate (Fig. 5) for the presupernova star, which was as high as $\sim 10^{-4}$ $M_\odot$ yr^{-1} approximately 1000 years before explosion and decreased to $\sim 10^{-5}$ $M_\odot$ yr^{-1} just before explosion. Recently Fransson & Björgsson (1998) have shown that the observed behavior of the free-free absorption for SN 1993J could alternatively be explained in terms of a systematic decrease of the electron temperature in the circumstellar material as the SN expands. It is not clear, however, what the physical process is which determines why such a cooling occurs efficiently in some SNe and not in others.

3.2. *Clumpiness of the presupernova wind*

In their study of the radio emission from SN 1986J, Weiler, Panagia, & Sramek (1990) found that the simple Chevalier (1982a,b) model could not describe the relatively slow turn-on. They therefore added terms described mathematically by τ' in Eqs. 2.1 and

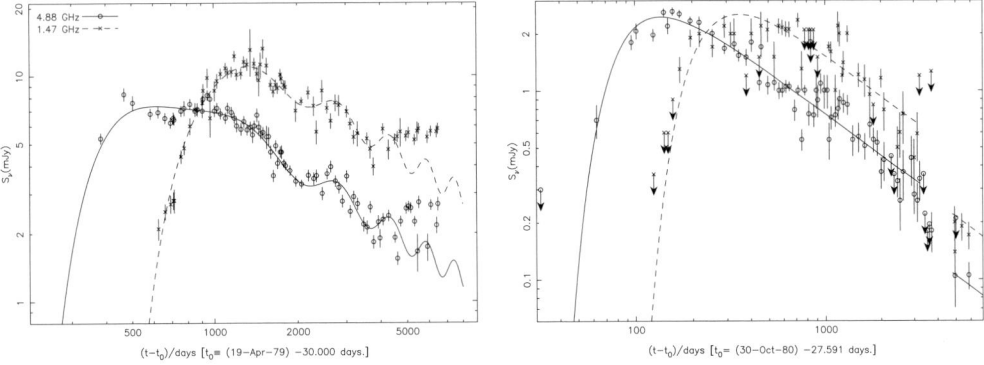

FIGURE 4. Shown are the radio light curves of Type II SN 1979C (Fig. 4a, left) at 6 cm (*circles*) and 20 cm (*crosses*) as observed over a period of about 18 years together with the best-fit model curves (*solid* and *dashed* curves, respectively). The model light curves for SN 1979C include the effects due to a possible binary companion in an eccentric orbit (Weiler et al. 1991, Weiler et al. 1992a). Note the *increase* in flux density of SN 1979C, with respect to the expected, continuing decline, starting at $\sim$ 4300 days (Montes et al. 2000). Also shown, with the same notation at the same frequencies, are the radio light curves of Type II SN 1980K (Fig. 4b, right). Note the sudden *drop* in the flux density of SN 1980K, with respect to the expected, continuing decline, which occurs at $\sim$ 3700 days (Montes et al. 1998).

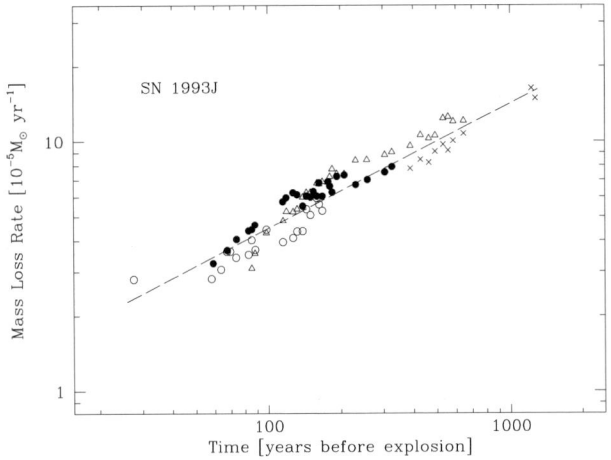

FIGURE 5. Mass-loss rate of the presumed red supergiant progenitor to SN 1993J *vs.* time before the explosion (Van Dyk et al. 1994).

2.3. This extension greatly improved the quality of the fit and was interpreted by Weiler, Panagia, & Sramek (1990) to represent the possible presence of filamentation or clumpiness in the CSM.

Such a clumpiness in the wind material was again required for modeling the radio data from SN 1988Z (Van Dyk et al. 1993) and SN 1993J (Van Dyk et al. 1994). Since that time, evidence for filamentation in the envelopes of SNe has also been found from optical and UV observations (Filippenko, Matheson, & Barth 1994, Spyromilio 1994).

3.3. Binary systems

In the process of analyzing a full decade of radio measurements from SN 1979C, Weiler et al. (1991, 1992a) found evidence for a significant, quasi-periodic, variation in the amplitude of the radio emission at all wavelengths of $\sim$ 15% with a period of 1575 days or ~ 4.3 years (see Fig. 4a at age < 4300 days). They interpreted the variation as due to a minor ($\sim 8\%$) density modulation, with a period of ~ 4000 years, on the larger, relatively constant presupernova stellar mass-loss rate. Since such a long period is inconsistent with most models for stellar pulsations, they concluded that the modulation may be produced by interaction of a binary companion in an eccentric orbit with the stellar wind from the presupernova RSG. This concept was strengthened by more detailed calculations for a binary model from Schwarz & Pringle (1996). Since that time, the presence of binary companions has been suggested for the progenitors of SN 1993J (Podsiadlowski et al. 1993) and SN 1994I (Nomoto et al. 1994), indicating that binaries may be common in presupernova systems.

3.4. HII along the line-of-sight

A reanalysis of the radio data for SN 1978K from Ryder et al. (1993) clearly shows flux density evolution characteristic of normal Type II SNe. Additionally, the data indicate the need for a time-independent, free-free absorption component along the line-of-sight. Montes, Weiler, & Panagia (1997) interpret this constant absorption term as indicative of the presence of HII along the line-of-sight to SN 1978K, perhaps a part of an HII region or a distant circumstellar shell associated with the SN progenitor. SN 1978K had already been noted for its lack of optical emission lines broader than a few thousand km s^{-1} since its discovery in 1990 (Ryder et al. 1993), indeed suggesting the presence of slowly moving circumstellar material.

To determine the nature of this absorbing region, a high-dispersion spectrum of SN 1978K at the wavelength range 6530–6610 Å was obtained by Chu et al. (1999). The spectrum shows not only the moderately broad Hα emission of the supernova ejecta, but also narrow nebular Hα and [N II] emission. The high [N II] 6583/Hα ratio of 0.8–1.3 suggests that this radio absorbing region is a stellar ejecta nebula. The expansion velocity and emission measure of the nebula are consistent with those seen in ejecta nebulae of luminous blue variables. Previous low-dispersion spectra have detected a strong [N II] 5755 Å line, indicating an electron density of $(3\text{--}12) \times 10^5$ cm^{-3}. These data suggest that the ejecta nebula detected towards SN 1978K is probably part of the pre-shock dense circumstellar envelope of SN 1978K. Another possible example of this type of system may be SN 1997ab, which looks in its optical spectrum like a young version of SN 1978K.

4. Rapid pre-supernova stellar evolution

SN radio emission that preserves its spectral index while deviating from the standard model is taken to be evidence for a change of the average circumstellar density behavior from the canonical r^{-2} law expected for a pre-SN wind with a constant mass-loss rate, $\dot{M}$, and a constant wind velocity, w. Since the radio luminosity of a SN is proportional to $(\dot{M}/w)^{(\gamma-7+12m)/4}$ (Chevalier 1982a) or, equivalently, to the same power of the circumstellar density (since $\rho_{\text{CSM}} \propto \dot{M}/w$), a measure of the deviation from the standard model provides an indication of deviation of the circumstellar density from the r^{-2} law. Monitoring the radio light curves of RSNe also provides a rough estimate of the time scale of deviations in the presupernova stellar wind density. Since the SN shock travels through the CSM roughly 1000 times faster than the stellar wind velocity which established the CSM ($v_{\text{shock}} \sim 10,000$ km s^{-1} vs. $w_{\text{wind}} \sim 10$ km s^{-1}) one year of radio light curve monitoring samples roughly 1000 years of stellar wind mass-loss history.

4.1. Radio evidence for CSM structure

SN 1979C (Type IIL) prior to 1991 (age < 4300 days; ~ 12 years) follows a standard, albeit sinusoidally modulated, declining radio emission (see §3.3). However, for age > 4300 days a slow increase in the radio light curve occurs at all wavelengths (see Fig. 4a). By day ~ 7100, this change in evolution implies an *excess* in flux density by a factor of ~ 1.7 with respect to the standard model, or a density enhancement by a factor of ~ 1.34 over the expected density at that radius. This may be understood as a change of the average CSM density profile from the r^{-2} law which was applicable until day ~ 4300, to an appreciably flatter behavior of ~ $r^{-1.4}$ (Montes et al. 2000).

SN 1980K (Type IIL) prior to epoch ~ 3700 days (~ 10 years) is also well behaved. However, more recent measurements show a steep *decline* in flux density at all wavelengths by a factor of ~ 2 occurring between day ~ 3700 and day ~ 4900 (see Fig. 4b). Such a sharp decline in flux density implies a *decrease* in ρ_{CSM} by a factor of ~ 1.6 below that expected for a r^{-2} CSM density profile (Montes et al. 1998).

SN 1988Z (Type IIn), similarly to SN 1980K, shows a sharp drop in its flux density with respect to its expected radio evolution at an age of a few thousand days (several years). Although the parameters of the change are yet to be quantified, it appears to also have evolved rapidly in the last several thousand years before explosion (Lacey et al. 2000).

SN 1987A (Type II) is the best studied RSN because its proximity makes it easily detectable even at very low radio brightness. The progenitor to SN 1987A was in a blue supergiant (BSG) phase at the time of explosion and had ended a RSG phase some ten thousand years earlier. After an initial, very rapidly evolving radio outburst (Turtle et al. 1987) which reached a peak flux density at 6 cm ~ 3 orders-of-magnitude fainter than other known Type II RSNe (presumably due to sensitivity limited selection effects), the radio emission declined to a low radio brightness within a year. However, at an age of ~ 3 years the radio emission started increasing again and continues to increase at the present time (see Ball et al. 1995, Gaensler et al. 1997). Although its extremely rapid development resulted in the early radio data at higher frequencies being almost non-existent, the evolution of the initial radio outburst is roughly consistent with the models described above in Eqs. 2.1–2.4 (i.e. a shock front expanding into a spherically symmetric circumstellar envelope). The density implied by such modeling is appropriate to a pre-SN mass-loss rate of ~ 10^{-7} $M_\odot$ yr^{-1} for a wind velocity of ~ 150 km s^{-1}. Because the *HST* can actually image the denser regions of the CSM around SN 1987A, we know that the current rise in radio flux density is caused by the interaction of the SN shock with the diffuse material at the inner edge of the well known inner circumstellar ring (Gaensler et al. 1997). Since the density increases as the SN shock interaction region moves deeper into the main body of the optical ring, the flux density is expected to continue to increase steadily at all wavelengths. Recently, increases at optical and X-ray have also been reported (Garnavich, Kirshner, & Challis 1997, Hasinger, Aschenbach, & Truemper 1996). Best estimates are that the shock/CSM interaction will reach a maximum by ~ 2003.

4.2. Discussion of CSM structure

For at least four supernovae, namely SN 1979C, SN 1980K, SN 1988Z, and SN 1987A, we have significant changes in radio flux density occurring a few years after the explosion. Since the SN shock is moving about 1000 times faster than the wind material of the RSG progenitor (i.e. ~ 10,000 km s^{-1} *vs.* ~ 10 km s^{-1}), such a time interval implies a significant change in the pre-supernova stellar wind properties several thousand years before the explosion. Such an interval is short compared to the lifetimes of typical

RSN progenitors (say, 10–30 Myrs) but is a sizeable fraction of its red supergiant phase ($t_{\rm RSG} \sim 2$–5×10^5 yrs), suggesting that a significant transition occurs in the evolution of pre-supernova stars just before the final explosive event.

Since the radio emission is determined by the mass-loss-rate to stellar-wind-velocity ratio ($\dot{M}/w$), one of these quantities, or both are required to change by as much as a factor of 2 over the last few thousand years before the SN explosion. Such a time is too short (for H and He burning), or too long (for C and heavier element burning) to correspond to any of the known nuclear burning phases and, therefore, it is unlikely that the stellar luminosity (which determines the mass-loss rate, $\dot{M} \propto L^{1-1.5}$), can vary on a time scale needed to account for the observed changes.

On the other hand, the wind velocity, w, is roughly proportional to the square of the effective temperature ($w \propto T_{\rm eff}^2$, e.g. Panagia & Macchetto 1982) so that change of a factor of ~ 2 in w requires a change of a factor of only ~ 1.4 in $T_{\rm eff}$, e.g. from $\sim 3,500$ K to $\sim 5,000$ K or, correspondingly, a change from an early M to an early K supergiant spectrum. Such a transition would define a loop in the HR diagram reminiscent of the blue loops which are characteristic of the evolution of moderately massive stars (e.g. Brocato & Castellani 1993, Langer & Maeder 1995). However, the apparent transition implied by these CSM density changes cannot be classical blue loops, since classical blue loops are much slower and more extreme processes occurring several $\times 10^5$ years before the terminal stages of an RSG and involving temperature excursions from $\sim 3,500$ K to $> 10,000$ K.

The smaller temperature changes which we infer from the radio data require a star to change only from a very red to a moderately red spectrum, and back, corresponding to a transition in the HR diagram which is more appropriately dubbed a *"pink loop."* The cause of such loops is not obvious, but may be similar to the not-so-well understood phenomenon that caused the SN 1987A progenitor to move in the HR diagram from being a red supergiant to a blue supergiant some 10^4 years before explosion.

Another possibility for explaining these implied CSM density changes around at least some presupernova stars derives from a recent study by Panagia and Bono (this Conference). They find from modeling that the pulsational instability of stars in the mass range 10–20 $M_\odot$ may, in some cases, be of suitable period and magnitude to account for the changes of the pre-supernova mass-loss rates implied by radio observations of RSNe.

5. Peak radio luminosities and distances

Our long-term monitoring of the radio emission from supernovae shows that the radio "light curves" evolve in a systematic fashion with a distinct peak flux density (and thus, in combination with a distance, a peak spectral luminosity) at each frequency and a well-defined time from explosion to that peak. Studying these two quantities at 6 cm wavelength, peak spectral luminosity ($L_{\rm 6\ cm\ peak}$) and time after explosion date (t_0) to reach that peak ($t_{\rm 6\ cm\ peak} - t_0$), we find that they appear related (Fig. 6; see also Weiler et al. 1998). In particular, based on twelve objects, Type II supernovae appear to obey a relation

$$L_{\rm 6\ cm\ peak} \simeq 5.5 \times 10^{23} \left(t_{\rm 6\ cm\ peak} - t_0\right)^{1.4}\ {\rm erg\ s^{-1} Hz^{-1}} \qquad (5.6)$$

with time measured in days. Thus, if this relation is supported by further observations, it provides a means for determining distances to supernovae, and thus to their parent galaxies, from purely radio continuum observations.

Although there are still relatively few objects to which this technique can be applied, RSNe could eventually provide a powerful and independent technique for investigating

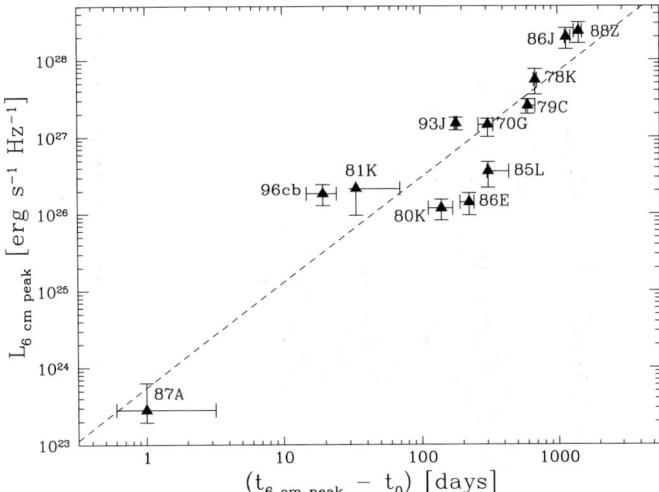

FIGURE 6. Peak 6 cm luminosity, $L_{6\text{ cm peak}}$, of RSNe vs. time, in days, from explosion to peak 6 cm flux density $(t_{6\text{ cm peak}} - t_0)$ for Type II SNe. The *dashed line* (given by Eq. 5.6 in the text) is the unweighted, best fit to the 12 available Type II RSNe. Error bars are based on best estimates. Where no error or only a stub of a line is shown, the error in that direction is indeterminate (from Weiler et al. 1998).

the long-standing problem of distance estimates in astronomy. With such intrinsically bright Type II RSNe as SN 1988Z and SN 1986J, the technique can be applied to distances of at least 100 Mpc with current VLA technology. With future sensitivity improvements and planned, new, more sensitive radio telescopes, the technique could be extended to large distances, even for less luminous RSNe. For example, with a sensitivity of 1 μJy, a SN of the same class as SNe 1988Z and 1986J could be detected to a redshift of ~ 1, while a relatively radio faint Type II SN, such as SN 1980K, could be studied to a redshift of > 0.1.

6. Sphericity of an SN explosion

It has often been suggested that SN explosions are non-spherical, and there is evidence in a number of stellar systems for jets, lobes, and other directed mass-loss phenomena. Also, the presence of polarization in the optical light from SNe (including SN 1993J) has been interpreted for non-sphericity (see, e.g. Hoeflich et al. 1996) and probably the most obvious evidence for non-spherical structure in an SN system is the very prominent inner ring around SN 1987A. However, our most direct evidence for the structure of at least the shock wave from an SN explosion and the CSM with which it is interacting is from VLBI measurements on SN 1993J. A series of images taken by Marcaide and co-workers (Fig. 7; Marcaide et al. 1997) over a period of two years from 1994 September through 1996 October show only a very regular ring shape indicative of a relatively spherical shock wave expanding into a relatively uniform CSM. The cause of such apparently conflicting results is still to be resolved.

7. Summary of RSN studies

Arising from one of the most energetic phenomena in the Universe, the radio emission from supernovae appears to be relatively well understood in terms of shock interac-

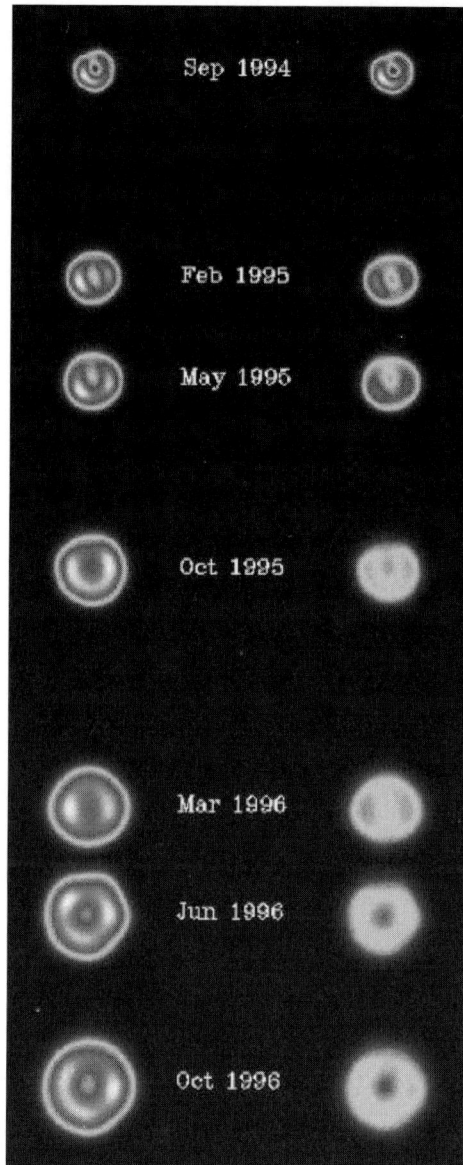

FIGURE 7. VLBI radio images of SN 1993J at 6 cm wavelength. Images on the left-hand side are normalized to the same peak brightness to emphasize structural changes. Images on the right-hand side are on a single brightness scale to illustrate the decrease in brightness with time (from Marcaide et al. 1997).

tion with a structured circumstellar medium as described by the Chevalier (1982a,b) model and its modifications (Weiler et al. 1986, Weiler, Panagia, & Sramek 1990, and Weiler et al. 1991). With this modeling, the radio emission can be used to estimate the circumstellar density, the pre-SN mass loss rate and changes therein, to show the existence of filamentation in the pre-SN stellar wind, to indicate the possible presence of binary companions, and to measure the symmetry of the explosion and the CSM with which it is interacting. More speculatively, radio observations may also lead to a more physical classification system for SNe and provide a new technique for estimating distances to SNe and their parent galaxies.

However, the recent suggestion of an association of the γ-ray burst GRB 980425 with the Type Ic supernova SN 1998bw provides evidence for yet a new phenomenon which may arise in at least some types of SN explosions. Since SN 1998bw is a strong radio emitter, it is important to consider how it compares with the known properties of RSNe discussed above.

8. Gamma-ray bursts and SN 1998bw

8.1. *Gamma-ray burst afterglows*

Gamma-ray bursts (GRBs) are "mysterious" flashes of high-energy radiation that appear from random directions in space and typically last a few seconds. They were first discovered by U.S. Air Force Vela satellites in the 1960s and, since then, numerous theories of their origin have been proposed. NASA's Compton Gamma-Ray Observatory (CGRO) satellite has detected several thousand bursts so far, with an occurrence rate of approximately one per day. The uniform distribution of the bursts on the sky has led theoreticians to suggest that their sources are either very near, and thus uniformly distributed around the solar system, are in an unexpectedly large halo around the Galaxy, or are at cosmological distances—not very restrictive proposals.

The principal limitation to understanding the origin of the bursts has been the difficulty in pinpointing their direction on the sky in order to obtain the multi-wavelength observations necessary to constrain physical models. Gamma-rays are exceedingly difficult to focus on to a position sensitive detector, and the bursts' short duration exacerbates the problem. Only with the launch of the Italian/Dutch satellite BeppoSAX in 1996 has it been possible to couple a quick response pointing system with relatively high precision position sensitive detectors for γ-rays and hard X-rays. This quick response, high accuracy position information has finally permitted rapid and accurate follow-up observations with the world's powerful ground-based and space-based telescopes, and has led to the discovery of long-lived "afterglows" of the bursts in soft X-rays, visible and infrared light, and radio waves. Although the γ-ray bursts generally last only seconds, their afterglows have, in a few cases, been studied for minutes, hours, days, or even weeks after discovery. These longer wavelength observations have allowed observers to probe the immediate environment of γ-ray burst sources and to assemble clues as to their nature.

The first GRB related optical transient was identified for GRB 970228 by Groot et al. (1997) and followup with the Hubble Space Telescope (*HST*) by Sahu et al. (1997) demonstrated that the GRB was associated with a faint (thus probably distant) late-type galaxy. A few months later Fruchter & Bergeron (1997) (see also, Pian et al. 1998) imaged the afterglow of another γ-ray burst, GRB 970508, with the *HST WFPC2* finding this source to be associated with a late type galaxy at a redshift of $z = 0.835$. GRB 970508 was also the first GRB to be detected in its radio afterglow (Frail & Kulkarni 1997).

Heise et al. (1997) detected the very energetic γ-ray burst GRB 971214 on 1997 December 14 with the BeppoSAX satellite and provided sufficient positional accuracy for Halpern et al. (1997) to identify a visible light afterglow using the KPNO 2.1 m telescope. As the visible light from the burst afterglow faded, an extremely faint galaxy ($R = 25.6 \pm 0.2$) was detected at its position. Using the 10-meter Keck II telescope on Mauna Kea, Hawaii, Kulkarni et al. (1998a) measured a redshift of $z = 3.42$. Subsequent images taken with the *HST* (Odewahn et al. 1998) confirmed the association of the burst afterglow and the redshift of this faint galaxy. For such a large distance, Odewahn et al. (1998) estimated that the amount of energy released in the γ-ray flash was extremely high, $\sim 3 \times 10^{53}$ erg. Thus, it appears that at least some GRB sources are very energetic explosions (hypernovae?; Paczynski 1998, Iwamoto et al. 1998) occurring at cosmological distances.

8.2. GRB 980425 and SN 1998bw

8.2.1. Background

While still generally accepted that "most" GRBs are extremely distant and energetic, the discovery of GRB 980425 (Soffitta et al. 1998) on 1998 April 25.90915 and its possible association with a bright supernova, SN 1998bw at RA(J2000) = $19^h35^m03^s.31$, Dec(J2000) = $-52°50'44''.7$ (Tinney et al. 1998), in the relatively nearby spiral galaxy ESO 184–G82 at $z = 0.0085$ (distance ~ 38 Mpc for $H_0 = 65$ km s^{-1} Mpc^{-1}) (Galama et al. 1998, Lidman et al. 1998, Tinney et al. 1998, Sadler et al. 1998), has introduced the possibility of multiple origins for GRBs. The estimated explosion date of SN 1998bw in the interval 1998 April 21–27 (Sadler et al. 1998) corresponds rather well with the time of GRB 980425. Iwamoto et al. (1998) feel that they can restrict the core collapse date for SN 1998bw even more from hydrodynamical modeling of exploding C + O stars and, assuming that the SN 1998bw optical light curve is generated by ^{56}Ni as in Type Ia SNe, they then restrict the coincidence between the core collapse of SN 1998bw to within $+0.7/-2$ days of the detection of GRB 980425.

Classified initially as an SN optical Type Ib (Sadler et al. 1998), then Type Ic (Patat & Piemonte 1998), then peculiar Type Ic (Kay, Halpern, & Leighly 1998, Filippenko 1998), then later, at an age of 300–400 days, again as a Type Ib (Patat et al. 1999), SN 1998bw presents a number of optical spectral peculiarities which strengthen the suspicion that it may be the counterpart of the γ-ray burst.

However, some doubt remains concerning the association of GRB 980425 with SN 1998bw. When the more precise BeppoSAX NFI was pointed at the BeppoSAX error box 10 hours after the detection of GRB 980425, two X-ray sources were present (Pian et al. 1999). One of these, named S1 by Pian et al. (1999), is coincident with the position of SN 1998bw and declined slowly between 1998 April and 1998 November. The second X-ray source, S2, which was $\sim 4'$ from the position of SN 1998bw, was not (or at best only marginally with a $< 3\sigma$ possible detection six days after the initial detection) detectable again in follow up observations in 1998 April, May, and November (Pian et al. 1999). Even though the *a posteriori* statistics indicate a very low probability($\sim 10^{-4}$) of a GRB being nearly coincident in space and time with a SN outburst, the concern remains that the Pian source S2 was the brief afterglow from GRB 980425 rather than the Pian source S1 associated with SN 1998bw.

8.2.2. Radio emission

Since the peak absolute magnitude of Type Ib/c SN 1998bw, while bright at $M_B \sim -19.5$, is not that much brighter than Type Ib SN 1966J ($M_B \sim -19$, converted from $H_0 = 75$ km s^{-1} Mpc^{-1} to $H_0 = 65$ km s^{-1} Mpc^{-1}, Miller, & Branch 1990; N.B.: SN 1966J may have been misclassified. Van Dyk, Hamuy, & Filippenko 1996 indicate in a *Note Added in Proof* that D. Branch has reclassified SN 1966J as a Type Ia.) and less luminous than the Type Ic SN 1992ar with $M_B \sim -20$ (Hamuy et al. 1992), much of the argument for the unusual nature of SN 1998bw rests on the radio observations.

The radio emission from SN 1998bw reached an unusually high 6 cm spectral luminosity at peak of $\sim 8 \times 10^{28}$ erg s^{-1} Hz^{-1}, i.e. about 3 times higher than any of the relatively well studied radio RSNe and ~ 30–40 times higher than a typical Type Ib/c SN at peak (see Table 2 and Weiler et al. 1998). Unfortunately, it is not clear how sig-

SN Name	Type	Peak 6cm Luminosity (erg s^{-1} Hz^{-1})
1983N	Ib	1.4×10^{27}
1984L	Ib	2.6×10^{27}
1990B	Ic	5.6×10^{26}
1994I	Ic	1.4×10^{27}
1997X	Ic	2.6×10^{27}
1998bw	Ib/c	7.9×10^{28}

TABLE 2. Peak Radio Luminosities of Type Ib/c Supernovae)

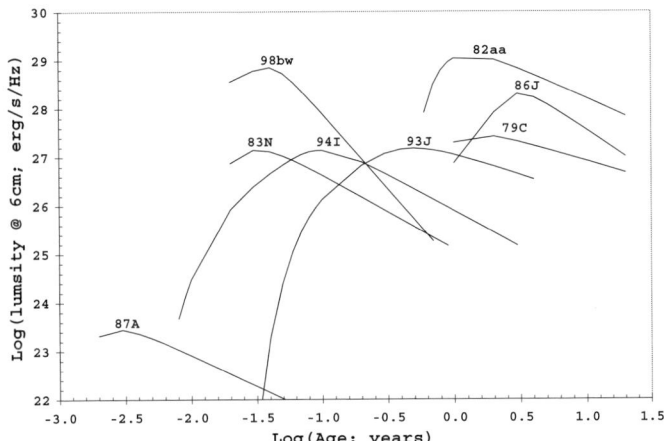

FIGURE 8. Approximate model 6 cm luminosity vs. time "light" curves for radio supernovae (RSNe). Derived from best fits to observations, these model curves smooth out the many "bumps and wiggles" seen in actual data to show the general character of the time evolution rather than the details. Both Type Ib/c (SN 1983N, SN 1994I, SN 1998bw) and Type II (SN 1979C, SN 1982aa, SN 1986J, SN 1987A, SN 1993J) SNe are shown.

nificant this "excess" radio luminosity is since only 6 other examples of radio emitting Type Ib/c SNe are known (see Table 1). Also, one must keep in mind that, contrary to the often stated opinion that SN 1998bw is "the most luminous radio supernova ever observed," SN 1998bw is still exceeded in peak 6 cm spectral luminosity by the poorly studied, presumed supernova, SN 1982aa (Green 1994) in the starburst galaxy NGC 6052 (Markarian 297). SN 1982aa is estimated to have peaked at a 6 cm spectral luminosity of $\sim 1 \times 10^{29}$ erg s^{-1} Hz^{-1}. Although SN 1982aa was not optically identified and, therefore, has no optical spectral type classification, its radio evolution strongly resembles that of Type II RSNe (Yin 1994).

As a further check for unusual characteristics of SN 1998bw, we can compare a smoothed 6 cm light curve for SN 1998bw with model 6 cm light curves for other RSNe. This is illustrated in Fig. 8. Inspection of the figure shows that SN 1998bw is unusual in its radio emission, but not extreme. For example, the time from explosion to peak 6 cm luminosity for both SN 1987A and SN 1983N was faster and, as mentioned above, the 6 cm spectral luminosity of SN 1998bw at peak is exceeded by that of SN 1982aa. However, as is clear from Fig. 8, SN 1998bw is the most luminous Type Ib/c RSN ever observed by a factor of ~ 30 and it reaches a higher radio luminosity earlier than any RSN known.

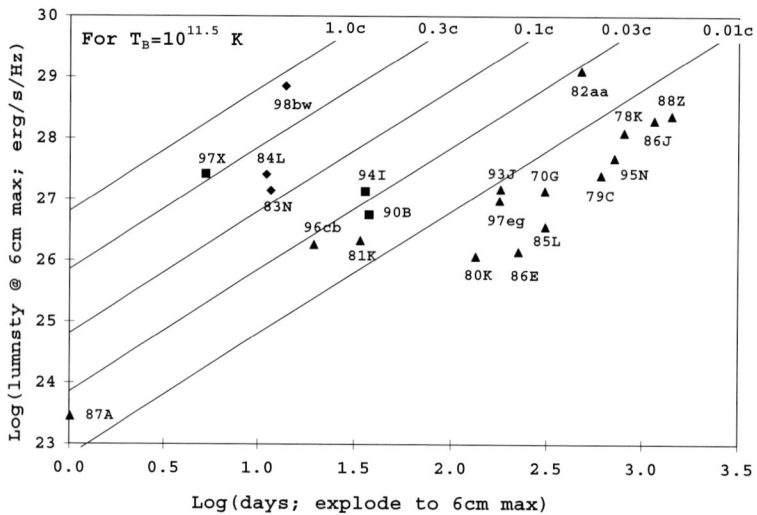

FIGURE 9. A plot of the 6 cm luminosity at peak vs. the time to reach that peak for all of the known RSNe. The individual RSNe are labeled and their types are indicated by their symbols; Type Ib, filled squares, Type Ic, filled diamonds, and Type II filled triangles. The lines of constant expansion velocity are shown for a brightness temperature of $10^{11.5}$ K.

8.2.3. Brightness temperature

Although unique in neither the speed of evolution nor in radio luminosity, SN 1998bw is certainly unusual in the combination of these two factors—very radio luminous very soon after explosion. Kulkarni et al. (1998b,c) have used these observed qualities, together with the lack of interstellar scintillation at early times, brightness temperature estimates, and physical arguments to conclude that the shock wave from SN 1998bw giving rise to the radio emission must have been expanding relativistically. On the other hand, Waxman & Loeb (1999) argue that a sub-relativistic shock can generate the observed radio emission. However, both sets of authors agree that a very high expansion velocity ($\geq 0.3c$) is required for the radio emitting region under a spherical geometry.

Simple arguments confirm this high velocity since, to avoid the well known Compton Catastrophe, Kellermann & Pauliny-Toth (1969) have shown that $T_B < 10^{12}$ K must hold and Readhead (1994) has better defined this limit to $T_B < 10^{11.5}$ K. From geometrical arguments, such a limit requires the radiosphere of SN 1998bw to have expanded at $\geq 200,000$ km s^{-1}, at least during the first few days after explosion. While such a value is still only mildly relativistic ($\gamma \sim 1.5$), it seems very high. However, measurements by Gaensler et al. (1997) have demonstrated that the radio emitting regions of SN 1987A have expanded at an *average* velocity of $\sim 35,000$ km s^{-1} over the 3 years from 1987 February to mid-1990 so that, in a very low density environment such as one finds around Type Ib/c SNe, very high shock velocities may be possible.

This is illustrated graphically in Fig. 9 where, for an adopted brightness temperature of $10^{11.5}$ K, the lines of constant expansion velocity are shown on a plot of the 6 cm spectral luminosity at peak *vs.* the time to reach that peak for a number of RSNe. Although expected, it is noteworthy that all Type II SNe, which have relatively dense circumstellar envelopes, have the lowest expansion velocities while the Type Ib/c SNe, with their relatively less dense CSM, have significantly higher expansion rates. Once again, SN 1998bw appears to represent a more extreme form of Type Ib/c RSNe.

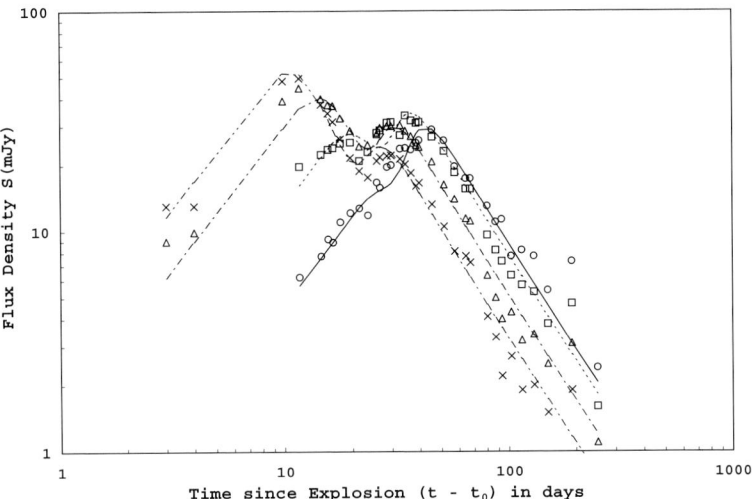

FIGURE 10. The radio light curves of SN 1998bw at 8.64 GHz (3.5 cm; *cross, dash-double dot line*), 4.80 GHz (6.3 cm; *open triangle, dash-dot line*), 2.50 GHz (12 cm; *open square, dotted line*) and 1.38 GHz (21.7 cm; *open circle, solid line*). The curves are derived from a best fit model described by the equations in §2.1 and the parameters listed in Table 3.

8.2.4. *Radio light curves*

A final, obvious comparison of SN 1998bw with other RSNe is the evolution of its radio flux density at multiple frequencies. The radio data for the first 80 days are plotted in Kulkarni et al. (1998c), up to day 250 on *http://www.narrabri.atnf.csiro.au/public/grb /grb980425/*, and have been kindly supplied to us in tabular form by D. A. Frail (private communication). SN 1998bw shows, at higher frequencies ($\nu \geq 2.5$ GHz) an early peak which reaches a maximum as early as day 10–12 at 8.64 GHz, a minimum almost simultaneously for the higher frequencies at day ~ 20–24, then a secondary, somewhat lower peak at later times after the first dip. An interesting characteristic of this "double humped" structure is that it dies out at lower frequencies and is relatively inconspicuous in the 1.4 GHz radio measurements (see Fig. 10). Such a structure is not as prominent for other known radio supernovae on such a short time scale, although the Type Ic SN 1994I (S. D. Van Dyk, private communication) shows a dip in its 15 GHz flux density, also at an age of ~ 20 days.

Li and Chevalier (2000) propose an initially synchrotron self-absorbed (SSA), rapidly expanding shock wave in a $\rho \propto r^{-2}$ circumstellar wind model to describe the radio light curve. This is in many ways similar to the Chevalier (1998) model for Type Ib/c SNe, which also included SSA, and the structure described in §2.1 which includes only thermal free-free initial absorption. However, to improve the quality of the fit, Li and Chevalier (2000) introduce the additional free parameters of a boost of shock energy by a factor of ~ 2.8 on day ~ 22 in the observer's time frame. This provides the energy necessary to produce the second peak in the radio light curves.

While is not possible to rule out such complexities, we have attempted to describe the radio light curve by a relatively straightforward, two-component model very similar to those used successfully for previous SNe described above in §2.1. Such an approach has no requirement for shock re-acceleration. Also, although an attempt was made to include the effects of SSA at early times, the fit was not improved by such an addition. The two component model fit which we obtained using only synchrotron emission and thermal, free-free absorption is shown as the lines in Fig. 10, and its parameter values are listed

Parameter	Component 1	Component 2
α	-1.08	-0.80
β	-3.13	-1.55
K_1	3.8×10^5	6.2×10^3
K_2	0	0
δ	–	–
K_3	2.4×10^5	8.4×10^{12}
δ'	-4.43	-8.80
K_4	4.3×10^{-2}	3.4×10^{-2}

TABLE 3. Parameters for a two component, thermal model fit (see §2.1)

in Table 3. The "double humped" structure of the radio light curves for SN 1998bw is reproduced by a single energy shock wave encountering regions of differing CSM density as it travels rapidly outward.

Since Li & Chevalier (2000) do not give a quantitative indication of the quality of their fit, it is impossible to compare the two models directly. However, a visual comparison of Fig. 10 here with their Fig. 9 indicates that the fits are of similar quality at late times and low frequencies, where the initial absorption mechanisms have become less important. This is not surprising, since they are based on the same models from Chevalier (1982a,b). Both our model and the Li & Chevalier (2000) model have difficulty adequately describing the dip at all frequencies near day 20. However, at early times and high frequencies the free-free absorption model shown in Fig. 10 gives a clearly superior fit to the data to that of Li & Chevalier (2000).

Note that our fit requires essentially no diffuse/uniform absorption ($K_2 = 0$; see Table 3 and Eqs. 2.1–2.4) so that all of the free-free absorption either is due to a clumpy medium or is internal to the emitting zones (which has the same mathematical form), both of which contribute to a K_3 term in Table 3 and Eqs. 2.1–2.4. These results, combined with the estimate of a high shock velocity, suggest that the CSM around SN 1998bw is highly clumped in numerous and dense knots with little, if any, intra-knot gas. The clump filling factor has to be high enough to intercept a considerable fraction of the shock energy and still be low enough to let radiation escape from any given knot without being appreciably absorbed by any other knot. This can be somewhat quantified by postulating that the covering factor (i.e. the fraction of solid angle occupied by clumps) has to be less than $\sim 1/4$ and, perhaps, greater than $\sim 1/10$. In this scenario the shock front can easily move at a speed which is a significant fraction of the speed of light because it is moving essentially in a vacuum. However, one still has strong energy dissipation and relativistic electron acceleration at the knot surfaces facing the SN explosion center. If the knots are dense and relatively opaque to radio emission, we mainly observe radiation produced at the surfaces of the knots in the CSM on the far-side of the SN in which some internal absorption is also occurring. The radiation from near-side knots is probably absorbed internally by the knots themselves and, therefore, lost from the signal we detect.

The presence of two components, as evidenced by the presence of two peaks in the radio light curve, implies the existence of two regimes within the clumpy CSM that surrounds the SN progenitor with the transition between the two regimes occurring at about 10–20 light-days from the star, i.e. at about 3×10^{16} cm or ~ 3000 AU from the stellar progenitor. This distance may be somewhat speculatively interpreted as the separation of the exploding primary from a lower mass binary companion, which is reminiscent of the binary structure inferred for SN 1979C (Weiler et al. 1992a, Boffi & Panagia 1996). In the case of SN 1998bw a mass at explosion of about ~ 14 M$_\odot$ and an original progenitor mass of ~ 30–40 M$_\odot$ has been proposed (Danziger et al., this Conference). A binary hypothesis is also the best explanation for the large mass loss that the progenitor star must have suffered to expose the H-free layers typical of Type Ib/c supernovae before exploding. Within this framework, one may argue that the CSM within the SN progenitor Roche lobe was mostly determined by the progenitor mass-loss itself, whereas at larger distances the CSM was largely determined by the interaction of the progenitor wind with the companion wind, thus leading to a two component CSM structure.

Clearly, there is much more complexity in the physics of the GRB phenomenon and the structure of the SN 1998bw system than can be described by simple models, and none proposed is completely satisfactory as yet. However, while the γ-ray emission itself appears to require relativistic boosting, the radio emission from the best-studied possible example, SN 1998bw, need not be interpreted as proof of relativistic beaming, highly relativistic shock waves, or other esoteric phenomena. Only mildly relativistic processes and thermal absorption by the ionized CSM appear at least satisfactory and the arguments for a SN 1998bw/GRB 980425 connection are correspondingly weakened.

8.2.5. Summary of SN 1998bw/GRB 980425 relations

On balance, SN 1998bw appears to be a relatively normal, if rather over-bright example of the SN Type Ib/c phenomenon. Many observers tend to accept that it is associated with GRB 980425 and it is impossible to disprove that postulate. On the other hand, neither the optical nor radio emission are sufficiently unusual to firmly establish a connection and the Pian et al. (1999) X-ray source S2 in the GRB field remains a distinct possibility for the location of GRB 980425. We can only hope that both the continued study of the GRB phenomenon and its (sometimes) associated afterglows and the slowly increasing statistics of observations of Type Ib/c SNe will be able to establish the true nature of both phenomena and any possible relation.

CKL, KWW, & MJM wish to thank the Office of Naval Research (ONR) for the 6.1 funding supporting this research.

REFERENCES

BALL, L., CAMPBELL-WILSON, D., CRAWFORD, D. F., & TURTLE, A. J. 1995 *ApJ* **453**, 864.
BARTEL, N., ET AL. 1985 *Nature* **318**, 25.
BOFFI, F. R. & PANAGIA, N. 1996. In *Radio Emission from the Stars and the Sun* (eds. A. R. Taylor & J. M. Paredes). Astr. Soc. Pacific Conf. Series, vol. 93, p. 153.
BONO, G. & PANAGIA, N. 2000, in preparation.
BROCATO, E. & CASTELLANI, V. 1993 *ApJ* **410**, 99.
CHEVALIER, R. A. 1982a *ApJ* **259**, 302.
CHEVALIER, R. A. 1982b *ApJL* **259**, L85.
CHEVALIER, R. A. 1998 *ApJ* **499**, 810.
CHEVALIER, R. A. & FRANSSON, C. 1994 *ApJ* **420**, 268.
CHU, Y.-H., CAULET, A., MONTES, M. J., PANAGIA, N., VAN DYK, S. D., & WEILER, K. W. 1999 *ApJ* **512** L51.
FILIPPENKO, A. 1998 *IAUC* 6969.
FILIPPENKO, A., MATHESON, T., & BARTH, A. 1994 *AJ* **108**, 222.
FRAIL, D. A. & KULKARNI, S. R. 1997 *IAUC* 6662.
FRANSSON, C. & BJÖRGSSON, C.-I. 1998 *ApJ* **509**, 861.

FRUCHTER, A. & BERGERON, L. 1997 *IAUC* 6674.

GAENSLER, B. M., MANCHESTER, R. N., STAVELEY-SMITH, L., TZIOUMIS, A. K., REYNOLDS, J. E., & KESTEVEN, M. J. 1997 *ApJ* **479**, 845.

GALAMA, T. J., ET AL. 1998 *IAUC* 6895.

GARNAVICH, P., KIRSHNER, R., & CHALLIS, P. 1997 *IAUC* 6710.

GREEN, D. W. E. 1994 *IAUC* 5953.

GROOT, P. J., ET AL. 1997 *IAUC* 6584.

HALPERN, J., ET AL. 1997 *IAUC* 6788.

HASINGER, G., ASCHENBACH, B., & TRUEMPER, J. 1996 *A&A* **312**, 9.

HEISE, J.,ET AL. 1997 *IAUC* 6787.

HOEFLICH, P., WHEELER, J. C., HINES, D. C., & TRAMAELL, S. R. 1996 *ApJ* **459**, 307.

HAMUY, M., ET AL. 1992 *IAUC* 5574.

IWAMOTO, K., ET AL. 1998 *Nature* **395**, 672.

KAY, L. E., HALPERN, J. P., & LEIGHLY, K. M. 1998 *IAUC* 6969.

KELLERMANN, K. I. & PAULINY-TOTH, I. I. K. 1969 *ApJ* **155**, L71.

KULKARNI, S., ET AL. 1998a *Nature* **393**, 35.

KULKARNI, S. R., BLOOM, J. S., FRAIL, D. A., EKERS, R., WIERINGA, M., WARK, R., & HIGDON, J. L. 1998b *IAUC* 6903.

KULKARNI, S. R., FRAIL, D. A., WIERINGA, M. H., EKERS, R. D., SADLER, E. M., WARK, R. M., HIGDON, J. L., PHINNEY, E. S., & BLOOM, J. S. 1998c *Nature* **395**, 663.

LACEY, C. K., WEILER, K. W., SRAMEK, R. A., PANAGIA, N., & VAN DYK, S. D. 2000, in preparation.

LANG, K. R. 1986. In *Astrophysical Formulae*, p. 47. Springer-Verlag.

LANGER, N. & MAEDER, A. 1995 *A&A* **295**, 685.

LI, Z.-Y. & CHEVALIER, R. A. 2000, in press (also astro-ph/9903483).

LIDMAN, C., ET AL. 1998 *IAUC* 6895.

MARCAIDE, J. M., ET AL. 1997 *ApJ* **486**, L31.

MILLER, D. L. & BRANCH, D. 1990 *AJ* **100**, 530.

MONTES, M. J., WEILER, K. W., & PANAGIA, N. 1997 *ApJ* **488**, 792.

MONTES, M. J., VAN DYK, S. D., WEILER, K. W., SRAMEK, R. A., & PANAGIA, N. 1998 *ApJ* **506**, 874.

MONTES, M. J., WEILER, K. W., VAN DYK, S. D., SRAMEK, R. A., PANAGIA, N., & PARK, R. 2000 *ApJ*, in press (also astro-ph/9911399).

NOMOTO, K., YAMAOKA, H., POLS, O. R., VAN DEN HEUVEL, E., IWAMOTO, K., KUMAGAI, S., & SHIGEYAMA, T. 1994 *Nature* **371**, 227.

ODEWAHN, S. C., ET AL. 1998 *ApJ* **509**, 5.

PACZYNSKI, B. 1998 *ApJ* **494** L45.

PANAGIA, N. & MACCHETTO, F. 1982 *A&A* **106**, 266.

PATAT, F., CAPPELLARO, E., RIZZI, L., TURATTO, M., & BENETTI, S. 1999 *IAUC* 7215.

PATAT, F. & PIEMONTE, A. 1998 *IAUC* 6918.

PIAN, E., ET AL. 1998 *ApJ* **492** L103.

PIAN, E., ET AL. 1999 *A&ASS* **138**, 463.

PODSIADLOWSKI, P., HSU, J., JOSS, P., & ROSS, R. 1993 *Nature* **364**, 509.

READHEAD, A. C. S. 1994 *ApJ* **426**, 51.

RYDER, S., STAVELEY-SMITH, L., DOPITA, M., PETRE, R., COLBERT, E., MALIN, D., & SCHLEGEL, E. 1993 *ApJ* **417**, 167.

SADLER, E. M., STATHAKIS, R. A., BOYLE, B. J., & EKERS, R. D. 1998 *IAUC* 6901.

SAHU, K. C., ET AL. 1997, *ApJ* **489** L127.

SCHWARZ, D. H. & PRINGLE, J. E. 1996 *MNRAS* **282**, 1018.

SOFFITTA, P., ET AL. 1998 *IAUC* 6884.

SPYROMILIO, J. 1994 *MNRAS* **266**, 61.

TINNEY, C., STATHAKIS, R., CANNON, R., & GALAMA, T. 1998 *IAUC* 6896.

TULLY, R. B. 1988. In *Nearby Galaxies Catalogue*. Cambridge Univ. Press.

TURTLE, A. J., CAMPBELL-WILSON, D., BUNTON, J. D., JAUNCEY, D. L., KESTEVEN, M. J., MANCHESTER, R. N., NORRIS, R. P., STOREY, M. C., & REYNOLDS, J. E. 1987 *Nature* **327**, 38.

VAN DYK, S. D., SRAMEK, R. A., WEILER, K. W., & PANAGIA, N. 1993 *ApJ* **419** L69.

VAN DYK, S. D., WEILER, K. W., SRAMEK, R. A., RUPEN, M., & PANAGIA, N. 1994 *ApJ* **432** L115.

VAN DYK, S. D., HAMUY, M., & FILIPPENKO, A. V. 1996 *AJ* **111**, 2017.

WAXMAN, E. & LOEB, A. 1999 *ApJ* **515**, 721.

WEILER, K. W., SRAMEK, R. A., PANAGIA, N., VAN DER HULST, J. M., & SALVATI, M. 1986 *ApJ* **301**, 790.

WEILER, K. W., PANAGIA, N., & SRAMEK, R. A. 1990 *ApJ* **364**, 611.

WEILER, K. W., VAN DYK, S. D., PANAGIA, N., SRAMEK, R. A., & DISCENNA, J. 1991 *ApJ* **380**, 161.

WEILER, K. W., VAN DYK, S. D., PRINGLE, J., & PANAGIA, N. 1992a *ApJ* **399**, 672.

WEILER, K. W., VAN DYK, S. D., PANAGIA, N., & SRAMEK, R. A. 1992b *ApJ* **398**, 248.

WEILER, K. W., VAN DYK, S. D., MONTES, M. J., PANAGIA, N., & SRAMEK, R. A. 1998 *ApJ* **500**, 51.

YIN, Q. F. 1994 *ApJ* **420**, 152.

Models for Ia Supernovae and evolutionary effects

By P. HÖFLICH[1] AND I. DOMINGUEZ[2]

[1] Dept. of Astronomy, University of Texas, Austin, TX 78712

[2] Dept. of Astronomy, University of Granada, Granada, Spain

From the spectra and light curves it is clear that SNIa events are thermonuclear explosions of white dwarfs (WD). However, details of the explosion are highly under debate. Here, we present detailed models which are consistent with respect to the progenitor evolution, explosion mechanism, the optical and infrared light curves (LC), and the spectral evolution. This leaves the description of the burning front and the structure of the white dwarf as the only free parameters. The explosions are calculated using one-dimensional Lagrangian codes including nuclear networks. Subsequently, optical and IR-LCs are constructed. Detailed NLTE-spectra are computed for several instants of time using the density, chemical and luminosity structure resulting from the LCs (§2).

Theoretical models allow for two basic approaches: Test of explosion scenarios by comparison with existing observations and the study of the influence of variations in the underlaying models, e.g. in the metallicity, on the observables.

Different models for the thermonuclear explosion are discussed including detonations, deflagrations, delayed detonations (DD), pulsating delayed detonations (PDD) and helium detonations (HeD, or sub-Chandrasekhar mass explosions) (§3). Comparisons between theoretical and observed LCs and spectra provide an insight into details of the explosion and nature of the progenitor stars. The combined analysis of optical and IR data provides evidence for a layered chemical structure in 'normal' bright SNe Ia. At the very outer layers, we may have some unburned C/O, there is evidence for layers with have undergone explosive carbon layers, followed by a layer of incomplete Si burning and an inner region ($M \leq 0.6...1.0$ M$_\odot$) of complete burning up to NSE (§§4–5). We try to answer several related questions relevant for cosmology. How can we understand the brightness/decline relation (§4)? Can we determine H_o independently from primary distance indicators, and how do the results compare with empirical methods (§6)? What do we learn about the progenitor evolution and its metallicity? What are the systematic effects for the determination of the cosmological parameters Ω_M and Λ and how can we recognize this potential 'pitfall' and correct for evolutionary effects (§7)?

1. Introduction

During the last few years it became evident that Type Ia supernovae are a less homogeneous class than previously believed (e.g. Phillips et al. 1987). In particular, the observation of subluminous SN 1991bg (Filippenko et al. 1992, Leibundgut et al. 1993) raised questions on the SN-rate and whether we have missed a huge subgroup. The possible impact on our understanding of supernovae statistics and, consequently, the chemical evolution of galaxies must be noted.

With respect to the use of Type Ia Supernovae as distance indicators, for nearby supernovae ($z \leq 0.1$), different schemes have been developed and tested to cope with the problems of deducing the intrinsic brightness based on theoretical models or observed correlations between spectra or light curves and the absolute brightness using primary distance indicators (e.g. Norgaard-Nielsen et al. 1989; Branch & Tammann 1992; Sandage & Tammann 1993; Müller & Höflich 1994; Hamuy et al. 1996; Riess, Press & Kirshner 1995; Höflich & Khokhlov 1996; HK96 thereafter). These methods have been tested locally and provide consistent results. Due to the new recalibrations of the brightness of SNe Ia by Cepheid variables in nearby galaxies (e.g. Saha et al. 1997 and references therein), a common agreement could be reached on the value of H_o. In a next step, new telescopes including the extensive use of the *Hubble Space Telescope* and observational programs put SNe Ia at large redshifts well within reach up to a redshift of 1.3. The Berkeley group has discovered more than 50 SNe Ia up to a redshift of 1.3 (e.g. Perlmutter et al. 1995, 1999). The CfA/CTIO/ESO/MSSSO collaboration has found a similar number of supernovae (e.g. Garnavich et al. 1998). A major concern with the use of SNe Ia are systematic errors due to evolutionary effects (Höflich, Wheeler & Thielemann 1998, HWT98 thereafter). This is especially true for statistical methods which are calibrated only on local SNe Ia.

Despite the success of purely empirical methods, theoretical work on SNe Ia is critical to provide an independent test for the distance scale, to get an insight into the underlying physics of the explosion, it provides a tool to investigate benefits of new wavelength ranges not yet well explored (i.e. the IR) and it allows to test for the importance (or unimportance) of systematic effects when it comes to the study of distant SNe Ia. The goal of this paper is to provide a brief overview of the different aspects just mentioned.

2. Numerical methods

A consistent treatment of the explosion, light curves and spectra is critical to provide a tight coupling between the explosion model and the observational quantities. Therefore, we start with a brief description of the numerical methods. For details, see the corresponding papers cited below.

2.1. *Stellar evolution*

One example shown below is based on the stellar evolution calculation by Umeda et al. (1999). Other models have been calculated using the code FRANEC of the Italian group (Chieffi & Straniero, 1989; Straniero et al. 1997) up to the end of the helium burning. The actual calculations have been performed by the second author. Subsequently, the evolution of the C/O core is calculated by accreting H/He rich material for a given accretion rate on the core solving the standard equations for stellar evolution using a Henye scheme. Nomoto's equation of state is used (Nomoto et al. 1982). Cristallization is neglected. For the energy transport, conduction (Itoh et al. 1983), convection in the mixing length theory, and radiation is taken into account. Radiative opacities for free-free and bound free transitions are approximated in Kramer's approximation and by free electrons. A nuclear network of 35 species up to ^{24}Mg is taken into account.

2.2. *Hydrodynamics*

The explosions are calculated using a one-dimensional radiation-hydro code, including nuclear networks (Höflich & Khokhlov 1996 and references therein). This code solves the hydrodynamical equations explicitly by the piecewise parabolic method (Collela & Woodward 1984) and includes the solution of the frequency averaged radiation transport implicitly via moment equations, expansion opacities (see below), and a detailed equation of state. Nuclear burning is taken into account using a network which has been tested in many explosive environments (see Thielemann et al. 1996 and references therein).

2.3. *Description of the burning front*

In the hydrodynamical models, the nuclear burning may propagate either as a deflagration or detonation front, or a combination in which a deflagration turns into a detonation.

In case of a detonation, its speed can be calculated directly by the hydrodynamic equations and the nuclear energy release. This mode is realized in models for the explosion of sub-Chandrasekhar Mass White dwarfs (WD) and, likely, in the outer layers of massive WDs. In massive WDs, the burning front starts as a deflagration which, later on, may turn into a detonation. Intrinsically, combustion is a three-dimensional problem problem which, still, is not completely solved. Here, we have to use a parameterized description and it may be useful study the sensitivity of the results of the calculations below on this critical assumption, because the explosions models for massive WDs provide the best agreement with observations (see below).

For the description of the deflagration front, we have tested three different cases (Dominguez & Höflich 1999):

Case 1) $v_{\text{burn}} = const.\ v_{\text{sound}}$. In our previous investigations, $const = 0.03$ has been found to give the best fits to observations. This corresponds to the fractal dimension $D = 2$ in the description of Woosley & Weaver (1994a) who suggested $D = 2$ to 2.5.

Case 2 & 3) Here we assumed that $v_{\text{burn}} = max(v_t, v_l)$ where v_l and v_t are the laminar and turbulent velocities, respectively.

Turbulent combustion is driven on large scales by the buoyancy of the burning products. The turbulent cascade penetrates down to very small scales, and makes the rate of deflagration independent of the microphysics. Turbulent combustion in a uniform gravitational field and static conditions singles out the propagation of the flame against gravity g. Chemical combustion experiments have been performed in confined environments, so called Combustion chambers. These experiments can be reproduced by numerical simulations. The propagation speed can be described by

$$v_t = C_1 \sqrt{\alpha_T\ g\ L_f};\ \ C_1 = 0.5; \alpha_T = (\alpha - 1)/(\alpha + 1),\ \alpha = \rho^+(r_{\text{burn}})/\rho^-(r_{\text{burn}})\ , \quad (2.1)$$

where α_T is the Atwood number, L_f is the characteristic length scale, and ρ^+ and ρ^- are the densities in front and behind the front, respectively. However, despite the success in terrestrial experiments, the basic assumptions of both a uniform gravitational field and static conditions are violated in the rapidly expanding envelopes of SNe Ia. The main effect of expansion is the freeze out of the turbulence on scales L_f where the turbulent velocity due to Rayleigh Taylor instabilities is comparable to the differential expansion velocities on those scales, i.e.

$$v_t \approx v_{exp} = L_f/\tau_{ex}\ . \quad (2.2)$$

Based on this idea, Khokhlov et al. (1997b) suggested to use the average turbulent velocity (eq. 2.1), use α for uniform, static conditions, and to use the mean expansion time scale determined by one dimensional simulations $\tau_{exp} \approx dt/d\ ln\ R_{\text{WD}}$. He found for the propagation speed of the turbulent burning front

$$v_t = 0.0474 * \sqrt{(g\ L_f)}\ . \quad (2.3)$$

As third case for the description, we followed the recipe of Khokhlov but did some modifications by taking α, L_f and τ_{exp} directly from the hydro at the location of the burning front. Freeze-out was assumed when the radius of a mass element has doubled after being burned. C_1 in equation 2.1 has been varied. Note that a variation in C_1 is equivalent to scaling the relative length scale for the freeze out. We varied C_1 in the range to cover a parameter space which includes both the descriptions suggested by Khokhlov et al. (1997b) and Niemeyer & Woosley (1997).

Within the scenario of models where the deflagration front turns into a detonation, the so called delayed detonations (DD), we find that the results of the explosion are rather insensitive to details of the description of the deflagration front, even if its speed during

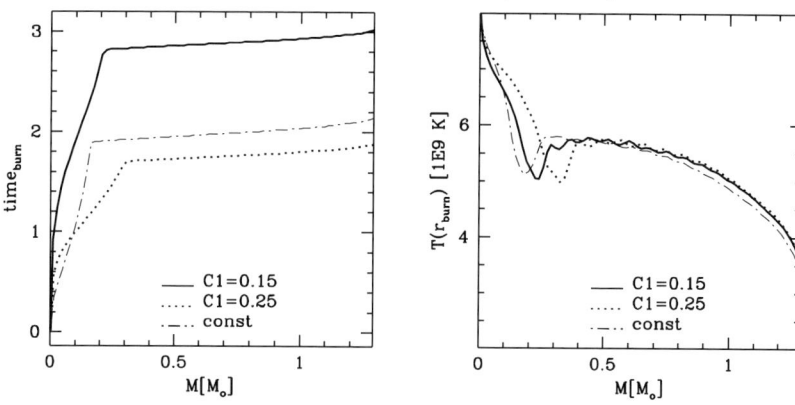

FIGURE 1. Burning conditions in a typical delayed detonation model for various description for the deflagration speed. Time at which a mass element is burned and, as example, the temperature behind the burning front is given for deflagration speeds according eq. 2.1 with various C_1, and with constant deflagration speed with $v = 0.03 \times v_{\rm sound}$. The central density of the initial WD is $2. \times 10^9$ g cm^{-3} and the transition from deflagration to detonation was triggered at 2.4×10^7 g cm^{-3}. Apparently, there are little changes in the burning conditions.

the deflagration phase differs by almost a factor of 2 because the conditions are rather uneffected under which burning takes place (see Fig. 1). The same should also apply for pure deflagration models in which the front accelerates rapidly after an initially slow phase as suggested by Hillebrand (1999, private communication). Qualitatively, this insensitivity of the final outcome of the explosion on the details of the flame propagation during the (slow) deflagration phase can be understood as follows: For plausible variations in the speed of the turbulent deflagration, the duration of this phase is several times longer than the sound crossing time in the initial WD. Therefore, the energy produced during the early nuclear burning can be redistributed over the entire WD causing a slow preexpansion. In this intermediate state the WD is still bound but its binding energy is reduced by the amount of nuclear energy. The expansion ratio depends mainly on the total amount of burning during the deflagration phase. Consequently, the conditions are very similar under which nuclear burning takes place during the subsequent detonation phase. An exception to the similarity are the innermost layers of ≈ 0.03 to 0.05 M$_\odot$. Still, nuclear burning is in nuclear statistical equilibrium (NSE) but the rate of electron capture is larger if the flame propagates close to the laminar speed. Consequently, the production of very neutron rich isotopes is increased. We note that, by using parameterized description, we cannot study the influence of environmental changes on the properties of the nuclear burning itself. 3-dimensional calculations are needed to get an inside into the relations between model parameters such as central density and properties of the deflagration front, its relation to the transition density between deflagration and detonation, and to make use of information on asphericity that is provided by polarization measurements.

2.4. Light curves

Based on the explosion models, the subsequent expansion and bolometric as well as monochromatic light curves are calculated using a scheme recently developed, tested and widely applied to SN Ia (HWT98 and references therein). The code used in this phase is similar to that described above, but nuclear burning is neglected and γ ray transport is included via a Monte Carlo scheme. In order to allow for a more consistent treatment of scattering, we solve both the (two lowest) time-dependent, frequency averaged radiation moment equations for the radiation energy and the radiation flux, and a total energy equation. At each time step, we then use $T(r)$ to determine the Eddington factors and mean opacities by solving the frequency-dependent radiation transport equation in the comoving frame and integrate to obtain the frequency-averaged quantities. About one thousand frequencies (in one hundred frequency groups) and about five hundred depth points are used. The averaged opacities been calculated under the assumption of local thermodynamical equilibrium. Both the monochromatic and mean opacities are calculated using the Sobolev approximation. The scattering, photon redistribution and thermalization terms used in the light curve opacity calculation are calibrated with NLTE calculations using the formalism of the equivalent-two-level approach (Höflich 1995).

2.5. Spectral calculations

Our non-LTE code (Höflich 1995, HWT98 and references therein) solves the relativistic radiation transport equations in comoving frame. The energetics of the supernova are calculated. The evolution of the spectrum is not subject to any tuning or free parameters.

The non-LTE spectra are computed for various epochs using the chemical, density and luminosity structure and γ-ray deposition resulting from the light curve code providing a tight coupling between the explosion model and the radiative transfer. The effects of instantaneous energy deposition by γ-rays, the stored energy (in the thermal bath and in ionization) and the energy loss due to the expansion are taken into account. The transport equations are solved consistently with the statistical equations and ionization due to γ rays for the most important elements (C, O, Ne, Na, Mg, Si, S, Ca, Fe, Co, Ni). Besides about 20,000 lines treated in full non-LTE, about 10^6 additional lines are included assuming LTE-level populations and an equivalent-two-level approach for the source functions.

3. Hydrodynamical models

3.1. Explosions of massive White Dwarfs

A first group consists of massive carbon-oxygen white dwarfs (WDs) with a mass close to the Chandrasekhar mass which accrete mass through Roche-lobe overflow from an evolved companion star (Nomoto & Sugimoto 1977). The explosion is triggered by compressional heating. The key question is how the flame propagates through the white dwarf. Several models of SNe Ia have been proposed in the past, including detonation (Arnett 1969, Hansen & Wheeler 1969), deflagration (e.g. Nomoto, Sugimoto & Neo 1976) and the delayed detonation model, which assumes that the flame starts as a deflagration and turns into a detonation later on (e.g. Khokhlov 1991ab, Woosley & Weaver 1994ab).

Our sample includes detonations (DET1/2), deflagrations (W7, DF1), delayed detonations (M35-39, DD13-27, DD200) and pulsating delayed detonations (PDD1-9). The deflagration speed is parameterized as $D_{def} = \alpha a_s$, where a_s is the local sound velocity ahead of the flame and α is a free parameter. The speed of the detonation wave is given by the sound-speed behind the front. For delayed detonation models, the transition to a detonation is given by another free parameter ρ_{tr}. When the density ahead of the deflagration front reaches ρ_{tr}, the transition to a detonation is forced by increasing α to 0.5 over 5 time steps bringing the speed well above the Chapman-Jouguet threshold for steady deflagration. For pulsating delayed detonation models, the initial phase of burning fails to release sufficient energy to disrupt the WD. During the subsequent contraction phase, compression of the mixed layer of products of burning and C/O formed at the dead deflagration front would give rise to a detonation via compression and spontaneous ignition (Khokhlov 1991ab). In this scenario, ρ_{tr} represents the density at which the

Model	$M_\star$ [$M_\odot$]	ρ_c [10^9 g cm^{-3}]	C/O	α	ρ_{tr} [10^7 g cm^{-3}]	M_{Ni} [$M_\odot$]
DET1	1.4	3.5	1.	—	—	0.92
DF1	1.4	3.5	1.	0.30	—	0.50
W7	1.4	2.0	1.	n.a.	—	0.59
M35	1.4	2.8	1.	0.03	3.0	0.67
M36	1.4	2.8	1.	0.03	2.4	0.60
M37	1.4	2.8	1.	0.03	2.0	0.51
M39	1.4	2.8	1.	0.03	1.4	0.34
DD200c	1.4	2.0	1.	0.03	2.0	0.61
DD13c	1.4	2.6	1.	0.03	3.0	0.79
DD21, 24–27c	1.4	2.6	1.	0.03	2.7	0.69
DD23c	1.4	2.6	2/3	0.03	2.7	0.59
PDD3	1.4	2.1	1.	0.04	2.0	0.49
PDD5	1.4	2.7	1.	0.03	0.76	0.12
PDD8	1.4	2.7	1.	0.03	0.85	0.18
PDD7	1.4	2.7	1.	0.03	1.1	0.36
PDD9	1.4	2.7	1.	0.03	1.7	0.66
PDD6	1.4	2.7	1.	0.03	2.2	0.56
PDD1a	1.4	2.4	1.	0.03	2.3	0.61
PDD1c	1.4	2.4	1.	0.03	0.71	0.10
HeD2	0.6+0.22	.013	1.	—	—	0.43
HeD4	1.0+0.18	.150	1.	—	—	1.07
HeD6	0.6+0.172	.0091	1.	—	—	0.252
HeD10	0.8+0.22	.036	1.	—	—	0.75
HeD12	0.9+0.22	.083	1.	—	—	0.92
DET2env2	1.2 + 0.2	0.04	1.	—	—	0.63
DET2env4	1.2 + 0.4	0.04	1.	—	—	0.63

TABLE 1. Some quantities are given for the models discussed in the sections 3-6 (see text).

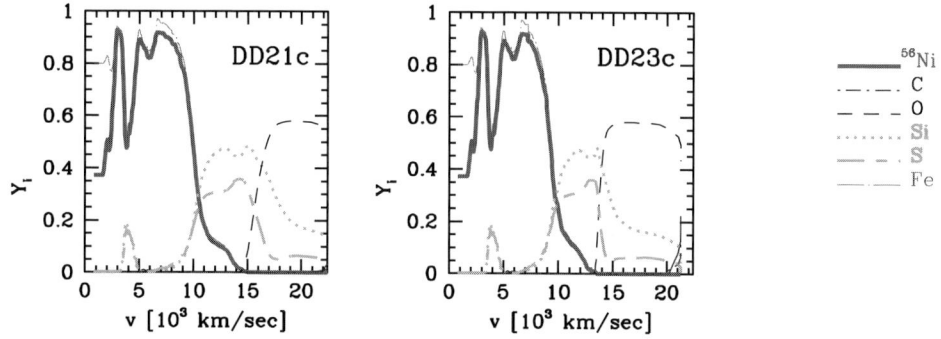

FIGURE 2. Abundances as a function of the expansion velocity for two delayed detonation models (see Table 3.1).

detonation is initiated after the burning front dies out. Besides the description of the burning front, the central density of the WD at the time of the explosion is another free parameter. For white dwarfs close to the Chandrasekhar limit, it depends sensitively on the chemistry and the accretion rate $\dot{M}$ at the time of the explosion.

In addition, the C/O ratio of the initial WD has been varied. In general, it has been assumed to be 1:1 unless otherwise quoted in brackets after the name. A significantly lower C/O ratio can be expected if the WD originates from a progenitor of more than ≈ 2 $M_\odot$ (see §7.2). The initial metallicity for $Z \geq 20$ is assumed to be solar, but for

DD24c, DD25c, DD26c and DD27c for which 1/3, 3, 0.1 and 10 times solar abundances are used, respectively. Otherwise, these models are identical to DD21c.

As example, typical chemical structures are given in Fig. 2. Since the burning time scales to NSE or partial NSE are shorter than the hydrodynamical time scales for all but the very outer layers, the final products depend mainly on the density at which burning takes place. With decreasing transition density, lesser ^{56}Ni is produced and the intermediate mass elements expand at lower velocities because the later transition to a detonation allows for a longer pre-expansion of the outer layers (DD21c vs. DD13c). Similarly, with increasing C/O ratio in the progenitor, the specific energy release during the nuclear burning is reduced (DD23c vs. DD21c) and the transition density at the burning front reached later in time, resulting in a larger preexpansion of the outer layers. This may allow to determine the main sequence mass of the progenitor.

To test the influence of the metallicity Z (i.e. nuclei beyond Ca) we have constructed models with parameters identical to DD21c but initial Z between 0.1 and 10 times solar. The energy release, the density and velocity structure are virtually identical. The main difference is an increase of more neutron-rich Fe group nuclei, namely ^{54}Fe, in the outer layers because, there, Y_e is inherented from the progenitor. Most remarkable is the change in the ^{54}Fe production which is the dominant contributor to the abundance of iron group elements at these velocities since little cobalt has yet been decayed near maximum light. While, for 1/3 solar metallicity, hardly any ^{54}Fe is produced at high velocities, ^{54}Fe is as high as 5% by mass fraction if we start with ten times the solar metallicity (see §7). Note that the temperature in the inner layers is sufficiently high during the explosion to allow electron capture which determines Y_e and, in those layers, the initial metallicity has no influence on the final burning product (HWT98).

3.2. Merging White Dwarfs

The second group of progenitor models consists of two low-mass white dwarfs in a close orbit which decays due to the emission of gravitational radiation and this, eventually, leads to the merging of the two WDs (e.g. Iben & Tutukov 1984). After the initial merging process, one low density WD is surrounded by an extended envelope (e.g. Benz, Thielemann & Hills 1989). This scenario is mimicked by our envelope models DET2env2...6 in which we consider the detonation in a low mass WD surrounded by a compact envelope between 0.2 and 0.6 $M_\odot$.

3.3. Explosions of Sub-Chandrasekhar Mass White Dwarfs

Another class of models—double detonation of a C/O-WD triggered by detonation of helium layer in low-mass WDs—was explored by Nomoto (1980), Woosley & Weaver (1980), and most recently by Woosley and Weaver (1994a, hereafter WW94), Livne & Arnett (1995) and HK96. For these models, the mass of the C/O core and of the He-layers given separately.

For ease of comparison, we have used parameters close to those suggested in WW94. To prevent repetition, we refer to the latter work for a detailed discussion of this class of models. Helium detonations show a qualitatively different structure in comparison to all models with a Chandrasekhar mass WD. The intermediate mass elements are sandwiched by Ni and He/Ni rich layers at the inner and outer regions, respectively. Generally, the density smoothly decreases with mass because partial burning produces almost the same amount of kinetic energy as the total burning, but a moderate shell-like structure is formed just below the former Helium layers. Observationally, a distinguishing feature of this scenario is the presence of Helium and Ni with expansion velocities above 11,000 to

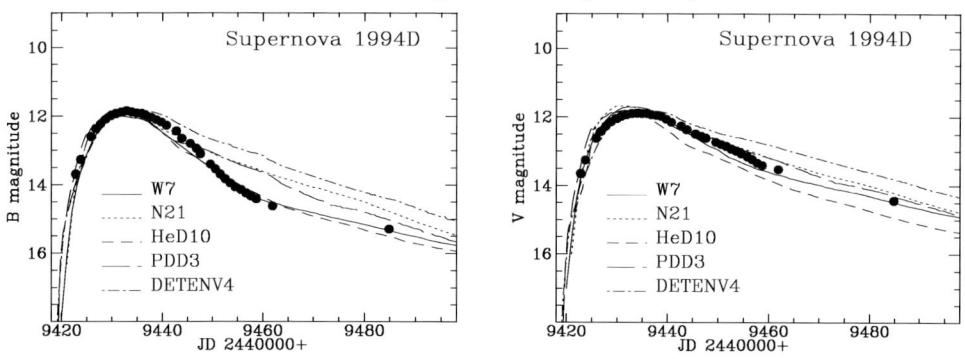

FIGURE 3. Theoretical LCs of the deflagration W7, the delayed detonation N21, the helium detonation HeD10, the pulsating delayed detonation PDD3, and the envelope model DET2env4. For comparison, the observation for SN1994D are given.

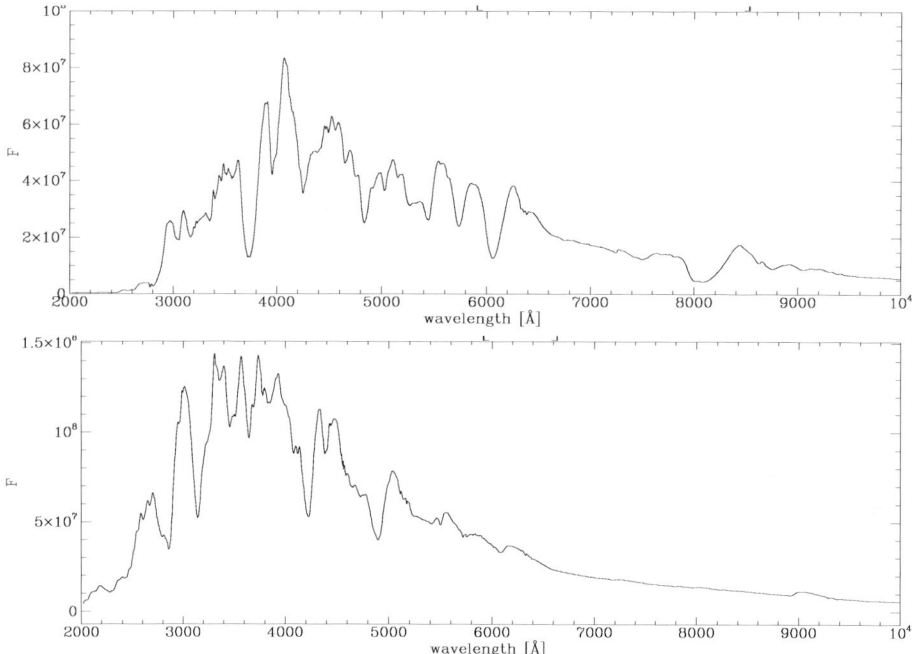

FIGURE 4. Comparison of spectra at about maximum light of the typical delayed detonation model DD200 and the normal bright helium detonation model HeD10 (lower graph).

14,000 km s^{-1}. Typically 0.07 to 0.13 $M_\odot$ of Ni are produced in the outer layers, mainly depending on the mass of the Helium shell.

4. General Properties of the Explosion Models

The different explosion scenarios can generally be distinguished based on differences in the slopes of the early monochromatic LCs (Fig. 3) and spectra (Fig. 4), e.g. by the expansion velocities indicated by various elements. Note that the differences between the slopes of the LCs are much larger than the uncertainties imposed by the model assumptions (see §2).

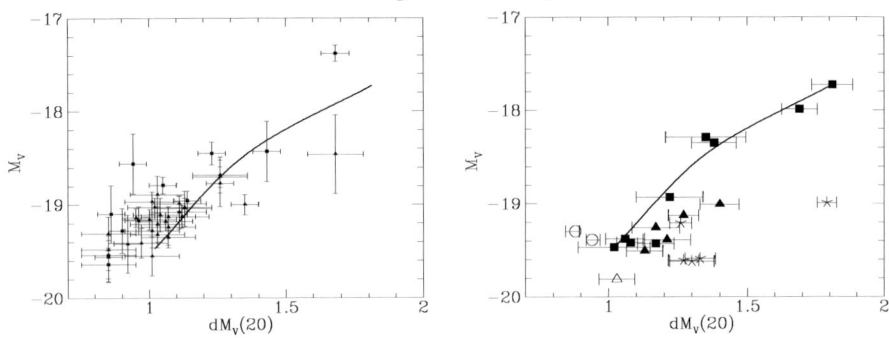

FIGURE 5. (Left panel) Observed light curve maximum brightness-decline rate relation. M_V is presented as a function of the decline from maximum at 20 days. (Right panel) The predicted relation for an array of models of SNe Ia representing delayed detonations (open triangles), pulsating delayed detonations (filled circles), merging models (open circles) and helium detonations (asterisks) (Höflich et al. 1996).

For all models with a ^{56}Ni production ≥ 0.4 M$_\odot$, M_V ranges from -19.0 to -19.7 (HK96). As a general tendency, the post-maximum declines are related to M_V, but there is a significant spread in the decline rate among models with similar brightness. For all models but the Helium detonations, the colors become very red for small M_{Ni} consistent with the observations (Hamuy et al. 1996; Höflich et al. 1996).

The maxima of subluminous supernovae are more pronounced. For bright models, the opacity stays high and the photosphere recedes mainly by the geometrical dilution of matter. For models with a slower rise time or little ^{56}Ni, the opacity drops strongly at about maximum light. The photosphere receeds quickly in mass, and thermal energy can be released from a larger region. Because no additional energy is gained, the energy reservoir is exhausted faster, and the post-maximum decline becomes steeper.

The radioactive decay of the variable nickel mass is the dominant factor which gives a range in maximum brightness and ^{56}Ni is the most important factor which governs the light curve shape (Fig. 5). A given amount of nickel can be produced by different combinations of the model parameters and, from the models, we expect a spread of ≈ 0.3 to 0.5^m around the mean maximum brightness-decline relation. A similar spread of $\approx \pm 0.4^m$ ($\sigma \approx 0.18^m$ in B) in the local, observed relation between maximum brightness and decline (Hamuy et al. 1996) may also suggest that there are additional parameters needed in the empirical relations. Note that new observations and recalibrations of the old observations indicate a somewhat tighter relation ($\approx 0.12^m$, Jha et al. 1999). From the current status of theoretical models, this very narrow spread cannot be understood, but it cannot be ruled out either. Both reanalysis of new light curves and theoretical investigation of the coupling between the "free" parameters may help to answer this puzzle. In figure 5, we also show a line produced by a sequence of models in which we changed the transition density only. The difference in the slope may point towards an intrinsic coupling of the free model parameters. E.g. the transition density may be related to the progenitor structure or its central density at the time of the explosion. Note that this also implies that observable spreads in relations, e.g. between rise time and decline time, cannot be expected to change along the path given by the change of a particular free model parameter.

The blue and UV spectrum of the delayed detonation model and its variants is dominated by strongly blended lines due to Fe group elements. Strong features due to Si II at ≈ 6100 Å and Ca II at about ≈ 3800 & 8100 Å are consistent with observations. In contrast, the spectra of Helium detonations are very blue due to the heating by radioactive

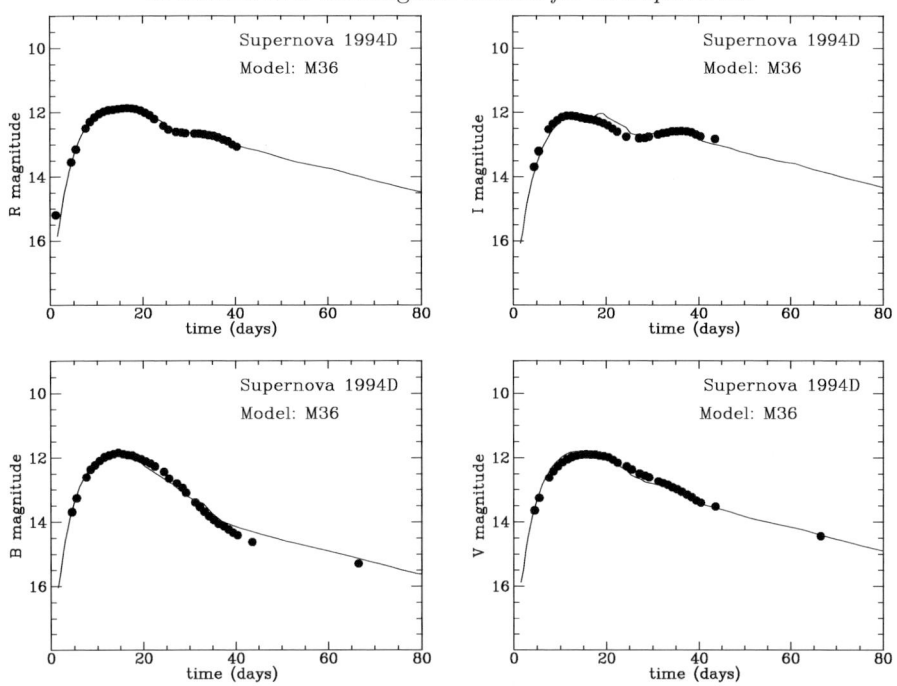

FIGURE 6. Observed LCs of SN 1994D in comparison with the theoretical LCs of M36

Ni both for normal and subluminous SNe Ia, and the spectra are dominated by strong Ni lines without showing a strong Si line. Already from the general properties, these HeDs can be ruled out as a general scenario.

5. Comparison between observation and models predictions

To get more information, detailed fitting of individual supernovae is required. The following 29 SNe Ia have been analyzed: SN 37C, SN 70J, SN 71G, SN 72E, SN 72J, SN 73N, SN 74G, SN 75N, SN 81B, SN 83G, SN 84A, SN 86G, SN 88U, SN 89B, SN 90N, SN 90T, SN 90Y, SN 90af, SN 91M, SN 91T, SN 91bg, SN 92G, SN 92K, SN 92bc, SN 92bo, SN 94D, SN 94M, SN 94S, SN 94T (HK96). For an example, see Figs. 6 & 7. To fit the light curves, we use a quantitative method for fitting data to models based on Wiener filtering (Rybicki and Press, 1995). The reconstruction technique is applied to the standard deviation from the theoretical LC to avoid problems with measurements distributed unevenly in time. By minimizing the error, the time of the explosion, the distance, and the reddening correction can be determined. We used models which cannot be ruled out within the uncertainties. This certainly implies that the restrictions depend on the quality of data which had been available in the early 90th. New data will provide more stringent restrictions.

According to our results, normal bright, fast SNe Ia (e.g. SN 71G, SN 94D) with rise times up to 20 days for the visual LC can be explained by delayed detonation with different densities ρ_{tr} for the transition from a deflagration to a detonation. For PDDs, the density ρ_{tr} stands for the density at which the detonation starts after the first pulsation. Typically, ρ_{tr} is about 2.5×10^7 g cm^{-3}. Central densities of the initial WDs range from 2.1 to 3.5×10^9 g cm^{-3}. As a tendency, models at the lower end of this range give better fits. Lower densities are also favored by recent observations of the rise times

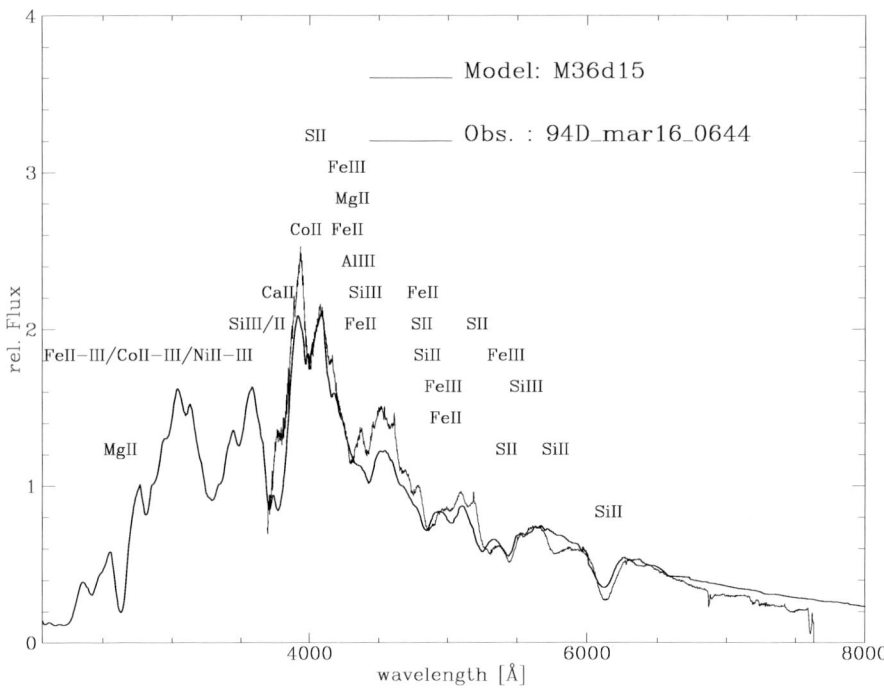

FIGURE 7. Synthetic spectrum at day 15 for M36 compared to the observations of SN 1994D at Mar. 16th.

by Goldhaber (1998). We note that the classical deflagration W7 (Nomoto et al. 1984) provides similar good fits of the optical in several cases because its structure resembles those of DD models. Distinguishing features between delayed detonation models and deflagration models such as W7 are the outer layers. As a generic feature in W7 and similar deflagration models, unburned C and O are present down to about 14,000 km/sec whereas, in DDs, the entire star is burned and only little C/O remains at the very high velocities (Fig. 2, $v \leq 22,000$ km s^{-1}). A couple of recent observations favor DDs or, alternatively, deflagration models where the deflagration speed increases to velocities close to the speed of sound as suggested by Hillebrandt (private communication, 1999). From the observations, Si lines are seen at velocities up to 22,000 km s^{-1}, e.g. in SN 1994D (Höflich 1995). This may be explained either by mixing or in situ burning. More insight is provided by IR spectra at about and before maximum light, although few IR data have been obtained in the past and the result still is hampered by small number statistics. At about 1.05 μm, a strong line can be seen in spectra which can be attributed to Mg II at expansion velocities at and above 15,000 to 16,000 km s^{-1}. Mg is a product of explosive carbon burning but it is depleted in explosive oxygen burning. This argues against strong mixing as an explanation for the presence of Si at high velocities (Wheeler et al. 1998). In addition, Fisher et al. (1997) found that unburned C may be visible in ordinary SN in early time spectra but only at velocities above 20,000 to 25,000 km s^{-1}.

Our "standard" explosion models are unable, however, to reproduce very slow rise times ($\geq$ 20–22 days) even within the uncertainties, unless the C/O ratio is rather low (see §7, HWT98), or by models with an envelope of typically 0.2 to 0.4 M$_\odot$. The envelope can be produced during a strong pulsation or during the merging of two WD. Distinguishing features between the alternatives of low C/O ratio and an envelope is the presence of unburned material and the absence of high velocity Si in the latter case, and a slow

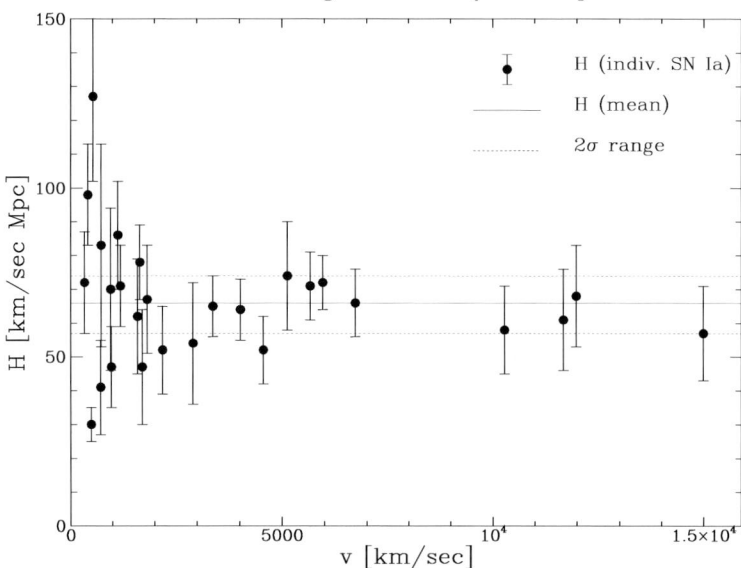

FIGURE 8. Hubble values H are shown based on individual distances based on fitting of the LCs and spectra (§5 & HK96).

decline rate compared to DDs (see also §7). Another, unique features of models with massive envelopes are very high photospheric expansion velocities ($v_{ph} \approx 16,000$ km s^{-1}) shortly before maximum light, which drop rapidly to an almost constant value between 9000 and 12,000 km s^{-1}. This "plateau" in v_{ph} lasts for 1 to 2 weeks depending on the envelope mass (Khokhlov et al. 1993). In fact, there is some evidence for the plateau in v_{ph} from the Doppler shift of lines for SNe Ia with a slow pre-maximum rise and post-maximum decline (e.g. SN 84A, SN 90N, Müller & Höflich 1994) and, for SN 1991T, there is evidence of unburned carbon down to velocities of 16,000 km s^{-1} (Fisher et al. 1999).

Qualitatively, strongly subluminous SNe Ia (SN 91bg, SN 92K, SN 92bc) can be explained within the framework of pulsating delayed detonation models with a low transition density which produce a very low mass envelope and very little ^{56}Ni but Si down to small expansion velocities (Höflich et al. 1995). Due to the small heating, these models become systematically redder and the post-maximum decline becomes steeper with decreasing brightness in agreement with observations. The evolution of the photospheric expansion velocity v_{ph} and, in particular, its steady decline, is consistent with observations (Fisher et al. 1996).

6. Distance determinations and H_o

Based on our LCs, we have also determined the individual distances of the parent galaxies of the analyzed SNe Ia (§5). Our method does not rely on secondary distance indicators and allows for a consistent treatment of interstellar reddening and the interstellar redshift. We find H_o to be 67 ± 9 km s^{-1} Mpc within a 95% confidence level (Fig. 8). This value agrees well with our previous analysis based on a subset of observations and models (66 ± 10 km Mpc^{-1} sec^{-1}, Müller & Höflich 1994). By using the brightness at maximum light alone (but including the individual reddening correction), our H_o decreases to 63 km Mpc^{-1} sec^{-1}.

Other determinations of H_o are based on independent, purely statistical methods and primary distance indicators. It may be encouraging that the result of different SN Ia based methods agree if SN Ia are not treated as as standard candles. Tammann & Sandage (1995) get 59, Hamuy et al. (1995) get 65 ± 5, Riess et al. (1995) give 67 ± 5, and Nugent et al. (1995) get a values of 60 ± 12. Taking the excellent agreement between all methods, one may conclude that the question of H_o has been settled at least within an 10% error range. From our models, both the empirical relations between $M_V/dM(15)$ like-relations and the ansatz to deselect subluminous SNe Ia seems to be justified, but we expect an individual dispersion of $\approx 20\%$ (Höflich et al. 1996).

7. Evolutionary effects with Redshift z

Time evolution is expected to produce the following main effects: (a) a lower metallicity will decrease the time scale for stellar evolution of individual stars by about 20% from Pop I to Pop II stars (Schaller et al. 1992) and, consequently, the progenitor population which contributes to the SNe Ia rate at any given time. The stellar radius also shrinks. This will influence the statistics of systems with mass overflow; (b) Evolutionary effects of the stellar population will change the mass function present at the time corresponding to a given redshift; (c) The initial metallicity Z will effect the nuclear burning during the explosion or, more precisely, the electron to nucleon fraction (§3); (d) Systems with a shorter life time may dominate early on and, consequently, the typical C/O ratio of the central region of the WD is reduced; (e) The metallicity may influence the structure of the initial WD. (f) In principle, a change of the metallicity may alter the R-M relation of a WD but this issue is not a major concern (HWT98); (g) The properties of the interstellar medium may change.

Before discussing the *possible(!)* implications for cosmology in more detail, we address the influence of changes in the initial composition on individual light curves and spectra, i.e. (c), (d) and (d). The influence of the initial composition on light curves and spectra has been studied for the example of a set of delayed detonation models with DD21c being the reference model (Table 3.1). Because the time evolution of the composition is not well known and because we have not considered the entire variety of possible models, the values given below *do not(!)* provide a basis for quantitative corrections of existing observations. The goal is to get a first order estimate of the size of the systematic effects to be expected, to demonstrate how evolutionary effects in a real data sample can be recognized and how one may be able to compensate.

7.1. *Influence of the chemical structure of the WD*

The influence of the initial metallicity and C/O ratio on light curves and spectra has been studied for the example of a set of delayed detonation models with the basic as follows: Central density of the WD $\rho_c = 2.6 \; 10^9$ g/cm^{-3}, $v_{\rm burn} = \alpha * v_{\rm sound}$ with $\alpha = 0.03$ during the deflagration phase and a transition to detonation occurred at $\rho_{tr} = 2.7 \times 10^7$ and 3.0×10^7 g cm^{-3} for DD21c to DD27c and DD13c, respectively (see table 1). The parameters are close to those which reproduce both the spectra and light curves reasonably well (Nomoto et al. 1984; Höflich 1995).

7.1.1. *Light curves*

Reducing the ratio C/O from 1 to 2/3 in the WD reduces the ^{56}Ni production and the kinetic energy in DD23c compared to DD21c. The smaller expansion due to the smaller E_{kin} causes a slower rise to maximum light by about 3 days. The smaller ^{56}Ni results in a steeper decline after maximum light because the photosphere recedes faster and the

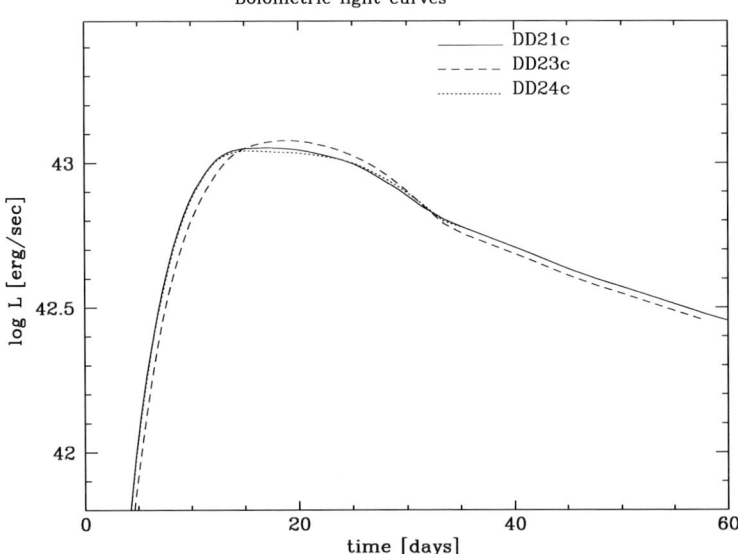

FIGURE 9. Comparison of bolometric LCs of the delayed detonation models DD21c, DD23c and DD24c with otherwise identical parameters but with different C/O ratios and metallicity relative to solar (C/O; R_Z) of (1;1), (2/3;1) and (1;0.3), respectively.

luminosity after about day 35 is smaller by about 10% (Fig. 9). Both the bolometric and monochromatic light curves are effected by a similar amount because the energetics of the explosion changes. Overall, the change of the C/O ratio from 1/1 to 2/3 has a similar effect on the colors, light curve shape and the distribution of elements as a 10% reduction in the transition density or central density in delayed detonations. However, for a given peak magnitude to tail ratio, a model produced by a reduced C/O ratio will be brighter than a one obtained by varying the transition density (Höflich 1995). The rise time and the mean expansion rate provide a way to determine the C/O ratio and the transition density independently (compare Khokhlov et al., 1993, Höflich 1995, HK96).

Changing the initial metallicity Z has very little influence on the bolometric light curve because the ^{56}Ni production and energy release vary by only 4% if Z is varied between 0.1 and 10 times solar. In addition, diffusion time scales are mainly determined by deeper layers and, and there, the electron capture during burning determines Y_e and not the initial composition (§3).

7.1.2. *Spectra at maximum light*

Spectral changes due to the C/O ratio produce a variation in the pattern of the most abundant elements similar to a change in the transition density and the central density of the WD. Consequently, the variation of the spectra is similar. However, this degeneracy can be overcome if we combine the information from the LC with the expansion velocity measured by the line shifts.

More interesting is the influence of the initial metallicity because it influences the iron group elements even outside the region which underwent incomplete Si-burning during the explosion. As an example, the spectra of the delayed detonation models with solar and 1/3 solar metallicity are given in Fig. 10. At maximum light (≈ 17 days), the line forming region ($\tau \approx 0.1 \ldots 1.$) extends between 1 and 2×10^{15} cm in the optical, corresponding to expansion velocities between 8000 and 16,000 km s^{-1}. Consequently, the upper edge of the ^{56}Ni rich layers ($v \approx 12,000$ km s^{-1}) become visible at maximum

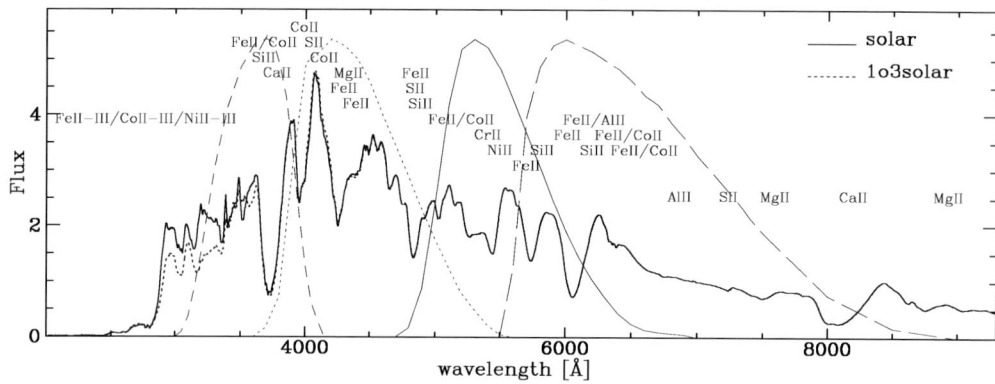

FIGURE 10. Comparison of synthetic NLTE spectra at maximum light for initial compositions of solar and 1/3 of solar, respectively. The standard Johnson filter functions for UBV, and R are also shown.

light and all but the far UV is formed in regions where at least partial burning to Si took place. The spectrum is scarcely effected by metallicity effects except for wavelengths below 4200 Å which are dominated by the layers with ^{54}Fe.

By 2 to 3 weeks after maximum, the spectra are completely insensitive to the initial Z because the spectrum is formed in even deeper layers where none of the important abundances are effected by the metallicity. Thus, for two similar bright SNe with similar expansion velocities, a comparison between the spectral evolution can provide a method to determine the metallicity difference or may be used to detect evolutionary effects for distant SNe Ia if high quality spectra are available.

Although small, the changes in the spectrum with metallicity have an important effect on the colors of SN at high redshifts where they are shifted into other bands. For local SNe Ia, a change of the metallicity by a factor of 3 implies a variation in color of two to three hundredths of a magnitude. For V–R and R–I, the effect remains small for redshifts $z \leq 0.5$ and 0.7, respectively, but it increases to the order of 0.2 to 0.3^m at redshifts relevant for the determination of cosmological parameters (HWT98). The amplitude of this effect and hence the uncertainty in color, reddening and brightness with metallicity is again comparable to the brightness change imposed by cosmological deceleration.

7.2. Influence of the stellar evolution on the WD structure

Up to now, we have neglected the influence of the metallicity and the mass of the progenitor on the structure of the initial WD. Such dependencies may become of important if supernovae at large distance are observed. On cosmological distance scales, the metallicity must be expected to be correlated with redshift when the system has been formed and more systems with a shorter life time must be expected to contribute to the SNe Ia population. At the time of the explosion, the WD masses are close to the Chandrasekhar limit as it has grown by accretion of H/He and subsequent burning from the mass of the central core of a star with less than ≈ 7 M$_\odot$. In the accreted layers, the C/O-ratio is close to 1. However, the initial mass of the C/O WD is given by the results of stellar evolution of the main sequence star. Its size depends on the main sequence mass of the progenitor and its metallicity. An extended set of models with various parameters have been calculated up to the first thermal pulses (see Table 2 for a selection). Here, we want to discuss the consequences mainly with respect to changes in the light curves and spectra of SNe Ia. For more details on the stellar evolution, see Dominguez et al. (1998).

For a given metallicity, the size of the core is growing with mass. Consequently, the

M	R_Z	M_{CO}	X_C	C/O	Rate	M	R_Z	M_{CO}	X_C	C/O	Rate
3.0	0.0	0.74	0.35	0.82	CF85	3.0	0.02	0.54	0.25	0.79	CF85
4.0	0.0	0.83	0.36	0.78	CF85	4.0	0.02	0.74	0.31	0.78	CF85
5.0	0.0	0.89	0.36	0.73	CF85	5.0	0.02	0.85	0.33	0.77	CF85
6.0	0.0	0.94	0.35	0.70	CF85	6.0	0.02	0.90	0.34	0.73	CF85
8.0	0.0	1.06	0.33	0.60	CF85	7.0	0.02	0.98	0.33	0.68	CF85
3.0	0.001	0.72	0.30	0.77	CF85						
4.0	0.001	0.84	0.32	0.74	CF85	5.0	0.0	0.89	0.36	0.73	CF85
5.0	0.001	0.90	0.34	0.73	CF85	5.0	0.0	0.88	0.33	0.69	BuH
6.0	0.001	0.96	0.35	0.69	CF85	5.0	0.0	0.92	0.69	1.68	BuL
7.0	0.001	1.06	0.34	0.63	CF85						

TABLE 2. Basic parameters of the CO-cores of low mass stars. The main sequence mass M in $M_\odot$, the metallicity R_Z the size of the the C/O-core in $M_\odot$, and the central mass fraction of carbon X_C, and the final C/O ratio after growing to the Chandrasekhar mass. In addition, we give the source of the $^{12}C(\alpha,\gamma)$ rate (CF85: corresponds to Caughlan & Fowler (1985); and BuH and BuL to the upper and ower limits according Buchmann (1997). The helium mass fraction Y is 0.23 for Z=0.001 and Z=0, and Y=0.285 for Z=0.02.

total mass fraction of the C/O ratio in the WD at the time of explosion will decrease, making the corresponding SN Ia more luminous at maximum light but with a steeper $M_V(dM_{20})$-relation and a slightly slower expansion velocities as indicated by spectra.

Increasing metallicity effects mainly the size of the C/O core and the central carbon fraction X_C its because its influence on the convection during the stellar evolution. Lower metallicity results in smaller convection and, consequently, less alpha capture on Carbon and a higher X_C. However, as a competing effect, the size of the C/O core decreases for solar metallicity. As a result, the overall effect of a decreasing metallicity is similar to those of an increasing mass. However, we want to note that the latter tendency depends sensitively on the assumed $^{12}C(\alpha,\gamma)^{16}O$ rate. Whereas the upper limit for this rate hardly effects the results compared to the use of Caughlan & Fowler (1985) using the lower limit for this rate as given by Buchmann (1997), reverses this trend.

In addition, we want to stress that the trends are given by assuming that the ignition density of the WD and the propagation of the WD is assumed to be uneffected. In reality, both the propagation of the burning front and the central density of the WD at time of the ignition may be effected by the chemical structure of the WD. These questions need further investigation.

Worth noting is the following trend: For realistic cores, the mean M_C/M_O tends to be smaller than the canonical value of 1 used in all calculations prior to 1998 (e.g. Nomoto et al. 1984, Woosley & Wheaver 1994ab, HK96). Consequently, as a general trend, the rise times are about 1-5 days slower compared to the models published with C/O ratios of 1 (see §3). This slower slower rise time is also consistent with the finding of Goldhaber et al. (1998) and suggests that the progenitors of SNe Ia originate from stars with masses $\geq$ 3-4 $M_\odot$.

7.2.1. Example: Structure and light curves of a progenitor with 7 $M_\odot$

Here, we want to discuss the size of the metallicity effect at the example of a 7 $M_\odot$ with $Z = 0.02$ and 0.004. Z mainly effects the convection during the stellar Helium burning and, consequently, the size of the C/O core and the central C/O ratio (Höflich et al. 1999).

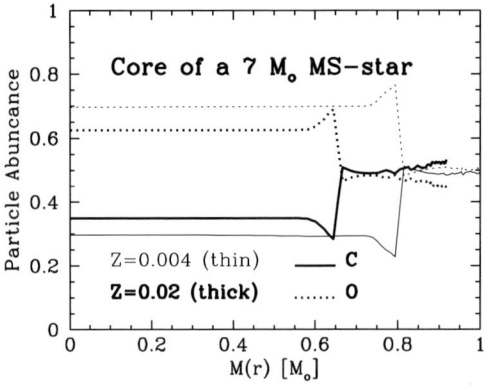

FIGURE 11. Final chemical profile of the C/O core of a star with $M_{MS} = 7$ M$_\odot$.

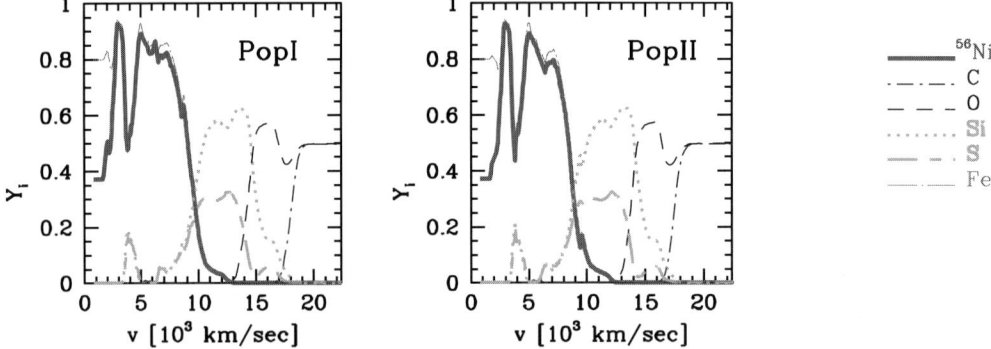

FIGURE 12. Abundances as a function of the final expansion velocity for progenitors of 7 M$_\odot$ with $Z = 0.02$ and $Z = 0.004$.

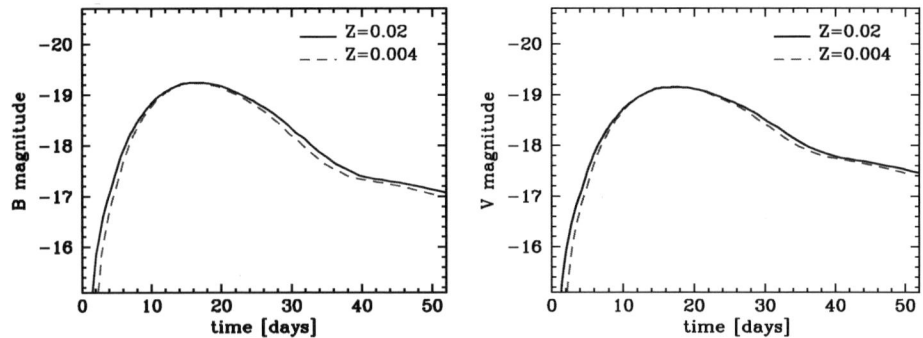

FIGURE 13. Comparison of light curves in B (left) and V (right) of the delayed detonation models with metallicities with $Z = 0.02$ and 0.004.

In agreement with §3.1, the total C/O masses determines the explosion energy and, consequently, has the main effect on the light curves. After accretion on the initial core, the total mass fraction of M_C/M_{Ch} is 0.75 and 0.61 for the models with $Z = 0.04$ and 0.02 respectively. At the time of the explosion, $\rho_c = 2.4E9$ g cm^{-3}. For the burning, $\alpha = 0.02$ and $\rho_{tr} = 2.4 \times 10^7$ g cm^{-3}) (Fig. 11).

Monochromatic light curves are shown in Fig. 13. As expected from the last sections, the main effect on the light curves is caused by the different expansion ratio caused by the change in the integrated C/O ratio. The change in the maximum brightness remains small $(M_V(Z = 0.02) - M_V(Z = 0.004) = -0.03^m)$ and the rise times are by about 1 day $(t_B((Z = 0.02) = 16.4d$ vs. $t_B(Z = 0.04) = 17.6\ d)$. The most significant effect is a steeper decline ratio and a reduced ^{56}Ni production for the model with solar Z, mainly due to the slower expansion ratio. This translates into a systematic offset of $\approx 0.1^m$ in the maximum brightness decline ratio (Hamuy et al. 1996). Using either the stretching method or the LCS-method gives similar offsets.

7.3. *Implications for cosmology*

Systematic errors that must be taken into account in the use of SNe Ia to determine cosmological parameters include technical problems, changes of the environment with time, changes in the statistical properties of the SNe Ia, and changes in the physical properties of SNe Ia.

In the first class, we put corrections for redshift. If standard filter systems are used, they can be well calibrated to local standards but the k-correction is of some concern. This problem can be overcome if redshifted "standard" filters are used. Another technical problems may arise then from the fact that the transmission functions must be identical to those resulting from the redshift because no direct calibration can be applied by using a comparison star.

In the second class of problems an important example is that the properties of dust may change at high redshift. In the first place, the element abundances in the ISM can change. In addition, important donors of dust such as low mass stars during the red giant phase cannot contribute because their evolutionary time is comparable to or longer than the age of the universe at $z \approx 0.5$ to 1. Another problem related to the correction for extinction is that the extinction law to be applied depends on the redshift of the absorbing dust cloud (HK96).

In the third class of problems is the fact that the contribution of different progenitor types may change with redshift. For local supernovae, it is likely that a large variety of binary star properties (total mass of donor stars, separation, etc.) with very different evolutionary life times account for the variety of SNe Ia observed (Canal et al. 1997, HWT98). Given this variety, we expect a time evolution of the statistical properties of the progenitors because, early on, progenitors with a short life time will dominate the sample. Among other changes, more massive, shorter lifetime progenitors will have a lower C/O ratio in the center. Changes in the statistical properties are expected to increase with redshifts larger than 0.7 to 0.8 when the age of the universe becomes comparable or shorter than the suspected progenitor life times. Other evolutionary effects have already been mentioned: The ZAMS life time changes with metallicity, Roche lobe radii change with metallicity and the lower limit for accretion and steady burning of hydrogen on the surface of the WD changes by a factor of 2 for Pop I and Pop II stars, respectively (Nomoto et al. 1982).

Finally, the physical properties of a typical SNe Ia may change. If more massive stars contribute to the supernovae population, we will see a smaller mean C/O ratio and, consequently, the explosions are less energetic (§7.2). Changing the C/O ratio from 1 to 2/3 with otherwise identical parameters (see §7.1), will result in a smaller ^{56}Ni production and, consequently, a lower bolometric luminosity at late times. On the other hand, the slower expansion causes less adiabatic cooling during early times and, thus, the luminosity at maximum light is slightly larger. This implies that the peak to tail luminosity ratio changes and produces a systematic offset in the brightness decline relation of up to a few tenth of a magnitude.

As a secondary effect, a change of the metallicity in the progenitor on the main sequence influences the central helium burning and, consequently, the size of the C/O core which becomes a C/O WD and its C/O ratio. Consequently, the metallicity changes the C/O structure of the exploding white dwarf. This effect works similar as a change of the typical main sequence mass of the progenitor.

A change of the C/O ratio reveals itself mainly by the change in the ratio between rise time and decline. This change may be caused either by a change of Z or, alternatively, by a change of the main sequence mass of the progenitor. When using statistical methods to determine cosmological parameters, both a change in the distribution of progenitor masses and the metallicity of the sample may cause systematic evolutionary effects. Fortunately, it does not matter whether a change of C/O is due to a change in the metallicity or the typical mass of the progenitors. In both cases, we find that

$$\Delta M \approx \Delta t_{\rm rise} \quad , \tag{7.4}$$

where ΔM is the offset in the maximum brightness/decline relation in magnitudes and Δt is measured in days (compare HWT98).

The effect of progenitor mass and metallicity can be untangled by simultaneous analysis of both spectra and light-curves. In principle, this allows a method to get a handle on the progenitor mass. With the current level of modeling such analysis is restricted to differential comparisons between individual supernovae.

Recently, a comparison of rise times between local and distant supernovae has been made (Riess et al. 1999). They find a difference in the rise times between the local and the distant sample of 2.5 ± 0.4 days if all observed supernovae are normalized to the stretch factor s of 1. This result is based on a preliminary analysis of the high redshift data by Goldhaber (1998). It may also be noted that both the local and the distant sample show an spread of ± 1 day in the rise time/decline relation without overlap which is hard to understand as it would imply that the progenitors are distinctive although a comparison of SNe Ia in local spirals and ellipticals indicates that, locally, we see both progenitors with a very short and long life time (Wang et al. 1998). However, if confirmed and in light of our analysis, this would imply a systematic offset $\approx 0.25^m$ in the brightness decline relation. A change of this order would not alter the basic conclusion of the high z searches that we live in a a low Ω universe but, then, a low Ω universe with $\Omega_M = 0.2$ and $\Omega_\Lambda = 0$ may be consistent with SN data as well.

Even strong variations in the metallicity will hardly produce selection effects with respect to the counting rates for SNe Ia at large red-shifts. Variations of the expected size are critical if, in future, SNe Ia may be used to measure large scale scalar-fields because the metallicity may show large local variations during the early phases when individual SNe explosions govern the metallicity.

Finally, note that the transition density from deflagration to detonation may depend on the energy release during the deflagration phase and, hence the transition density could be a function of the C/O ratio, both effects could alter the LC and should must be explored in the future.

This research was supported in part by NASA Grant LSTA-98-022. The calculations for the explosion and light curves were done in parts on a cluster of workstations financed by the John W. Cox Fund of the Department of Astronomy at the University of Texas.

REFERENCES

ARNETT, W. D. 1969 *ApJS*, **5**, 180.

BENZ, W., THIELEMANN, F. K., & HILLS, J. G. 1989 *ApJ*, **342**, 986.

BRANCH, D. & KHOKHLOV, A. M. 1995 *Phys. Rep.*, **256**, 53.

BRANCH D. & TAMMANN G. A. 1992 *ARA&A*, **30**, 359.

BUCHMANN, L. 1997 *ApJ*, **479**, L153.

CANAL, R. 1995 in *Les Houches Lectures* (eds. J. Audouze et al.). Elsevier, in press.

CAUGHLAN, G.R. & FOWLER, A .M. 1985 *ARA&A*, **21**, 165.

CHIEFFI, A. & STRANIERO, O. 1989 *ApJ*, **71**, 47.

COLLELA, P. & WOODWARD, P. R. 1984 *J. Comp. Phys.*, **54**, 174.

DOMINGUEZ, I., HÖFLICH, P., STRANIERO, O., & WHEELER, C. J. 1998 in *Nuclei in the Cosmos V* (eds. Prantzos & Harissopulos), p. 48. Editions Frontieres.

DOMINGUEZ, I. & HÖFLICH, P. 1999 *ApJ*, in press.

FILIPPENKO, A. V., ET AL. 1992 *AJ*, **104**, 1543.

FISHER, A., BRANCH, D., HATANO, K., & BARON, E. 1999 *ApJ*, **304**, 67.

FISHER, A., BRANCH, D., HÖFLICH, P., KHOKHLOV, A. 1996 *ApJ*, **447**, 83.

FISHER, A., BRANCH, D., NUGENT, P., BARON, E. 1997 *ApJ*, 481, L89.

GARNAVICH, P., ET AL. 1998 *ApJ*, **509**, 74.

GOLDHABER, G. 1998 *BAAS*, **193**, 4713.

HAMUY, M., PHILLIPS, M. M., MAZA, J., SUNTZEFF, N. B., SCHOMMER, R. A., & AVILES, A. 1996 *AJ*, **112**, 2438.

HANSEN, C. J. & WHEELER, J. C. 1969 *ApJS*, **3**, 464.

HÖFLICH, P. 1995 *ApJ*, **443**, 89.

HÖFLICH, P. & KHOKHLOV, A. 1996 *ApJ*, **457**, 500, [HK96].

HÖFLICH, P., KHOKHLOV, A., & MÜLLER, E. 1991 *A&Ap*, **248**, L7.

HÖFLICH, P., KHOKHLOV, A., & MÜLLER, E. 1992 *A&Ap*, **259**, 243.

HÖFLICH, P., MÜLLER, E., & KHOKHLOV, A. 1993 *A&Ap*, **268**, 570.

HÖFLICH, P., KHOKHLOV, A., WHEELER, J. C., PHILLIPS, M. M., SUNTZEFF, N. B., & HAMUY, M. 1996 *ApJ*, **472**, L81.

HÖFLICH, P., WHEELER, J. C., & THIELEMANN, F. K. 1998 *ApJ*, **495**, 617, [HWT98].

HÖFLICH, P., WHEELER, J. C., NOMOTO, K., UMEDA, H. 1999 *ApJ*, in press.

IBEN, I., JR. & TUTUKOV, A. V. 1984 *ApJS*, **54**, 335.

ITOH, N., MITAKE, S., IYETOMI, H., & ICHIMARU, S. 1983 *ApJ*, **273**, 774.

JHA, S., ET AL. 1999 *ApJS*, submitted (astro-ph/9906220).

KARP, A. H., LASHER, G., CHAN, K. L., SALPETER E. E. 1977 *ApJ*, **214**, 161.

KHOKHLOV, A., MÜLLER, E., & HÖFLICH, P. 1993 *A&Ap*, **270**, 23.

KHOKHLOV, A. 1991ab *A&Ap*, **245**, & 114 & L25.

KHOKHLOV, A. 1995 *ApJ*, **449**, 695.

KHOKHLOV, A., ORAN, E. S., WHEELER, J. C. 1997a *ApJ*, **478**, 678.

KHOKHLOV, A., ORAN, E. S., WHEELER, J. C. 1997b in *Thermonuclear Supernovae* (eds. Canal, et al.), **486**, p. 475. Kluwer.

KURUCZ, R. 1993 *Atomic Data for Opacity Calculations*, Cambridge, CfA.

LEIBUNDGUT, B., ET AL. 1993 *AJ*, **105**, 301.

LIVNE, E. & ARNETT, D. 1995 *ApJ*, **452**, L62.

MÜLLER, E., HÖFLICH, P. 1994 *A&Ap*, **281**, 51.

NIEMEYER, J.C. & WOOSLEY, S. 1997 *ApJ*, **475**, 740.

NOMOTO, K. 1980 in *IAU-Sym. 93* (eds. D. Sugimoto, D Q. Lamb & D. Schramm), p. 295. Reidel.

NOMOTO, K., YAMAOKA, H., SHIGEYAMA, T., IWAMOTO, K. 1995 in *Supernovae* (ed. R. A. McCray) Cambridge University Press.

NOMOTO, K., THIELEMANN, F.-K., YOKOI, K. 1984 *ApJ*, **286**, 644.

NOMOTO, K. & SUGIMOTO, D. 1977 *PASP*, **29**, 765.

NOMOTO, K. 1982 *ApJ*, **253**, 798.

NOMOTO, K., SUGIMOTO, D., & NEO, S. 1976 *ApJS*, **39**, 137.

NORGAARD-NIELSEN, H. U., HANSEN, L., HENNING, E. J., SALAMANCA, A. A., ELLIS, R. S., & WARRICK, J. C. 1989 *Nature*, **339**, 523.

NUGENT, P., ET AL. 1995 *Phys. Rev. Let.*, **75, 394**, & 1974E

OLSSON, G. L., AUER, L. H., BUCHLER, J. R. 1986 *JQSRT*, **35**, 431.

PENNYPACKER, C., ET AL. 1991 *IAUC*, 5207.

PERLMUTTER, C., ET AL. 1995 *ApJ*, **440**, L95.

PERLMUTTER, C., ET AL. 1999 *ApJ* **517**, 565.

PHILLIPS, M. M., ET AL. 1987 *PASP*, **90**, 592.

RIESS, A. G., PRESS, W. H., & KIRSHNER, R. P. 1995 *ApJ*, **438**, L17.

RIESS, A., FILIPPENKO, A., WEIDONG, L., & SCHMIDT, B. 1999 *AJ*, submitted & astro-ph/9907038.

RIESS, A. G., ET AL. 1999 *AJ*, **118**, 2675.

RYBICKI, G. B. & PRESS, W. H. 1995 *Phys. Rev. Let.*, **74**, 1060.

SAHA, A., SANDAGE, A., LABHARDT, L., TAMMANN, G.A., MACCHETTO, F. D., & PANAGIA, N. 1997 *ApJ*, **486**, 1.

SANDAGE, A. & TAMMANN, G. A. 1993, *ApJ*, **415**, 1.

SCHALLER, G., SCHAERER, D., MEYNET, G., & MAEDER, A. 1992 *A&AS*, **96**, 269.

SCHMIDT, B. P, ET AL. 1996 *BAAS*, **189**, 1089.

SPYROMILIO, J., PINTO, P. A., & EASTMAN, R. G. 1994 *MNRAS*, **266**, L17.

STRANIERO, O., CHIEFFI, A., & LIMONGI, M. 1997 *ApJ*, **490**, 425.

THIELEMANN, F. K., NOMOTO, K., & HASHIMOTO, M. 1996 *ApJ*, **460**, 408.

UMEDA, H., NOMOTO, K., YAMAOKA, H., & WANAGO, S. 1999 *ApJ*, **513**, 861.

WANG, L., HÖFLICH, P., WHEELER, C. J. 1998 *ApJ*, **487**, 29.

WHEELER, J. C., HÖFLICH, P., HARKNESS, R. P., SYPROMILIO, J. 1997, *ApJ*, **496**, 908.

WOOSLEY, S. E. & WEAVER, T. A. 1994a in *Supernovae*, p. 423. Elsevier.

WOOSLEY, S. E. & WEAVER, T. A. 1994b *ApJ*, **423**, 371 [WW94].

WOOSLEY, S. E., WEAVER, T. A., & TAAM, R. E. 1980 in *Type I Supernovae* (ed. J. C. Wheeler), p. 96. U. Texas.

Deflagration to detonation

By A. M. KHOKHLOV

Laboratory for Computational Physics and Fluid Dynamics, Naval Research Laboratory, Washington, DC

Thermonuclear explosions of Type Ia supernovae (SNIa) involve turbulent deflagrations, detonations, and possibly a deflagration-to-detonation transition. The physics of these processes is discussed. A phenomenological delayed detonation model of SNIa successfully explains many observational properties of SNIa including monochromatic light curves, spectra, brightness—decline and color—decline relations. Observed variations among SnIa are explained as a result of varying nickel mass synthesized in an explosion of a Chandrasekhar mass C/O white dwarf. Based on theoretical models of SNIa, the value of the Hubble constant $H_o \simeq 67$ km s^{-1} Mpc^{-1} was determined without the use of secondary distance indicators. The cause for the nickel mass variations is still debated. It may be a variation of the initial C/O ratio in a supernova progenitor, rotation, or other effects.

1. Introduction

Type Ia supernovae (SNIa) are important astrophysical objects which are increasingly used as distance indicators in cosmology. SNIa appear to be a rather well behaving group of objects. There are deviations in maximum brightness of $\sim 2m$ among SNIa, but they correlate with variations in the shape of SNIa light curves: less bright supernovae tend to decline faster. This is often expressed as a correlation between m and dm_{15}, where dm_{15} is the decrease in magnitude 15 days since maximum. Another correlation exists between SNIa color at maximum and postmaximum decline, $(B-V)$-dm_{15}—less bright supernovae tend to be more red (Phillips 1993). These two correlations can be used to account for variations in brightness of SNIa and for interstellar absorption. Using these has led to improved determinations of H_o (Riess et al. 1996) and to new findings concerning Ω_m and Ω_Λ (Shmidt et al. 1998; Perlmutter et al. 1999).

Are there exact and unique maximum brightness—postmaximum decline and color—postmaximum decline relations among SNIa? Are these relations the same for nearby and cosmological supernovae? Before these questions so important for Ω_m and Ω_Λ can be addressed from theoretical grounds, we would like the theory of SNIa to answer more general questions:
(1) Why do SNIa differ from each other?
(2) Why do some of SNIa characteristics correlate?

2. Pre-supernovae

It is believed that SNIa are thermonuclear explosions of carbon-oxygen (CO) white dwarfs (WD). However, evolutionary paths leading to SNIa are still a bit of a mystery (Livio 2000). Three major scenarios have been considered based on the evolution of binary stellar systems: (1) a CO-WD accreting mass through Roche-lobe overflow from an evolved companion star (accretion scenario). The explosion is triggered by compressional heating near the WD center when the WD approaches the Chandrasekhar mass. (2) Merging of two low-mass WDs caused by the loss of angular momentum due to gravitational radiation (Webbink 1984; Iben & Tutukov 1984). Resulting merged configuration consists of a massive WD component surrounded by the rotationally supported envelope made of less massive, disrupted WD (Benz et al. 1990). If ignition takes place at low densities, near the base of the rotating envelope, it will probably lead to slow burning and subsequent core collapse (Mochkovich & Livio 1990). Otherwise, gradual redistribution of angular momentum may lead to a growth of a massive CO core which then ignites near the center when its mass approaches the Chandrasekhar limit. The exploding configuration will resemble an isolated $\sim 1.4 M_\odot$ CO-WD, but with rotation and surrounded by an extended CO envelope. (3) a CO-WD accreting mass through Roche-lobe overflow as in (1), but the explosion is triggered by the detonation of an accumulated layer of helium before the total mass of the configuration reaches the Chandrasekhar mass (Nomoto 1980; Woosley et al. 1980). Only the first two models appear to be viable. The third, the sub-Chandrasekhar WD model, has been ruled out on the basis of predicted light curves and spectra.

3. Phenomenological models of SNIa explosion

Many ingredients of the SNIa explosion physics such as equation of state and nuclear reaction rates are known well. However, flame propagation in a supernova is difficult to model from first principles due to an enormous disparity of spatial and temporal scales involved. One is forced to make assumptions about regimes of burning (detonation or deflagration), and about the speed with which the flame propagates in case of turbulent deflagration. Once the assumptions are made, an outcome can be calculated by solving the equations of fluid dynamics coupled with the (prescribed) nuclear energy release terms, and with terms describing self-gravity of the star.

Three major models of the explosion of a Chandrasekhar mass CO-WD have been considered: (1) detonation model (Arnett 1969; Hansen & Wheeler 1969), (2) deflagration model (Nomoto et al. 1976), and (3) delayed-detonation (DD) model (Khokhlov 1991; Khokhlov 1991; Yamaoka et al. 1992; Woosley & Weaver 1994). The most detailed computations of SNIa explosion to date involve a hydrodynamic calculation of the thermonuclear explosion that includes a nucleosynthesis computation, a time-dependent radiation-transport computation that gives the light curve, including mechanisms for γ-ray and positron deposition, the effects of expansion opacity, and scattering, and NLTE spectra computations (Höflich & Khokhlov 1996; Nugent et al. 1997 and references therein).

It was found that purely detonation models do not fit observations because they do not produce intermediate mass (Si-group) elements which are so prominent in the spectra of SNIa around maximum light. Deflagration models produce intermediate mass elements, but typically in a too narrow velocity range, and also have difficulty explaining the variety of SNIa. Delayed detonation models are successful in reproducing the main features of SNIa, including multi-wavelength light curves, the spectral behavior, and the brightness—decline and color—decline correlations (Höflich & Khokhlov 1996; Höflich et al. 1995; Höflich, et al. 1996; Wheeler et al. 1998, and references therein).

Delayed detonation models assume that burning starts as a subsonic deflagration and then turns into a supersonic detonation. The deflagration speed, S_{def}, and the moment of deflagration-to-detonation transition (DDT) are free parameters. The moment of DDT is conveniently parameterized by introducing the transition density, ρ_{tr}, at which DDT happens. Initial central density and initial composition (C/O ratio) of the exploding WD must also be specified. To reproduce observations, deflagration speed should be a rather small fraction of the speed of sound a_s, say, $S_{\text{def}} < 0.1 a_s$. Physical arguments why S_{def} is small are discussed in the next section. The models are very sensitive to the variations of ρ_{tr}, but to a much lesser extent on the exact assumed value of the deflagration speed, initial central density of the exploding star, and the initial chemical composition.

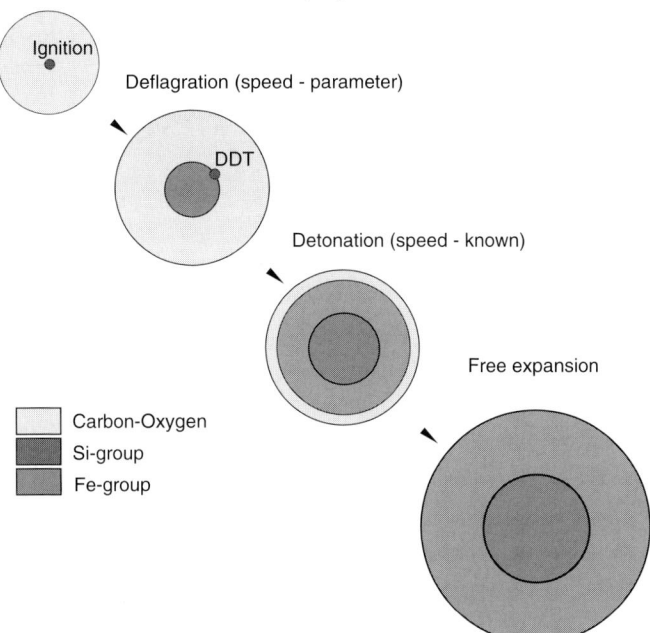

FIGURE 1. Schematics of the delayed detonation explosion. Ignition takes place in a dense, Chandrasekhar mass carbon-oxygen white dwarf. Flame propagates from the center as a subsonic turbulent deflagration. Deflagration-to-detonation transition (DDT) takes place in a significantly expanded star after only a small fraction of mass has been burned. Detonation incinerates the rest of the white dwarf. The resulting configuration consists of the inner core of Fe-group elements including ^{56}Ni surrounded by a massive envelope of Si-group elements.

A delayed detonation explosion is schematically illustrated in Figure 1. Because the speed of deflagration is less than the speed of sound, pressure waves generated by burning propagate ahead of the deflagration front and cause the star to expand. As a result, deflagration propagates through matter which density continuously decreases with time. After deflagration turns into a detonation, detonation wave incinerates the rest of the WD left unburned during the deflagration phase. Detonation produces Fe-group elements if it occurs at densities greater than $\rho \simeq 10^7$ g cm^{-3}. At lower densities it produces intermediate mass elements. At even lower densities around $\sim 10^6$ g cm^{-3} only carbon has time to burn. The outermost layers of a supernova will consist of products of explosive carbon burning such as O, Ne, Mg, etc. To reproduce observations, $\rho_{\rm tr}$ must be selected in the range $\rho_{\rm tr} \simeq (1-3) \times 10^7$ g cm^{-3}. Virtually no intermediate mass elements will be produced for larger values of $\rho_{\rm tr}$. For lower $\rho_{\rm tr}$, the WD expands so much that a detonation cannot be sustained. With $\rho_{\rm tr}$ in the right range, the inner parts of the exploded star consist of Fe-peak elements and contain radioactive ^{56}Ni. Outer parts contain intermediate group elements and products of explosive carbon burning (Figure 1).

The amount of ^{56}Ni produced during the explosion is very sensitive to $\rho_{\rm tr}$. Varying $\rho_{\rm tr}$ in the range $(1-3) \times 10^7$ g cm^{-3} gives nickel mass in the range $\simeq 0.1$–0.7 M$_\odot$, respectively. The reason for such a sensitivity is the combination of an exponential temperature dependence of reaction rates and the dependence of the specific heat of a degenerate matter on density. Small differences in density at which burning takes place translate into small differences in burning temperature. These, however, translate into large differences in reaction rates, and into qualitative differences in the resulting chemical composition. The kinetic energy of the explosion, on the other hand, is very

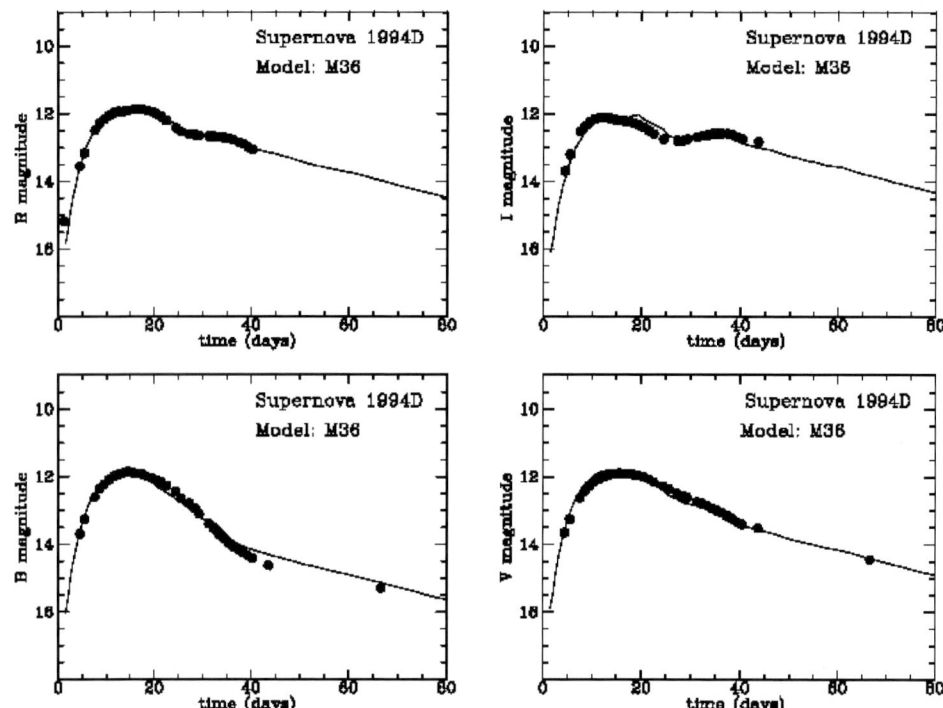

FIGURE 2. Comparison of observed (SN1994D) and theoretical (M36) B, V, R, I light curves (Höflich 1995).

insensitive to $\rho_{\rm tr}$. It depends on the total amount of burned material (Fe-group and Si-group together). This is because the difference in binding energies of Fe-group and Si-group nuclei is relatively small compared to the difference between binding energies of both Fe- and Si-group elements and the initial CO mixture. Thus, the delayed detonation model thus predict SNIa with significantly varying nickel mass but with almost the same kinetic energy and expansion velocities.

The above property of the delayed detonation model is perhaps the key to the explanation of the brightness—decline and color—decline relations among Type Ia supernovae. All delayed detonation supernovae expand with approximately the same velocity. Explosions with more nickel give rise to brighter supernovae. Also, because of more nickel decays, envelopes of these supernovae are heated better and stay hot and opaque. The result is a slow post-maximum decline and a blue color. Explosions with less nickel give rise to dim supernovae. Envelopes of these supernovae are cool and transparent because they contain less nickel. The result is a fast post-maximum decline and a red color.†

As a representative example, Figure 2 shows results of numerical modeling of the bright SNIa 1994D (Höflich 1995). This SN is especially interesting because it was relatively blue and rather bright for its rate of decline and did not fit exactly into the "average" one-parameter brightness-decline relation. The light curves of SN1994D are fit with the light curves of the best fit delayed detonation model M36, one of the models of the series of (Höflich & Khokhlov 1996) with the initial central density $\rho_c = 2.7 \times 10^9$ g cm^{-3} and

† In deflagration models, the amount of nickel and kinetic energy of the explosion are tightly related. Supernovae with more nickel expand and cool faster, while supernovae with less nickel are expanding slowly. Deflagration models predict that light curves of brighter supernovae should decline faster, which is contrary to observations.

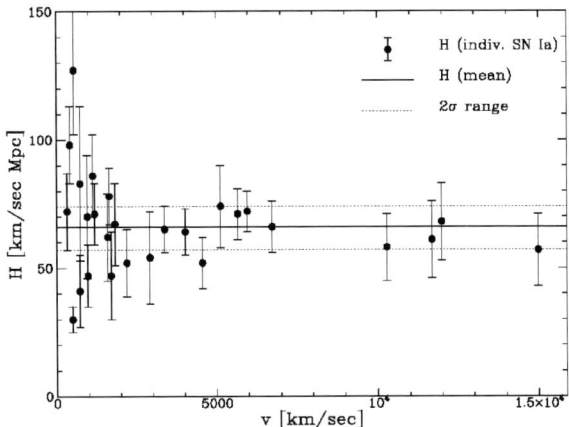

FIGURE 3. Direct determination of the Hubble constant ($H_o = 67 \pm 9$ km s^{-1} Mpc^{-1}) using delayed detonation models (Höflich & Khokhlov 1996). Values of H_o are plotted for individual SNIa based on distances determined by fitting their light curves and spectra with theoretical models.

$S_{\text{def}} = 0.04 a_s$. For the M36 model, the transition density was $\rho_{\text{tr}} = 2.4 \times 10^7$ g cm^{-3}. As can be seen, both optical and IR light curves are fit by M36 rather well, including the secondary maximum in R and I typical of normally bright SNIa. Models with other values of ρ_{tr} led to much worse fits to observations.

Delayed detonation models have been used to predict a purely theoretical (without using Cepheid distances) value of the Hubble constant (Müller & Höflich 1994; Höflich & Khokhlov 1996). The idea is to fit a supernova with the model which best reproduces its light curves and spectra. The model then gives the absolute brightness of a supernova. The method takes into account both brightness variations among SNIa and possible interstellar absorption. The result is shown in Figure 3. Values of H_o determined from individual supernovae show a large spread for close SNIa but converge to $H_o \simeq 67 \pm 9$ km s^{-1} Mpc^{-1} with increasing z. This value is in agreement with H_o found using Cepheid variables (Sandage 2000; Riess et al. 1998; Ferrarese et al. 1999).

4. Three-dimensional SNIa

Three-dimensional effects in the propagation of turbulent flames and deflagration-to-detonation transition (DDT) must play a key role in SNIa in determining the actual speed of the flame propagation, energetics, and nucleosynthesis, and also are likely to translate initial differences in presupernova structure into the observed differences among SNIa.

4.1. Deflagration

Laminar flame in a WD is driven by heat conduction due to degenerate electrons and propagates very subsonically with the speed $S_{\text{lam}} < 0.01 a_s$ (Timmes & Woosley 1992). Such a slow flame cannot account for the explosion properties of SNIa. However, in the presence of gravity, the flame speed will be enhanced by the Rayleigh-Taylor instability (Nomoto et al. 1976). Whether the Rayleigh-Taylor instability can itself sufficiently increase the flame speed to cause the explosion or whether deflagration just serves to pre-expand the star which is incinerated later by a supersonic detonation, has been a subject of numerous studies and 3D simulations (Khokhlov 1995; Livne & Arnett 1993; Arnett & Livne 1994; Khokhlov et al. 1995; 32 and references therein).

Simple scaling arguments show that turbulent flame subjected to a uniform gravity acceleration in a vertical column must propagate with a speed (Khokhlov 1995)

$$S_{\text{def}} \simeq \alpha \left(gL \frac{\rho_0 - \rho_1}{\rho_0 + \rho_1} \right)^{1/2}, \qquad (4.1)$$

where g is local acceleration, L is the width of the column, ρ_0 and ρ_1 are densities ahead and behind the flame, respectively, and $\alpha < 1$ is a constant which depends on the column's geometry and boundary conditions. Formula (1) is valid, of course, only when S_{def} is much larger than S_{lam}. It tells that when the characteristic RT speed $\simeq (gL)^{1/2}$ is greater than S_{lam}, the flame speed is determined by the turbulence on the largest scale, L, independent of details of flame propagation on smaller scales. The reason for this behavior is self-similarity of the flame. Turbulent flame speed is the product of the area A of the flame surface and the laminar flame speed, $S_{\text{def}} = A \cdot S_{\text{lam}}$. Turbulence tends to increase A whereas intersections of different portions of the flame front tend to diminish A. The latter effect is proportional to S_{lam}. In equilibrium, the two effects balance each other, and $A \propto 1/S_{\text{lam}}$. The product of A and S_{lam} remains constant.† Three-dimensional numerical simulations with varying g, L and varying laminar flame speed confirm equation (4.1) including its independence of S_{lam} and indicate $\alpha \simeq 0.5$ (Khokhlov 1995; Khokhlov et al. 1996). The results are consistent with high-gravity combustion experiments that used a centrifuge to study premixed turbulent flames at various g.

Equation (4.1) tells several important things. Near the WD center $g \simeq 0$. Thus, in the beginning the deflagration speed $S_{\text{def}} \simeq S_{\text{lam}}$ should be small. The speed will then tend to increase as the flame goes away from the center and the gravity increases. When the intensity of turbulence increases, the flame speed will become independent of the physics of burning on small scales. The latter conclusion is very important since it gives us a hope that SNIa explosions can be modeled in three dimensions without resolving all small spatial and temporal scales. However, equation (4.1) is missing an important piece of physics. It is valid only in a uniform gravitational field and only when there is no global expansion of matter.

In a supernova explosion, burning causes a global expansion of a star. Equation (4.1) may be valid only on scales where the expansion velocity is less than the characteristic RT speed. On larger scales, expansion will tend to freeze the turbulence out. The net result will be a substantial decrease of the turbulent flame speed. A crude estimate of the scale L_f at which the turbulence becomes frozen and of the effective deflagration speed limited by expansion can be obtained as follows. First, carry out a one-dimensional simulation assuming no turbulence freeze-out, that is, with the flame speed given by equation (4.1) with L equal to the flame radius R_f. This gives the expansion rate. Then estimate L_f as a scale at which the expansion velocity becomes comparable with the characteristic RT velocity. Finally, estimate the effective deflagration speed from equation (4.1) using $L = L_f$ (Khokhlov et al. 1995). The estimates are $L_f \simeq$ (a few) $\times 10^7$ cm, and

$$S_{\text{def}} \simeq 1.5 \times 10^7 \text{ cm s}^{-1} \left(\frac{g}{10^9 \text{ cm s}^2} \right)^{1/2} \left(\frac{L_f}{10^8 \text{ cm}} \right)^{1/2}. \qquad (4.2)$$

Equation (4.2) shows that in conditions typical of the exploding white dwarf, a turbulent

† There is a close analogy here with the self-similarity of an ordinary Kolmogorov cascade. In the Kolmogorov cascade, changes in the fluid viscosity lead to changes in the viscous microscale, but do not influence the amount of energy dissipated into heat. The rate of dissipation depends only on the intensity of turbulent motions on the largest scale. S_{lam} plays the role of viscosity in a turbulent flame.

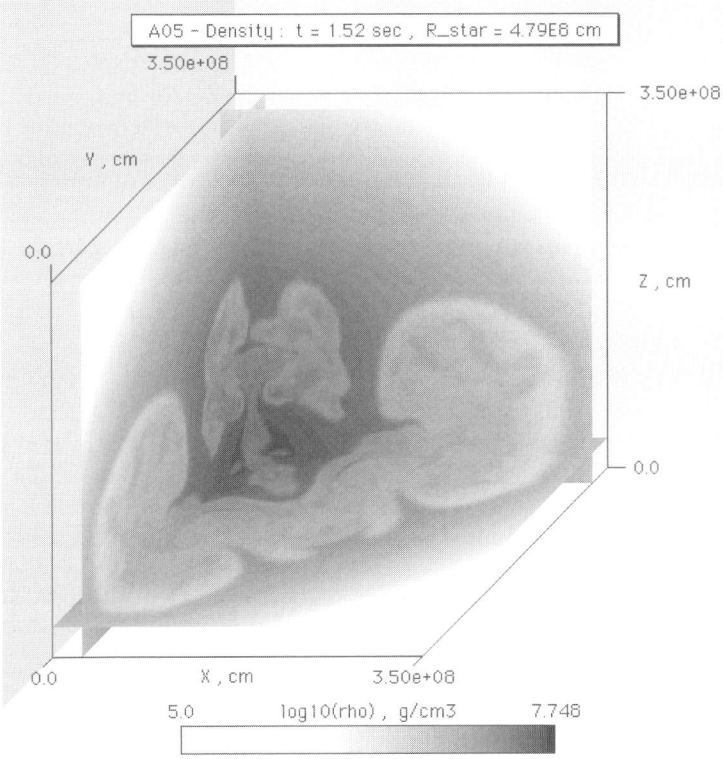

FIGURE 4. Three-dimensional simulation of an explosion of a Chandrasekhar mass CO-WD (Khokhlov 1995). The figure shows density distribution during the deflagration phase of the explosion.

burning speed is a few percent of the sound speed $a_s \simeq 5 \times 10^8$ cm s^{-1}. This is not enough to cause a powerful explosion. An additional effect that further limits the rate of deflagration is a deviation from the steady-state turbulent burning regime. A certain time is required for a turbulent flame to reach a steady-state. This time is larger for larger scales. Scales of the order of R_f might never reach a steady-state during the explosion.

Figure 4 shows some results of a three-dimensional numerical simulation of the entire Chandrasekhar mass CO-WD exploding as a supernova. In this simulation, equation (4.1) has been used for the turbulent flame velocity on scales not resolved numerically. For regions of the "average" flame front not oriented "upwards" against gravity this formula most probably overestimates the local turbulent flame speed. Despite this, only $\simeq 5\%$ of the mass has been burned by the time the star has expanded and quenched the flame, and the white dwarf has not even become unbound. These results show that spherical expansion is indeed important and that burning on large scales does not reach a steady state. Big blobs of burned gas rise and penetrate low density outer layers, whereas unburned matter flows down and reaches the stellar center. The model experienced an almost complete overturn. This has obviously important implications for nucleosynthesis and may cause an element stratification incompatible with observations if composition inhomogeneities are not smeared out during the subsequent detonation stage of burning. The results of 3D modeling indicate that the deflagration alone is not sufficient to cause an explosion. To make a powerful explosion, the deflagration must somehow make a transition to a detonation (delayed detonation model).

4.2. Deflagration-to-detonation transition

In terrestrial conditions detonation may arise from a non-uniform explosion of a region of a fuel with a gradient of reaction (induction) time via the Zeldovich gradient mechanism (Zeldovich et al. 1970; Lee et al. 1978). The region may be created by mixing of fresh fuel and hot products of burning, as in jet initiation, or it may be created by multiple shocks, etc. The same gradient mechanism can operate in supernovae (Blinnikov & Khokhlov 1986; Khokhlov et al. 1997; Niemeyer & Woosley 1997). There exist a minimum, critical size of the region capable of generating a detonation, L_i. This parameter is determined by the equation of state and nuclear reaction rates and is mainly a function of the density of the material. L_i is much less than the size of a WD for all but very low densities $\rho < 10^7$ g cm^{-3} (Khokhlov et al. 1997).

Why then DDT does not happen in supernovae at high densities? Why does it have to wait until the WD expands significantly? The explanation may be this. The critical size L_i, however small, is still several orders of magnitude larger than the thickness of a laminar flame. To mix fresh fuel with products of burning, the surface of the flame must be disrupted. But this is difficult to achieve unless the turbulence on a scale of a flame front is larger than the laminar flame speed. Only at very low densities, where reactions slow down, the width of the laminar flame becomes very large, and its speed becomes very small, the turbulence may have a chance to create the right conditions for DDT (Khokhlov et al. 1997; Niemeyer & Woosley 1997).

In addition to mixing fuel and products inside an active deflagration front, another mechanism for creating the right conditions for DDT may be as follows. As mentioned above, turbulence in a SNIa will be limited by the expansion. The conditions for DDT during the expansion of a star may not be fulfilled at all. But when deflagration speed is small, deflagration quenches due to expansion before the WD becomes unbound. This happens, in particular, in the simulation shown in Figure 4. The star will then experience a pulsation and collapse back. During the expansion and contraction phases of the pulsation, the high entropy ashes of dead deflagration front will mix with the fresh low entropy fuel again to form a mixture with the reaction time gradients. Mixing will be facilitated during the contraction phase due to the increase of turbulent motions from the conservation of angular momentum (like an ice skater increases his rotation by squeezing his arms). The estimate of the mixing region formed during pulsation is $\simeq 10^6$–10^7 cm, much larger than L_i. Is was also shown that as soon as only a few percent of hot ashes are mixed with a cold fuel, the mixture cannot be compressed to densities higher than $\simeq$ (a few) $\times 10^7$ g cm^{-3}. Further compression will lead to a burnout on time scales much shorter than the pulsation time scale. As soon as this mixture returns to high enough densities $\simeq 10^7$ and re-ignites, the detonation will be triggered (Khokhlov 1991; Arnett & Livne 1994; Khokhlov et al. 1997).

It should be noted at this point that three-dimensional theory of flame propagation and DDT in supernovae is far from being finished, and remains a subject of an active research. In particular, it was speculated recently that DDT may be caused by a sudden acceleration of a quasi-spherical deflagration front, due to the Landau-Darrheus or some other yet unknown internal instabilities of the flame; that a suddenly accelerated deflagration might keep propagating with the speed of sound without turning into a detonation, etc. (Niemeyer 1999). Whether any of these can actually happen should be either tested in appropriately scaled terrestrial experiments or demonstrated in three-dimensional simulations. Further work is required, and it will undoubtedly improve our understanding of SNIa explosions.

It may also be possible to distinguish between different multi-dimensional explosion mechanisms on the basis of observations. One of the amazing properties of SNIa is their apparently small deviation from spherical symmetry. We do not expect all three-dimensional models to have this property. For example, pure deflagration models are expected to be clumpy (Figure 4) and asymmetric with large blobs of Si and Fe group elements embedded in the unburned CO envelope. Delayed detonation models, on the other hand, should be more symmetric. A supersonic detonation mode of burning that follows deflagration will tend to homogenize the ejecta. Rotation of the progenitor may impose a global, low order asymmetry on the ejecta. Viable models can be limited by computing the polarization of the emerging radiation and comparing the predictions with the existing (Wang et al. 1997) and planned observations.

5. Discussion

We described a phenomenological delayed detonation model of SNIa based on the explosion of a Chandrasekhar mass carbon-oxygen white dwarf. The model assumes that the explosion starts as a subsonic deflagration and then turns into a supersonic detonation mode of burning. The model is successful in reproducing the main features of SNIa, including multi-wavelength light curves, the spectral behavior, and the brightness—decline and color—decline correlations. It was argued that an apparently low deviation of SNIa from spherical symmetry (low polarization of SNIa) may be attributed in delayed detonation models to the homogenizing effect of the detonation phase of an explosion.

The model interprets existing brightness—decline and color—decline relations among SNIa as a result of varying nickel mass synthesized during the explosion. Major free parameters of the model are the deflagration speed S_{def}, the transition density ρ_{tr} at which deflagration turns into a detonation, and also initial density and composition (C/O ratio) of the exploding WD. The variation of nickel mass in the model is caused by the variation of ρ_{tr}. Strong sensitivity of the nickel mass to ρ_{tr} is probably the basis of why, to first approximation, SNIa appear to be a one-parameter family. Nonetheless, variations of the other parameters also lead to some relatively small variations of the predicted properties of SNIa, which indicate that the assumption of a one-parameter family may not be strictly valid.

To fit observations, the delayed detonation model requires low values of $S_{\text{def}} < 0.1 a_s$ and low values of $\rho_{\text{tr}} \simeq (1-3) \times 10^7$ g cm^{-3}. In Section 4 it was argued that slow deflagration is the result of an expansion of a star caused by the deflagration itself. The expansion tends to freeze the turbulence and, thus, limits the deflagration speed. The actual rate of deflagration in a supernova is determined by the competition of the Rayleigh-Taylor instability which is the turbulence driving force, the turbulent cascade from large to small scales, and the turbulence freeze-out. Two possible mechanisms that lead to a low ρ_{tr} were discussed—one related to the disruption of an active deflagration front by the existing turbulence, and the other related to quenching of deflagration, mixing of the low-entropy fuel with high-entropy burning products, and its subsequent compression.

It may seem strange that the two apparently different mechanisms predict almost the same low values for ρ_{tr}, these same low values that are required to fit observations. Note, however, that predictions of both mechanisms and the very reason why low ρ_{tr} is needed to fit observations steam from the same two fundamental facts: (1) specific heat of matter in supernovae depends on density; (2) nuclear reactions depend on temperature exponentially. Numbers are such that at densities above 10^7 g cm^{-3} nuclear burning timescales are much shorter than the sound crossing time ($\simeq$ explosion timescale) in a

WD. At densities below 10^7 g cm^{-3} the timescales become much longer the sound crossing time. The dependence of burning timescales on density is very steep. That is why a laminar flame front can widen and be disrupted by turbulence only at approximately $\leq 10^7$ g cm^{-3}. That is also why a mixture of cold fuel and hot products cannot be compressed to densities much higher than $\leq 10^7$ g cm^{-3}—it will react faster than it is being compressed. And that is also why intermediate mass elements can be synthesized in an SNIa only at densities around $\sim 10^7$ g cm^{-3}.

What may cause the variations of ρ_{tr} among SNIa? There are several possibilities. One is differences in C/O ratio. If less carbon is present in a WD, the energy released by burning is less, and this will affect both the buoyancy of burning products and the rate of expansion of a WD. This, in turn, will affect the speed of deflagration, and will lead to different conditions for DDT. Variations of initial C/O ratio among SNIa has been recently studied in the framework of one-dimensional phenomenological delayed detonation models in (Dominguez & Höflich 1999; Höflich & Dominguez 2000; Umeda et al. 1999; Höflich et al. 1999). The effect of varying C/O ratio may result in small but noticeable variations in the rise time to maximum and in some other variations in light curve behavior. This is a potential source of systematic evolutionary effects, and has obvious implications for using SNIa in cosmology. However, in one-dimensional models one has to assume how changes in C/O influence the deflagration speed and ρ_{tr}, and the conclusions then depend on these assumptions. Three-dimensional modeling is required in order to predict the actual influence of C/O ratio on the outcome of the explosion. Another possibility is the influence of rotation if SNIa are the result of a merger of low-mass CO-WD. Rotation will undoubtedly influence the turbulent deflagration phase which, in turn, will affect DDT. Merger configurations may also differ in their mass, so that slightly super-Chandrasekhar mass WD explosions are probable. Could they be responsible for unusually bright SN1991T-like events? An extended CO envelope around a merger WD may manifest itself in SNIa light curves and spectra. Further work is needed to answer these questions.

I thank David Branch, Peter Höflich, Elaine Oran and Craig Wheeler for discussions. This research was supported in part by the NASA Grant NAG-52888 and by the Office of Naval Research.

REFERENCES

Arnett, D. & Livne, E. 1994 *ApJ*, **427**, 330.

Arnett, W. D. 1969 *Ap. Space Sci.*, **5**, 280.

Arnett, W. D. & Livne, E. 1994 *ApJ*, **427**, 315.

Benz, W., Bowers, R. L., Cameron, A. G. W., & Press, W. H. 1990 *ApJ*, **348**, 647.

Blinnikov, S. I. & Khokhlov, A. M. 1986 *Soviet Astron. Lett.*, **12**, 131.

Dominguez, I. & Höflich, P. A. 1999 *ApJ*, in press, astro-ph/9908204.

Ferrarese, L., et al. 1999 *ApJ*, in press.

Hansen, C. J. & Wheeler, J. C. 1969 *Ap. Space Sci.*, **3**, 464.

Iben, I. Jr. & Tutukov, A. V. 1984 *ApJS*, **54**, 335.

Höflich, P. 1995 *ApJ*, **459**, 307.

Höflich, P. A. & Dominguez, I. 2000, this volume.

Höflich, P. & Khokhlov, A. M. 1996 *ApJ*, **457**, 500.

Höflich, P., Khokhlov, A., & Wheeler, J. C. 1995 *ApJ*, **444**, 831.

HÖFLICH, P., KHOKHLOV, A., WHEELER, C. J., PHILLIPS, M. M., SUNZEFF, N. B., HAMUY, M. 1996 *ApJ*, **472**, L81.

HÖFLICH, P. A., NOMOTO, K., UMEDA, H., & WHEELER, J. C. 1999 *ApJ*, in press.

KHOKHLOV, A. M. 1991 *AA*, **245**, 114.

KHOKHLOV, A. M. 1991 *AA*, **245**, L25.

KHOKHLOV, A. M. 1995 *ApJ*, **449**, 695.

KHOKHLOV, A. M., ORAN, E. S., & WHEELER, J. C. 1995 in *Type Ia Supernovae*.

KHOKHLOV, A. M., ORAN, E. S., & WHEELER, J. C. 1996 *Combustion & Flame*, **105**, 28.

KHOKHLOV, A. M., ORAN, E. S., & WHEELER, J. C. 1997 *ApJ*, **478**, 678.

LEE, J. H. S., KNYSTAUTAS, R., & YOSHIKAWA, N. 1978 *Acta Astron.*, **5**, 971.

LIVIO, M. 2000, this volume.

LIVNE, E. & ARNETT, W. D. 1993 *ApJ*, **415**, L107.

MOCHKOVICH, R. & LIVIO, M. 1990 *A&A*, **236**, 378.

MÜLLER, E. & HÖFLICH, P. A. 1994 *A&A*, **281**, 51.

NIEMEYER, J. C. 1999 *ApJ*, **523**, L57.

NIEMEYER, J. C. & WOOSLEY, S. E. 1997 *ApJ*, **475**, 740.

NOMOTO, K. 1980 *ApJ*, **248**, 798.

NOMOTO, K., SUGIMOTO, S., & NEO, S. 1976 *ApSS*, **39**, L37.

NUGENT, P., ET AL. 1997 *ApJ*, **485**, 812.

PERLMUTTER, S., ET AL. 1999 *ApJ*, **517**, 565.

PHILLIPS, M. M. 1993 *ApJ*, **413**, L105.

RIESS, A., NUGENT, P., FILIPPENKO, A. V., KIRSHNER, R., & PERLMUTTER, S. 1998 *ApJ*, **504**, 935.

RIESS, A. G., PRESS, W. H., & KIRSHNER, R. P. 1996 *ApJ*, **473**, 588.

SANDAGE, A. 2000, this volume.

SHMIDT, B., ET AL. 1998 *ApJ*, **507**, 46.

TIMMES, F. X. & WOOSLEY, S. E. 1992 *ApJ*, **396**, 649.

UMEDA, H., NOMOTO, K., KOBAYASHI, C., HACHISU, I., & KATO, M., 1999, *ApJL*, in press

WANG, L., WHEELER, J. C., & HÖFLICH, P. 1997 *ApJ*, **476**, 27.

WEBBINK, R. F. 1984 *ApJ*, **277**, 355.

WHEELER ET AL. 1998 *ApJ*, **496**, 908.

WOOSLEY, S. E. & WEAVER, T. A. 1994 in *Les Houches, Session LIV, Supernovae* (eds. S. A. Bludman, R. Mochkovich, & J. Zinn-Justin), p. 63. North-Holland.

WOOSLEY, S. E., WEAVER, T. A., & TAAM, R. E. 1980 in *Type I Supernovae* (ed. C. Wheeler), p. 96. U. Texas.

YAMAOKA, H., NOMOTO, K., SHIGEYAMA, T., & THIELEMANN, F.-K. 1992 *ApJ*, **393**, 55.

ZELDOVICH, YA. B., LIBROVICH, V. B., MAKHVILADZE, G. M., & SIVASHINSKY, G. L. 1970 *Acta Astron.*, **15**, 313.

Universality in SN Iae and the Phillips relation

By DAVID ARNETT

Steward Observatory, University of Arizona, Tucson, AZ 85721

The use of supernovae of Type Ia for the determination of accurate distances rests upon the empirical Phillips relation, in which the brightest events are the broadest in time. Implications of new data upon the homogeneity of light curves under the operation of a stretch in time, of the parabolic luminosity increase at the earliest times, and of the time from explosion to maximum light are discussed. The early luminosity is in excellent agreement with the predictions of Arnett (1982), and the lack of prominent higher modes of diffusion constrain progenitor and explosion models. Difficulties with reproducing the observed rise time are restricted to radiative transfer models (e.g. Höflich & Khokhlov 1996), and probably due to an overestimate of thermal photon escape due to inadequate line lists. Because of the strong dependence of luminosity on ^{56}Ni mass, some simple models can give a Phillips relation of the correct sense.

1. Introduction

Supernovae of Type Ia (SN Iae) have been found at high redshift (up to $z \approx 1$) (Perlmutter et al. 1997b; Schmidt et al. 1999; Perlmutter et al 1999a). Being about 10^6 times brighter than Cepheids, SN Iae can be seen about 10^3 times farther, and consequently provide a more interesting possibility for determining truly cosmic distances. Although the SN Iae are *not* identical, so that each event is not strictly like the standard one, Phillips (1993) found that the brighter events are broader in time (the Phillips relation). Calibration of this variation, with a set of relatively nearby supernovae, the Calan-Tololo sample of Hamuy et al. (1995), allows individual events to be placed upon a standard scale (Riess, Press & Kirshner 1995), and so doing reduces the scatter in the Hubble diagram (a plot of redshift versus calibrated brightness).

To properly use this breakthrough, we must understand the underlying physics of Phillip's empirical relation. Several recent observational developments, namely

(*a*) the explicit demonstration, by Perlmutter et al. 1997a and Perlmutter (1999a), that the application of a time stretch operation to SN Iae and a luminosity normalization, give a universal light curve shape, and

(*b*) measurement of the rise to maximum from low luminosities (Riess et al. 1999a) gives the detailed structure of the early light curve,

provide interesting information on the nature of the supernova event. The fact that SN Ia light curves, except for a few odd events that can be easily recognized by their spectra, represent a single parameter family of shapes which can be "stretched" to a single universal shape, needs to be explained. It is this universality that allows the nature of the individual event to be identified, independent of distance, and their use as distance indicators to be precise.

2. Perspective on analytic solutions

The luminosity of SN Iae is provided almost entirely by the decay of ^{56}Ni and ^{56}Co; other radioactive sources become important well after maximum light, and thermal emission from shocks is small. Determining the shape of the light curve, that is, the luminosity as a function of time, involves three separate issues:

(a) The transfer of radioactive energy by gamma-rays and positrons to the thermal energy of the plasma, properly including energy that escapes in the form of positrons, gamma and x-rays,

(b) The work done on the expanding plasma by the thermal radiation (or alternatively, the reduction in radiation energy by the accumulated redshift of the trapped photons), and,

(c) The energy escaping as thermalized radiation, which forms the supernova spectrum observed from the infra-red through the visible to ultra-violet (UVOIR).

The analytic solutions used approximate the last item, the escape probability, by a grey diffusion operator, parameterized by an effective opacity. The actual physics of the escape of thermal radiation is complex, and this complexity has obscured some simple and general features of the light curves. In particular, *a reasonable reproduction of the observed spectrum cannot guarantee a reasonable representation of the thermal photon escape probability.*

Here we will use the simple model to illustrate some general points (a more accurate discussion is in preparation with P. Nugent and P. Pinto). Note that the escape operator for thermal photons might be better thought of as diffusion down in energy space as well as out in radius; see Pinto & Eastman (1999). It will be an interesting challenge to develop a better approximation to the thermal escape operator, which is simple enough to allow analytic solutions.

In this analysis we will use only the solutions presented in Arnett (1982) and earlier to emphasize the phenomena already contained in these early efforts, which have only now had observational confirmation.

3. The shape of the light curve

The bolometric luminosity may be written as

$$L = \epsilon_{Ni} M_{Ni} \Lambda(x, y). \tag{3.1}$$

Here M_{Ni} is the mass of ^{56}Ni; ϵ_{Ni} is the energy of radioactive decay of ^{56}Ni per unit mass, divided by the mean lifetime τ_{Ni}. Actually ^{56}Co contributes as well. This may be included in the Λ function, with the effect that the peak shifts higher in luminosity and later in time. In either case, $\epsilon_{Ni} M_{Ni}$ is a convenient scale factor for the luminosity.

The shape of the light curve, which is independent of distance, is contained in $\Lambda(x, y)$, where

$$x = t/\tau_m, \tag{3.2}$$

and

$$y = \tau_m/2\tau_{Ni}. \tag{3.3}$$

The effective escape time in a diffusing and expanding medium is a logarithmic average of the expansion time τ_h and the diffusion time τ_0,

$$\tau_m = \sqrt{2\tau_h \tau_0}, \tag{3.4}$$

see Arnett (1982). Note that xy is simply time in units of two mean lives of ^{56}Ni, or about 17.6 days. Although y was defined within the particular analytic context of simple diffusion in an expanding medium, it may be more generally interpreted as a measure of the probability of escape of energy by photon transport.

The general character of these light curves is shown in Figure 1, which represents the solutions for ^{56}Ni decay alone. For the observationally interesting case of maxima

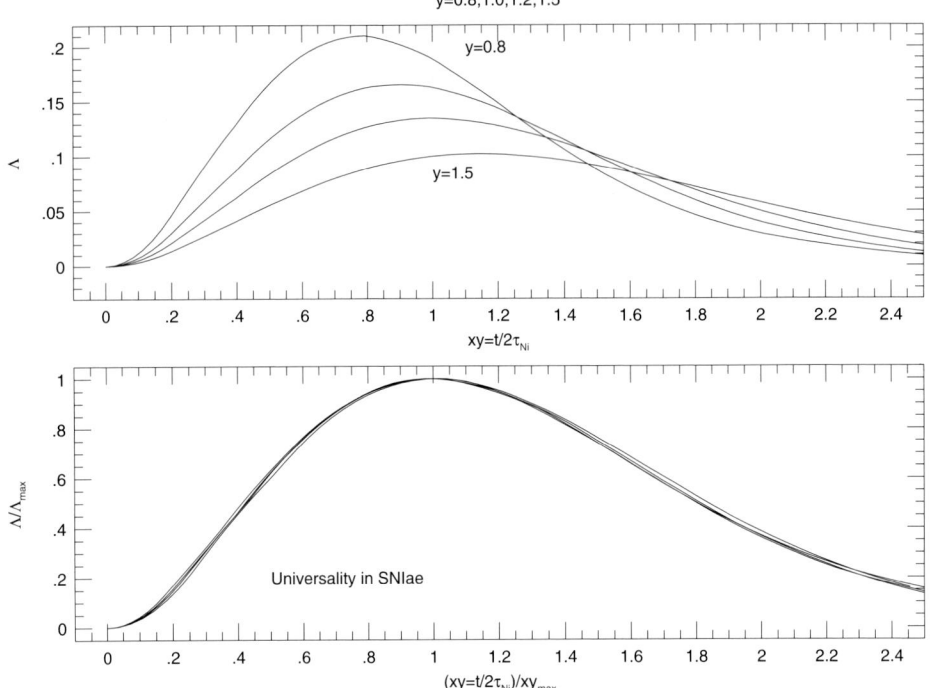

FIGURE 1. Lambda vs. time (panel 1), normalized (panel 2).

occurring around $xy \approx 1$ as shown, we have

$$\Lambda_{\max} \approx 0.165/y, \tag{3.5}$$

and

$$t_{\max}/2\tau_{\mathrm{Ni}} \approx 0.42 + 0.48y. \tag{3.6}$$

These approximations apply to the simple case in which only the ^{56}Ni heating is included, not that of ^{56}Co decay (the more general case was calculated but not tabulated in Arnett, 1982). Near maximum light ($t \approx 2\tau_{\mathrm{Ni}}$), the heating from ^{56}Co is equal to that from ^{56}Ni, and dominates at later times. This additional heating will increase both $t_{\max}$ and $\Lambda_{\max}$ relative to the values given by these approximations, but make no qualitative change. The lower panel shows the effect of (1) renormalizing the luminosity ($\Lambda \to \Lambda/\Lambda_{\max}$), and (2) stretching the time scale to line up the maxima ($t \to t/t_{\max}$ was used here). The curves lie almost on top of each other, so that in this sense, the shape is "universal."

For theoretical light curves the time of explosion is easily defined, which is not true observationally. The observational stretch includes a shift in time as well, $t \to (t - t(0))/t_{\max}$.

The light curves shown are those first presented by Arnett (1982). However, it was only after the discovery by Perlmutter et al. (1997a) that the observational data could be mapped into a universal curve by a normalization of luminosity and a stretch of time scale relative to the time of peak luminosity, that the analytic solutions were plotted in this form. For this simple case, the analytic solutions have this same property of (approximate) universality as the data.

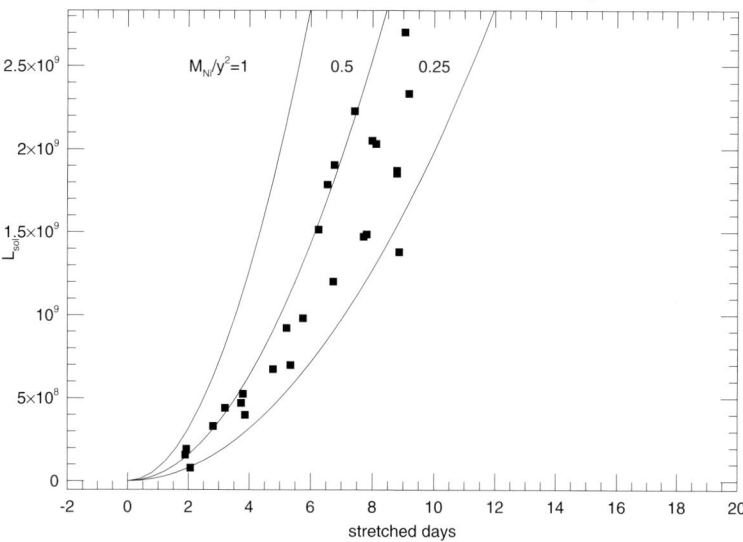

FIGURE 2. Scaled Luminosity of SN Iae vs. stretched time.

4. The early light curve

Figure 1 also shows that all the light curves have a parabolic dependence upon time during the earliest times after the explosion. Riess et al. (1999a) present measurements of the earliest detections of nearby SN Iae, which delineate the rise behavior for 18 to 10 days before maximum.

According to Riess et al. (1999a), Goldhaber has proposed a method of determining the rise time of SN Iae which is based upon the "stretch" method of Perlmutter et al. (1997a). Riess et al. (1999a) and Riess et al. (1999b) have applied a similar approach to the B-band light curves of a number of SN Iae. Goldhaber proposed that we describe the young SNIa as a homologously expanding fireball whose luminosity is most sensitive to its increasing radius, rather than effective temperature. The luminosity is then

$$L = \alpha(t_{\max} + t_r)^2, \qquad (4.7)$$

where $t_{\max}$ is the time elapsed relative to maximum, t_r is the rise time, and α is the "speed" of the rise.

Figure 2 shows the data from Riess et al. (1999a). The squares represent their ten SN Iae; upper limits have not been plotted. The time coordinate has been stretched by their stretch factors, and shifted so that $t = 0$ corresponds to the onset of the explosion. Thus, after stretching, the new time coordinate is $t = t_{\max} + t_r$. Their B magnitude has been rudely converted to solar luminosities by simply ignoring bolometric corrections (this is roughly correct, see Riess et al. 1999a). In this linear plot, the quadratic nature of the time dependence is obvious.

This behavior was predicted in Arnett (1982). Ten days before maximum is roughly ten days after explosion, at which time the Co luminosity is about 0.3 of that of Ni, and should not yet make a qualitative difference. Note that $xy = t/2\tau_{\text{Ni}}$, so for early times $(t \ll \tau_{\text{Ni}})$, $\Lambda \approx x^2 = (t/2\tau_{\text{Ni}})^2/y^2$, so that

$$L = \epsilon_{\text{Ni}} M_{\text{Ni}} (t/2\tau_{\text{Ni}} y)^2. \qquad (4.8)$$

The luminosity scale is set by the mass of ^{56}Ni and the shape parameter y, the time dependence is quadratic in t, and

$$\alpha = \epsilon_{\mathrm{Ni}} M_{\mathrm{Ni}} / (2\tau_{\mathrm{Ni}} y)^2. \tag{4.9}$$

This identifies Goldhaber's α with the solution parameters, $2\tau_{\mathrm{Ni}}^2 y^2 = \kappa M / 2\beta c v_{\mathrm{sc}}$, where κ is the effective opacity and $\beta \approx 13.7$ (see Arnett 1980, Table 2), and M_{Ni} is the mass of ^{56}Ni.

The solid lines in Figure 2 represent this solution for $(M_{\mathrm{Ni}}/M_\odot)/y^2 = 0.25$, 0.5 and 1.0. At early times $(M_{\mathrm{Ni}}/M_\odot)/y^2 \approx 0.4$. For a popular estimate of $(M_{\mathrm{Ni}}/M_\odot) = 0.6$, we have $y = 1.2$. As we shall see, this is a plausible value.

5. Higher modes

The behavior shown in Figure 1 and Figure 2 is based on a theoretical model which assumes that the higher modes in the spatial solution of the diffusion equation are small (Arnett 1980). These higher modes can be driven by

(a) a distribution of ^{56}Ni which is different from the distribution of energy in the fundamental mode for diffusion, (Pinto & Eastman 1999),

(b) a time dependence in the opacity (effective escape parameter y), or

(c) the interaction of the exploding star with surrounding matter or a companion.

Such overtones modify the shape of the light curve, and in principle can be detected as a distance independent characteristic of SN Iae. The Riess et al. (1999a) data in Figure 2 place limits on these effects.

6. The risetime

Most theoretical models of SN Iae predict significantly shorter risetimes than are found (Vacca & Leibundgut 1996; Riess et al. 1999a). For example, Höflich & Khokhlov (1996) give risetimes to visual maximum of 9 to 16 days, with an average value of 14 days for single white dwarf explosions. The same difference is seen by Pinto & Eastman (1999). Riess et al. (1999a) state:

"If these models are otherwise accurate, we concur with the conclusion of Vacca & Liebundgut (1996) that the model atmospheric opacity has been significantly underestimated. Past work suggests that deficient resonance line lists may be the culprit. By increasing the number of resonance lines from 500 to 100,000, the risetime for models by Harkness (1991) increased by 8 days."

By including the additional lines, Harkness decreased the escape probability for thermal photons, which in our language means increasing y. This means not only *atmospheric* opacity, but opacity at all depths. Because the opacity is strongly frequency dependent (Pinto & Eastman 1999), the atmosphere does not have a sharply defined radius. This makes the escape probability view a clearer one. If maximum light occurs at 19.5 days, about 3.2 half-lives, the ^{56}Ni has dropped to about 0.10 of its initial value, and most of the decayed Ni is in the form of Co. The line lists for Co and Ni are less extensive than for Fe, although these elements have comparably complex atomic and ionic states. While this may not affect spectral synthesis, which is often more sensitive to strong lines (which are likely to be in even poor line lists), it would affect the thermal escape probability.

If the spectral synthesis modelling of SN Iae is deficient in this way, then attempts to use these models to infer global properties of SN Iae will inevitably be biased, and their use for detailed inferences concerning distances and evolution of SN Iae is suspect. This argues for the relevance of the simpler approach pioneered by Dave Branch (see

Nugent et al. 1995), which focuses on the atmosphere, and for the systematic effort to understand the physics of the complex models in order to make them adequately robust (Baron et al. 1999; Pinto & Eastman 1999).

If we use a value of $y = 1.2$ (see above), then we expect $\Lambda_{\max} \approx 0.138$ and $t_{\max} \approx 17.5$ days. This is less than the value of 19.5 of Riess et al. (1999a), but it is an underestimate because heating from ^{56}Co decay will shift the maximum luminosity to later times. The luminosity at maximum is then $L/L_\odot = 32.0 \times 10^9$, or a B magnitude of -18.35. The additional Co luminosity brightens this by about -0.75 to -19.1. This is to be compared to $M_B = -19.45$ of Riess et al. (1999a), which is encouragingly consistent, given the crudeness of our analysis.

7. The Phillips relation

As is clear from the top panel in Figure 1, the Λ curves which peak earlier (the ones with smaller y), have larger values at peak. Thus, to the extent that the nickel mass M_{Ni} does not change with y (from event to event), we have an *anti-Phillips* relation (Phillips 1993). How should $M(Ni)$ change with y?

Let us begin by examining a carbon-oxygen white dwarf of near Chandrasekhar mass, which makes a transition to detonation after expanding to some lower central density ρ_{tran}. The material will be heated to a temperature which depends primarily upon its current density, and the energy available from burning. We may divide the star into layers, depending upon whether its peak temperature allows explosive carbon burning, oxygen burning, or silicon burning (see Woosley, Arnett, & Clayton 1973, Arnett 1996). We will ignore deviations of the mass-density structure from that of an $n = 3$ polytrope. The transition to detonation may occur in a violent deflagration or in the compressional phase following a mild deflagration. Details may vary from the simple model we use, depending upon the explosion mechanism and the progenitor characteristics.

The top panel of Figure 3 shows the final composition expected for a white dwarf, as a function of the central density it has when it detonates. At low density there is no burning, and the initial CO abundances are preserved. Carbon burning produces mostly O and Mg at these explosive temperatures. At still higher density, oxygen burning makes Si, S, Ar, and Ca (SiCa). At the highest density, nickel is the dominant ash.

If all these changes in composition, throughout the white dwarf, are converted into an implied explosion energy, we may relate the explosion energy to the mass of nickel produced. This is shown in the lower panel. At the lowest densities for detonation, the burning does not proceed through silicon burning to make Ni, but energy is released from burning up to SiCa. There is an abrupt rise to about 0.7 foe (10^{51} ergs), and then a gradual increase to over 1.3 foe, so

$$E_{\text{SN}}/(10^{51}\text{erg}) \approx 0.7 + (6/7)(M_{\text{Ni}}/M_\odot). \tag{7.10}$$

This is related to the velocity scale through the distribution of post-explosion velocity with density;

$$E_{\text{SN}} = \frac{1}{2}M<v^2>. \tag{7.11}$$

If the composition has no effect on the opacity, or more precisely, the thermal photon escape time, then the y parameter depends upon $M(Ni)$ through the velocity scale, or equivalently, the explosion energy. Thus,

$$y \propto \sqrt{\kappa M/v_{\text{sc}}} \to M^{3/4}/E_{\text{SN}}^{1/4}, \tag{7.12}$$

which is a relatively weak dependence on E_{SN}.

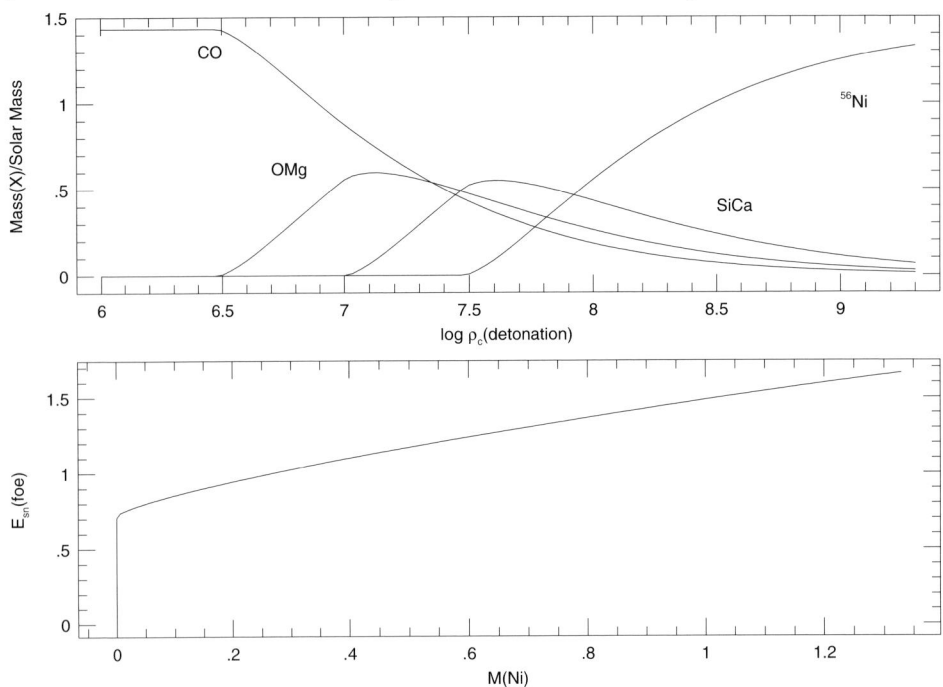

FIGURE 3. Polytropes $n = 3$ yield vs. central density at detonation.

Collecting results,

$$L = \epsilon_{\text{Ni}} M_{\text{Ni}} \Lambda_{\max}$$
$$\propto M_{\text{Ni}}/y$$
$$\propto y^3 \qquad (7.13)$$

where the last result assumes $E \gg 0.7$ foe. This gives the observed sense of the Phillips relation: brighter SNe have broader light curves. However, there is a potential difficulty here (Pinto & Eastman 1999): if the ^{56}Ni is distributed as a central sphere of pure Ni, then the average distance to the surface increases with M_{Ni}, and so does the probability of leakage from gamma and x-ray escape. This changes the light curve shape, and may destroy the sense of the Phillips relation. Alternatively, if the Ni distribution is not a strong function of M_{Ni} at a time several days after the explosion, the previous result holds. The production of a Phillips relation that agrees with observation depends upon the nature of the explosion assumed, but may not be difficult to get for some simple and attractive models.

8. Conclusions

The new data allow us to independently determine the parameters in the analytic models of SN Iae. The predictions for the premaximum behavior of the light curves are confirmed, and it will be possible to place new constraints on the nature of the progenitors and explosions. Even at the crude level sketched here, it is possible to get self consistent values, and a more accurate computation following this logic is warranted. In contrast, the radiative transfer models of Höflich & Khokhlov (1996) fail to give the correct rise time, probably due to incomplete line lists, and are likely to give biased results when applied to the description of such global properties of SN Iae, at least until this weakness is corrected.

Enlightening discussions with participants in this workshop and the Aspen Center for Physics workshop on Type Ia Supernovae are gratefully acknowledged. In particular, Pat Nugent, Phil Pinto, Paolo Mazzali, Adam Riess, Eddy Baron, Brian Schmidt, and Saul Perlmutter made specific comments which affected the development above. This work was supported in part by DOE grant DE-FG03-98DP00214/A001.

REFERENCES

ARNETT, D. 1980 *ApJ*, **237**, 541.

ARNETT, D. 1982 *ApJ*, **253**, 785.

ARNETT, DAVID 1996 *Supernovae and Nucleosynthesis*, Princeton University Press.

BARON, E., BRANCH, D., HAUSCHILDT, P. H., FILIPPENKO, A. V., & KIRSHNER, R. P. 1999 *ApJ*, in press.

HAMUY, M., PHILLIPS, M. M., MAZA, J., SUNTZEFF, N. B., SCHOMMER, R. A., & AVILES, R. 1995 *A.J.*, **109**, 1.

HARKNESS, R. P. 1991 in *SN1987A and Other Supernovae*, (eds. I. J. Danziger & K. Kjar). ESO.

HÖFLICH, P. & KHOKHLOV, A. 1996 *ApJ*, **457**, 500.

NUGENT, P., BRANCH, D., BARON, E., FIGHER, A., & VAUGHAN, T. 1995 *Phys. Rev. Lett.*, **75**, 394.

PERMUTTER, S., ET AL. 1997a in *Thermonuclear Supernovae*, (eds. P. Ruiz-Lapuente, J. Isern, & R. Canal), p. 749. Kluwer Publishers.

PERMUTTER, S., ET AL. 1997b *ApJ*, **483**, 565.

PERMUTTER, S., ET AL. 1999a *ApJ*, **517**, 565.

PERMUTTER, S. 1999b, this conference.

PHILLIPS, M. M. 1993 *ApJ*, **413**, L105.

PINTO, P. A. & EASTMAN, R. 1999 *ApJ*, in press.

RIESS, A. G., PRESS, W. H., & KIRSHNER, R. P. 1995 *ApJ*, **438**, L17.

RIESS, A. G., FILIPPENKO, A. V., LI, W., TREFFERS, R. R., SCHMIDT, B. P., QUI, Y., HU, J., ARMSTRONG, M., FARANDA, C., & THOUVENOT, E. 1999a *ApJ*, in press.

RIESS, A. G., FILIPPENKO, A. V., LI, W., & SCHMIDT, B. P. 1999b *ApJ*, in press.

SCHMIDT, B., ET AL. 1998 *ApJ*, **5507**, 46.

VACCA, W. D. & LEIBUNDGUT, B. 1996 *ApJ*, **471**, L37.

WOOSLEY, S. E., ARNETT, W. D., & CLAYTON, D. D. 1973 *ApJ Suppl.*, **26**, 231.

Abundances from supernovae

By F.-K. THIELEMANN,[1,2] F. BRACHWITZ,[1]
C. FREIBURGHAUS,[1] S. ROSSWOG,[1,3]
K. IWAMOTO,[4] T. NAKAMURA,[5]
K. NOMOTO,[5] H. UMEDA,[5]
K. LANGANKE,[6] G. MARTININEZ-PINEDO,[6]
D. J. DEAN,[2] W. R. HIX[2] AND M. S. STRAYER[2]

[1]Department of Physics and Astronomy, University of Basel, CH-4056 Basel, Switzerland

[2]Physics Division, Oak Ridge National Laboratory, Oak Ridge, TN 37831-4576, USA

[3]Center for Parallel Computing, University of Cologne, D-50931 Köln and German Aerospace Center (DLR), D-51147 Köln, Germany

[4]Department of Physics, Nihon University, Tokyo 101-8308, Japan

[5]Department of Astronomy and Research Center for the Early Universe, University of Tokyo, Tokyo 113-0033, Japan

[6]Institute of Physics & Astronomy, University of Aarhus, DK-8000 Aarhus C, Denmark

Supernovae of both types are responsible for essentially all of the element abundances between O and the Fe-group, and possibly for about 1/2 of the heavy elements beyond Fe. Therefore, they represent the most important nucleosynthesis source for the evolution of galaxies. Unfortunately our theoretical understanding is still burdened with many uncertainties. The present overview tries especially to clarify the link between model uncertainties and their nucleosynthesis imprint.

Type II supernovae (SNe II) are linked to massive stars (M > 8 $M_\odot$) at the end of their evolution. Gravitational collapse results in a central hot proto-neutron star which cools via neutrino emission. Neutrino opacities and transport determine the neutrino emission luminosity, which is responsible for the heating of outer layers, expected to cause the explosion and ejection of matter. Nucleosynthesis products are effected via two to three main uncertainties, (i) the location of the mass cut between the ejecta and the remaining neutron star, (ii) the total explosion energy responsible for explosive nucleosynthesis, and (iii) the possibility that explosion assymetries may cause severe effects on the composition of the innermost ejected layers. Observations of individual supernovae can constrain these quantities. The full theoretical understanding how (i)–(iii) depend on the progenitor mass is still missing. Connections to the chemical evolution of Mn, Cr, Co, Ni, and Fe can provide constraints.

Type Ia supernovae (SNe Ia) are explained by exploding white dwarfs in binary stellar systems. The favored systems are accreting white dwarfs, approaching the (maximum stable) Chandrasekhar mass before contraction and central ignition. Their major uncertainties are related (i) to the the accretion rate in the binary system which determines the carbon ignition density, (ii) the flame speed after central ignition, (iii) a possible transition from a subsonic deflagration to a supersonic detonation, and (iv) a possible dependence of (i)–(iii) on the metallicity of the object or the accreted matter. SNe Ia represent the major source of Fe-group nuclei.

Observations of r-process nuclei show already in low metallicity stars a solar r-process pattern (beyond Ba). Recent oservations of a non-solar r-process pattern for e.g. Ag, I, and Pd in these objects indicate the need for a second r-process component in the nuclear mass range A ≈ 80–120. Unfortunately, the stellar origins are not yet unambiguously defined. However, the existence of the heavy component at metallicities of −3 requires an origin related to evolution times comparable to those of massive stars. Theoretical options are SNe II, possibly even the recently discovered hypernovae (related to a specific class of gamma ray bursts), as well as the ejection of neutron star matter in binary mergers. The latter require delay times of 10^7–10^8 y in early galactic evolution.

1. Introduction

The present review focuses on nucleosynthesis contributions from supernovae and attempts to give a realistic presentation of our current knowledge and a critical assessment of the inherent uncertainties. We will not discuss here in detail the nuclear input for astrophysical calculations, i.e. experimental cross sections in hydrostatic burning stages of stellar evolution, electron captures in late burning stages and in stellar collapse, the nuclear equation of state and its influence on the supernova core collapse, neutrino transport and opacities in dense nuclear matter, the precision of cross section predictions for explosive burning, the properties (e.g. masses and half-lives) of nuclei far from beta-stability and resulting effects on abundance features of heavy nuclei. These topics are covered in recent reviews by e.g. Baraffe et al. (1997) and Käppeler et al. (1998); for an independent view see also Arnould & Takahashi (1999). Only if directly of major importance for the abundance yields, we will refer here to such aspects.

The discussion of supernova yields can (in principle) be performed on two levels. (a) Self-consistent, first-principle calculations of supernovae of both types which suffer, however, from inherent uncertainties in e.g. neutrino luminosities of core collapse supernovae, 2D vs. 3D treatment and the resolution of convective turnover, numerical methods of neutrino transport, neutrino opacities in nuclear matter, the equation of state at multiples of the nuclear matter density, and the progenitor systems of both massive stars (SNe II) and binary systems for SNe Ia. In SNe Ia instabilities and the multidimensional treatment of conductive and convective burning fronts determine the speed of deflagrations, as well as the possible transition to detonations. There has been major progress in recent years on all these topics, but present day self-consistent models still face problems in causing SNe II to explode and in understanding SN Ia progenitor systems. (b) Thus, being interested in the outcome and nucleosynthesis yields of such events, leads typically to a more pragmatic approach. SN II explosions are artificially induced via pistons or thermal bombs with predetermined explosion energies, mass cuts between the central neutron star and ejecta, and the assumption of spherical symmetry [which still seems to be a good description for the not innermost nucleosynthesis products]. The uncertainties in a realistic nuclear equation of state at high densities, which would automatically determine a maximum neutron star mass, is replaced by constraints on the maximum progenitor mass for successful SN II explosions. Similarly, ignition densities, the burning front speed, and the deflagration/detonation transition density are treated in SNe Ia as free parameters.

Under such assumptions (approximations) one can predict nucleosynthesis yields, but they are functions of the parameters employed (explosion energies and mass cuts in SNe II, or ignition densities, burning front speed, and transition densities in SNe Ia). Therefore, such parameters have to be calibrated in the best possible way. The most promising way would be the direct comparison with individual, observed supernova properties. However, often this is not possible and the test has to be performed via chemical evolution calculations. Both methods provide the chance that resulting abundance compositions from explosive burning can give clues to the detailed working of supernovae.

2. Nuclear burning in supernovae and their progenitors

Nuclear burning can in general be classified into two categories: (1) hydrostatic burning stages on timescales dictated by stellar energy loss and (2) explosive burning due to hydrodynamics of the specific event. Both are of importance for supernovae.

Hydrostatic burning stages are characterized by temperature thresholds, permitting Maxwell-Boltzmann distributions of (charged) particles (nuclei) to penetrate increasing Coulomb barriers: H-burning [conversion of ^{1}H into ^{4}He via pp-chains, initiated by ^{1}H$(p,e^+\nu)^2$H, or the CNO-cycle with the slowest reaction ^{14}N$(p,\gamma)^{15}$O], He-burning [^{4}He$(2\alpha,\gamma)^{12}$C (triple-alpha) and ^{12}C$(\alpha,\gamma)^{16}$O], C-burning [^{12}C$(^{12}$C$,\alpha)^{20}$Ne], and O-burning [^{16}O$(^{16}$O$,\alpha)^{28}$Si]. The alternative is that photodisintegrations start to play a role when $30kT \approx Q$ (the Q-value of the inverse capture reaction). This ensures sufficient photons with energies $> Q$ in the Planck distribution and leads to Ne-Burning [^{20}Ne$(\gamma,\alpha)^{16}$O, ^{20}Ne$(\alpha,\gamma)^{24}$Mg] at $T > 1.5 \times 10^9$ K (preceding O-burning) due to a small Q-value of ≈ 4 MeV and Si-burning at temperatures in excess of 3×10^9 K [initiated like Ne-burning by photodisintegrations]. The latter ends in a chemical equilibrium with an abundance distribution around Fe (nuclear statistical equilibrium, NSE), as Q-values of 8–10 MeV along the valley of stability permit photodisintegrations at these temperatures as well as the penetration of the corresponding Coulomb barriers.

In such an NSE the abundance of each nucleus $Y_{(Z,A)}$, with atomic weight A and charge Z, is only governed by chemical potentials and thus only dependent on temperature T, density ρ, its nuclear binding energy B, and partition function $G(T)$, while fulfilling mass conservation $\sum_i A_i Y_i = 1$ and charge conservation $\sum_i Z_i Y_i = Y_e$ (the total number of protons equals the net number of electrons). Y_e is changed by weak interactions (beta-decays and electron captures) on longer timescales. While still approaching NSE, different nuclear mass regions, usually separated by closed neutron or proton shells and small Q-values, can already be in equilibrium with the background of free neutrons, protons and alphas, but such quasi-equilibrium (QSE) clusters have total abundances which are offset from their NSE values (Thielemann & Arnett 1985, Woosley & Weaver 1995, Hix & Thielemann 1996, 1999).

Many of the hydrostatic burning processes occur also under explosive conditions at higher temperatures and on shorter timescales, but often the beta-decay half-lives of unstable products are longer than the timescales of the explosive processes under investigation. This requires in general the additional knowledge of nuclear cross sections for unstable nuclei (Rauscher et al. 1997, Thielemann et al. 1998, Rauscher & Thielemann 1999). Extensive calculations of explosive C, Ne, O, and Si burning have been performed for many years (Arnett 1995). The fuels for explosive nucleosynthesis consist mainly of $N = Z$ nuclei like ^{12}C, ^{16}O, ^{20}Ne, ^{24}Mg, or ^{28}Si, resulting in heavier nuclei, again with $N \approx Z$, unless densities are high enough to ensure substantial electron captures due to energetic, degenerate electrons.

Explosive Si-burning differs strongly from its hydrostatic counterpart and can be devided into three different regimes: (i) incomplete Si-burning and complete Si-burning with either (ii) a normal (high density, low entropy) or (iii) an alpha-rich (low density, high entropy) freeze-out of charged-particle reactions. At high temperatures or during a "normal" freeze-out, the abundances are in a full NSE. An alpha-rich freeze-out is caused by the inability of the triple-alpha reaction ^{4}He$(2\alpha,\gamma)^{12}$C, transforming ^{4}He into ^{12}C, and the ^{4}He$(\alpha n,\gamma)^9$Be reaction, to keep light nuclei like n, p, and ^{4}He, and intermediate mass nuclei beyond $A = 12$ in an NSE during declining temperatures, when the densities are small. This causes a large alpha abundance after freeze-out, which shifts abundances in QSE groups to heavier nuclei, transforming e.g. ^{54}Fe, ^{56}Ni, ^{57}Ni, and ^{58}Ni into ^{58}Ni, ^{60}Zn, ^{61}Zn, and ^{62}Zn. It leads also to a slow supply of ^{12}C still during freeze-out, leaving traces of alpha nuclei, ^{32}S, ^{36}Ar, ^{40}Ca, ^{44}Ti, ^{48}Cr, and ^{52}Fe, which did not fully make their way up to ^{56}Ni. This effect, most pronounced for SNe II, is a function of remaining alpha-particle mass fraction (or entropy).

In general, the initial hydrostatic composition also influences the ejected yields of explosively processing layers. A moderate influence on the fuels and Y_e is caused by the original metallicity of the stellar object. CNO isotopes are converted in H-burning to ^{14}N, this changes to ^{22}Ne in He-burning via alpha-captures and one beta-decay. As ^{22}Ne consists of 12 neutrons and 10 protons, we have here a surplus of neutrons, affecting slightly the products of explosive burning and acting also as neutron source via an (α, n)-reaction in hydrostatic core burning. This aspect can have an important impact on SN Ia products, originating from C/O white dwarfs, i.e. the ashes of He-burning. Later burning stages produce β^+-unstable fusion products which can further decrease Y_e (Thielemann & Arnett 1985) and affect the composition of explosive SN II ejecta (see section 3).

r-process nucleosynthesis (rapid neutron capture) relates to subsets of explosive Si-burning, either with low or high entropies and thus experiencing a normal or alpha-rich freeze-out. The requirement of a neutron/seed ratio of 10 to 150 after charged particle freeze-out translates for a normal freeze-out into a $Y_e = <Z/A>$ as low as 0.12–0.3, as found in neutron star matter (Rosswog et al. 1999ab, Freiburghaus et al. 1999b). Another option is an extremely alpha-rich freeze-out with a moderate $Y_e > 0.40$ (Takahashi et al. 1994, Woosley et al. 1994, Freiburghaus et al. 1999a). The inclusion of non-standard neutrino properties (McLaughlin et al. 1999) may also achieve low Y_es for intermediate entropies and a solar r-process pattern (beyond Ba). Recent oservations of a non-solar r-process pattern for e.g. Ag, I, and Pd in these objects indicate the need for a second r-process component in the nuclear mass range $A \approx 80$–120.

In SNe II the outer ejected layers are unprocessed by the explosion and contain results of prior H-, He-, C-, and Ne-burning in stellar evolution. If there exist differences in nuclear reaction rates between two stellar evolution calculations, they will enter directly into reaction flows and abundance differences. The interior parts of SNe II contain products of explosive Si, O, and Ne burning. A recent reaction rate sensitivity analysis by Hoffman et al. (1999) noticed that, due to chemical equilibria attained in explosive burning, the dependence on cross section uncertainties is strongly reduced in explosive O and Si-burning. The largest deviation, of however only 22% (while the reaction rates involved differed by factors), was encountered for ^{44}Ti, a nucleus which is strongly affected by an alpha-rich freeze-out. Thus, if there exist differences in products of explosive burning between calculations of different groups, they trace back in most cases to model uncertainties but less to reaction rate uncertainties in explosive burning. The same holds true for SNe Ia which contain essentially only explosive burning products with vanishing amounts of unburned outermost layers. If observed via very-early time spectra, they might give important clues about the metallicity of accreted matter. There exists one exception to this rule. Due to the high density of matter in the innermost central regions of SNe Ia, which burns explosively and is ejected, electron captures can play an important role for the Y_e (the proton to nucleon ratio) of these mass zones and thus the isotopic composition of the Fe-group. Therefore, we will discuss in the SN Ia section model uncertainties together with existing uncertainties in electron capture rates on Fe-group nuclei.

The different sites discussed above for r-process nuclei, being either related to SNe II or neutron stars in binary systems, would cause a clearly different dependence on metallicity in galactic history and consequently have to be scrutinized further. The large scatter of the r/Fe ratio in low metallicity stars (see e.g. CS 22892–052, Sneden et al. 1996) raises doubts about a (typical) SN II origin from which one would also expect an r/(Si-Ca) and r/Fe ratio close to the SNe II average, unless there exist very strong variations from supernova to supernova.

3. Type II supernovae

All stars with main sequence masses M > 8 $M_\odot$ (Nomoto 1987, Hashimoto et al. 1993, Weaver & Woosley 1993) produce a collapsing core after the end of their hydrostatic evolution, which proceeds to nuclear densities. The total energy released, 2–3×10^{53} erg, equals the gravitational binding energy of a neutron star. Because neutrinos are the particles with the longest mean free path, they are able to carry away that energy in the fastest fashion as seen for SN1987A in the Kamiokande, IMB and Baksan experiments (Burrows 1990). The apparently most promising mechanism for supernova explosions is based on neutrino heating beyond the hot proto-neutron star via the dominant processes $\nu_e + n \to p + e^-$ and $\bar{\nu}_e + p \to n + e^+$ with a (hopefully) about 1% efficiency in energy deposition. The neutrino heating efficiency depends on the neutrino luminosity, which in turn is affected by neutrino opacities (Schinder 1990, Mezzacappa & Bruenn 1993, Keil & Janka 1995, Reddy & Prakash 1997, Burrows & Sawyer 1998, 1999, Reddy et al. 1999, Yamada et al. 1999, Pons et al. 1999). The explosion via neutrino heating is delayed after core collapse for a timescale of seconds or less. The exact delay time t_{de} and other aspects of the explosion mechanism are still uncertain and depend on Fe-cores from stellar evolution, the supranuclear equation of state and maximum neutron star mass, related to the total amount of gravitational binding energy release of the collapsed protoneutron star, the resulting total amount of neutrinos, and the time release (luminosity), dependent on neutrino transport via the correct numerical treatment, convective transport, and opacities (Herant et al. 1994, Janka & Müeller 1995, 1996, Burrows et al. 1995, 1996, Mezzacappa et al. 1998, Messer et al. 1998, Mezzacappa & Messer 1999, Liebendörfer et al. 2000).

The observational fact that many SNe Ibc or SNe II show polarized light emission (Wang et al. 1996, 1999) is an indication of a nonspherical explosion mechanism, which is either a complication of the existing spherical models—without changing the basic mechanism as such—or even a sign of the working of another core collapse origin (Khokhlov et al. 1999).

3.1. Induced explosion calculations

Although there are positive signs that with full GR, Boltzmann neutrino transport, and adequate hydrodynamics techniques (Liebendörfer et al. 2000) successful, self-consistent supernova explosions can be obtained, the still existing uncertainties of theoretical models suggest to make use of the fact that typical kinetic energies of 10^{51} erg are observed. Light curve as well as explosive nucleosynthesis calculations can be performed by introducing a shock of appropriate energy in the pre-collapse stellar model (Weaver & Woosley 1993, Woosley & Weaver 1995, Thielemann et al. 1996, Nomoto et al. 1997a). Present explosive nucleosynthesis calculations for SNe II are still based on such induced supernova explosions by either depositing thermal energy or invoking a piston with a given kinetic energy of the order 10^{51} erg. Induced calculations (lacking self-consistency) utilize the constraint of requiring ejected ^{56}Ni-masses from the innermost explosive Si-burning layers in agreement with supernova light curves, being powered by the decay chain ^{56}Ni-^{56}Co-^{56}Fe, which can also serve as guidance to the supernova mechanism (Thielemann et al. 1996, Nomoto et al. 1997a). However, it should be clear that—even if ^{56}Ni ejecta from a variety of progenitor masses are known—this is not a simple one parameter problem. First of all, already the assumption of spherical symmetry is restrictive. Secondly the explosion energy can vary as a function of progenitor mass (or implicit parameters like rotation). Thirdly, if the explosion mechanism is dependent on the progenitor star mass, the delay time between core collapse and explosion can vary as well, thus implicitly influencing the Y_e of the ejecta. In the following we want to show this in more detail.

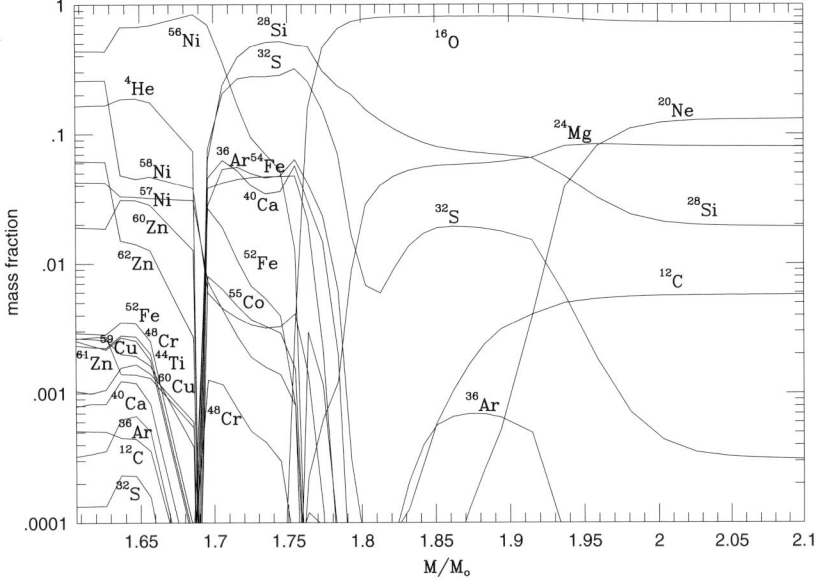

FIGURE 1. Isotopic composition for a core collapse supernova from a 20 $M_\odot$ progenitor star with a 6 $M_\odot$ He-core and a net explosion energy of 10^{51} erg, remaining in kinetic energy of the ejecta. The exact mass cut in $M(r)$ between neutron star and ejecta depends on the details of the delayed explosion mechanism. At $M(r) = 1.69$ $M_\odot$ one notices the transition from an alpha-rich freeze-out to incomplete Si-burning. For the relation between the abundance changes at $M(r) = 1.63$ $M_\odot$ and $Y_e(r)$ see Figure 2.

Fig. 1 shows the composition after explosive processing for a changing Y_e from 0.4989 to 0.494 in the inner ejecta, which experience incomplete and complete Si-burning. This Y_e originates from the pre-explosive hydrostatic fuel in these layers (see Fig. 2). Huge changes occur in the Fe-group composition for mass zones below $M(r) = 1.63$ $M_\odot$, where the abundances of ^{58}Ni and ^{56}Ni become comparable. All neutron-rich isotopes increase (^{57}Ni, ^{58}Ni, ^{59}Cu, ^{61}Zn, and ^{62}Zn), the even-mass isotopes (^{58}Ni and ^{62}Zn) show the strongest effect. One can also recognize the increase of ^{40}Ca, ^{44}Ti, ^{48}Cr, and ^{52}Fe with an increasing alpha-rich freeze-out, but a reduction of these $N = Z$ nuclei in the inner more neutron-rich layers.

In Fig. 2 we display the corresponding Y_e-distribution of a 20 $M_\odot$ star and the position of the outer boundary of explosive Si-burning with complete Si exhaustion, up to where predominantly ^{56}Ni is produced. This position changes as a function of the delay time t_{de} (here 0, 0.3, 0.5, 1, and 2s) between core bounce and explosion, because matter is accreted onto the proto-neutron star. Observed ^{56}Ni ejecta indicate the corresponding neutron star masses. The upper mass limit of neutron stars, due to the nuclear equation of state (Prakash et al. 1997), causes therefore also the upper mass limit of stars which undergo successful supernova explosions before the formation of central black holes sets in. Different maximum stable masses between the initially hot and a cold neutron star, possibly related to kaon condensates (Brown & Bethe 1994, Prakash et al. 1997), could result in a supernova explosion *and* afterwards the formation of a central black hole.

3.2. Observational constraints and the progenitor mass connection

There is limited direct observational information from optical, UV, and X-ray spectra, supernova lightcurves, as well as gamma-ray lines following nuclear decay for individual supernovae with known progenitors (SN 1987A, 1993J, 1997D, 1996N, 1994I), possible

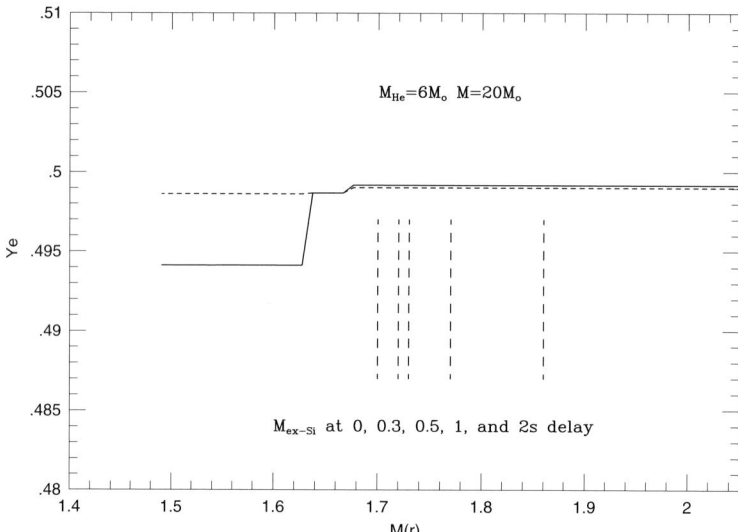

FIGURE 2. The Y_e-distributions in a 20 $M_\odot$ star (solid line) and the position of the outer boundary of complete Si-burning, as a function of the delay/accretion period t_{de}. A required amount of Ni-ejecta and the delay time t_{de} determine Y_e and the composition in the innermost ejected material. The steep drop in Y_e corresponds to the edge of core O-burning.

hypernovae (SN 1997ef, 1998bw), and supernova remnants (e.g. Cas A, G292.0+1.8, N132D) (e.g. Shigeyama & Nomoto 1990, Iwamoto et al. 1994, 1998, 2000, Turatto et al. 1998, Sollerman et al. 1998, Kozma & Fransson 1998, Bouchet et al. 1991, Suntzeff et al. 1992, Hughes & Singh 1994, Blair et al. 1994, Thielemann et al. 1996, Diehl & Timmes 1998). In those cases element abundances, like C, O, Si, Cl, Ar, Co, and Ni, and isotopic abundances, like ^{56}Ni, ^{57}Ni, ^{44}Ti, and ^{26}Al, are provided, which constrain the products of hydrostatic burning stages like C, O, Ne and Mg and the composition close to the mass cut between the central neutron star and the supernova ejecta (^{56}Co, ^{57}Co, ^{44}Ti). The latter are a measure of temperature, entropy, and Y_e close to the mass cut and give a consistent picture for neutrino heating delay times of 0.3–0.5s, explosion energies of the order 10^{51} erg, and a mass cut close to or outside the O-burning shell with minute permitted admixtures of deeper layers so that only matter with $Y_e \geq 0.497$–0.498 is ejected. Such constraints will have to be met by self-consistent hydro calculations.

The yields of ^{44}Ti, a $Z/A = 0.5$ nucleus (Iyudin et al. 1994, Dupraz et al. 1997, Kozma & Fransson 1998), based on recent half-life determinations (Norman et al. 1998, Görres et al. 1998, Ahmad et al. 1998), although dominantly depending on entropy, are also a function of Y_e. A change from 0.4989 to 0.4915 decreases its abundance by a factor of 2. Thus, as discussed above, it is also very important to know the correct Y_e in the ejecta. One might be tempted to explain larger ^{44}Ti abundances for too low Y_e-values with higher entropies alone (see e.g. the recent induced 2D simulations Nagataki et al. 1997, 1998, 1999), neglecting that the Ni isotopic composition changes with Y_e (^{58}Ni by a factor of 10, see Fig. 1). Here one should make use of the fact that the stable Ni/Fe ratio from optical and IR observations (mainly ^{58}Ni/^{56}Ni) as well as the gamma-ray observations (CGRO) of ^{57}Co/^{56}Co (from ^{57}Ni/^{56}Ni) are strong indicators that only ejecta with Y_e of the order 0.497–0.498 are allowed (e.g. Kumagai et al. 1991, Thielemann et al. 1996). Multi-D calculations are very important in future investigations, but they should be analyzed in comparison to *all* available observations.

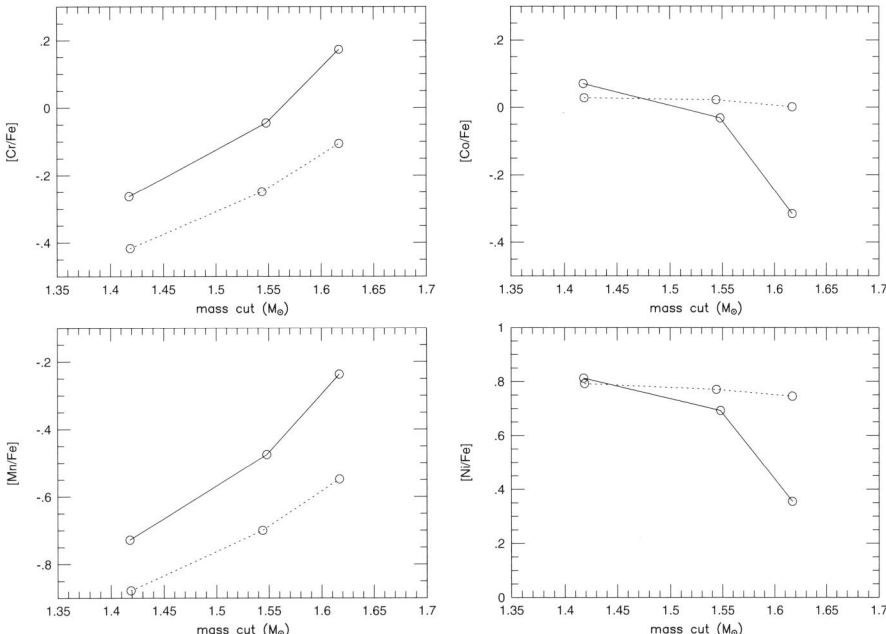

FIGURE 3. Dependence on mass cut. Three models with different mass cuts are used for each line. The solid line indicates a 6 $M_\odot$ He-core (corresponding to a 20 $M_\odot$ star), the dotted line represents a 8 $M_\odot$ He-core (corresponding to a 25 $M_\odot$ star). The position of the mass cut between the remaining neutron star (baryonic mass) and the ejecta is given on the abzissa. In both cases the explosion energy is assumed to be 10^{51} erg. The innermost Y_e affecting the ejecta for deep mass cuts is 0.4940 in case of the 20 $M_\odot$ star and 0.4985 for the 25 $M_\odot$ star.

These general features have probably been discussed in most detail for a 20 $M_\odot$ progenitor star corresponding to SN1987A (Hashimoto et al. 1989, Thielemann et al. 1990). We have to investigate how these features might depend on the supernova progenitor mass and possibly other parameters. Fig. 1 indicates in addition to the aspects discussed above that ^{55}Co (decaying to ^{55}Mn) and ^{52}Fe (decaying to ^{52}Cr) are products of incomplete Si-burning, while ^{59}Cu (decaying to ^{59}Co) is resulting from alpha-rich freeze-out. As these nuclei are the only (or dominant) stable isotopes of the respective element, this fact could be related to the Mn, Cr, Co, and Fe abundances observed in very low metallicity stars (McWilliam 1997), if interpreted as a change in mass cut position as a function of progenitor mass (Nakamura et al. 1999). Fig. 3 shows how a change in mass cut can affect the ratio of ejected (stable) Mn/Fe, Cr/Fe, Co/Fe, and Ni/Fe for two models related to 20 and 25 $M_\odot$ supernova explosions. The effect is very similar for both masses, as the nucleosynthesis layering is in all cases similar to that of Fig. 1. The difference for Co and Ni is related to the fact that the nuclei responsible (^{59}Co and ^{58}Ni) are neutron-rich and the Y_e structure for both stellar models differs somewhat. Mn and Cr show clearly the expected sign observed in galactic evolution via abundances in low metallicity stars, if the mass cut as a function of stellar progenitor mass leads to (at least initially) increasing Fe (^{56}Ni) ejecta. This requires the additional assumption that galactic evolution involves instantaneous mixing of ejecta or that more massive supernovae mix their ejecta with a larger amount of interstellar matter (see for details Nakamura et al. 1999).

Fig. 4 gives an idea what we know observationally about this fact from direct supernova observations and interpretations via synthetic spectra. We expect for low mass SNe II

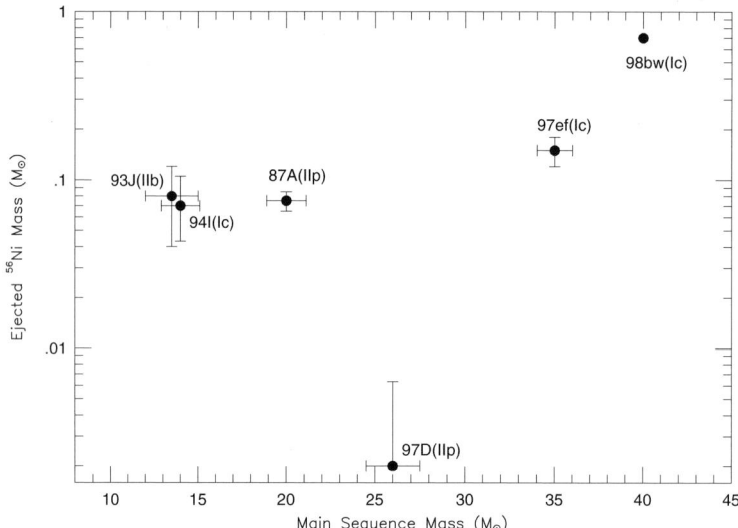

FIGURE 4. Ejected ^{56}Ni mass as a function of main sequence mass, as estimated for SN 1997A (Shigeyama & Nomoto 1990), SN 1993J (Shigeyama et al. 1994), SN 1994I (Iwamoto et al. 1994), SN 1997D (Turatto et al. 1998), SN 1997ef (Iwamoto et al. 2000) and SN 1998bw (Iwamoto et al. 1998).

(see e.g. the Crab where essentially only He enhancements are observed) small amounts of Ni(Fe) ejecta (not shown in Fig. 4 which only contains recent SN observations, where some clues about progenitor masses are available). Then a rise of Ni(Fe) ejecta to full blown SNe II occurs, before a decline sets in when massive stars cause a large amount of fall back via inverse shocks and large gravitational potentials. Such supernova explosions (SN 1997D) seem to avoid barely the creation of a black hole. With SN 1998bw and SN 1997ef, which appear in optical spectra like SNe Ic, the transition to hypernovae is encountered (failed supernovae which produce black holes and are probably powered by black hole accretion and jet ejection, see Iwamoto et al. 1998, 2000, Iwamoto 1999, MacFadyen & Woosley 1999). Their nucleosynthesis products have not yet been seriously analyzed nor included in chemical evolution studies.

Fig. 3 indicates that the above given explanation of Mn/Fe, Cr/Fe, and Co/Fe ratios faces the largest problems for Co (one needs a declining Co/Fe ratio, opposite to Cr/Fe and Mn/Fe for explaining low metallicity observations), unless Y_e depends strongly on the mass cut position. Such an interpretation encounters, however, also the ^{58}Ni overabundance discussed already above for low Y_e-values. Probably the Co contribution from hydrostatic burning and its dependence on metallicity has to be analyzed as well (Woosley & Weaver 1995, Hoffman et al. 1999).

3.3. A practioner's guide to SNe II abundances and their uncertainties

Here we want to give a short summary of SNe II nucleosynthesis results and discuss their key features, together with an assessment of the uncertainties by comparing independent calculations (Woosley & Weaver 1995, Thielemann, Nomoto, & Hashimoto 1996, Nomoto et al. 1997a). The synthesized elements form three different classes which are sensitive to different aspects of stellar models and the supernova explosion mechanism, being (1) either fully determined by stellar evolution alone, (2) by stellar evolution plus the explosion energy, or (3) by details of the explosion mechanism, which includes aspects of stellar evolution determining the size of the collapsing Fe-core. These three classes are discussed in the following in more detail.

1: The abundances of C, O, Ne, and Mg originate from the unaltered (essentially only hydrostatically processed) C-core and from explosive Ne/C-burning. They are mainly dependent on the structure and zone sizes of the pre-explosion models resulting from stellar evolution. Due to large variations in stellar zones, the amount of ejected mass varies strongly over the progenitor mass range. O, Ne, and Mg change by a factor of 10–20 between a 13 $M_\odot$ and a 25 $M_\odot$ progenitor star. This behavior can vary with the treatment of stellar evolution and is strongly related to the amount and method of mixing in unstable layers. Woosley & Weaver (1995) employed the Ledoux criterion with semiconvection for Schwarzschild-unstable but Ledoux-stable layers. Nomoto & Hashimoto (1988) made use of the Schwarzschild criterion for convection (neglecting composition gradients) which ensures mixing over more extended regions than the Ledoux criterion. The Schwarzschild criterion causes larger convective cores (see also Chieffi et al. 1998, Straniero et al. 1999) which leads to larger ^{16}O, ^{20}Ne, and ^{24}Mg yields, the latter being also dependent on the ^{12}C(α,γ) rate (Langer and Henkel 1995). In addition, it is important to know the mixing velocity in unstable regions. Recent calculations by Umeda et al. (2000) within the diffusion approximation for mixing (Spruit 1992, Saio & Nomoto 1998), but with a remaining free paramter permitted to vary between 0 and 1, show that the the Weaver & Woosley results can be reproduced when choosing a rather small value of 0.05, which minimizes the extent of mixing. A correct treatment of rotation induces mixing as well via rotational (hydrodynamic) instabilities (Langer et al. 1997, Talon et al. 1997, Meynet & Maeder 1997, Heger et al. 1999). This can bring the models making use of the Ledoux criterion closer to those evolved with the Schwarzschild criterion and the compositions closer to those obtained with instantaneous mixing (high mixing velocities). Thus, the amount of mixing (being influenced by the mixing criterion utilized, the mixing velocity, and rotation) determines the size of the C/O core in stellar evolution. While the yield of O can be fixed with a combination of the still uncertain ^{12}C(α,γ) rate (Buchmann 1996) and a mixing description, the yields of Ne and Mg depend on the extent of mixing. Recent galactic evolution calculations (Thomas et al. 1998, Matteucci et al. 1999, Chiappini et al. 1999) prefer apparently a larger extent of mixing (caused by either of the effects mentioned above) in order to reproduce the observed Mg in low metallicity stars.

2: The mass of the elements S, Ar, and Ca, originating from explosive O- and Si-burning, is very similar for all massive stars in the Thielemann et al. (1996) models. This behavior is different from the strong progenitor mass dependence of C, O, Ne, and Mg. Si has some contribution from hydrostatic burning and varies by a factor of 2–3. Thus, the first set of elements (C, O, Ne, Mg) tests the stellar progenitor models, while the second set (Si, S, Ar, Ca) tests the progenitor models and the explosion energy, because the amount of explosive burning depends on the structure of the model plus the energy of the shock wave which passes through it. Present models make use of an artificially induced shock wave (see section 3.1) with a given kinetic energy. In our models this is an average energy of 10^{51} erg, known from remnant observations, which does not reflect possible explosion energy variations as a function of progenitor mass. The apparent underproduction of Ca seen in some chemical evolution calculations (e.g. Thomas et al. 1998, Matteucci et al. 1999, Chiappini et al. 1999) could apparently be solved by a progenitor mass dependent explosion energy.

3: The amount of Fe-group nuclei ejected (which includes also one of the so-called alpha elements, i.e. Ti) and their relative composition depends directly on the explosion mechanism, connected also to the size of the collapsing Fe-core. Three types of uncertainties affect the Fe-group ejecta, related to (i) the total amount of Fe (group) nuclei ejected (mostly measured by ^{56}Ni decaying via ^{56}Co to ^{56}Fe) and the mass cut between neutron star and ejecta, (ii) the total explosion energy which influences the entropy of the ejecta and the degree of alpha-rich freeze-out from explosive Si-burning which determines the abundances of radioactive ^{44}Ti as well as ^{48}Cr (^{48}Cr decays to ^{48}Ti and is responsible for elemental Ti), and (iii) finally the neutron richness or $Y_e = <Z/A>$ of the ejecta, depending on stellar structure and the delay time between collapse and explosion. Y_e influences strongly the ratio of the isotopes 57/56 in Ni(Co,Fe) and the overall elemental Ni/Fe ratio, the latter being dominated by ^{58}Ni and ^{56}Fe. The position of the mass cut, besides determining the total amount of ^{56}Ni, influences also the ratio of abundances from alpha-rich freeze-out and incomplete Si-burning. In this way the abundances of the elements Mn (^{55}Co decay), Cr (^{52}Fe decay) and Co (^{59}Cu decay) are affected as discussed in Nakamura et al. (1999).

Because at present, explosive nucleosynthesis calculations cannot yet rely on self-consistent explosion models, the position of the mass cut is in all cases an assumption and has mostly been normalized to observations of SN 1987A. Whether there is a decline in Fe-ejecta as a function of progenitor mass (as assumed in Thielemann et al. 1996) or actually an increase (Woosley & Weaver 1995) or a more complex rise, maximum and decline (Nakamura et al. 1999) is not really understood (see also Fig. 4). Thus, the results by Thielemann et al. show the correct IMF integrated behavior of e.g. Si/Fe, but one has to keep in mind that e.g. O/Fe, Mg/Fe, Si/Fe, Ca/Fe yields of individual supernovae could be quite uncertain and even show an incorrect progenitor mass dependence or a larger scatter than (yet unknown) realistic models. Ratios within the Fe-group (like e.g. Ni/Fe) have been obtained by mass cut positions which reproduce the solar ratios. Thus, the theoretical yields might show already the average values and a much smaller scatter than some observations (see e.g. Henry 1984, Argast et al. 1999). Later work attempted to choose mass cuts in order to represent in galactic evolution calculations some specific element trends like e.g. in Cr/Fe, Co/Fe or Mn/Fe (Nakamura et al. 1999).

In general we should keep in mind that as long as the explosion mechanism is not completely and quantitatively understood yet, one has to assume a position of the mass cut. Dependent on that position, which is a function of explosion energy and the delay time between collapse and final explosion, the total amount of Fe-group matter can vary strongly, Ti-yields can vary strongly due to the attained explosion energy and entropy, and the ejected mass zones will have a variation in neutron excess which changes relative abundances within the Fe-group, especially the Ni/Fe element ratio.

4. Some words on the r-process

SNe II have long been expected to be the source of r-process elements. Some recent calculations seemed to be able to reproduce the solar r-process abundances well in the high entropy neutrino wind, emitted from the hot proto-neutron star after the SN II explosion (Takahashi et al. 1994, Woosley et al. 1994). If the r-process originates from supernovae, a specific progenitor mass dependence has to be assumed in order to reproduce the r-process abundances in low metallicity stars as a function of [Fe/H] (Mathews et al. 1992, Wheeler et al. 1998), which shows a delayed emergence of r-process matter in galactic evolution (McWilliam et al. 1995, McWilliam 1997), requiring SNe II with small progenitor mass as r-process source. Such a "hypothetical" r-process yield curve for SNe II has been constructed by Tsujimoto & Shigeyama (1998), in agreement with ideas of Ishimaru & Wanajo (1999) and Travaglio et al. (1999). However, we should keep in mind that present-day supernova models have difficulties to reproduce the entropies required for such abundance calculations. In addition, they could exhibit the incorrect abundance features of lighter r-process nuclei (Freiburghaus et al. 1999a). The inclusion of non-standard neutrino properties (McLaughlin et al. 1999) may perhaps achieve low enough Y_es for intermediate entropies to correct for such unwanted features. However, recent observations shed some doubts on the supernova origin. On average SNe II produce Fe to intermediate mass elements in ratios within a factor of 3 of solar (Thielemann et al. 1996). If they would also be responsible for the r-process, the same limits should apply. But the observed bulk r-process/Fe ratios vary widely in low metallicity stars. In CS 22892−052 the r/Fe ratio is 30 times solar (Sneden et al. 1996)!

The previous discussion underlined that the question whether we understand fully all astrophysical sites leading to an r-process is not a settled one. From meteoritic abundances and observations in low metallicty stars we know by now that at least two r-process sources have to contribute to the solar r-process abundances (Wasserburg et al. 1996, Qian et al. 1998, Cowan et al. 1999b). The observed non-solar r-process pattern for e.g. Ag, I, and Pd in some objects indicates the need for a second r-process component in the lower mass range A $\approx$ 80–120, in addition to the main process which provides a solar r-process pattern beyond Ba. It is not exactly clear which of the two processes is related to SNe II and which one is related to possible other sources.

Neutron star mergers or still other low entropy sites have been debated in the past (Lattimer et al. 1977, Meyer 1989, Eichler et al. 1989). Recent calculations (Rosswog et al. 1999ab, Ruffert & Janka 1999) show that on average about 10^{-2} $M_\odot$ of neutron-rich matter are ejected from such objects while preliminary calculations with assumptions on Y_e (Freiburghaus et al. 1999b) predict a solar-type r-process pattern for nuclei beyond A = 130. Lighter nuclei are depleted due to a long duration r-process with a large neutron supply in such neutron-rich matter, leading also to fission cycling. This seems (accidentally?) in accordance with the observed main r-process component. Given the frequency and amount of ejected matter, this component alone could be responsable for the heavy solar r-process pattern and also explain the large scatter of r/Fe elements found in low metallicity stars.

4.1. Constraints from solar and low metallicity abundance patterns

The observations of stellar spectra of low metallicity stars are all consistent with a solar r-abundance pattern for elements heavier than Ba, and the relative abundances among heavy elements do apparently not show any time evolution (Sneden et al. 1996, Cowan et al. 1997). This suggests that all contributing astrophysical events produce the same relative r-process abundances for the heavy masses, although a single astrophysical site will still have varying conditions in different ejected mass zones, leading to a superposition of individual components. High temperature r-process sites are characterized by quasi-equilibrium clusters (QSE, see section 2) which consist of isotopic chains in equilibrium with neutron reactions. A component is given by a combination of neutron number density n_n and temperature T (defining the neutron separation energy S_n where the maximum abundances in each isotopic chain can be found, i.e. the r-process path) for a duration time τ; or more physically for an adiabatic expansion: entropy, Y_e, and an expansion timescale τ. The physical conditions must vary smoothly, as expected from a single astrophysical site.

The site-independent classical analysis of Kratz et al. (1988, 1993), based on n_n, T, and τ, led to the conclusion that the r-process experienced a fast drop from equilibrium in order not to wash out the observed odd-even staggering in the lower mass range via slow freeze-out effects. A continuous superposition of components with neutron separation energies in the range 4–1 MeV on timescales of 1–2.5 s provided a good overall fit (for random superpositions and mathematical fits see Goriely & Arnould 1996, 1997). For the heavier elements beyond A = 130 this reduces to about $S_n = 3$–1 MeV. The beta-decay properties along contour lines of constant S_n towards heavy nuclei are responsible for the resulting abundance pattern. These are predominantly nuclei not accessible in laboratory experiments to date. Exceptions exist in the $A = 80$ and 130 peaks (Kratz et al. 1988, 1993, 1999) and continuous efforts are underway to extend experimental information in these regions of the closed shells $N = 50$ and 82 with radioactive ion beam facilities. Such classical r-process studies were extended to deduce necessary requirements for nuclear properties like masses, half-lives, and deformation (Thielemann et al. 1994, Chen et al. 1995, Pfeiffer et al. 1997, Kratz et al. 1999). A recent detailed analysis of the A = 206–209 abundance contributions to Pb and Bi isotopes from alpha-decay chains of heavier nuclei permitted for the first time also to predict abundances of nuclei as heavy as Th with reasonable accuracy (Cowan et al. 1999a, see also Qian et al. 1999). One of the major conclusions was the quest for the quenching of shell effects far from stability in order to avoid abundance deficiencies. Similar conclusions were drawn from more realistic, entropy-based calculations (Freiburghaus et al. 1999a) and are not changed due to neutrino interactions with matter (Qian et al. 1996, Meyer et al. 1998).

4.2. Stellar r-process sites

A different question is related to the actual astrophysical realization of such conditions. An r-process requires 10 to 150 neutrons per seed nucleus (in the Fe-peak or somewhat beyond) which have to be available to form all heavier r-process nuclei by neutron capture. For a composition of Fe-group nuclei and free neutrons that translates into a $Y_e = <Z/A>$ of 0.12–0.3. Such a high neutron excess for material in NSE and a normal freeze-out from Si-burning is only possible at high densities in neutron stars under beta equilibrium ($e^- + p \leftrightarrow n + \nu$, $\mu_e + \mu_p = \mu_n$) due to the high electron Fermi energies which are comparable to the neutron-proton mass difference (Meyer 1989).

Another option is an extremely alpha-rich freeze-out in complete Si-burning with moderate $Y_e > 0.40$, which was outlined in Takahashi et al. (1994), Woosley et al. (1994), Hoffman et al. (1997), and Freiburghaus et al. (1999a). However, explaining the r-process by ejecta of SNe II faces two difficulties: (i) whether the required entropies can really be attained in supernova explosions has still to be verified, (ii) the mass region 80–120 cannot be reproduced adequately. Whether the inclusion of specific non-standard neutrino properties (McLaughlin et al. 1999) which apparently can cure both difficulties is a realistic one has to be scrutinized.

The neutron star matter option mentioned first requires a mechanism to eject such material. A possible site is related to neutron star mergers. Interest in a scenario where a binary system, consisting of two neutron stars (ns-ns binary), looses energy and angular momentum through the emission of gravitational waves comes from various sides. Such systems are known to exist (Thorsett 1996). The measured orbital decay gave the first evidence for the existence of gravitational radiation (Taylor 1994). Further interest in the inspiral of a ns-ns binary arises from the fact that it is a prime candidate for the detection by gravitational wave detector facilities. A merger of two neutron stars can also lead to the ejection of neutron-rich material and r-process elements. It is possible that such mergers account for all heavy r-process matter in the Galaxy (Lattimer et al. 1977, Eichler et al. 1989, Rosswog et al. 1999ab). The decompression of cold neutron star matter has been studied in the past (Lattimer et al. 1977, Meyer 1989). However, a hydro calculation coupled with a preliminary r-process study was only undertaken recently (Freiburghaus et al. 1999b).

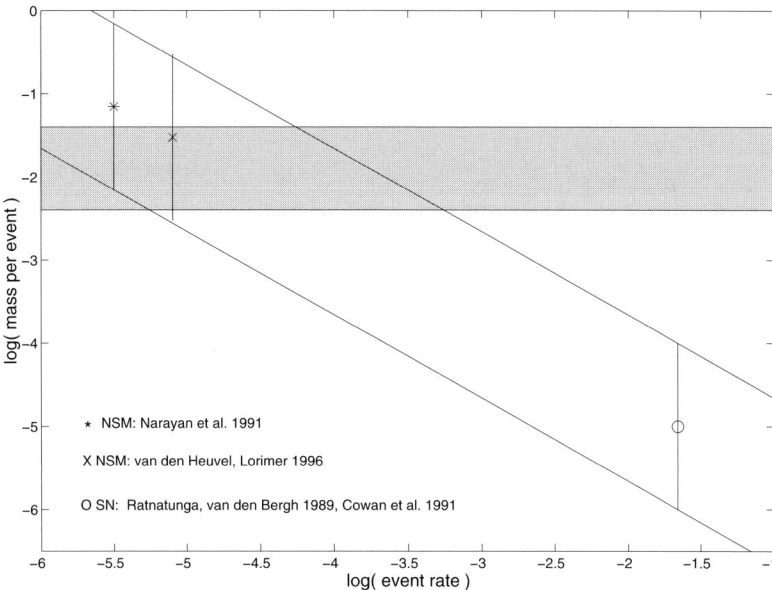

FIGURE 5. For a given galactic frequency of a specific r-process producing event, the total solar r-process abundances can be explained with ejecta masses within the two parallel lines. We include SNe II and neutron star mergers (with two frequency estimates). The shaded region corresponds to the predicted ejecta masses by Rosswog et al. (1999ab). Hypernovae, if producing high enough entropies and r-process matter, could be included as well.

The rate of neutron star mergers has been estimated to be of the order 10^{-5}y^{-1} per galaxy (Eichler et al. 1989, Narayan et al. 1992, van den Heuvel & Lorimer 1996, Bethe & Brown 1998). Simulations of neutron-star mergers should include 3D hydrodynamics, general relativitistic effects, a realistic equation of state, neutrino transport and neutrino cooling, and possible nuclear reactions (Davies et al. 1994, Janka & Ruffert 1996, Lombardi & Rasio 1997, Baumgarte et al. 1997, Ruffert & Janka 1998, 1999). Rosswog et al. (1999ab) have performed extensive ns-ns merger calculations (however only without GR effects and neutrino transport) and find typically ejecta of the order 10^{-2} M$_\odot$. Fig. 5 shows that this is sufficient to reproduce the solar system r-process abundances, provided that this matter is converted into an r-process abundance pattern. Alternatively, SNe II with their given galactic occurance frequency, would need to eject $\sim 10^{-5}$ M$_\odot$ per event.

There are several uncertainties related to the Y_e of the ejected and expanding matter in ns-ns mergers. In the co-rotating case, the ejected matter originates from the outermost layers of the neutron star. This can evolve of the order of 1% of the total neutron star mass of crust material with a Y_e up to ≈ 0.5 (Pethick & Ravenhall 1995, Haensel & Zdunik 1990a, Haensel & Zdunik 1990b). On the other hand the outer core layers (excluding the crust) are expected to have a $Y_e < 0.05$. Thus, the exact Y_e of the ejected matter is not fully determined unless a very fine resolution of the hydro calculation includes the correct properties of the neutron star surface. On the other hand, calculations with varying spins of the ns binaries also permit to eject matter from deeper layers, where a typical $Y_e \approx 0.1$ can be found. A part of this material gets also compressed and is heated when the stars come into contact. Thus, due to electron/positron capture reactions that become very effective at high temperatures, very different Y_e-histories might be encountered for different initial spins. These cases demand further careful investigations of the resulting abundance distributions. Finally our hydro calcu-

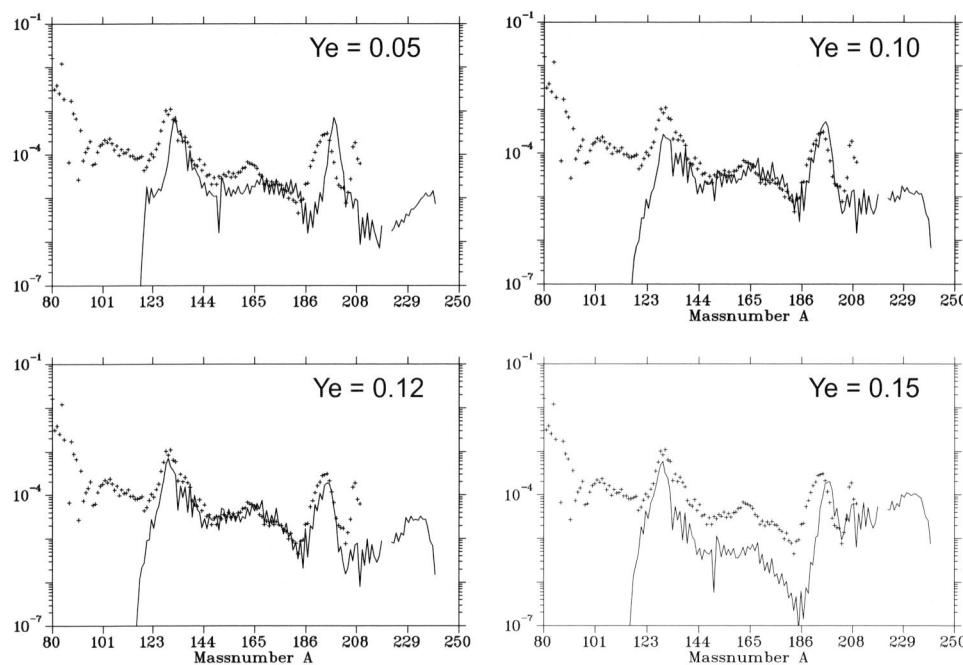

FIGURE 6. Calculated r-process distribution for different Y_es. In general one obtains useful contributions for $0.08 < Y_e < 0.15$. A further discussion is given in the text. Y_e determines the total neutron/seed ratio, which is an indication of the strength of the r-process. It affects also the combination of n_n and T, i.e. the r-process path, and therefore the position of peaks. Finally, fission cycling is responsible for the drop of abundances below A = 130, but only an improved incorporation of fission yields will provide the correct abundance distribution in this mass range.

lation did not yet include neutrino transport or neutrino captures, also having an effect on the Y_e evolution.

For that reason we used in Freiburghaus et al. (1999b) Y_e as a free parameter in a reasonable range (0.05–0.20; see Ruffert et al. 1997). The value of Y_e has basically two effects: (i) It determines the neutron to seed ratio which is responsible for the maximum nucleon number A produced, and (ii) it also determines the location (neutron separation energy) of the r-process path and thus the beta-decay half-lives encountered which influence the process speed and the nuclear energy release. Y_e values of 0.08 to 0.15 yield an almost perfect agreement with the observed r-process abundance distribution for $A > 130$ (see Fig. 6). Higher or lower values shift the abundance peaks to incorrect positions. High Y_e-values are not sufficient to achieve a full r-process. Also very low Y_e values (like e.g. 0.05) do not necessarily produce perfect (solar) r-process abundances, although a very strong r-process is encountered. In such a case the neutron/seed ratio is so large that neutrons are not exhausted until they decay with their natural half-life. Due to the decreasing density ρ during the expansion, the neutron number density $n_n = \rho N_A Y_n$ declines and leads to r-process paths approaching stability. This causes abundance peaks very close to (solar) s-process rather than r-rocess peaks, similar to calculations in the inhomogeneous big bang of Rauscher et al. (1994). Thus, neutron star mergers bear the possibility of a successful r-process site, but more detailed and self-consistent calculations are required to ensure the correct prediction of the Y_e of ejected matter.

Another interesting feature is that the Y_e-range which reproduces the peaks nicely also causes some amount of fission cycling, exhausting the seed abundances and the initial mass flow below $A = 130$, which leads to an underproduction of the low end of the solar r-process distribution. This, however, coincides exactly with the recent finding of observations in low metallicity stars that Ag and other abundances with $A < 130$ are underproduced, while abundances beyond Ba show a regular solar r-process pattern (Cowan et al. 1999b). This is consistent with evidence from meteoritic abundances that two r-process sites exist and one of them is responsible for all nuclei with $A > 130$ (Wasserburg et al. 1996, Qian et al. 1998). Neutron star merger ejecta could just play this role, but a a more self-consistent set of coupled hydro and r-process nucleosynthesis calculations is needed together with an implementation of improved (beta-delayed) fission rates and fission abundance yields.

5. Type Ia supernovae

There are strong observational and theoretical indications that SNe Ia are thermonuclear explosions of accreting white dwarfs (Wheeler et al. 1995, Höflich & Khokhlov 1996, Nomoto et al. 1997b, Livio 1999, Nomoto et al. 2000). Theoretical considerations lead to two options, the so-called single degenerate or double degenerate scenarios, involving either one or two C+O white dwarf(s) in a binary stellar system. The double degenerate scenario is related to the merging of two C+O white dwarfs with a combined mass exceeding the Chandrasekhar limit (Iben & Tutukov 1984). The single degenerate scenario relates to accretion of H or He via mass transfer from the binary companion (e.g. Nomoto 1982a). In the case of He accretion at low rates, He detonates at the base of the accreted layer before the system reaches the Chandrasekhar mass (Nomoto 1982b, Woosley & Weaver 1994, Livne & Arnett 1995, Arnett 1996). High accretion rates of H cause quasistable H-burning as well as subsequent He-burning in the accreted matter. This leads to an increase of the C+O white dwarf mass close to the Chandrasekhar mass limit before contraction and central ignition of C sets in. Both, the latter Chandrasekhar mass white dwarf models and sub-Chandrasekhar mass models have been considered so far (Nomoto et al. 1984, Canal 1997, Höflich et al. 1998). The Chandrasekhar vs. sub-Chandrasekhar mass issue has recently experienced some progress. Observational features of SNe Ia in early phases clearly indicate that Chandrasekhar models give a more consistent picture than the sub-Chandrasekhar models of helium detonations (Höflich & Khokhlov 1996, Nugent et al. 1997, Livio 1999).

Details of the Fe-group composition in Chandrasekhar mass models, however, depend on variations in the central ignition densities, subsonic deflagration front speeds, and possible transitions from deflagrations to supersonic detonations (Iwamoto et al. 1999) which can provide some improvements over the traditional model W7 (Nomoto et al. 1984). This relates to the overall agreement with solar abundances based on a mix of SNe II and averaged SNe Ia contributions. Luckily, different Fe-group isotopes depend in different ways on (i)–(iii). Direct abundance determinations in early and late time spectra as well as remnants of individual SNe Ia can also give constraints on the variation of these parameters. One interesting feature is e.g. the observed Ni/Fe ratio (Ruiz-Lapuente 1997, Liu et al. 1997, Mazzali et al. 1998, Dupke & White 1999), connected mostly to the deflagration speed. Ignition densities are related to accretion rates. Recent findings of supersoft X-ray sources make them potential progenitors of SN Ia events with high accretion rates and low ignition densities (Nomoto & Kondo 1991, Hachisu et al. 1996, 1999ab, Li & van den Heuvel 1997). The best available electron capture rates for Fe-group (pf-shell) nuclei should be utilized in order to relate ignition densities and the

Fe-group composition (Brachwitz et al. 1999). In addition, variations of SN Ia properties like maximum brightness and lightcurve decay time (Phillips et al. 1990, Hamuy et al. 1995, Riess et al. 1999, Perlmutter et al. 1999) have to be understood within this framework (Höflich & Khokhlov 1996, Dominguez & Höflich 1999). This also includes metallicity effects on the white dwarf (IMF and C/O ratio) and the accreted matter, possibly affecting galactic evolution as well cosmological use (Höflich et al. 1998, 1999, Umeda et al. 1999ab, Yoshi et al. 1996, Kobayashi et al. 1998, 2000). Very early time spectra could possibly help to identify the metallicity (Hatano et al. 1999, Lentz et al. 1999).

5.1. Ignition and burning front propagation

For Chandrasekhar mass models carbon ignition in the central region leads to a thermonuclear runway. High accretion rates cause a higher central temperature and pressure, favoring lower ignition densities. A flame front then propagates at a subsonic speed as a *deflagration wave* due to heat transport across the front (Nomoto et al. 1984). Here the most uncertain quantity is the flame speed which depends on flame front instabilities of various scales (Dominguez & Höflich 1999). Multi-dimensional hydro simulations of the flame propagation have been attempted by several groups, though the results are still preliminary (Khokhlov 1995, Niemeyer & Hillebrandt 1995, Niemeyer & Woosley 1997, Reinecke et al. 1999, Niemeyer 1999, Niemeyer et al. 1999, Livne 1999). These simulations have suggested that a carbon deflagration wave might propagate at a speed v_{def} as slow as a few percent of the sound speed v_{s} in the central region of the white dwarf. The nucleosynthesis consequences witness the actual burning front velocities and can thus serve as a constraint. After an initial deflagration in the central layers, the deflagration is assumed to turn into a detonation at lower densities (Khokhlov 1991, Woosley & Weaver 1994, Livne 1999). We will later discuss independent nucleosynthesis constraints for this deflagration/detonation transition.

In the deflagration wave, electron captures enhance the neutron excess. The amount of electron capture depends on both v_{def} (influencing the time duration of matter at high temperatures, and with it the availability of free protons) and the central density of the white dwarf ρ_{ign} (increasing the electron chemical potential or Fermi energy). The resultant nucleosynthesis in slow deflagrations (Khokhlov 1991) has some distinct features compared with faster deflagrations like W7 (Nomoto et al. 1984), thus providing important constraints on these two parameters. Initially slower deflagrations cause an earlier expansion of the outer layers with respect to the arrival of the burning front (information of the central ignition propagates with sound speed) and lead to lower burning densities for the outer deflagration and detonation layers. Our main aim is here to find the "average" SN Ia conditions responsible for their nucleosynthesis contribution to galactic evolution, especially for the Fe-group composition (Iwamoto et al. 1999).

This situation is complicated by the fact that electron capture on intermediate mass and Fe-group nuclei plays an important role in explosive burning. Electron capture affects the central electron fraction Y_e, which determines the composition of the ejecta from such explosions. Up to present, astrophysical tabulations based on shell model matrix elements were only available for light nuclei in the sd-shell (Fuller et al. 1980, 1982, 1985, Oda et al. 1994, Aufderheide et al. 1994). Recently new Shell Model Monte Carlo (SMMC) and large-scale shell model diagonalization calculations have also been performed for pf-shell nuclei (Dean et al. 1998, Martinez-Pinedo et al. 1999). These lead in general to a reduction of electron capture rates in comparison with previous, more phenomenological, approaches. Making use of the new shell model based rates, first results were reported by Brachwitz et al. (1999) for the composition of Fe-group nuclei

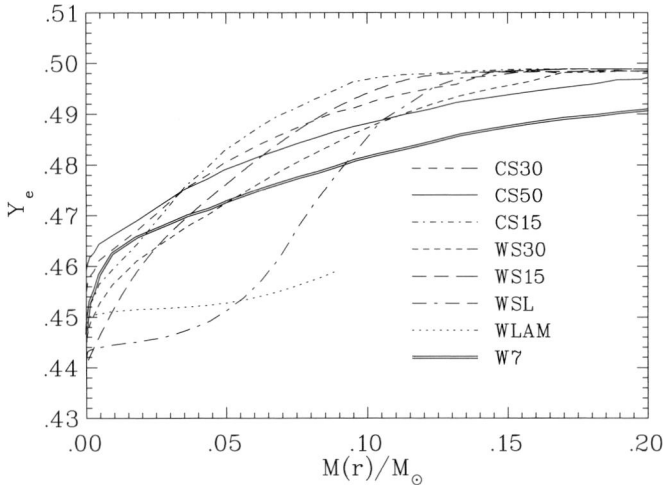

FIGURE 7. Y_e after freeze-out of nuclear reactions measures the electron captures on free protons and nuclei. Small burning front velocities lead to steep Y_e-gradients which flatten with increasing velocities (see the series of models CS15, CS30, and CS50 or WS15, WS30, and W7). Lower central ignition densities shift the curves up (C vs. W), but the gradient is the same for the same propagation speed. Only when the Y_e from electron captures is smaller than for stable Fe-group nuclei, subsequent β^--decays will reverse this effect (WSL and WLAM).

produced in the central regions of SNe Ia in combination with possible changes in the constraints on model parameters like ignition densities ρ_{ign} and burning front speeds $v_{\rm def}$.

The transition from a deflagration to a detonation (delayed detonation model) leads to a change in the ratio of incomplete Si-burning and complete Si-burning with either normal freeze-out or alpha-rich freeze-out. The latter is less alpha-rich than in SNe II due to the smaller entropies or higher densities which leaves an imprint on the Fe-group composition. We know from low metallicity stellar observations that the average Fe-group yields of SNe II differ from those of SNe Ia (e.g. in the elemental Cr/Fe, Mn/Fe or also O-Ca/Fe ratio). We will see that this fact gives the strongest constraints on the transition density ρ_{tr} from deflagrations to detonations.

5.2. The Fe-group composition in slow deflagrations

In order to show the effect of ignition densities we adopt two models with central densities of $\rho = 1.37$ (C) and 2.12×10^9 g cm^{-3} (W) at the onset of thermonuclear runaway and assume that a slow (S) deflagration propagates with speeds $v_{\rm def}/v_{\rm sound} = 0.015$ (WS15, CS15), 0.03 (WS30,CS30) or 0.05 (CS50) and consider also two extreme cases of fully and initially laminar flame fronts (WLAM,WSL).

Fig. 7 from Iwamoto et al. (1999b) summarizes the major options to change Y_e values in the central part of SNe Ia explosions: (i) the burning front speed $v_{\rm def}$, determining the Y_e gradient and (ii) the central ignition density ρ_{ign}, being inversely proportional to Y_e. These features hold true as long as Y_e values in explosive burning do not drop below 0.44, when competing β^--decays have also to be taken into account (case WSL and WLAM). Y_e values of 0.47–0.485 lead to dominant abundances of ^{54}Fe and ^{58}Ni, values between 0.46 and 0.47 produce dominantly ^{56}Fe, values in the range of 0.45 and below are responsible for ^{58}Fe, ^{54}Cr, ^{50}Ti, ^{64}Ni, and values below 0.43–0.42 are responsible for ^{48}Ca. The intermediate Y_e-values 0.47–0.485 exist for all models of the set shown in Fig. 7, but the related mass zones which experience these conditions depend on the Y_e-

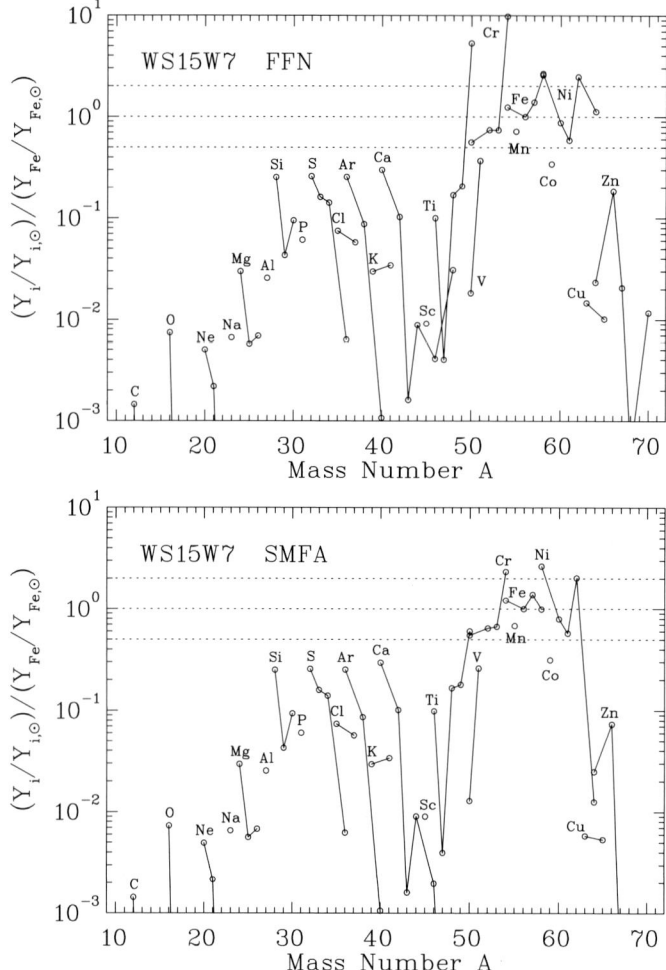

FIGURE 8. Ratio of abundances to solar predicted in model WS15 for different electron capture rate sets. Isotopes of one element are connected by lines. The ordinate is normalized to ^{56}Fe. Intermediate mass elements exist, but are underproduced by a factor of 2–3 for SNe Ia models in comparison to Fe-group elements. For FFN rates the Fe-group does not show a composition close to solar. Especially ^{54}Cr and ^{50}Ti are strongly overproduced by more than a factor of 3. The change from FFN rates (top) to SMFA (bottom) reduces the overproduction over solar strongly.

gradient and thus v_{def}. Whether the lower values with $Y_e < 0.45$ are attained, depends on the central ignition density ρ_{ign}. Therefore, ^{54}Fe and ^{58}Ni are indicators of v_{def} while ^{58}Fe, ^{54}Cr, ^{50}Ti, ^{64}Ni, and ^{48}Ca are a measure of ρ_{ign}.

If SNe Ia are responsible for at least 50% of the Fe-group nuclei in galactic evolution (see e.g. the discussion given in Iwamoto et al. 1999), the averaged SN Ia Fe-group contribution should in principle not exhibit isotopic overabundances with respect to ^{56}Fe by more than a factor of 2, independent of the theoretical models which try to explain SNe Ia. Observational uncertainties might lead to larger deviations. However, recent findings reported by Dupke & White (1999) of Ni/Fe ratios in SN Ia remnants close to three times solar have to be analyzed with possible selection effects in mind, perhaps not showing a metallicity average. If we follow the above given arguments on averaged type

Ia yields, and permit a limiting factor of 3 due to model and nuclear uncertainties, we can obtain limits for the model parameters v_{def} and ρ_{ign}.

^{54}Fe and ^{58}Ni can be produced within the factor of 3 uncertainty limits for the cases WS15, WS30, CS15, CS30, i.e. favoring v_{def} of 1.5–3% of the sound speed. ^{54}Cr is produced within a factor of two in CS15 and CS30, but apparently overproduced in WS15 and WS30, due to the too low central Y_es. ^{50}Ti is produced close to solar values for CS15, slightly underproduced by CS30, well produced for WS30, and clearly overproduced in WS15 (for full details see Iwamoto et al. 1999 and the top of Fig. 8). This exercise argues for CS15 and CS30 to be better models than WS15, WS30, W7 or CS50, in terms of avoiding overproduction of neutron-rich elements. The conclusions to be drawn from these results are that (i) v_{def} in the range 1.5–3% of the sound speed is preferred (cases 15 and 30 over 50), and (ii) ignition densities $\rho_{ign} < 2 \times 10^9$ g cm^{-3} provide apparently a better agreement with the solar abundances of very neutron-rich species (case C rather than W). The latter interpretation seems to change, however, with improved electron capture rates on pf-shell nuclei (Dean et al. 1998, Martinez-Pinedo et al. 1999) and which permit ignition densities slightly in excess of $\rho_{ign} = 2 \times 10^9$ g cm^{-3} (see bottom of Fig. 8). In either case, ignition densities in the range 1.5–2×10^9 g cm^{-3} are in agreement with identifying the recently discovered supersoft X-ray sources as SN Ia progenitor systems with hydrogen accretion rates close to $\dot{M} = 10^{-7}$ M$_\odot$ (Li & van den Heuvel 1997).

5.3. The effect of improvements in Fe-group electron capture rates

As the electron gas in white dwarfs is degenerate, characterized by high Fermi energies for the high density regions in the center, electron capture on intermediate mass and Fe-group nuclei plays an important role in explosive burning. Electron capture affects the central electron fraction Y_e, which determines the composition of the ejecta from such explosions. Recently new Shell Model Monte Carlo (SMMC) and large-scale shell model diagonalization calculations have also been performed for pf-shell nuclei (Dean et al. 1998, Martinez-Pinedo et al. 1999). These lead in general to a reduction of electron capture rates in comparison with previous, more phenomenological, approaches. Making use of these new shell model based rates, we present the first results from Brachwitz et al. (1999) for the composition of Fe-group nuclei produced in the central regions of SNe Ia and possible changes in the constraints on model parameters like ignition densities ρ_{ign} and burning front speeds v_{def}.

Four rate sets were employed: (i) The original FFN rates by Fuller et al. (1980, 1982) as a benchmark for further comparisons; (ii) inclusion of the electron captures rates calculated within the SMMC method by Dean et al. (1998), replacing the corresponding FFN rates. SMMC rates were used for the parent nuclei ^{45}Sc, ^{48,50}Ti, ^{51}V, ^{50,52}Cr, ^{55}Mn, $^{54-56,58}$Fe, and ^{55,57,59}Co, ^{56,58,60}Ni, otherwise rates were taken from FFN; (iii) to simulate potential modification of the rates not provided by the SMMC method, we also multiplied the FFN electron capture and beta-decay rates within the Fe-group nuclei by factors, derived from comparison between FFN and rates from large-scale shell model diagonalization calculations by Martinez-Pinedo et al. (1999, labeled with SMFA). (iv) A further option is to treat even-even (ee), odd-A (oa), and odd-odd (oo) nuclei in different ways, in order to test the sensitivity of the models and the importance of the rates of particular nuclei. Such calculations are denoted by SMFA with the corresponding extension ee, oa, oo or by combinations, e.g. ee+oa. With these modifications of the electron capture rates we recalculated the nucleosynthesis for the SN Ia models WS15 and CS15.

The resulting Y_e-curves (Fig. 9b) display a small Y_e shift between SMMC and SMFAee+oa, and a larger Y_e-shift between SMFAee+oa and SMFA. Therefore, the inclu-

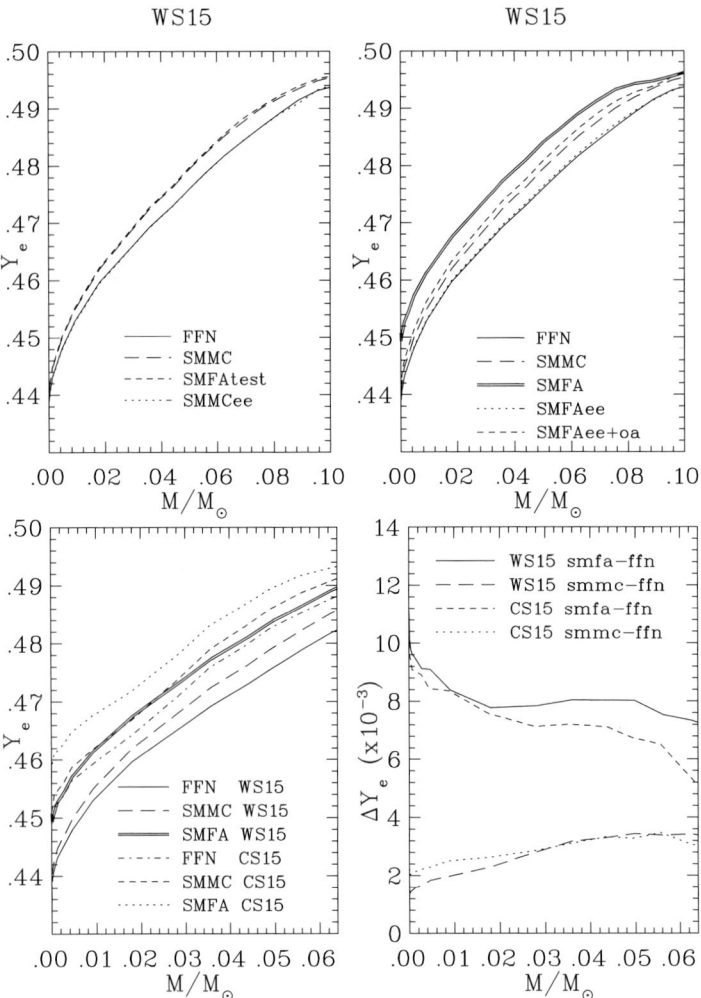

FIGURE 9. Y_e, the total proton to nucleon ratio and thus a measure of electron captures on free protons and nuclei, after freeze-out of nuclear reactions, as a function of radial mass for different models and electron capture rates. Also the Y_e-difference ΔY_e between various cases is shown at the bottom right (d). A detailed discussion of the changes with different electron capture rate sets [FFN = Fuller et al. (1980, 1982), SMMC = Dean et al. (1998), and SMFA = Martinez-Pinedo et al. (1999)] is given in the text.

sion of odd-odd nuclei has the largest influence on the Y_e difference between SMFA and SMMC. Thus, the rate change for odd-A nuclei is mostly responsible for the Y_e-shift between FFN and SMMC, and the inclusion of odd-odd nuclei causes the largest part of the Y_e-shift between SMMC and SMFA. This makes clear that the changes in the electron capture rates for odd-A and odd-odd nuclei are responsible for the Y_e difference between SMFA and FFN, while the contribution of even-even nuclei is negligible, an assertion which was directly tested by case SMFAee (Fig. 9b).

Notice, however, that the changes for a given model (here WS15 and CS15) lead to almost parallel Y_e-curves in the intermediate Y_e-range responsible for the major abundances of ^{54}Fe and ^{58}Ni. This can also be seen in the close to constant ΔY_e-curves in (Fig. 9d). Thus, a change in electron capture rates does (to first order) not affect the Y_e-gradient of a model. Iwamoto et al. (1999) showed that the Y_e gradient is determined by v_{def}. Therefore, we can conclude that the consequences for the permitted range of burning front speeds remain the same. In Iwamoto et al. (1999) we determined this range v_{def}/v_s to be of the order 0.015–0.03. The central neutronization is dependent on ρ_{ign}, as shown in the previous subsection, and on the set of electron capture rates employed (see Fig. 9c). If the trend as experienced between models 'C (CS15 $\rho_{ign} = 1.37 \times 10^9$ g cm^{-3}) and W (WS15, $\rho_{ign} = 2.12 \times 10^9$ g cm^{-3}) continues in a similar way, we would expect a central Y_e-value comparable to that of WS15 with FFN rates for $\rho_{ign} = 2.6 \times 10^9$ g cm^{-3} when utilizing SMFA. This corresponds to a permitted ignition density increase by about a factor of 1.24 when shifting from FFN to SMFA rates.

5.4. Deflagration to detonation transition

Finally, if a deflagration turns into a detonation, the transition density ρ_{tr} affects the total amount of ^{56}Ni, the intermediate mass elements Si-Ca, and the ratio of different explosive Si-burning regimes like alpha-rich freeze-out to incomplete Si-burning (see Fig. 10 taken from Iwamoto et al. 1999). The ^{58}Ni and ^{54}Fe plateaus indicate the regions of alpha-rich freeze-out vs. incomplete Si-burning. The most obvious consequence of choosing different transition densities is the amount of ^{56}Ni produced in a SN Ia event (DD2 and DD1 stand for deflagration/detonation transitions occurring at densities of 2.2 and 1.7×10^7 g cm^{-3}). Höflich & Khokhlov (1996) found from light curve modeling and spectra that the typical ^{56}Ni mass should be in the range 0.5–0.7 M$_\odot$. This agrees with the original fast deflagration model W7 (Nomoto et al. 1984, Thielemann et al. 1986). Among the DD models it would ask for a value somewhere between DD1 and DD2 (closer to DD2). The amount of Si-Ca in comparison to Fe is too large in DD1 models in order to compensate during galactic evolution for the well known overproduction of Si-Ca in SNe II. Si/Fe ratios in SN Ia models put constraints on permitted Ia/(II+Ib) frequencies in order to obtain a solar mix combined with SN II contributions. DD2 seems to be closest to the present observational limits for this ratio by Cappelaro et al. (1997, see also van den Bergh & Tammann 1991).

Small transition densities favor larger amounts of matter which experience incomplete Si-burning. Low metallicity constraints (McWilliam 1997) require some overproduction of Mn (and Cr) in SNe Ia. These elements are mostly made as ^{55}Co and ^{52}Fe (decaying to Mn and Cr), which are favorably produced in incomplete Si-burning and would also require a deflagration/detonation transition between DD1 and DD2. (One should, however, realize that a fast deflagration like W7 can simulate this as well and in addition that these numbers would have to be rescaled or reinterpreted in multi-D calculations, Livne 1999). Thus, combining all requirements on transition densities from total ^{56}Ni-yields, Si/Fe and Ia/(II+Ib) ratios, as well as specific elements favored in incomplete Si-burning, we would argue for a transition density close to 2×10^7 g cm^{-3}, i.e. between models DD1 and DD2. One should, however, be careful with these constraints based on spherically symmetric approximations of the burning front. Full 3D calculations could possibly produce the required ratio of matter from incomplete Si-burning and complete Si-burning with alpha-rich freeze-out in a different way (Livne 1999).

6. Conclusions and outlook

This overview concentrated on nucleosynthesis processes in supernovae and their yields for galactic chemical evolution. Given the still existing problems with fully self-consistent calculations, parametrized explosion models were introduced for SNe II as well as SNe Ia and compared with a number of constraints from observations of individual supernovae as well as low metallicity stars which witness the past galactic evolution.

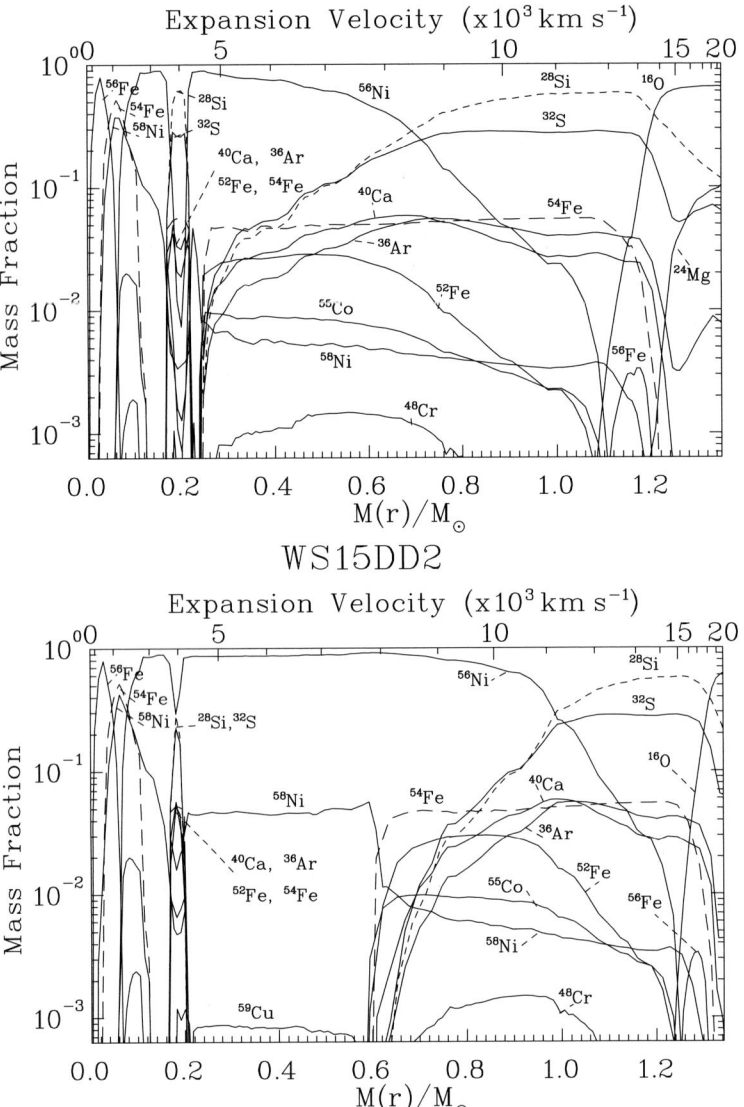

FIGURE 10. Composition of delayed detonation models WS15DD1 and DD2 as a function of expansion velocity or radial mass $M(r)$. In the series DD1–DD2 we see a decrease in the total amount of intermediate mass elements (Si-Ca), an increase in ^{56}Ni, and a change of the ratio between matter experiencing an alpha-rich freeze-out (indicated by the ^{58}Ni-plateau) and incomplete Si-burning (^{54}Fe-plateau).

SNe II yields have to pass the following tests. (i) One of the major issues is whether the average SNe II yields can reproduce the alpha element to Fe ratios ([x/Fe], x = O, Ne, Mg, Si, S, Ar, Ca) at metallicities [Fe/H] < −1. O, Ne, and Mg put severe constraints on the treatment of stellar evolution, S, Ar, and Ca measure also the SN explosion energy. (ii) The Fe-group elements witness more details of the explosion mechanism, like (a) mass cuts between the proto neutron star and ejecta, (b) the neutron excess in the innermost ejected layers, and (c) explosion energies. Three specific obervables of supernova explosions ^{56,57}Ni (^{56,57}Co) and ^{44}Ti [via light curve and gamma-ray observations] play a key role in the understanding of the explosion mechanism as a function of the progenitor mass. (iii) Variations in the above mentioned parameters (a–c) change the predictions for these unstable isotopes, observable in individual supernova explosions, but they might also explain the interesting trend of other Fe-group element abundance ratios, like [Cr/Fe], [Mn/Fe], [Ni/Fe], and [Co/Fe], of halo stars in the very low metallicity range $-4 <$ [Fe/H] < -2.5. They might be explained by a changing mix of nucleosynthesis products from explosive Si-burning with an alpha-rich freeze-out vs. incomplete Si-burning. (iv) Finally, we raised the question of the production of heavy elements in the r-process up to Th and U and possible connections to supernovae or other possible stellar sources.

Type Ia supernovae (SNe Ia) represent the major source of Fe-group nuclei, i.e. they have to explain the "solar" Fe-group composition in combination with SN II yields, in particular abundances of nuclei like e.g. ^{50}Ti, ^{52,54}Cr, ^{55}Mn, ^{54,56}Fe, and ^{58}Ni. Major uncertainties involved in Chandrasekhar mass models are: (i) the companion star of the accreting white dwarf (which determines the accretion rate and the carbon ignition density ρ_{ign}), (ii) the flame speed of the deflagration front after central C-ignition v_{def}, and (iii) the density ρ_{tr} at which a possible transition from deflagrations to detonations occurs. Nucleosynthesis calculations for relatively slow deflagrations with a variety of deflagration speeds and ignition denisities, combined with delayed detonations, can provide several constraints for these parameters as they apply to "average" progenitor systems. The Si-Ca/Fe ratio is another measure of the deflagration/detonation transition density and is observationally related to the SN Ia/SN II ratio.

We have tested the effect of improved electron capture rates for Fe-group (pf-shell) nuclei on the conclusions drawn from the considerations listed above. A systematic analysis of metallicity effects on SN Ia events and their outcome is still pending.

This work has been supported in part by the Swiss Nationalfonds (2000-53798.98), the U.S. Department of Energy (DOE contracts DE-AC05-96OR22464 and DE-FG02-96ER40983), the Danish Research Council, the grant-in-Aid for COE research (07CE2002) of the Ministry of Education, Science, and Culture in Japan, and a fellowship of the Japan Society for the Promotion of Science for Japanese Junior Scientists (6728). Some of us (KN and FKT) thank the Aspen Center for Physics for hospitality and inspiration during the 1999 Type Ia supernova program.

REFERENCES

AHMAD, I., ET AL. 1997 *Phys. Rev. Lett.*, **80**, 2550.

ARGAST, D., SAMLAND, M., GERHARD, O., THIELEMANN, F.-K. 1999 *A&A*, in press.

ARNETT, W. D. 1995 *Ann. Rev. Astron. Astrophys.*, **33**, 115.

ARNETT, W. D. 1996, in *Nucleosynthesis and Supernovae*, Princeton Univ. Press.

ARNOULD, M. & TAKAHASHI, K. 1999 *Rep. Prog. Phys.*, **62**, 395.

AUFDERHEIDE, M., FUSHIKI, I., WOOSLEY, S.E., & HARTMANN, D. 1994 *ApJS*, **91**, 389.

BARAFFE, I., ET AL. 1997, in *NuPECC Report on Nuclear and Particle Astrophysics*, http://quasar.physik.unibas.ch/nupecc/.

BAUMGARTE, T. W., COOK, G. B., SCHEEL, M. A., & SHAPIRO, S. L. 1997 *Phys. Rev. Lett.*, **79**, 1182.

BLAIR, W. P., RAYMOND, J. C., & LONG, K. S. 1994 *ApJ*, **423**, 334.

BOUCHET, P., DANZIGER, I. J., & LUCY, L. B. 1991 *AJ*, **102**, 1135.

BRACHWITZ, F., DEAN, D. J., HIX, W. R., IWAMOTO, K., KISHIMOTO, N., LANGANKE, K., MARTÍNEZ-PINEDO, G., NOMOTO, K., STRAYER, M.S., & THIELEMANN, F.-K. 1999 *ApJ*, in press.

BETHE, H. A. & BROWN, G. E. 1998 *ApJ*, **506**, 780.

BROWN, G. E. & BETHE, H. A. 1994 *ApJ*, **423**, 659.

BUCHMANN, L. 1996 *ApJ*, **468**, L127.

BUCHMANN, L. 1997 *ApJ*, **479**, L153.

BURROWS, A. 1990 *Ann. Rev. Nucl. Part. Sci.*, **40**, 181.

BURROWS, A., HAYES, J., & FRYXELL, B. 1995 *ApJ*, **450**, 830.

BURROWS, A. 1996 *Nucl. Phys.*, **A606**, 151.

BURROWS, A. & SAWYER, R. F. 1998 *Phys. Rev.*, **C58**, 554.

BURROWS, A. & SAWYER, R. F. 1999 *Phys. Rev.*, **C59**, 510.

CANAL, R. 1997 in *Thermonuclear Supernovae* (eds. P. Ruiz-Lapuente, R. Canal, J. Isern), p. 257. Kluwer Academic Publishers.

CAPPELLARO, E., TURATTO, M., TSVVETKOV, D. YU., BARTUNOV, O. S., POLLAS, C., EVANS, R., & HAMUY, M. 1997 *A&A*, **322**, 431.

CHEN, B., DOBACZEWSKI, J., KRATZ, K.-L., LANGANKE, K., PFEIFFER, B., THIELEMANN, F.-K., & VOGEL, P. 1995 *Phys. Lett.*, **B355**, 37.

CHIAPPINI, C., MATTEUCCI, F., BEERS, T. C., & NOMOTO, K. 1999 *ApJ*, **515**, 226.

CHIEFFI, A., LIMONGI, M., & STRANIERO, O. 1998 *ApJ*, **502**, 737.

COWAN, J. J., MCWILLIAM, A., SNEDEN, C., & BURRIS, D. L. 1997 *ApJ*, **480**, 246.

COWAN, J. J., PFEIFFER, B., KRATZ, K.-L., THIELEMANN, F.-K., BURLES, S., TYTLER, D., & BEERS, T. C. 1999 *ApJ*, **521**, 194.

COWAN, J. J., SNEDEN, C., IVANS, I., BURLES, S., BEERS, T. C., & FULLER, G. 1999 *Bull. AAS*, **194**, 67.04.

DAVIES, M. B., BENZ, W., PIRAN, T., & THIELEMANN, F.-K. 1994 *ApJ*, **431**, 742.

DEAN, D. J., ET AL. 1998 *Phys. Rev.*, **C58**, 536.

DIEHL, R. & TIMMES, F. X. 1998 *PASP*, **110**, 637.

DOMINGUEZ, I. & HÖFLICH, P. 1999 *ApJ*, in press, astro-ph/9908204.

DUPKE, R. A. & WHITE, R. E. 1999 *ApJ*, in press, astroph/9907343.

DUPRAZ, C., WESSOLOWSKI, U., OBERLACK, U., GEORGII, R., LICHTI, R., IYUDIN, A., SCHÖNFELDER, V., STRONG, A. W., BLOEMEN, H., MORNS, D., RYAN, J., WINKLER, C., KNÖDLSEDER, J., & VON BALLMOOS, P. 1997 *A&A*, **324**, 683.

EICHLER, D., LIVIO, M., PIRAN, T., & SCHRAMM, D. N. 1989 *Nature*, **340**, 126.

FREIBURGHAUS, C., REMBGES, F., RAUSCHER, T., KOLBE, E., THIELEMANN, F.-K., KRATZ, K.-L., PFEIFFER, B., & COWAN, C. 1999 *ApJ*, **516**, 381.

FREIBURGHAUS, C., ROSSWOG, S., & THIELEMANN, F.-K. 1999 *ApJ*, **525**, L121.

FULLER, G. M., FOWLER, W. A., & NEWMAN, M. 1980 *ApJS*, **42**, 447.

FULLER, G. M., FOWLER, W. A., & NEWMAN, M. 1982 *ApJS*, **48**, 279.

FULLER, G. M., FOWLER, W. A., & NEWMAN, M. 1985 *ApJ*, **293**, 1.

GÖRRES, J., ET AL. 1998 *Phys. Rev. Lett.*, **80**, 2554.

GORIELY, S. & ARNOULD, M. 1996 *A&A*, **312**, 327.

GORIELY, S. & ARNOULD, M. 1997 *A&A*, **322**, L29.

HACHISU, I., KATO, M., & NOMOTO, K. 1996 *ApJ*, **470**, L97.

HACHISU, I., KATO, M., & NOMOTO, K. 1999a *ApJ*, **519**, 314.

HACHISU, I., KATO, M., NOMOTO, K., & UMEDA, H. 1999b *ApJ*, **522**, 487.

HAENSEL, P. & ZDUNIK, J. L. 1990a *A&A*, **227**, 431.

HAENSEL, P. & ZDUNIK, J. L. 1990b *A&A*, **229**, 117.

HAMUY, M., PHILLIPS, M. M., MAZA, J., SUNTZEFF, R. A., & AVILES, R. 1995 *AJ*, **109**, 1.

HASHIMOTO, M., NOMOTO, K., & SHIGEYAMA, T. 1989 *A&A*, **210**, L5.

HASHIMOTO, M., IWAMOTO, K., & NOMOTO, K. 1993 *ApJ*, **414**, L105.

HATANO, K., BRANCH, D., FISHER, A., BARON, E., & FILIPPENKO, A. V. 1999 *ApJ*, in press.

HEGER, A., LANGER, N., & WOOSLEY, S. E. 1999 *ApJ*, in press, astro-ph/9904132.

HENRY, R. B. C. 1984 *ApJ*, **281**, 644.

HERANT, M., BENZ, W., HIX, W. R., FRYER, C. L., COLGATE, S. A. 1994 *ApJ*, **435**, 339.

HIX, W. R. & THIELEMANN, F.-K. 1996 *ApJ*, **460**, 869.

HIX, W. R. & THIELEMANN, F.-K. 1999 *ApJ*, **511**, 862.

HÖFLICH, P. & KHOKHLOV, A. 1996 *ApJ*, **457**, 500.

HÖFLICH, P., WHEELER, J. C., & THIELEMANN, F.-K. 1998 *ApJ*, **495**, 617.

HÖFLICH, P., NOMOTO, K., UMEDA, H., & WHEELER, J. C. 1999 *ApJ*, in press, astro-ph/9908226.

HOFFMAN, R. D., WOOSLEY, S. E., & QIAN, Y.-Z. 1997 *ApJ*, **482**, 951.

HOFFMAN, R. D., WOOSLEY, S. E., WEAVER, T. A., RAUSCHER, T., & THIELEMANN, F.-K. 1999 *ApJ*, **521**, 735.

HUGHES, J. P. & SINGH, K. P. 1994 *ApJ*, **422**, 126.

IBEN, I., JR. & TUTUKOV, A. V. 1984 *ApJS*, **54**, 335.

ISHIMARU, Y. & WANAJO, S. 1999 *ApJ*, **511**, L33.

IWAMOTO, K., NOMOTO, K., HÖFLICH, P., YAMAOKA, H., KUMAGAI, S., & SHIGEYAMA, T. 1994 *ApJ*, **437**, L115.

IWAMOTO, K., ET AL. 1998 *Nature*, **395**, 672.

IWAMOTO, K. 1999 *ApJ*, **512**, L47, **517**, L67.

IWAMOTO, K., BRACHWITZ, F., NOMOTO, K., KISHIMOTO, N., UMEDA, H., HIX, W. R., & THIELEMANN, F.-K. 1999 *ApJS*, **125**, 439.

IWAMOTO, K., NAKAMURA, T., NOMOTO, K., MAZZALI, P. A., DANZIGER, I. J., GARNAVICH, P., KIRSHNER, R., JHA, S., BALAM, D., & THORSTENSEN, J. 2000 *ApJ*, in press, astro-ph/9807060.

IYUDIN, A. F., DIEHL, R., BLOEMEN, H., HERMSEN, W., LICHTI, G. G., MORNS, D., RYAN, J., SCHÖNFELDER, V., STEINLE, H., VARENDORFF, M., DE VRIES, C., & WINKLER, C. 1994 *A&A*, **284**, L1.

JANKA, H.-T. & MÜLLER, E. 1995 *Phys. Rep.*, **256**, 135.

JANKA, H.-T. & MÜLLER, E. 1996 *A&A*, **306**, 167.

JANKA, H. T. & RUFFERT, M. 1996 *A&A*, **307**, L33.

KÄPPELER, F., THIELEMANN, F.-K., & WIESCHER, M. 1998 *Ann. Rev. Nucl. Part. Sci.*, **48**, 175.

KEIL, W. & JANKA, H.-T. 1995 *A&A*, **296**, 145.

KHOKHLOV, A. M. 1991 *A&A*, **245**, 114; **245**, L25.

KHOKHLOV, A. M. 1995 *ApJ*, **449**, 695.

KHOKHLOV, A. M., HÖFLICH, P. A., ORAN, E. S., WHEELER, J. C., WANG, L., & CHTCHELKANOVA, A. YU. 1999 *ApJ*, **524**, L107.

KOBAYASHI, C., TSUJIMOTO, T., NOMOTO, K., HACHISU, I., & KATO, M. 1998 *ApJ*, **503**, L155.

KOBAYASHI, C., TSUJIMOTO, T., & NOMOTO, K. 2000 *ApJ*, in press, astro-ph/9908005.

KOZMA, C. & FRANSSON, C. 1998 *ApJ*, **497**, 431.

KRATZ, K.-L., BITOUZET, J.-P., THIELEMANN, F.-K., MÖLLER, P., & PFEIFFER, B. 1993 *ApJ*, **402**, 216.

KRATZ, K.-L., PFEIFFER, B., & THIELEMANN, F.-K. 1998 *Nucl. Phys.*, **A630**, 352c.

Kratz, K.-L., Pfeiffer, B., Thielemann, F.-K., & Walters, W. B. 1999 in *Hyperfine Interactions*, in press, astro-ph/9907071.

Kumagai, S., Shigeyama, T., Hashimoto, M., & Nomoto, K. 1991 *A&A*, **243**, L13.

Langer, N., Fliegner, J., Heger, A., & Woosley, S. E. 1997 *Nucl. Phys.*, **A621**, 457c.

Langer, N. & Henkel, C. 1995 in *Nuclei in the Cosmos III* (eds. M. Busso, R. Gallino, & C. M. Raiteri), p. 413. AIP Press.

Lattimer, J. M., Mackie, F., Ravenhall, D. G., & Schramm, D. N. 1977 *ApJ*, **213**, 225.

Lentz, E. J., Baron, E., Branch, D., Hauschildt, P.H., & Nugent, P.E. 1999 *ApJ*, submitted.

Li, X. D. & van den Heuvel, E. P. J. 1997 *A&A*, **322**, L9.

Liebendörfer, M., et al. 2000 Ph.D. thesis, Univ. Basel, and to be publ.

Livio, M. 1999 in *Type Ia Supernovae: Theory and Cosmology*, Cambridge Univ. Press, in press.

Livne, E. 1999 *ApJ*, in press, astro-ph/9910471.

Livne, E. & Arnett, W. D. 1995 *ApJ*, **452**, 62.

Liu, W. H., Jeffery, D.J., & Schultz, D. R. 1997 *ApJ*, **483**, L107.

Lombardi, J. C. & Rasio, F. A. 1997 *Phys. Rev. D*, **56**, 3416.

MacFadyen, A. & Woosley, S. E. 1999 *ApJ*, **524**, 262.

Martinez-Pinedo, G., Langanke, K., & Dean, D. J. 1999 *ApJ*, in press.

Mathews, G. J., Bazan, G., & Cowan, J. J. 1992 *ApJ*, **391**, 719.

Matteucci, F., Romano, D., & Molaro, P. 1999 *A&A*, **341**, 458.

Mazzali, P., Cappelaro, E., Danziger, I. J., Turatto, M., Benetti, S. 1998 *ApJ*, **499**, L49.

McLaughlin, G. C., Fetter, J. M., Balantekin, A. B., & Fuller, G. M. 1999 *Phys. Rev.*, **C59**, 2873.

McWilliam, A. 1997 *Ann. Rev. Astron. Astrophys.*, **35**, 503.

McWilliam, A., Preston, G. W., Sneden, C., & Searle, C. 1995 *AJ*, **109**, 2757.

Messer, O. E. B., Mezzacappa, A., Bruenn, S. W., & Guidry, M. W. 1998 *ApJ*, **507**, 353.

Meyer, B. S. 1989 *ApJ*, **343**, 254.

Meyer, B. S., McLaughlin, G. C., & Fuller, G. M. 1998 *Phys. Rev.* **C58**, 3696.

Meynet, G. & Maeder, A. 1997 *A&A*, **321**, 465.

Mezzacappa, A. & Bruenn, S. W. 1993 *ApJ*, **405**, 637.

Mezzacappa, A., Calder, A. C., Bruenn, S. W., Blondin, N. J. M., Guidry, M. W., Strayer, M. R., & Umar, A. S. 1998 *ApJ*, **495**, 911.

Mezzacappa, A. & Messer, O. E. B. 1999 *J. Comp. Appl. Math.* **109**, 281.

Nagataki, S., Hashimoto, M., Sato, K., & Yamada, S. 1997 *ApJ*, **486**, 1026.

Nagataki, S., Hashimoto, M., Sato, K., Yamada, S., & Mochizuki, Y. S. 1998 *ApJ*, **492**, L45.

Nagataki, S. 1999 *ApJ*, **511**, 341.

Nakamura, T., Umeda, H., Nomoto, K., Thielemann, F.-K., & Burrows, A. 1999 *ApJ*, **517**, 193.

Narayan, R., Paczynski, B., & Piran, T. 1992 *ApJ*, **395**, L83.

Niemeyer, J. C. & Hillebrandt, W. 1995 *ApJ*, **452**, 769.

Niemeyer, J. C. & Woosley, S. E. 1997 *ApJ*, **475**, 740.

Niemeyer, J. C., Bushe, W. K., & Ruetsch, G. R. 1999 *ApJ*, **524**, 290.

Niemeyer, J. C. 1999 *ApJ*, **523**, L57.

Nomoto, K. 1982a *ApJ*, **253**, 798.

Nomoto, K. 1982b *ApJ*, **257**, 780.

NOMOTO, K. 1987 *ApJ*, **322**, 206.

NOMOTO, K. & HASHIMOTO, M. 1988 *Phys. Rep.* **163**, 13.

NOMOTO, K., HASHIMOTO, M., TSUJIMOTO, T., THIELEMANN, F.-K., KISHIMOTO, N., KUBO, Y., & NAKASATO, N. 1997a *Nucl. Phys.*, **A161**, 79c13.

NOMOTO, K., IWAMOTO, K., & KISHIMOTO, N. 1997b *Science*, **276**, 1378.

NOMOTO, K. & KONDO, Y. 1991 *ApJ*, **367**, L19.

NOMOTO, K., THIELEMANN, F.-K., & YOKOI, K. 1984 *ApJ*, **286**, 644.

NOMOTO, K., UMEDA, H., HACHISU, I., KATO, M., KOBAYASHI, C., & TSUJIMOTO, T. 2000 in *Type Ia Supernovae: Theory and Cosmology* (eds. J. Niemeyer & J. W. Truran). Cambridge Univ. Press, in press, astro-ph/9907386.

NORMAN, E. B., ET AL. 1998 *Phys. Rev.*, **C57**, 2010.

NUGENT, P., BARON, E., BRANCH, D., FISHER, A., & HAUSCHILDT, P. H. 1997 *ApJ*, **485**, 812.

ODA, T., HINO, M., MUTO, K., TAKAHARA, M., & SATO, K. 1994 *At. Data Nucl. Data Tables* **56**, 231.

PERLMUTTER, S., ET AL. 1999 *ApJ*, **517**, 565.

PETHICK, C. J. & RAVENHALL, D. G. 1995 *Ann. Rev. Nucl. Part. Sci.* **45**, 429.

PHILLIPS, M. M., HAMUY, M., MAZA, J., RUIZ, M. T., CARNEY, B. W., & GRAHAM, J. A. 1990 *PASP*, **102**, 299.

PFEIFFER, B., KRATZ, K.-L., & THIELEMANN, F.-K. 1997 *Z. Phys.*, **A357**, 235.

PONS, J. A., REDDY, S., PRAKASH, M., LATTIMER, J. M., & MIRALLES, J. A. 1999 *ApJ*, **513**, 780.

PRAKASH, M., ET AL. 1997 *Phys. Rep.* **280**, 1.

QIAN, Y.-Z., HAXTON, W. C., LANGANKE, K., & VOGEL, P. 1996 *Phys. Rev.* **C55**, 1532.

QIAN, Y.-Z., VOGEL, P., & WASSERBURG, G. J. 1998 *ApJ*, **494**, 285.

QIAN, Y.-Z., VOGEL, P., & WASSERBURG, G. J. 1999 *ApJ*, **524**, 213.

RAUSCHER, T., APPLEGATE, J. H., COWAN, J. J., THIELEMANN, F.-K., & WIESCHER, M. 1994 *ApJ*, **429**, 499.

RAUSCHER, T., THIELEMANN, F.-K., & KRATZ, K.-L. 1997 *Phys. Rev.* **C56**, 1613.

RAUSCHER, T. & THIELEMANN, F.-K. 1999 *At. Data Nucl. Data Tables*, in press.

REDDY, S. & PRAKASH, M. 1997 *ApJ*, **423**, 689.

REDDY, S., PRAKASH, M., LATTIMER, J. M., & PONS, J. A. 1999 *Phys. Rev.* **C59**, 2888.

REINECKE, M., HILLEBRANDT, W., & NIEMEYER, J. C. 1999 *A&A*, submitted.

RIESS, A. G., ET AL. 1999 *AJ*, **117**, 707.

ROSSWOG, S. K., LIEBENDÖRFER, M., THIELEMANN, F.-K., DAVIES, M. B., BENZ, W., & PIRAN, T. 1999a *A&A*, **341**, 499.

ROSSWOG, S. K., DAVIES, M. B., THIELEMANN, F.-K., & PIRAN, T. 1999b *A&A*, submitted.

RUFFERT, M., JANKA, H.-T., TAKAHASHI, K., & SCHÄFER, G. 1997 *A&A*, **319**, 122.

RUFFERT, M. & JANKA, H.-T. 1998 *A&A*, **338**, 535.

RUFFERT, M. & JANKA, H.-T. 1999 *A&A*, **344**, 573.

RUIZ-LAPUENTE, P. 1997 in *Thermonuclear Supernovae* (eds. P. Ruiz-Lapuente, R. Canal, & J. Isern), p. 681. Kluwer Academic Publishers.

SAIO, H. & NOMOTO, K. 1998 *ApJ*, **500**, 388.

SCHINDER, P. J. 1990 *ApJS*, **74**, 249.

SHIGEYAMA, T. & NOMOTO, K. 1990 *ApJ*, **360**, 242.

SHIGEYAMA, T., SUZUKI, T., KUMAGAI, S., NOMOTO, K., SAIO, H., & YAMAOKA, H. 1994 *ApJ*, **420**, 341.

SNEDEN, C., MCWILLIAM, A., PRESTON, G. W., COWAN, J. J., BURRIS, D. I., & ARMOSKY, B. J. 1996 *ApJ*, **467**, 819.

SOLLERMAN, J., LEIBUNDGUT, B., & SPYRIMILIO, J. 1998 *A&A*, **337**, 207.

STRANIERO, O., CHIEFFI, A., & LIMONGI, M. 1999 *ApJ*, in press.

SUNTZEFF, N. B., PHILLIPS, M. M., ELIAS, J. H., WALKER, A. R., & DEPOY, D. L. 1992 *ApJ*, **384**, L33.

SPRUIT, H. C. 1992 *A&A*, **253**, 131.

TAKAHASHI, K., WITTI, J., & JANKA, H.-T. 1994 *A&A*, **286**, 857.

TALON, S., ZAHN, J.-P., MAEDER, A., & MEYNET, G. 1997 *A&A*, **322**, 209.

TAYLOR, J. 1994 *Rev. Mod. Phys.*, **66**, 711.

THIELEMANN, F.-K. & ARNETT, W. D. 1985 *ApJ*, **295**, 604.

THIELEMANN, F.-K., HASHIMOTO, M., & NOMOTO, K. 1990 *ApJ*, **349**, 222.

THIELEMANN, F.-K., NOMOTO, K., & YOKOI, K. 1986 *A&A*, **158**, 17.

THIELEMANN, F.-K., KRATZ, K.-L., PFEIFFER, B., RAUSCHER, T., VAN WORMER, L., & WIESCHER, M. C. 1994 *Nucl. Phys.* **A570**, 329c.

THIELEMANN, F.-K., NOMOTO, K., & HASHIMOTO, M. 1996 *ApJ*, **460**, 408.

THIELEMANN, F.-K., ET AL. 1998 in *Nuclear and Particle Astrophysics* (eds. J. Hirsch, D. Page), p. 27. Cambridge Univ. Press.

THOMAS, D., GREGGIO, L., & BENDER, R. 1998 *MNRAS*, **296**, 119.

THORSETT, S. E. 1996 *Phys. Rev. Lett.*, **77**, 1432.

TRAVAGLIO, C., GALLI, D., GALLINO, R., BUSSO, M., FERRINI, F, & STANIERO, O. 1999 *ApJ*, **521**, 691.

TSUJIMOTO, T. & SHIGEYAMA, T. 1998 *ApJ*, **508**, L151.

TURATTO, M., ET AL. 1998 *ApJ*, **498**, L129.

UMEDA, H., NOMOTO, K., YAMAOKA, H., & WANAJO, S. 1999a *ApJ*, **513**, 861.

UMEDA, H., NOMOTO, K., KOBAYASHI, C., HACHISU, I., & KATO, M. 1999b *ApJ*, **522**, L43.

UMEDA, H., NOMOTO, K., & NAKAMURA, T. 2000 in *The First Stars* (eds. A. Weiss, et al.). Springer, in press, astro-ph/9912248.

VAN DEN BERGH, S. & TAMMANN, G. 1991 *ARA&A*, **29**, 363.

VAN DEN HEUVEL, E. P. J. & LORIMER, D. R. 1996 *MNRAS*, **283**, L37.

WANG, L., WHEELER, J. C., LI, Z. W., & CLOCCHIATTI, A. 1996 *ApJ*, **467**, 435.

WANG, L., WHEELER, J. C., & HÖFLICH, P. 1999 in *SN 1987A: Ten Years After* (eds. M. M. Phillips & N. Suntzeff). ASP, in press.

WASSERBURG, G., BUSSO, M., & GALLINO, R. 1996 *ApJ*, **466**, L109.

WEAVER, T. A. & WOOSLEY, S. E. 1993 *Phys. Rep.* **227**, 65.

WHEELER, J. C., COWAN, J. J., & HILLEBRANDT, W. 1998 *ApJ*, **493**, L101.

WHEELER, J. C., ET AL. 1995 *Phys. Rep.*, **53**, 221.

WOOSLEY, S. E. & WEAVER, T. A. 1994 *ApJ*, **423**, 371.

WOOSLEY, S. E. & WEAVER, T. A. 1995 *ApJS*, **101**, 181.

WOOSLEY, S. E., WILSON, J. R., MATHEWS, G. J., HOFFMAN, R. D., & MEYER, B. S. 1994 *ApJ*, **433**, 229.

YAMADA, S., JANKA, H.-T., & SUZUKI, H. 1999 *A&A*, **344**, 533.

YOSHI, Y., TSUJIMOTO, T., & NOMOTO, K. 1996 *ApJ*, **462**, 266.

SNe, GRBs and the global properties of the Universe

By BRIAN SCHMIDT

The Research School of Astronomy and Astrophysics, The Australian National University, Mount Stromlo Observatory, Private Bag, Weston Creek PO, ACT 2611 Australia; brian@mso.anu.edu.au

Supernovae and Gamma Ray Bursts are the most powerful events observed in the Universe. Both SN II and SN Ia are useful tools for measuring distances, providing some of the strongest results on the Hubble Constant, and SN Ia strongly indicate an accelerating Universe. In addition, there is mounting evidence that SNe and GRBs maybe connected, and the searches we use to discover SN Ia, can also shed light onto the field of GRB events. In this paper I will survey the use, limitations, and successes of SNe as distance indicators, and discuss the SN and GRB connection.

1. Supernovae, cosmology, and GRBs

Supernovae represent some of the most extreme physical situations in the Universe. Interest in their properties has crescendoed since the 1960s when it was realized that these events could be divided into two distinct classes, Type II and Type I; those objects with and without hydrogen, respectively. With this subdivision came a paradigm for the physical mechanism behind the explosions: Type II supernovae represent the core collapse of massive stars as their centers start to burn iron, whereas Type I supernovae seem to fit well with Chandrasekhar's predicted explosion of a 1.4 $M_\odot$ white dwarfs. This neat physical/observational separation was complicated during the mid 1980s when it was realized that the SN I class was polluted with core collapse explosions where a star had lost its hydrogen (and possibly helium) envelope. These stars are now called SN Ib (if they have a helium rich envelope) and SN Ic if (they have neither helium or hydrogen rich outer envelopes). These objects, despite evading detection for many decades, are usually easy to separate our from their SN Ia cousins, provided a good spectrum can be obtained (Filippenko 1997).

Type II Supernovae are well understood physical entities, and models (Eastman, Schmidt & Kirshner 1996) of these events can be used to predict their brightness based on a spectral time-series. Combining these models with observations provides a primary distance indicator, and work done by Schmidt, Kirshner & Eastman (1992) suggests it is possible to measure distances to individual SNe, provided they are observed regularly while young, with an accuracy of about 15%. This provides a value of the Hubble Constant independent of Cepheid distance scale, and in agreement with the current measurements based on the Hubble Space Telescope (Mould et al. 1999).

Type Ia supernovae have long been claimed as a panacea for the determining the extra galactic distance scale. Data obtained before 1985 indicated, when their maximum brightness was plotted against their hubble velocity, that all of the observed dispersion could be due solely to observational error; some authors even speculated that they might be a perfect standard candle. In the pre-CCD era, accurate supernova photometry was difficult to obtain, but recent samples (Riess et al. 1999a, Hamuy et al. 1996) of many objects have revealed that beneath the cloak of old observational uncertainty, lies a candle with a dispersion between 0.4 mag (B-band) and 0.2 mag (I-band), depending on the observed wavelength. Furthermore, these samples demonstrated, beyond a reasonable doubt, that a SN Ia's brightness is correlated with the shape of its light curve (Phillips 1993), and often with the strength of some spectral features (Nugent et al. 1995). After correction, SN Ia provide distances as good as any other available—about 0.12 mag (Jha et al. 1999). Their brightness, however, allows them to be used to redshifts greater than $z = 1$, and they occur in all galaxy types. So while not a panacea, SN Ia do provide one of the most powerful implements in the quiver of modern astronomers. These attributes have spawned two large teams who are using these objects to measure the expansion history of the Universe (Schmidt et al. 1998, Perlmutter et al. 1998). The conclusions reached by these two teams has provided astronomy with, at least from my perspective, an unexpected result; the Universe is accelerating! The key focus of research in this area is to now see if this result is right, or somehow based on a misguided interpretation of the SN data.

Over the past three years Gamma Ray Bursts (GRBs) have exploded onto the scene of astronomy as the largest bangs in the Universe since the big one. Observations of these objects are still poor, but improving rapidly. They paint a confused picture of a class of objects which seems to have many heads, but no body. Certainly there is strong evidence that some GRBs are unusual SNe explosions (Galama et al. 1998), but doubt remains outside the SNe community, and there are still many other viable progenitors. The study of SNe, both core collapse and SN Ia, can provide some clues to GRBs, but this is an area where future data should provide exciting breakthroughs.

In the following sections I will elaborate on the current state of each of these topics, providing commentary about possible problems and future avenues of work.

2. SNe as distance indicators

2.1. *Using Type II supernovae as distance indicators*

The Expanding Photosphere Method (EPM) assumes that the flux escaping from a SN II can be accurately modeled by the current generation of radiative transfer codes (Eastman, Schmidt & Kirshner 1996). These codes are still cumbersome, and it is difficult to model many individual events in a reasonable time frame. To help expedite using SN II as distance indicators, Schmidt, Kirshner & Eastman 1992 showed that SN II could be modeled as a spherical ball of gas which radiates like a dilute blackbody, where the dilution depends only on the photospheric temperature of the gas:

$$\theta = \frac{R}{D} = \sqrt{\frac{f_\nu}{\zeta(T)^2 \pi B_\nu(T)}}.$$

Here θ is the angular size of the SN, R is the radius of the SN's photosphere, D is the distance to the SN, $B_\nu(T)$ is the Planck function, f_ν is the observed flux density of the SN, and $\zeta(T)^2$ is a correction factor that accounts for the effects of flux dilution; for a blackbody $\zeta^2 = 1$. In a SN II, the stellar envelope undergoes free expansion (the expansion rate is more than 10 times the local escape velocity), and the radius of the photosphere, R, at time t, is just

$$R = v(t - t_0) + R_0.$$

The expansion velocity of the material that is instantaneously at the photosphere, v, can be determined from absorption minima in the supernova spectra (Eastman & Kirshner 1989). The radius of the SN progenitor at t_0, R_0, is usually negligible, and so we can combine these two equations to get

$$t = D\left(\frac{\theta}{v}\right) + t_0.$$

To determine θ, the temperature of the photosphere must be measured either from spectrophotometry or broad band photometry. This leaves two unknowns, D and t_0, which, with two or more epochs of observations, can be simultaneously determined.

Schmidt et al. (1994) have applied this technique to 14 SN II and find $H_0 = 73 \pm 12$ km s^{-1} Mpc. While this is not an amazingly precise measurement of the Hubble Constant, it is completely independent of all other distance measures, and agrees very well with the Cepheid scale as observed by the HST Key Project (Mould et al. 1999).

The physical platform on which EPM stands makes it an especially valuable method by which to measure the Hubble Constant. However, compared to other distance methods, it requires many hard to acquire observations, and is not terribly accurate (10–15% scatter in distance, possibly due to asphericity in SN II explosions). Therefore, this method is much harder (a factor of 10 to 100 times more effort is required) to use in research where statistical accuracy is essential. These include measuring the peculiar motions of galaxies or the acceleration of the Universe. However, its physical basis may make overcoming these hardships worth the effort in the future; EPM distances are unlikely to be hampered by evolution or other biases.

3. Using Type Ia supernovae as distance indicators

Kowal was the first to publish a Hubble diagram (observed magnitude as function of redshift) of SN I when, in 1968, he showed (Kowal 1968) that the scatter of these objects was approximately 30% in distance. We now realize that Kowal's diagram was tainted with SN Ib/c, which have a similar absence of hydrogen in their spectra (and remarkably similar light curves), but are significantly dimmer, on average, than SN Ia. Recent Hubble diagrams of SN Ia brightness show scatter about the Hubble line of about 15–25%, depending on the data source and the bandpass of the observations.

Modern detectors have enabled precise observations to be made of many nearby SN Ia, and using this data Phillips (1993) provided unequivocal proof that a SN Ia's brightness depends on the shape of its light curve: fainter SN Ia rise and fall faster than their brighter counterparts. While this feature sounds as though it may doom SN Ia as useful distance indicators, work by a team of astronomers based at the University of Chile and at Cerro Tololo showed otherwise.

Before the Calan/Tololo SNe Search, a systematic photographic search for SN Ia which used the Curtis Schmidt telescope at CTIO (Hamuy et al. 1993), there was essentially only photographic photometric information gathered of SN Ia beyond the Virgo Cluster. Between 1990 and 1993 this program detected and acquired high-quality spectroscopic and photometric follow-up for nearly 30 SN Ia at $0.01 > z > 0.1$. This data set (Hamuy et al. 1996) allowed the observational properties of SN Ia to be measured without hindrance from large observational uncertainties. The Calan/Tololo data set confirmed the relation seen by Phillips (1993), and showed that the brightness of SN Ia at maximum light had a scatter of 0.5, 0.3, 0.2 mags, respectively for B, V, and I band observations.

Although SN Ia exhibit a significant scatter, these objects also show a strong correlation between the rate at which their luminosity declines and their absolute magnitude (Hamuy et al. 1996). Using a technique based on that of Phillips (1993), who plotted the absolute magnitude of SN Ia versus the amount each SN decreased in brightness in the B band over the 15 days following maximum light ($\Delta m_{15}(B)$), Hamuy et al. (1996) showed the scatter in the Hubble diagram could be lowered to $\sigma \sim 0.15$ mag. Exploiting this correlation and data set, Riess et al. (1996) developed the Multicolor Light curve Shape Method (MLCS), which averages over the dataset and produces a series of light curve templates as a function of the absolute magnitude at maximum light. This technique can provide a statistically more rigorous estimate of the uncertainty in a SN distance measurement, and in principle, combines the observed data in an optimal manner. In practice the distance uncertainties of the two methods are indistinguishable, although the scatter between the methods' individual distance measurements is surprisingly large. Both methods also can be used to derive extinctions based on the color excess of the observations compared to non-reddened objects. Many other groups have since exploited this property of the SN Ia to create distance methods and while each has its own virtues, they all give a similarly small scatter for distances on the Hubble line. The bottom line here is that, not matter how you do it, SN Ia give good distances.

4. Measuring the global properties of the Universe

There are many methods which have been proposed to measure the global properties of the Universe. Essentially all make the fundamental assumptions that the Universe is isotropic and homogeneous on large scales, and obeys the equations of General Relativity. These methods can be broken up into three general classes:

• Angular Size Distances: The size of an object as a function of z depends on the matter content of the Universe. Observations of anisotropies in the Cosmic Microwave Background (CMB) (Hu 1996), extended Radio galaxies (Guerra, Daly & Wan 1998), and compact radio sources (Kellerman 1993) all utilize the angular size versus redshift relationship. These observations have not yet yielded results which the general community believe are individually definitive, but future observations and work have the potential to change this view rapidly.

• Volume versus redshift: The Volume of space as a function of z is sensitive to the matter content of the Universe. The number counts of galaxies as a function of z has the potential to trace the volume of space back in time, but these efforts have been hampered by galaxy evolution uncertainties (Shanks et al. 1984), although moving to the infrared may provide more leverage for this method (Yoshii & Peterson 1995). Using the frequency of gravitational lenses to measure cosmology also falls under this category, as this method exploits the sensitivity of number of lenses to the integrated volume of space. This method is primarily useful for putting limits on Λ (Kochanek 1996), because a large Λ produces a prodigious number of lenses. This method also requires a good understanding of the potentials of galaxies at $z > 0.5$.

• Brightness versus redshift: The apparent brightness of an object as a function of redshift has long been used to attempt to understand the global properties of the Universe (Sandage et al. 1961). Early investigations concentrated on Brightest Cluster Galaxies to $z < 0.5$, but these attempts were more abandoned when it was realized that the effects due to galaxy evolution were much larger than the differences due to cosmology (Tinsley 1972). Most distance methods used in the nearby Universe fall into this category, including type Ia and type II supernovae.

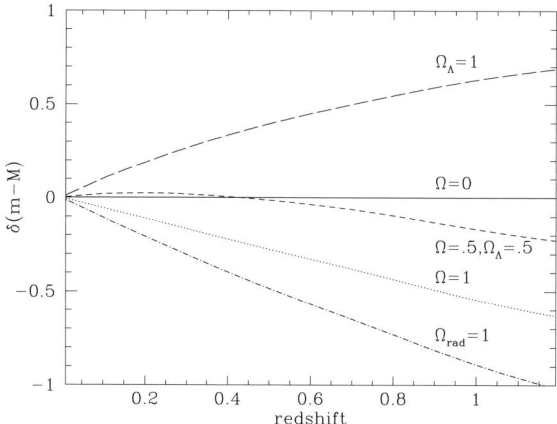

FIGURE 1. The difference in the $D_L : z$ relation for various cosmological models expressed as the difference in the distance modulus from an empty Universe, $\Omega_{\rm tot} = 0$. Ω: normal matter, $\Omega_{\rm rad}$: photons, Ω_Λ: Cosmological Constant

4.1. Luminosity distances

If we define the distance to an object such that $\mathcal{L}$ is the object's luminosity, and $\mathcal{F}$ is the flux of the object observed on Earth, the luminosity distance, D_L, is given by

$$D_L = \left(\frac{\mathcal{L}}{4\pi\mathcal{F}}\right)^{\frac{1}{2}}. \tag{4.1}$$

In a isotropic and homogeneous Universe, described by the Robertson-Walker metric and Friedmann equation, the luminosity distance to an object is given by

$$D_L H_0 = (1+z) |\kappa_0|^{-1/2} S\{|\kappa_0|^{1/2} \int_0^z dz' [\sum_i \Omega_i (1+z')^{3+3\alpha_i} - \kappa_0 (1+z')^2]^{-1/2}\}, \tag{4.2}$$

where $S\{x\} \equiv \sin(x)$, x, or $\sinh(x)$ for closed, flat, and open models ($k = 1, 0, -1$), respectively. H_0 is the current expansion rate of the Universe, given as a function of the scale factor of the Universe, a, $H_0 \equiv \dot{a}(t_0)/a(t_0)$. This equation assumes the matter content of the Universe is composed of a sum of components each having a fraction Ω_i of the current critical density $\rho_{\rm crit} \equiv 3H_0^2/8\pi G$ and various equations of state with density $\rho_i \propto ({\rm volume})^{-(1+\alpha_i)}$ [e.g., $\alpha = 0$ for normal matter (Ω_M), $\alpha = -1$ for a cosmological constant (Ω_Λ), $\alpha = +1/3$ for radiation ($\Omega_{\rm rad}$), $\alpha = -1/3$ for non-commuting strings (Ω_S)]. We have also utilized a parameter $\kappa_0 \equiv kc^2/[a(t_0)^2 H_0^2]$ which represents the scalar curvature in units consistent with the density parameters; the current physical radius of hypersphere curvature is $k^{-1/2}a(t_0) = \kappa_0^{-1/2} c H_0^{-1}$ and the definition of critical density gives $\kappa_0 = \sum \Omega_i - 1$.

It is conventional to define a "deceleration parameter" $q_0 \equiv -\ddot{a}(t_0) a(t_0)/\dot{a}^2(t_0)$, characterizing the low-redshift behavior, which can be expressed as

$$q_0 = \frac{1}{2} \sum_i \Omega_i (3 + 3\alpha_i) - \sum_i \Omega_i = \frac{1}{2} \sum_i \Omega_i (1 + 3\alpha_i). \tag{4.3}$$

When normal matter is the only contributor to Ω,

$$D_L H_0 = \frac{1}{q_0^2} \left[q_0 z + (q_0 - 1)(\sqrt{1 + 2q_0 z} - 1) \right]. \tag{4.4}$$

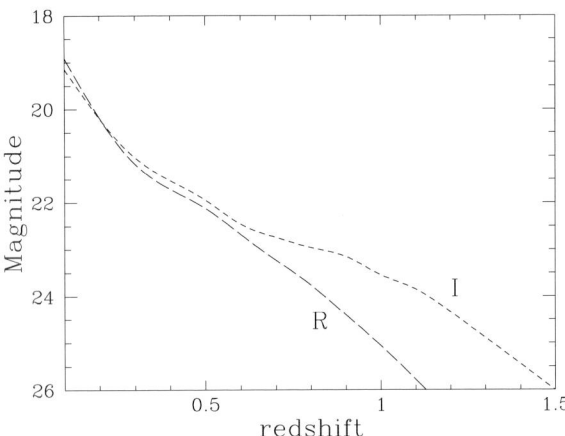

FIGURE 2. The magnitude at maximum of a typical SN Ia as a function of redshift in R_C and I_C for $\Omega_{\text{tot}} = 0$.

Alternatively, equation 4.2 can be expanded in z to give

$$D_L H_0 = z + z^2 \left(\frac{1-q_0}{2}\right) + \mathcal{O}(z^3), \tag{4.5}$$

which is valid for all models, but not terribly accurate for ($q_0 < 0$). The first term in this expansion is the Hubble law, and is the local rate of expansion of the Universe. In all plausible models, the expansion stays uniform to a few percent out to $z = 0.1$, and therefore objects closer than $z = 0.1$ are useful for measuring H_0.

Figure 1 shows the difference in magnitudes between different cosmologies as a function of redshift. If luminosity distances can be traced with 0.1 mag (5% error in distance) to $z \sim 0.5$ by averaging several objects, we should have every expectation of discerning the observations between the different curves shown on this diagram. In the next sections we will show that Type Ia supernovae can provide distances individually good to 0.15 mag (7% error in distance) and can be discovered to $z > 1$. And so with just a few objects, strong constraints can be put on the matter/energy content of the Universe. Type II SNe, with a scatter of 0.3 mags, also have the possibility of providing accurate enough distances to $z = 0.5$, but the degree of difficulty of the experiment is an order of magnitude harder: SN II provide distances half as accurate (providing 4 times less signal per object) as SN Ia, and are typically 5 to 10 times fainter.

4.2. Discovering High-Z supernovae

SN Ia are rare events, occurring once every 500 years in a typical galaxy (Cappellaro et al. 1997). However, through systematic searches over large piece of sky, it is possible to regularly uncover SN Ia, so that follow-up time can be pre-scheduled (Hamuy et al. 1993, Perlmutter et al. 1997). The first High-Z SN discovery (Norgaard-Nielsen et al. 1989) occurred in 1988 by a Danish team which spent nearly two years to find a single object; this effort was stifled by small detectors which provided only a small field of view. However, starting in 1994, large fields became available on 4m class telescopes, and high redshift SNe much easier to discover as demonstrated by the sudden success of the Supernova Cosmology Project after 5 years of hard work (Perlmutter et al. 1998). Our team, the High-Z SN Search team formed in 1994, when our members became convinced that it was possible to both use SN Ia at High-Z for measuring accurate distance (Hamuy et al. 1995), and discover them in interestingly large numbers (Perlmutter et al. 1997).

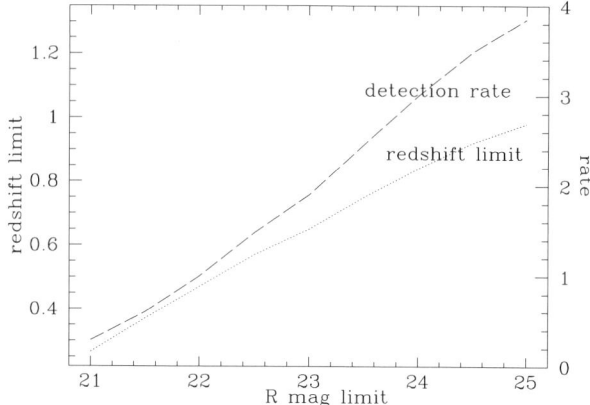

FIGURE 3. The number of SNe detected per square degree with a one month baseline in the R band, and the maximum redshift as a function of magnitude for $\Omega_{\rm tot} = 0$.

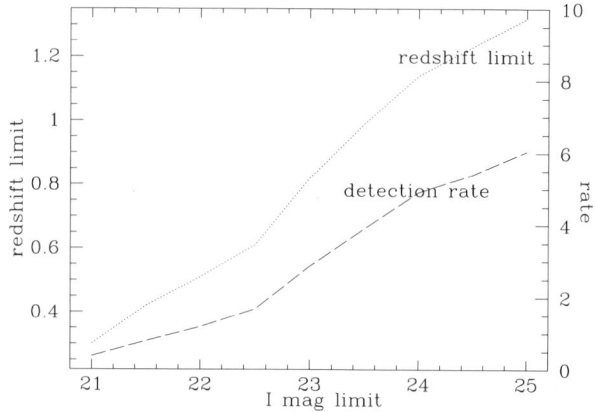

FIGURE 4. The number of SNe detected in I band per square degree with a one month baseline, and the maximum redshift as a function of magnitude for $\Omega_{\rm tot} = 0$.

To discover SN Ia we observe a patch of sky in two consecutive dark runs (Hamuy et al. 1993, Perlmutter et al. 1997) to ensure the SNe we discover are young, and to help eliminate the long period variability of Active Galactic Nuclei. In addition, at least two observations are made at each position to detect the motions of asteroids, eliminate cosmic rays, and remove chip defects. We usually make our observations near the celestial equator so that discoveries can be followed up in either hemisphere. Since we need to schedule significant amounts of telescope time to follow up our discoveries, it is essential not to be derailed by poor weather. During the period from December through March, the Chilean Atacama desert has virtually no time lost to clouds, and therefore we have concentrated our efforts at the only telescope which has had instrumentation that provides a large field of view at this geographic region; the Victor Blanco 4m telescope at CTIO. In recent times we have had the luxury of conducting a simultaneous search with the Canada-France-Hawaii Telescope. This combining of searches provides insurance for bad weather and instrument failure, while simultaneously allowing for us to target $z = 0.5$ objects (large areas to $m_R = 23.5$) and $z > 1$ objects (smaller areas to $m_I = 25$).

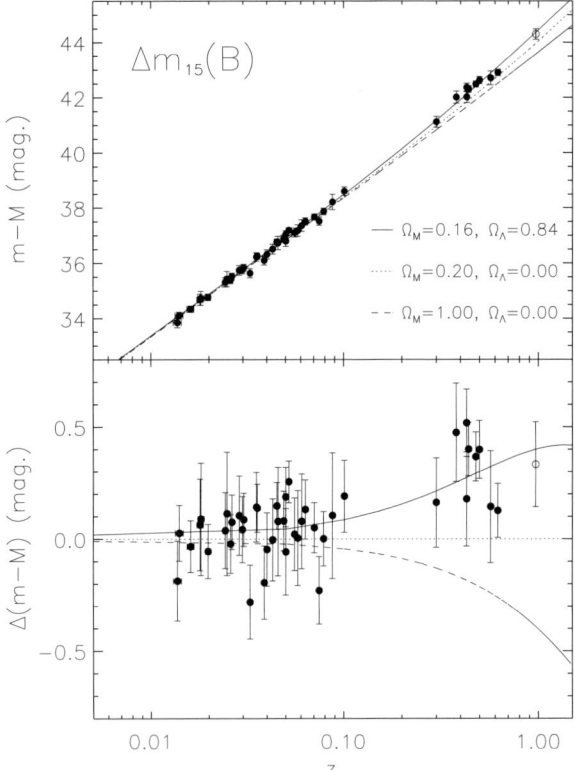

FIGURE 5. The $\Delta m_{15}(B)$ SN Ia data. The upper panel shows the Hubble diagram, and the lower panel, residuals relative to an $\Omega_M = 0.20$, $\Omega_\Lambda = 0$ Universe.

Figures 2, 3, and 4 show the approximate redshift and detection rate as a function of magnitude. With the "Big Throughput Camera" we are able to image 7 square degrees per night to $m_R = 23.5$, (50% more with the new NOAO 8k imagers) and this translates into finding 10–15 SNe per night with this instrument to $z = 0.85$. We also have interest in going deep, and finding SN Ia at $z > 1$. SN Ia are sufficiently faint and red that these objects can only be found with 4 meter (or larger) class telescopes, and I band observations with seeing better than 0.75 arcseconds. The Canada France Hawaii Telescope regularly provides images of this quality across a 1/2 square degree field of view with observations using the 12 K mosaic camera reaching $m_I = 25$ in 1 hour in $0.65''$ seeing.

Our SNe search is largely automated with the final decisions as to whether or not an object is a SN being made by eye. An automated script first aligns the 2nd epoch to the first epoch (template), and then computes a kernel which matches the point spread functions (PSFs) in the two images (Alard 1999). After convolution, the images are scaled in intensity, and then the template is subtracted from the 2nd epoch image. This procedure is applied to all 2nd epoch images, and these differenced images are averaged, rejecting any high pixels which are discordant by more than 3σ. This cleaned differenced image is searched for PSFs brighter than 4σ over the background, and this list sorted by magnitude, removing any object within a distance equal to the FWHM of the processed images from known bright stars. This process takes about 2 hours to run on a pair of 8192^2 images with a 270 Mhz Sun Ultrasparc computer, and typically leaves 50 candidates for inspection by eye, of which 1 in 10 is a SN. The remainder tend to

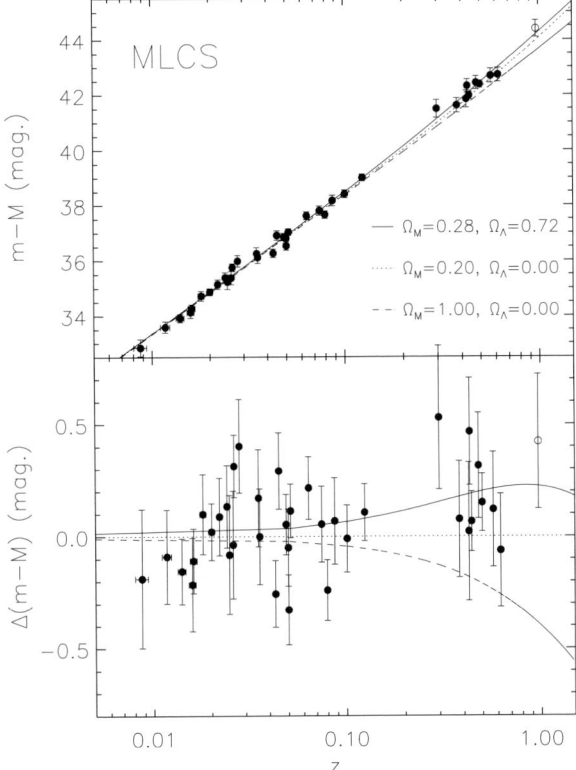

FIGURE 6. The MLCS SN Ia data. The upper panel shows the Hubble diagram, and the lower panel, residuals relative to an $\Omega_M = 0.20$, $\Omega_\Lambda = 0$ Universe.

be cosmic rays coincident with chip flaws, or poor subtractions of very bright galaxies. The vast majority of the area of the chip is subtracted to a noise level equal to the sky background. This procedure is discussed in detail by Schmidt et al. (1998).

4.3. Optical followup

Discovering High-Z Supernovae is not the ultimate goal of this program. Instead, we want to constrain the $D_L : z$ diagram as accurately as possible. Calculations into the optimal observing program show that to discriminate between an empty and matter-dominated, flat Universe, the maximum signal to noise ratio (S/N) per observing time is achieved by observing at $z \sim 0.5$. This occurs because as SNe become fainter with z (and we must move further and further to the red to observe them at the same rest wavelength), the S/N drops faster than the difference between these two cosmologies increases. In trying to distinguish between a matter-dominated Universe and a Λ-dominated Universe, this is even more true, because it is at a redshift between $z = 0.5$ that the difference between these two cosmologies is greatest (Figure 13).

To compare high and low redshift SN Ia accurately, it is best to observe them at a fixed wavelength in the rest frame. Kim et al. (1996) pointed out that at $z = 0.5$, R is a good match to rest frame B, and I a good match to rest frame V. We have designed a set of custom, high throughput filters which are the redshifted equivalents of B and V at $z = 0.35$ and $z = 0.45$. This minimizes the K-corrections (transformations to the restframe), and has the added advantage of providing significantly higher S/N per observing time than traditional filters.

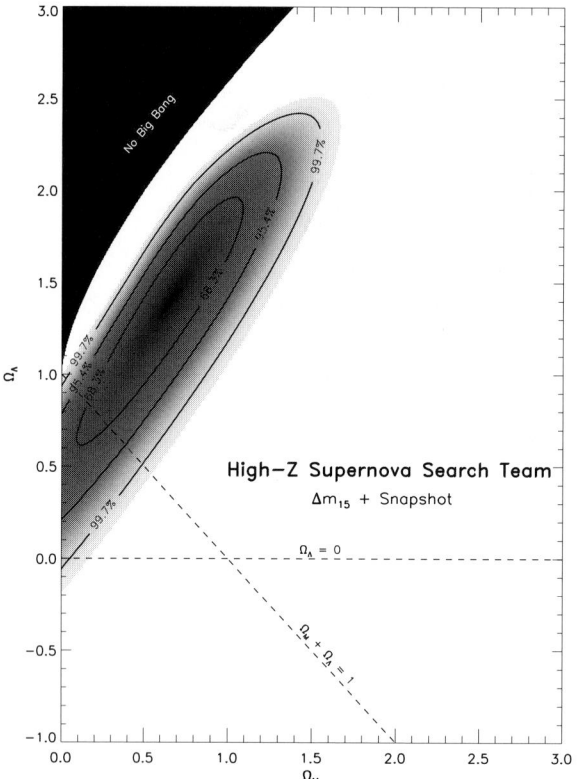

FIGURE 7. Confidence levels derived from the $\Delta m_{15}(B)$ SN Ia data.

4.4. Current High-Z Ia results

To date we have discovered over 100 SNe to $z = 1.23$, of which approximately 45 are SN Ia at appropriate redshifts, and with enough followup photometry/spectra, to be useful in measuring accurate distances. A total of 16 objects are fully analyzed and published (Schmidt et al. (1998), Garnavich et al. 1998a, Riess et al. 1998). These data are shown, combined with our nearby sample (Hamuy et al. 1996, Riess et al. 1999a), in Figures 5 and 6 Riess et al. (1998). By eye, it is clear that the data are some 0.25 mag fainter than would be expected for a matter-dominated Universe. The two distance methods, $\Delta m_{15}(B)$ and MLCS, show similar, although not identical, results; these differences are related to the subtle differences in how the two method exploit the light curve properties of SN Ia. Riess et al. (1998) have analyzed this data in detail and Figures 7 and 8 summarize these results. We find that $\Omega_\Lambda > 0$ at $> 3\sigma$ confidence levels for both distance methods, and a flat, matter-dominated Universe is ruled out at $> 7\sigma$.

The probability contours in Figures 7 and 8 can be used to estimate the dynamical age of the Universe, given by

$$t_0(H_0, \Omega_M, \Omega_\Lambda) = H_0^{-1} \int (1+z)^{-1}[(1+z)^2(1+\Omega_M z) - z(2+z)\Omega_\Lambda]^{-1/2} dz \qquad (4.6)$$

as shown in Figure 9. Using the HST Cepheid calibration of SN Ia host galaxies (Suntzeff et al. 1999, Saha et al. 1999, Jha et al. 1999, Gibson et al. 1999), this data set gives a value of $H_0 = 64 \pm 7$ km s^{-1} Mpc^{-1}, and a dynamical age of the Universe of 14.2 ± 1.5 Gyr.

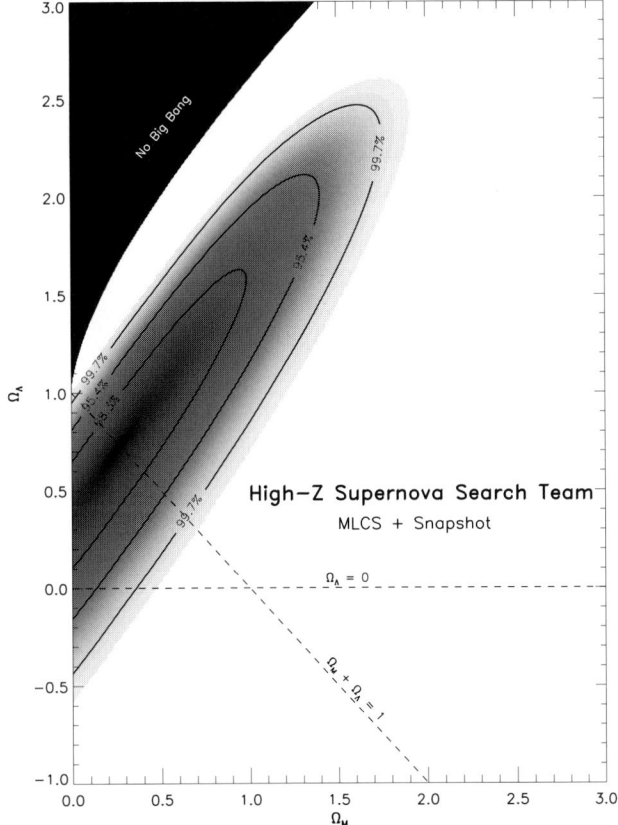

FIGURE 8. Confidence levels derived from the MLCS SN Ia data.

Garnavich et al. (1998b) has combined this data set with the CMB results White (1998) as shown in Figure 10, and find that for a Universe made up only of Λ and normal matter, that $\Omega_{\rm tot} = 0.94 \pm 0.26$. Furthermore, fitting for the equation of state of the unknown matter which is leading to the observed acceleration, yields confidence intervals shown in Figure 11, assuming the Universe is spatially flat, and only using the SN data. Parameterized as α (as per equation 4.2), the mean equation of state for the Universe, between $0 < z < 0.6$ must be < -0.55 with 95% confidence.

The conclusions presented here are echoed in the independent and simultaneously obtained Supernova Cosmology Project results (Perlmutter et al. 1998). The agreement between the two groups is both unprecedented and spectacular, and hopefully gives confidence that neither group has done something fatally wrong. However, presently, both groups share a very similar methodology, and an identical nearby SN Ia dataset. Consequently, the SN Ia results are not immune to systematic effects. Possible systematic effects include non-standard dust, gravitational lensing, photometric calibration and correction errors, and SN Ia evolution. These effects are discussed in detail by both groups (Schmidt et al. 1998, Riess et al. 1998, Perlmutter et al. 1998). There is still some indication of a discrepancy between the high-z and local SN Ia rise time (Riess et al. 1999c, Aldering et al. 2000) although not as large as initially indicated by (Riess et al. 1999b and Goldhaber 1998). In addition (Riess et al. 2000) have shown that in at least one case, even non-standard dust is not compatible with rest-frame I band observations

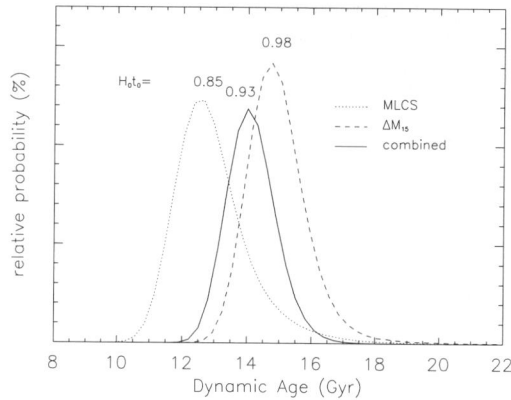

FIGURE 9. The probability density function for the dynamical age of the Universe given the SN Ia measurements. Lines are drawn from the both SN Ia distance methods.

(Figure 12). Instead, the SNe seem, if anything, too blue (which in itself is a cause for concern, and additional investigation).

4.5. *The future of High-Z SN Ia*

The $D_L : z$ relation has a serious degeneracy between Ω_M and Ω_Λ when observed at only two sets of redshifts, ($z \sim 0.05$ and $z \sim 0.5$ presently), as seen in Figures 7 and 8. This degeneracy, in principle, can be broken by observing over a wider range of redshifts. Unfortunately, our present observations are systematic limited. This is not to say that systematic error is dominating over statistical uncertainty, but rather that we cannot say this is not the case. This difficulty will continue, even as we move to larger redshift, because the effect of Λ goes away over time (Λ is overwhelmed by normal matter as one goes back in z), and any confining of the contours in Figures 7 and 8 will require the astronomical community to believe that the systematic errors in the SN Ia measurements are significantly smaller than the observed 0.25 magnitude effect we have already observed. This being said, moving to higher redshift can strengthen the case that our current measurements are not being driven by systematic effects. As Ω_Λ diminishes, the $D_L : z$ relationship should turn over around $z = 0.9$ (Figure 13), a prediction not expected to occur if the SNe results we are seeing are being caused by SN Ia evolution, dust, or most other systematic effects which increase with look-back time.

If SN Ia are evolving, it should be possible to see subtle difference in their spectra and light curves as a function of z. We are currently collecting high S/N spectra, matched with good light curves, so that we can make as careful a comparison between the two data sets as possible. While this can never prove the innocence of SN Ia against systematic effect (or even show the objects guilty of luminosity evolution), it can act as a warning sign for evolution which can be investigated with theoretical models, and targeted programs of both low and high redshift.

In the first year of the new millennium astronomy will have two new powerful tools for discovering and measuring High-Z SN Ia. These are Subaru with its wide-field imager, and HST's Advanced Camera. These two instruments should make discovering $z < 1.2$ SNe a breeze. The difficulty will be to extract redshifts, types, extinctions and distances from these objects, but I believe this will be an achievable challenge. The Supernova Cosmology Project's proposed satellite (SNAPsat) to do this same science will make

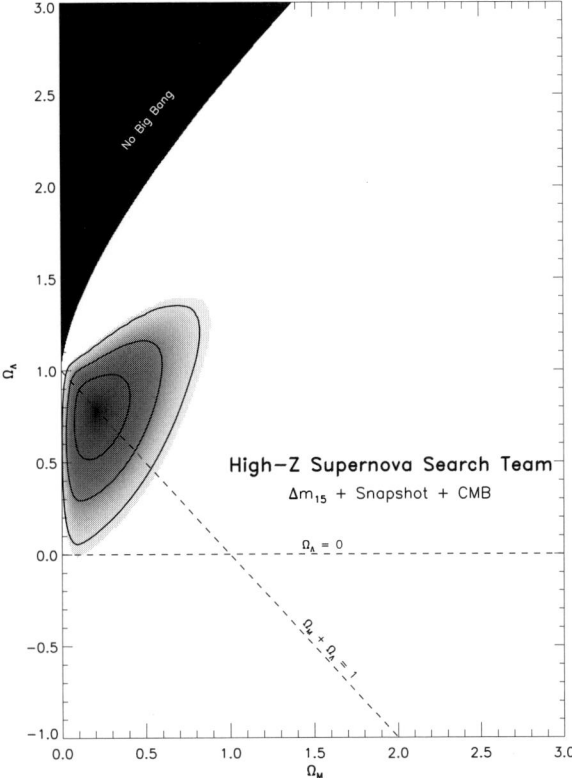

FIGURE 10. Confidence levels from combining $\Delta m_{15}(B)$ SN Ia distances with CMB measurements.

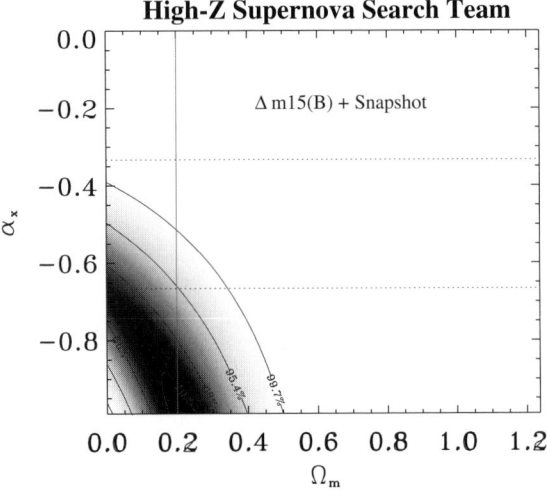

FIGURE 11. Confidence levels in measuring the equation of state parameter (α) which is causing the observed deceleration, using $\Delta m_{15}(B)$ SN Ia distances, and assuming a flat universe.

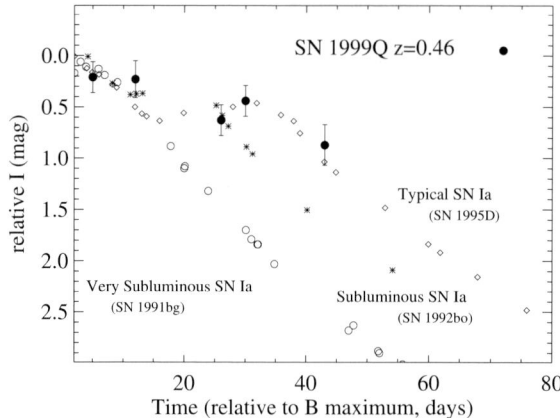

FIGURE 12. The light curve of SN 1999Q in rest frame I-band compared to a normal low-Z SN Ia

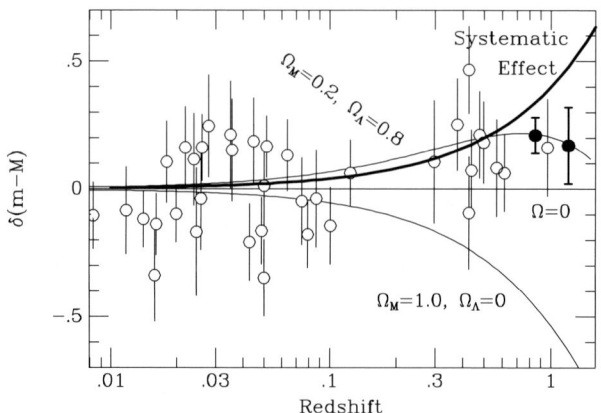

FIGURE 13. Our current $\Delta m_{15}(B)$ SN Ia distances, showing the measurement uncertainty of 10 SN Ia at $z = 1.2$.

all aspects look easy, albeit with a hefty price tag (more than the entire Australian astronomy budget for a decade).

As with any revolutionary result, the SN Ia results should be viewed with caution. In the end, it is only through confirmation through other methods that we can be sure the Universe is dominated by something than normal matter. I, for one, will be much more confident if the CMB demonstrates to high significance that the Universe is flat, but until a general consensus is reached with many methods that there exists a cosmological constant, there will still be a lingering doubt.

5. SNe, GRBs, and other transient objects

GRBs are amongst the most energetic and enigmatic phenomena studied in astronomy. The discovery of optical transients in association with these events and the evidence of their cosmological origin has demonstrated the enormous energies involved with these objects. However, the association of GRB980425 with SN 1998bw has added a new complexity to the subject. SN 1998bw, located at $z = 0.0085$, is the nearest GRB yet optically identified by a factor of 100, but is intrinsically fainter than previous GRB optical transients by several orders of magnitude. Its photometric and spectral resemblance to a type Ic SN, objects which are thought to result from the core collapse of massive stars whose outer envelopes have been stripped away either by binary interaction or a stellar wind, indicates that at least some GRBs are associated with this type of event. However, does SN 1998bw represent the same physical event as the optical counterparts to the more distant GRBs which are nearly 10^5 times brighter?

The coincidence of GRB980425 with a peculiar SN suggests that other such associations are likely. Unfortunately, the statistics are not usually compelling for singling out individual objects due to the large numbers of GRBs and their poor positional information. Wang & Wheeler (1998) have cross-correlated the SN and Burst and Transient Source Experiment (BATSE) catalogs and have identified a positive correlation with SN Ic. However, Kippen et al. (1998) have used a more sophisticated error model for the BATSE data and found no correlation.

As part of the Abell Cluster SN Search we have discovered SN 1997cy (Germany et al. 1999), the most luminous SN yet discovered. SN 1997cy has an unusual spectrum which is dominated by broad undulations in the blue, and $H\alpha$ in the red. Its spectrum is suggestive of circumstellar interaction, but is, taken in whole, unique to this object. An *a posteriori* statistical likelihood of a GRB/SN connection for this object is less than 1% (Germany et al. 1999), and while not overwhelming, the coincidence between this extreme SN and a GRB is suggestive. SN 1997cy's light curve is as bright as a GRB OTs, and exhibits a behavior, given the current state of observations, indistinguishable from GRB events at a similar age. Only future work will reveal, through similar associations— or lack there of, if our proposed connection is correct.

The disparity between SN 1998bw and other GRB OT luminosities can be naturally explained through beaming, but scenarios of GRBs arising from beamed SNe explosions predict that the underlying SN light should be visible in many cases a month after explosion (when the GRB has faded and the SN has reached peak). Indeed, there are several claims that the SN light has been seen, helping solidify support for this model. It is no surprise that this effect has not been seen in all events (even if they are all SN related events) because SN Ic come in all shapes and sizes, with some events observed to have absolute magnitudes nearly a 1000 times fainter than SN 1998bw in the UV.

Another consequence of a beaming scenario for GRBs should be the occurrence of a large number of optical transients not associated with GRBs. This argument does not apply if the optical radiation is also beamed (as is suggested for GRB990123). In any case, the high-z SN searches provide a reasonably strong constraint on intrinsically fast bright optical transients (FBOTs). The High-Z SN Search has now uncovered 5 FBOTs in searches where we have not biased against hostless transients—this translates to a rate of (very approximately) 1 FBOT per 6 square degrees per timescale. The timescale is not well constrained, but is longer than 10 minutes (our observing time between cosmic ray splits), but less than 3 days (these objects fade by more than a factor of 5 over this time). In one case, an object was observed to disappear within 3 hours, but this field was at low galactic latitude and it is possible the object was a distant flare star. In all but one case, the objects did not have obvious host galaxies.

While afficianados of the beamed GRB model may wish to claim this as the final proof of their point of view, I would like to propose that these are nothing other than the shock breakouts of ordinary SN II. Using the calculations of SN 1993J by Blinnikov et al. (1998), I have calculated the time evolution of this SN II as a function redshift. Shock breakouts of SN II should peak brighter than $m_R = 23.5$ at $z < 1.5$, and are probably observable to the edge of star formation, excluding the effects of dust, with HST. Using the ratio of the local SN Ia/II rate of 5, we can explain the observed rate of FBOTs from SN II at $z < 1$ if we assume a timescale of a few hours.

REFERENCES

ALARD, C. 1999; astro-ph/9903111.
ALDERING, G., KNOPF, R., & NUGENT, P. 2000; astro-ph/0001049.
BLINNIKOV, S., ET AL. 1998 *ApJ*, **496**, 454.
CAPPELLARO, E., ET AL. 1997, *A&A*, **322**, 431.
CARROLL, S. M, PRESS, W. H., & TURNER, E. L. 1992 *ARAA*, **30**, 49.
EASTMAN, R. G., SCHMIDT, B. P., & KIRSHNER, R. P. 1996 *ApJ*, **466**, 911.
EASTMAN, R. G. & KIRSHNER, R. P. 1989 *ApJ*, **347**, 771.
FILIPPENKO, A. V. 1997 *ARAA*, **35**, 309.
GALAMA, T. J., ET AL. 1998 *IAU Circ. 6895*.
GARNAVICH, P., ET AL. 1998 *ApJ*, **493**, L53.
GARNAVICH, P., ET AL. 1998 *ApJ* **509**, L74.
GERMANY, L., ET AL. 1999 *ApJ*, in press; astro-ph/9906096.
GIBSON, B. K., ET AL. 1999 *ApJ*, submitted.
GOLDHABER, G. 1998 in *Gravity: From the Hubble Length to the Planck Length*. Stanford Linear Accelerator Cent.
GUERRA, E. J., DALY, R. A., & WAN, L. 1998 *ApJ*, submitted, Astro-ph/9807249.
HAMUY, M., ET AL. 1993 *AJ*, **106**, 2392.
HAMUY, M., ET AL. 1995 *AJ*, **109**, 1.
HAMUY, M., ET AL. 1996 *AJ*, **112**, 2408.
HU, W. 1996, in *The Universe at High z, Large Scale Structure, and the Cosmic Microwave Background* (eds. E. Martnex-Gonzalez & J. L. Sanz), p. 207. Springer.
JHA, S., ET AL. 1999 *ApJS*, **125**, 73.
KELLERMAN, K. I. 1993 *Nature*, **364**, 134.
KIM, A., ET AL. 1996 *PASP*, **108**, 190.
KIPPEN, R. M., ET AL. 1998, *ApJL*, submitted; astro-ph/9806364.
KOCHANEK, C. 1996 *ApJ*, **466**, 638.
KOWAL, C. T. 1968 *AJ*, **73**, 1021.
MOULD, J. R., ET AL. 1999 *ApJ*, in press.
NORGAARD-NIELSEN, H., ET AL. 1989 *Nature*, **339**, 523.
NUGENT, P., ET AL. 1995 *ApJ*, **455**, L147.
PERLMUTTER, S., ET AL. 1997 *ApJ*, **483**, 565.
PERLMUTTER, S., ET AL. 1998 *ApJ*, in press; astro-ph/9812133.
PHILLIPS, M. M. 1993 *ApJ*, **413**, L105.
RIESS, A. G., PRESS, W. H., & KIRSHNER, R. P. 1996 *ApJ*, **473**, 88.
RIESS, A.G., ET AL. 1998 *AJ*, **116**, 1009.
RIESS, A. G., ET AL. 1999 *AJ*, **117**, 707.
RIESS, A. G. 1999 *AJ*, **118**, 2668.
RIESS, A. G. 1999 *AJ*, **118**, 2675.
RIESS, A. G. 2000 *ApJ*, accepted; astro-ph/0001384.
SAHA, A., ET AL. 1999 *ApJ*, in press; astro-ph/9904389.
SANDAGE, A. R. 1961 *ApJ*, **133**, 355.

SCHMIDT, B. P., ET AL. 1998 *ApJ*, **507**, 46.
SCHMIDT, B. P., KIRSHNER, R. P., & EASTMAN, R. G. 1992 *ApJ*, **395**, 366.
SCHMIDT, B. P., ET AL. 1994 *ApJ*, **432**, 42.
SHANKS, T., ET AL. 1984 *MNRAS*, **206**, 767.
SUNTZEFF, N., ET AL. 1999 *AJ*, in press; astro-ph/9811205.
TINSLEY, B. 1972 *ApJ*, **178**, 319.
WANG, L. & WHEELER, J. C. 1998 *ApJ*, submitted; astro-ph/9806212.
WHITE, M., ET AL. 1998 *ApJ*, **506**, 495.
YOSHI, Y. & PETERSON, B. A. 1995 *ApJ*, **444**, 15.

How good are SNe Ia as standard candles? A short history

By ALLAN SANDAGE,[1] G. A. TAMMANN,[2]
AND A. SAHA[3]
Representing the collaboration for the
HST SNe Ia Calibration Program for the Hubble Constant,
composed of A. Sandage, G. A. Tammann, A. Saha,
L. Labhardt, N. Panagia, and F. D. Macchetto

[1]The Observatories of the Carnegie Institution of Washington

[2]Astronomisches Institut der Universät Basel

[3]National Optical Astronomy Observatories

An abbreviated history is given of the evidence that type Ia supernovae are excellent standard candles with a small diversity of properties such that they are elite distance indicators. The evidence for both homogeneity and diversity is set out, beginning with the initial proof by Kowal (1968) of a relatively small dispersion of $<M(max)>_{\text{SNe Ia}}$, and ending with the modern corrections for diversity using various formulations of second-parameter effects for decay rate, intrinsic color, and parent galaxy types.

Most current investigations using the extant SNe Ia calibration give H_0 in the range of 58 to 63 km s^{-1} Mpc^{-1}, based on a Cepheid period-luminosity zero point with $(m - M)_o = 18.58$ for the LMC. The exception is $H_0 = 68$ offered by the "Key Project" consortium where they use a debatable elimination of several of the extant Cepheid-based calibrators plus controversial modifications of the absolute magnitudes of the remainder. In addition, they substitute other supernovae that are without direct Cepheid calibrations, based on an unproven premise that the Cepheid distance to the spiral NGC 1365 gives the distance to the compact E and S0 core of the Fornax cluster. The results of the substitutions and modifications are (1) too faint a mean absolute magnitude, $<M(max)>_{\text{SNe Ia}}$, and (2) too large a slope for the decay rate-absolute magnitude correlation for SNe Ia, with the consequent combined effects on H_0.

Nevertheless, the systematic and statistical errors of H_0 via the SNe Ia method remain at the level of $\sim 10\%$ in most of the current investigations, including ours. Because the differences at this level depend on precepts that will be difficult to adjudicate by observations in the short term, we have used three other independent methods to H_0 to test the SNe Ia method externally. These are (1) the route through the Virgo cluster tied to the remote (global) expansion frame, (2) the Tully-Fisher method using field galaxies corrected for obervational selection bias, and (3) the mean absolute magnitude, $<M>$, of Sb and Sc galaxies determined from the type-specific luminosity functions corrected for the systematic variation of $<M>$ with luminosity class. Each of these methods individually give $H_0 \sim 55$. Evidence (Fig. 15) is given that H_0(global) is smaller than 60 km s^{-1} Mpc^{-1}, based on the present SNe Ia calibrations with no second-parameter corrections.

The difference between these results and those of the "Key Project" consortium can be traced not only to differences in the use of the SNe Ia basic calibration data but also, independently, to the systematic differences in the distances and redshifts used for the Virgo and Fornax clusters (Fig. 16). Our distances average ~ 1.25 times larger than those given in the 1994 and 1997 summaries by "Key Project" authors, reducing their value of the Hubble constant from 70 ± 7 to $H_0 = 56 \pm 6$ km s^{-1} Mpc^{-1} (Fig. 16) based on the cluster distances alone, a route that is independent of the SNe Ia method. The agreement with the value of H_0 from SNe Ia as set out here supports the SNe Ia calibration in Table 1.

1. Prologue

1.1. *A mantra for the first approximation*

The organizers of this symposium assigned the title and defined the narrow boundaries for this account. Their hope was to test if each of us could still defend our original mantra that "SNe Ia are nearly perfect standard candles if we know what part of a complete sample to throw away," or, on the contrary, if we have been converted to a position that allowed more diversity than homogeneity in the absolute magnitude of SNe Ia at maximum. Have we experienced a real conversion or are we only apostate concerning the need for second-parameter corrections?

The difference between our conclusions concerning SNe Ia as standard candles and those who use the same data with different precepts concerns how to deal with *any* diversity and how to assess its impact. Does restricting membership in a sample according to some limits on particular characteristics (spectrum, decay rate, color at maximum, nearby redshifts, etc.) tilt the result, beg the question, and leave SNe Ia useless as distance indicators? Some astronomers have written that it does.

Others have taken a middle position that second-parameter corrections can be made to any complete SNe Ia sample even when no a priori restrictions have been imposed, restoring such supernovae as useful secondary distance indicators, albeit compromised.

The third position, and the one we used in the first several papers in the HST SNe Ia program for the Hubble constant (Sandage & Tammann 1982, 1993; Sandage et al. 1992, 1994; Saha et al. 1994, 1995, 1996a,b), is that the sample can be made remarkably homogeneous by using "appropriate" restrictions such that precise knowledge of second-parameter corrections could be made unimportant at better than the $\sim 10\%$ level. Indeed, if the mean decay rate (and any other second-parameter correction) of the calibrating SNe Ia is identical to that of the fiducial sample that defines the remote Hubble diagram, no corrections for diversity would be necessary.

We continue to assert that with this precept of "appropriate restrictions," the "normal" (defined below) SNe Ia remain the best primary distance indicator known. The purpose of this review is to discuss the issue.

But first we set out an abbreviated account of the progressive narrowing of the measured intrinsic dispersion in $<M(max)>_{\text{SNe Ia}}$ as the observational data improved, starting in the mid-1960s.

1.2. *The view from astronomy vs. the view from physics*

Astronomers have often used particular observed properties of astronomical objects to accomplish primary astronomical goals before the physics of objects were understood. On the other hand, the goal of the physicists is to understand the physics. They have often been suspicious of astronomical methods that are not tied to the underlying, initially unknown, physics. Many examples from the history of astronomy are known. Three of the most obvious from this century are part of astronomical literacy.

(1) Cepheid variables were used for distance determinations well before the pulsation mechanism was understood. Of course, what was necessary before the detailed physics was known were the astronomical proofs that a period-intrinsic luminosity relation exists, regardless of its cause. Furthermore, a calibration of its zero point in absolute magnitude could be made by astronomical methods without knowledge of the pulsation driver.

(2) Sequences in the HR diagram were used for stellar distance determinations via spectroscopic and photometric parallaxes before an understanding had been achieved of stellar structure and evolution that provided an explanation of the diagram.

(3) Redshifts were proved to be distance indicators for galaxies before an understanding of their cosmological meaning was clear (is the expansion real?). The astronomical test that ratios of redshifts are ratios of distance could be made by purely astronomical means, independent of cosmological theory.

The case discussed here is similar. The size of the dispersions in $<M(max)>_{\text{SNe Ia}}$ for both restricted and unrestricted supernovae samples can be measured in ways that are independent of the underlying supernova physics.

1.3. Homogeneity: A puzzle or a mystery?

The difference between a puzzle and a mystery is that a puzzle can be solved, whereas a mystery, by its definition, remains outside natural law and therefore outside of science. Is the homogeneity of $\sim 90\%$ of a flux-limited complete sample of SNe Ia a puzzle or a mystery? Clearly it is a puzzle; its solution even seems near at hand.

A valuable way to proceed with puzzles is to look at the extremes of any phenomenon to isolate the reason why the abnormal cases differ from the mean. In the end, the study of the diversities is often the means of bringing a subject, complete with its physics, nearer to solution.

In the yet uncompleted supernova problem, the astronomer attempts to proceed by simply mapping the extent and limits of the homogeneity. At this stage there is no need to understand the mechanisms. The observational astronomer proceeds by searching for instructions on how to treat particular samples of SNe I to produce increased homogeneity. To this end it is now known that the menagerie of "non-normal" supernovae of types Ib, Ic, hyper-supernovae etc. can be ignored. Purely astronomical methods have shown that their absolute magnitudes differ from those of the normals, and further, that such SN can be identified using particular observed characteristics. Nevertheless, the "abnormal" types that the practical astronomer throws away at the extremes of the distributions of luminosity, light curve shape, color, and frequency, are the most interesting for the theoretical physicist.

Throughout this conference the tension in the weight that each group puts on homogeneity vs. diversity, is evident; each according to his/her need, each according to his/her purpose.

1.4. Why such homogeneity?

It is now known that most of the supernovae of type I (defined as having no hydrogen in their spectra, cf. Minkowski 1964) in flux-limited samples are remarkably homogeneous in their characteristics at maximum. Such supernovae are now called "Branch normal" after the review of the characteristics of the sample known to 1992 (Branch, Fisher, & Nugent 1993). Their classifications are based on a long series of studies by Branch and others.

Khokhlov posed the question as "Why are we so lucky?" in the evident great homogeneity of the spectra and light curve shapes of this subset. Evidently the progenitor stars are sufficiently homogeneous so as to produce nearly identical standard thermonuclear bombs in their yield of Ni^{56}, and therefore in their maximum luminosity whose dispersion is less than 20%. Why?

It has long been supposed that one way to produce the homogeneity is to give a final nudge in the accretion of mass onto a white dwarf near the Chandrasekhar limit, driving that mass over the limit, causing catastrophic collapse and subsequent explosion. But if the process is so finely tuned, occurring immediately as the Chandrasekhar mass-limit is reached, why is there *any* dispersion at all in the explosion luminosity? Indeed, other models have been proposed for sub-Chandrasekhar masses, although these are widely believed to apply, if at all, to only a small fraction of any sample.

Then, just how lucky are we? By what stages did that knowledge of the extent of that luck develop?

2. Short history of the developing observational proof of homogeneity

2.1. The initial ex cathedra statement

It was Zwicky's position from the beginning in the 1930s until his reward in 1974 that supernovae were the only astronomical objects with which the Hubble constant could eventually be found. He made two points. (1) One must use a distance indicator so bright that it can be measured at large enough redshifts to be beyond any local velocity perturbation on the Hubble flow, and (2) the dispersion in absolute magnitude of the indicator must be small. He believed that supernovae were the only objects capable of such use.

Both in his lecture and in the question period after another paper at the 1961 Santa Barbara IAU Symposium 15 on *Problems of Extra-Galactic Research*, Zwicky (1962) stated:

"Preliminary investigations indicate that the absolute magnitudes of supernovae of Type I have a dispersion of less than a magnitude. If true, this fact will enable us to establish a relative, as well as an absolute, cosmic distance which is more reliable than any scales currently in use." "H from supernovae in the Cancer Cluster, the Coma Cluster, and the Local Group gives 175 km/sec (sic)."

However, the announcement was made ex cathedra. There were no supporting data, either on the size of the intrinsic dispersion or on his method of determining an absolute calibration of $<M(max)>$. Because of this, his statements were met with considerable skepticism.

The initial attempts to calibrate absolute magnitudes were the remarkable papers by Baade (1938) and Baade & Zwicky (1938). In the first study Baade, using fragmentary data from the literature, concluded that $<M(max)> = -14.3$ with a dispersion of 1.1 mag. However, his list of 18 supernovae had very fragmentary light curves (see e.g. the Basel Atlas; Leibundgut et al. 1991b, as cross checked with Baade's listing in his Table 1), and only 6 of the 18 are what we would now classify as SNe of type Ia. Baade himself comments on the very fragmentary nature of the data.

In addition, the two "modern" SN, at the time, whose light curves and absolute magnitudes were discussed by Baade & Zwicky, showed a very large difference in $<M(max)>$, with $<M(max)> = -16.6$ for what is now called SN 1937C in IC 4182, and -14.0 for SN 1937D in NGC 1003, not boding well for the small dispersion of 1.1 mag given in Baade's paper. Hence, Zwicky's statement at IAU Symposium 15 in 1961, where no new data were presented, had few believers.

The preliminary state of the calibration and a measure of the intrinsic dispersion in the early 1960s is seen from the review article by Zwicky (1958) and the summary by van den Bergh (1960) based on redshifts and an assumed Hubble constant.

2.2. The first convincing data

Zwicky's 1961 supposition changed into a probability when Kowal (1968) published his remarkable Hubble diagram (apparent magnitude at maximum vs. log redshift) of the SNe I that had "adequate" photometry to that date. He had assembled much of the literature, to which he added the results of the Palomar supernovae program, organized by Zwicky and with Kowal as one of the principal observers.

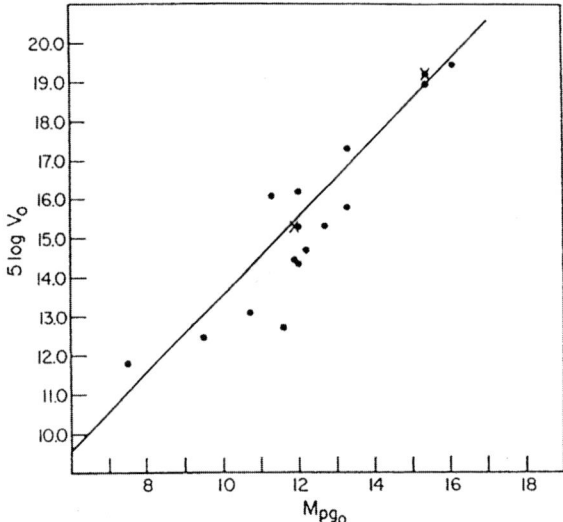

FIGURE 1. Kowal's (1968) initial Hubble Diagram for 22 SNe I that had "adequate" photometry at the time. The two crosses are the mean of 3 SN I in the Virgo and 5 SN I in the Coma clusters. The relative small scatter about the mean line of slope 5 was the first convincing proof that SNe I may have a small dispersion in absolute magnitude at maximum.

This diagram, reproduced here as Fig. 1, constituted the first convincing proof of a moderately small dispersion in $<M(max)>$ for SNe I. Note especially the positions of the two crosses that are the averages of 4 SNe I in the Virgo Cluster and 3 in the Coma Cluster. On the basis of this Hubble diagram Kowal remarks: "There is therefore considerable hope that the magnitudes of type I supernovae can be used as reliable distance indicators, presumably free from such complications as evolutionary effects, and visible at very great distances."

2.3. Decreasing dispersion with increasingly accurate photometry

2.3.1. The initial cases for homogeneity vs. diversity

By 1967 Pskovskii recognized that the light curves of SNe I were phenomenally similar, a point also known to Kowal & Minkowski (1964). In an important prescient summary paper Barbon, Ciatti, & Rosino (1973) combined all the photometric data available from the literature to 1972 and added much new data of their own from the supernova program at Asiago. They produced one of the first master-template light curves of SNe I using photometric data for 38 bona fide SNe I among the ~ 85 SNe I known at that time. Their subsample of 38 were those whose light curves were adequate to describe more than just fragmentary segments.

Their composite curve, found by shifting each individual curve to a developing template, had a dispersion at maximum of only 0.15 mag. Remarkably, they also suggested a division of the sample into two discrete groups of "fast" and "slow" events according to the rate of decline after maximum. With this discovery, following an initial discussion by Schmidt (1957) of possible internal differences in the light curves, they partially agreed with Pskovskii (1967, 1971, and later 1984) in his suggestion for differences in the slope of the light curve between maximum and the well defined inflection point ~ 40 days later. (Pskovskii had a continuum of variation whereas Barbon et al. had two discrete groups). There were also suggestions that the expansion velocities were correlated with

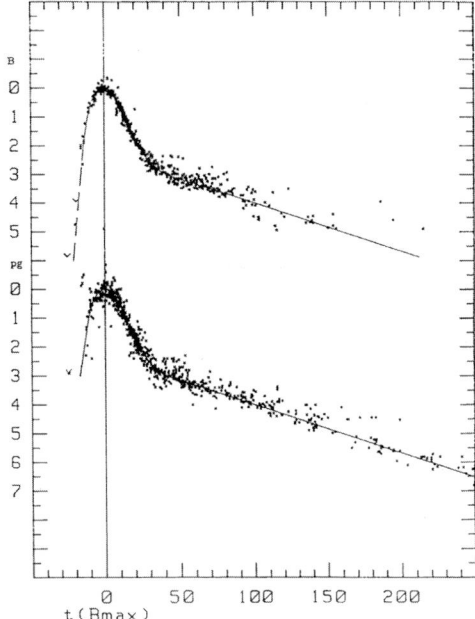

FIGURE 2. The master templates used for the Basel Atlas (Leibundgut et al. 1991b) using all data with adequate photometry up to 1986. Diagram is from Cadonau et al. (1985). Top curve is the composite template using the best 22 SNe I data in B. Bottom curve is the template using the best 16 SNe I curves in the earlier m_{pg} international magnitude system.

the absolute magnitudes and with the decay rates (Pskovskii 1971, 1984; Branch 1981, 1982).

With these discoveries, diversity was introduced amidst the attempt to maintain strict homogeneity. Taking a conservative view, maintaining a balance toward as much homogeneity as the extant data would allow in the presence of the still remaining significant photometric inaccuracies, some adopted a wait-and-see attitude toward the reality of significant second order differences relative to a universal template light curve (Tammann 1978, 1982; Cadonau, Sandage, & Tammann 1985; Leibundgut 1988, 1991; Branch & Tammann 1992).

The basis upon which our early skepticism on the severity of the inhomogeneity is described in the review by Branch & Tammann (1992) on "Supernovae as Standard Candles." They write:

"These differences [quoting Pskovskii 1977 and Branch 1981, 1982] became less convincing, however, when the analysis was restricted to the best observed SNe I (Tammann 1978, 1982). The infrared light curves also revealed an impressive uniformity (Elias et al. 1981).—A compilation of all optical magnitudes of SNe I in the literature (Cadonau & Leibundgut 1990) allowed the construction of template light curves in the UBV (Cadonau 1987) and JHK bands (Leibundgut 1988). Four particularly well observed SNe Ia have a scatter about the template light curve(s) for the six bands of only $\sigma = 0.06$–0.18 mag (Leibundgut 1990). From an atlas [the Basel Atlas] of the optical light curves of 75 SNe I (Leibundgut et al. 1991b) it is that most SNe I comply with the template within the photometric errors."

The master template derived by Cadonau and used for the Basel Atlas is shown in Figure 2, based on 16 of the best type I light curves in m_{pg} known to 1986, and on 22 of the best curves in B also to 1986. The appreciably smaller scatter in the upper curve in B than in the less accurate m_{pg} was one of the arguments we had used until the mid 1990s, quoted from Branch & Tammann above, for a small, if any, diversity in normal type I supernovae.

The case for "effective" homogeneity was also made by Branch & Tammann (1992) by showing the very close agreement in the observed apparent magnitudes of SNe in galaxies which had produced two events (NGC 1316, NGC 3913, and NGC 4753; their Table 1), and four galaxies in the Virgo complex, which however do show the depth effect (their Table 2).

However, beginning in 1985 new discoveries of a number of non-normal SNe I showed that the spectral classification of large samples could be subdivided into three groups, eventually called types Ia, Ib, and Ic. These discoveries of subtle spectral differences in the generic type I class (no hydrogen) were made soon after good spectra in large numbers were obtained en mass by the many new large telescopes everywhere, and used by the large numbers of new astronomers that were becoming supernovae experts, many of whom are at this meeting.

However, even as early as the mid-1960s Bertola (1964), Bertola & Sussi (1965), and Bertola et al. (1965) had noticed the lack of certain spectral features (no $\lambda 6150$ Å absorption, no P Cygni profiles) in a few SNe I, contrary to most of the class, but the discovery was not acted on further at that time. However, by 1985 the variations discovered by Bertola were shown to be general, and the SNe I class could be divided into types Ia and Ib (Wheeler & Levreault 1985; Uomoto & Kirshner 1985; Panagia 1985, Branch 1986; Harkness et al. 1987), with the class formally named by Elias et al. (1985) and formally defined by Porter & Filippenko (1987).

Diversity was soon again noticed even in the new Ib class. Harkness et al. (1987) and Harkness & Wheeler (1990) suggested that the subtype Ib could be divided again into two separate categories, with the new group called SNe Ic.

These diversities now seem crucial in providing data for an understanding of the physics of the explosion mechanisms for progenitors with presumed different masses. As emphasized at this meeting, it is now likely that the subtypes Ia, Ib, Ic, are from different types of progenitors leading to different explosion mechanisms and undoubtedly different mean absolute magnitudes at maximum. We are therefore concerned in what follows with the characteristics of SNe of type Ia.

Important summaries of the discoveries of three subtypes and their definitions are in Weiler & Sramek (1988), Branch et al. 1991, Branch & Tammann (1992), and Filippenko (1997).

2.3.2. Improved Hubble diagrams

But as the case became stronger for diversity in the generic type I class by the introduction of subtypes Ib and Ic, so also, at the same time, did the case become stronger for the remarkable homogeneity for the type Ia class itself. The data supporting homogeneity were the continuously decreased dispersion about the linear expansion line in the Hubble diagram as the photometric accuracy improved and the sample size increased.

An early improved Hubble diagram, made as a prelude to the Basel Atlas, was given by Sandage & Tammann (1982) in Paper VIII of the Steps series. The sample was restricted to SNe Ia that appeared only in E and S0 galaxies to overcome the pernicious problems of internal absorption.

The Hubble diagram in the B photometric band is shown in Figure 3. Data for six SNe Ia in the Virgo Cluster complex and five in the Coma Cluster are shown as the average points. Also shown are five additional SN in isolated E galaxies whose redshifts range from 3200 km s^{-1} to 12,000 km s^{-1}, believed then to be well into the unperturbed

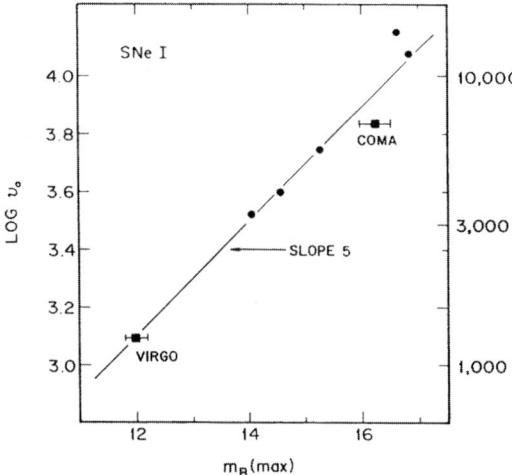

FIGURE 3. Hubble diagram for 16 SNe Ia in E and S0 galaxies with known $B(max)$ photometry. The data for six SNe I in Virgo and five in the Coma Cluster are averaged. Diagram from Sandage & Tammann (1982).

global Hubble flow. The line is the expected relation for a linear velocity-distance relation with a slope of $dM/d\log v = 5$.

The sample was expanded and data were added for the Hubble diagram in V by Sandage & Tammann (1993, section 2) where we also gave a review of the strong evidence at that time for homogeneity of SN Ia with a remarkably small dispersion of ~ 0.35 mag in $<M(max)>$. This small dispersion, that still contained the appreciable random photometric errors of the data available at the time, was consistent with the precept we had adopted for the HST calibration program that such SNe were sufficiently good standard candles that no second-parameter corrections were needed, at least not at the level of the factor of 1.6 between the short distance scale with $H_0 \sim 90$ (de Vaucouleurs 1979; Aaronson & Mould 1986; Jacoby et al. 1992; Pierce 1994) and the long scale with $H_0 \sim 55$ advocated by us from the earlier papers of the Steps series.

The resulting 1993 Hubble diagrams in m_{pg}, B, and V are shown in Fig. 4. These had been made in preparation for the initial determination of H_0 from the HST SNe Cepheid calibration program done in IC 4182 (Sandage et al. 1992; Saha et al. 1994) and NGC 5253 (Sandage et al. 1994; Saha et al. 1995) before repair of the HST's spherical aberration.

The data in Fig. 4 are from the Basel Atlas (Leibundgut et al. 1991b) and the discussion by Tammann & Leibundgut (1990). Five SNe Ia in the Virgo Cluster are included at the extreme lower left, using 1179 km s^{-1} for its global expansion velocity corrected for all local perturbations by the method of Sandage and Tammann (1990) and Jerjen & Tammann (1993). The redshifts, v_{220}, are the observed heliocentric velocities corrected to the Virgocentric kinematic frame using the "infall" 220 model of Kraan-Korteweg (1986).

By 1994, new high quality data were becoming available from the important discovery program for SN done by the Cerro Tololo/University of Chile collaboration begun in 1991. (The similarly important Harvard data were yet in the reduction stage [Riess 1996]). We combined 12 new data points from the Tololo/Calan collaboration with our master list used for Fig. 4 to give the Hubble diagrams in B and V in Fig. 5. We used this diagram in the first four papers after HST repair of the HST Cepheid SNe Ia calibration

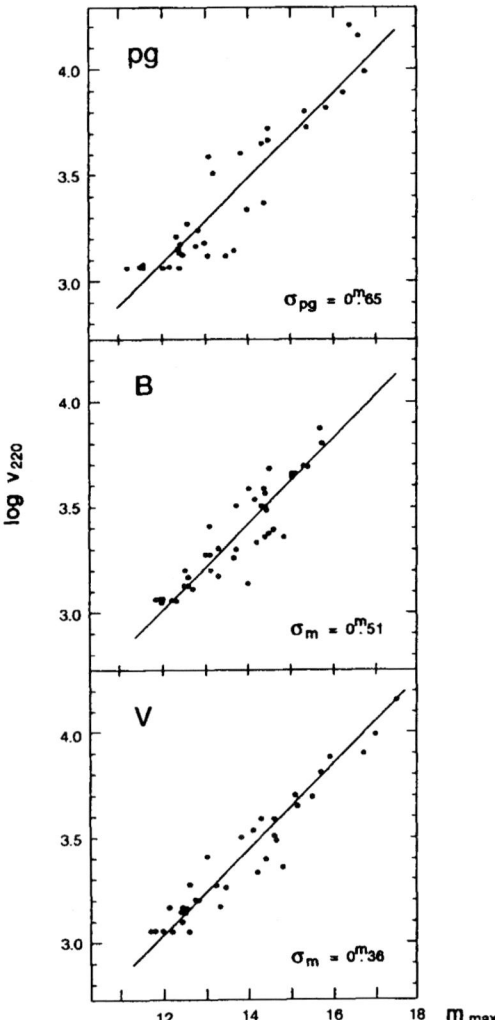

FIGURE 4. Hubble diagram in m_{pg}, B, and V from Sandage & Tammann (1993). The appreciably decreased scatter in the middle and lower panels compared with that in m_{pg} was a clue that observational errors still dominated the data at the $\sigma = 0.35$ mag level in the observed dispersion at the time.

program. The diagram was impressive, proving Kowal's discovery but with three times smaller dispersion. Although this was the tightest Hubble diagram in the literature at that time, it was soon replaced by diagrams that began to show fine structure, revealing particular second-parameter effects as the data became progressively even better.

By 1996 it had become evident that there was a small but systematic difference in the zero points of the Hubble diagram of E plus S0 galaxies compared with that of spirals. van den Bergh (1992) had already suggested a difference of 1.1 mag with the SNe Ia in spirals being brighter than those in E or S0 galaxies, but based on a highly scattered Hubble diagram to be shown as Fig. 7 in the next section. The data available by 1997, seen in the Hubble diagrams in Fig. 6 here, shows the small but significant (at the 3σ level) difference of 0.25 ± 0.08 mag.

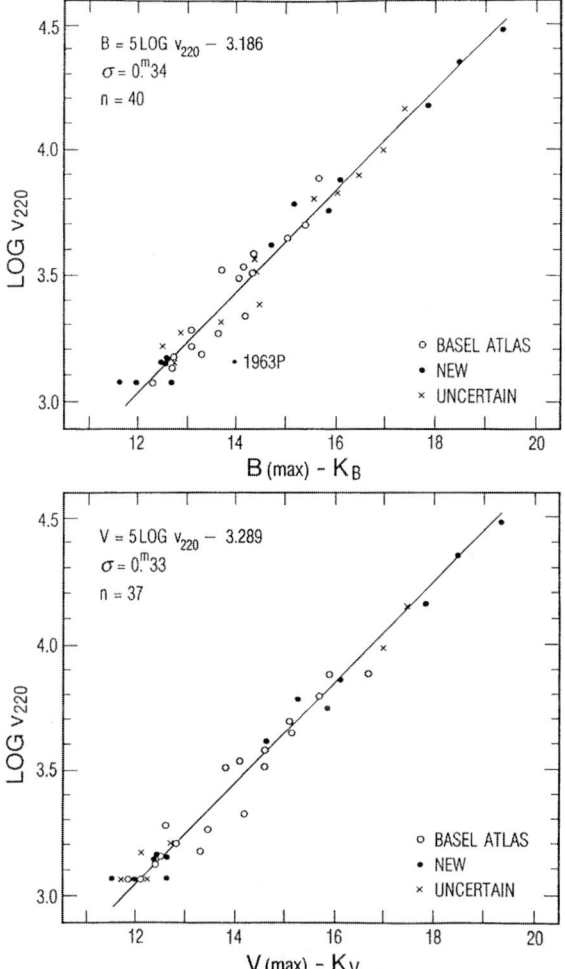

FIGURE 5. The Hubble diagram used in the first six HST SNe Ia calibration papers that measured Cepheid distances to parent galaxies that contained SNe Ia. This was the best Hubble diagram known at the time. The velocities are reduced to the Virgocentric kinematic frame. Diagram from Tammann & Sandage (1995).

SNe Ia in E and S0 galaxies are fainter than those in late type spirals. The reason for the difference is, of course, supposed to be due to a mass difference of the progenitors in spirals (higher mass) than those in E and S0 galaxies (lower mass) due to the different mean evolutionary state (age) between the stars that become white dwarfs in the different galaxy morphological types. This mass difference is presumed to produce a different mixture of carbon and oxygen in the white dwarf cores, and therefore different burning details once the progenitors ae nudged over the Chandrasekhar limit leading to explosion. Hence, one supposes that different amounts of ^{56}Ni are produced between E plus S0 and spiral SNe Ia (Branch, Kokholov, Wheeler, Woosley, and others in conversation at this meeting).

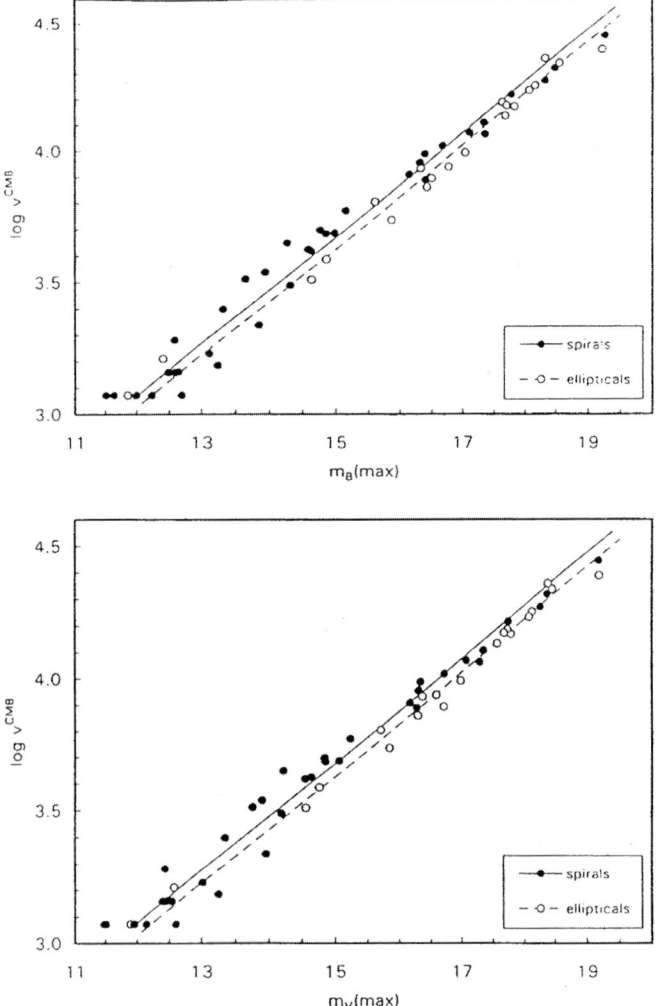

FIGURE 6. Hubble diagram showing the systematic offset of SNe Ia in E and S0 galaxies (open circles) from those in spirals (closed circles). The dashed lines for the E and S0 galaxies average 0.25 mag fainter at a given redshift than the solid line for the spirals. The velocities are reduced to the global frame of the microwave background. The data are from Leibundgut et al. (1991b), Patat et al. (1997), Hamuy et al. (1996), and Riess et al. (1996). Diagram is from Saha et al. (1997), and Sandage & Tammann (1997).

3. The 1992 conference at Viña del Mar, Chile, as the beginning of the debate on second-parameter difficulties

An important session on the problem of supernovae and the Hubble constant was held at Viña del Mar, Chile, in June 1992 as part of the 7th Interamerican Regional Reunion of the IAU. Our SNe Ia consortium for the Hubble constant had just obtained its first HST results (Sandage et al. 1992; Saha et al. 1994) for the Cepheid distance to IC 4182. This galaxy was parent to SN 1937C which was a classical Branch normal SN Ia that remains a prototype of the class. The supernova was one of the two studied for the light curves by Baade & Zwicky (1938), mentioned earlier.

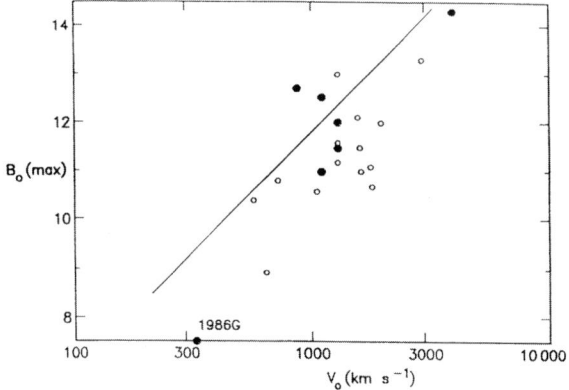

FIGURE 7. The Hubble diagram proposed by van den Bergh & Pazder (1992) with no restrictions on the nature of the SNe Ia sample and with "absorption" corrections, based on observed colors at maximum. Compare the scatter here with that of Figs. 4 and 5 which are restricted to SNe Ia that are bluer than $(B_{\max} - V_{\max}) = 0.2$ and that have Branch normal spectra.

Armed with the HST Cepheid data and the Hubble diagrams of Fig. 4 from our fiducial SNe Ia sample as it existed in 1992, one of us (AS) made the case for homogeneity and a value of H_0 near 50 km s^{-1} Mpc^{-1}. Clearly, those favoring the short distance scale were disappointed, leading to the lively discussions at the meeting, mostly unrecorded, on just how good SNe Ia were as reliable distance indicators.

We gave the evidence for homogeneity, similar to what is set out in section 2. The case seemed to be greatly strengthened in a lecture by Hamuy (1992) reporting on the Cerro Tololo/University of Chile discovery program (see also Hamuy et al. 1993). He emphasized strong homogeneity in the data as it had been analyzed at the time, and, although mentioning small differences in the light curve shapes, he did not suggest that these were so large as to invalidate our conclusion that $H_0 \sim 50$ based on SN 1937C with its HST Cepheid distance. Our result could not be wrong by the factor of 1.6 that would be required to support $H_0 \sim 90$ advocated by the short distance scale groups. Pleased at that point in the meeting by Hamuy's emphasis on homogeneity, our encomium for the result was short lived after two further reports later in the session, one by Phillips (1992) and the other by van den Bergh (1993).

Although Hamuy had emphasized homogeneity, Phillips using the same data, dwelt on the diversity by including data for non-Branch normal SNe and by emphasizing a range of decay rates, resurrecting the Pskovskii effect, but not yet in the form of his famous later diagram (Phillips 1993). Knowing of the tight Hubble diagrams of Fig. 3 and Fig. 4 where only Branch normal SN had been used, we asked in the discussion how many of the 20 new SNe that had been discovered in the Chile program were Branch normal. The reply was 19, meaning to us that whatever small effect the diversity of the light curve for non-Branch normal might have on the intrinsic dispersion of $<M(max)>$, it had to be smaller than the dispersion displayed by Fig. 4, known at the time. Hence, Phillips' emphasis on diversity, as presented at Viña del Mar, did not change our conclusions on H_0 via SN 1937C.

However, van den Bergh (1993) reached a different conclusion. He claimed that SNe Ia were such poor standard candles as to be worthless in the quest for H_0, despite the evidence from the tightness of the extant Hubble diagrams made using "appropriately" restricted samples. The thrust of his argument was that by including such abnormal SN as 1991bg, 1986G, and 1991T the Hubble diagram becomes highly scattered. In addition, making "absorption" corrections (van den Bergh & Pazder 1992) gave an even more widely scattered Hubble diagram. A representative Hubble diagram showing the effects of van den Bergh's precepts is given in Fig. 7, taken from van den Bergh & Pazder (1993).

Van den Bergh's (1993) conclusion was that the long distance scale with $H_0 \sim 50$, based on SNe Ia and the Hubble diagram of Fig. 4, was simply wrong. He maintained that SNe were not reliable standard candles at all; he would not agree to our precept that discarding both the abnormals and also SNe Ia whose colors at maximum were redder than $B - V = 0.2$ was justified despite the resulting tight Hubble diagram. We continued the disagreement in private at Viña del Mar but with no satisfaction to either position. The remarkable later paper (van den Bergh 1996), discussed in section 4, continued the disagreement.

4. The various Tololo/Calan formulations of the decay rate-absolute magnitude correlation—Our initial objections

Discussions of the efficacy of SNe Ia as distance indicators entered a new phase after the Viña del Mar conference. The influential paper by Phillips (1993) appeared soon thereafter, showing a range in $M(max)$ of ~ 2 magnitudes, with a step dependence on the Δm_{15} decay rate. Fig. 8 is the Phillips diagram from his 1993 paper. The range in absolute magnitude (on the distance scale adopted by Phillips) was so large that if the phenomenon also applied to the best Hubble diagrams known at the time (e.g. Figs. 3 and 4) we could not understand the evident small scatter. This was to become even more serious using the improved diagrams of Figs. 5 and 6, the preliminary versions of which were known earlier than their publication dates. For this and other reasons we were skeptical of both the steepness and the zero point of the 1993 Phillips diagram. Two of us wrote a dissension (Tammann & Sandage 1995).

Our principal argument was that four of the nine plotted points in Fig. 8 were abnormal SNe; three of the four (SN 1971I, 1986G, and 1991 bg) have the fastest decay rates with $\Delta m_{15} > 1.6$ mag. They also have the faintest luminosity of the group. They are also spectroscopically abnormal in the sense of Branch et al. (1993). These are the three points farthest to the right in Fig. 8. The fourth "non-Branch normal" is 1991T which is the extreme left data point. It is also the brightest in the diagram. All four would have been excluded from our Hubble diagrams by the restrictions on spectroscopic normality and the color cut at $B - V = 0.2$.

Eliminating the four points from Fig. 8 left even the existence of an effect suspect. The evidence is shown in Fig. 9 taken from Tammann & Sandage (1995, Fig. 3 there). The data for the filled dots are from redshift distances calculated with $H_0 = 52$. This was our calibration value from the first two SNe Ia/Cepheid calibration experiments (Saha et al. 1994, 1995). The three open squares in the diagram are these first calibrators. The small open circles and skipping-jack crosses show the data used by Phillips (1993) based on his adopted distances from the TF and SBF methods with their zero points as given in 1993.

Our second skepticism concerning Fig. 8 concerned the use of distances determined by these two methods. Our distrust of both the TF and the SBF data was based on the impossibly small distance modulus of the Virgo Cluster of $(m - M) = 30.96$ derived by Pierce & Tully (1992) using TF, and $(m - M) = 30.84$ by Tonry (1991) using SBF, whereas our modulus from a large number of different methods had consistently ranged from 31.5 to 31.8 (cf. Tammann 1996, 1997, 1998a,b; Sandage & Tammann 1997; Sandage, Tammann, and Saha 1998, for reviews). Furthermore, Pierce

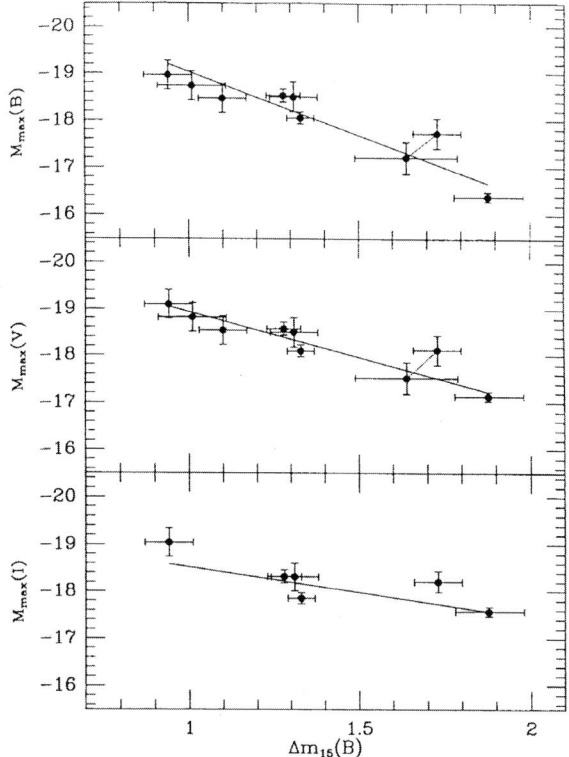

FIGURE 8. The original suggestion by Phillips (1993) for a steep variation of $M(max)$ with decay rate for nine SNe Ia for which he assumed the distance from the TF and the SBF methods. Four of the nine are not "Branch normal." The three most right hand data points are the spectroscopically peculiar (relative to the Branch et al. 1993 template) objects SN 1971I, SN 1986G, and SN 1991bg. The most left hand and brightest point is the superluminous and spectroscopically abnormal 1991T. Absolute magnitudes are on the faint zero points of the TF and the SBF methods as calibrated by Pierce & Tully (1992) and Tonry (1991). These average 0.7 mag fainter than the Cepheid zero point. Diagram from Phillips (1993).

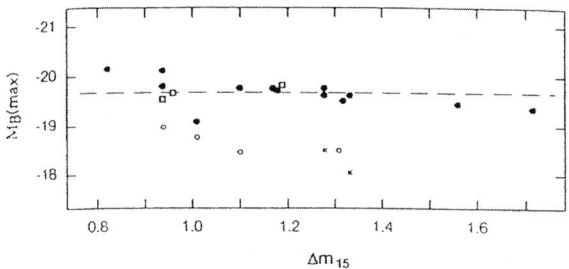

FIGURE 9. Decay rate-absolute magnitude correlation in the paper that argues against the steep slope in Fig. 8. The absolute magnitudes for the black circles are kinematic luminosities based on redshifts and $H_0 = 52$ for SNe Ia in the fiducial sample used for Fig. 5 for those SNe that also had known decay rates in 1995. The three open squares are based on the Cepheid distances of the parent galaxy from the first three calibrations of the HST calibration program (Saha et al. 1994, 1995 for IC 4182 and NGC 5253). The small open circles are from the Phillips (1993) paper based on TF distances. The two small skipping-jack crosses are also from Phillips, based on SBF distances. The mean line at $M_B = -19.65$ is from the extant Cepheid calibration known at the time. Based now on 9 calibrators (Saha et al. 1999), the mean has become fainter to $<M(max)>_V = -19.49 \pm 0.07$. Diagram from Tammann & Sandage (1995).

(1994) had derived $<M(max)>_{\text{SNe Ia}} = -18.74 \pm 0.14$ with his TF calibration of the SNe parent galaxies, whereas our first three SNe Ia calibrations using HST had, by 1995, given $<M(max)> \sim -19.5$ (Saha et al. 1994, 1995). One of the two calibrations of $<M(max)>_{\text{SNe Ia}}$ clearly contained a systematic error.

But even the *relative* distances derived using the TF method were suspect for us because of the known strong bias problems in both the calibration and the application of the Tully-Fisher method (Sandage 1988b, 1994b; Federspiel et al. 1994; Sandage et al. 1995) that distorts the linearity of the TF distance scale.

In addition, the statement by Tonry (1991) that "there is remarkable agreement between the fluctuation and the IRTF distances" showed that Tonry's zero point of the SBF method must also contain a systematic error of the same order as that from the TF relation used by Pierce.

For these reasons we distrusted both the zero points and the scale of the data that went into Fig. 8. These were the reasons we remained skeptical of the magnitude of the resurrected Pskovskii effect by Phillips. In the rebuttal paper (Tammann & Sandage 1995) we derived a slope to a magnitude-decay rate correlation that was four times smaller than that of Phillips (1993).

In the fullness of time, there is now high quality evidence that the distances used by Phillips (1993) *are* too small and also have a systematic scale error with distance. Cepheid distances using HST (Saha et al. 1996a for NGC 4536, 1997 for NGC 4639, and 1999 for NGC 3627) are presently known for three of the nine SNe in the diagram by Phillips. Inferred distances exist for two others in the Fornax Cluster (NGC 1316 and NGC 1380), based on the near identity of SN 1989B in NGC 3627 (for which a Cepheid modulus of 30.22 exists) with SN 1980N in NGC 1316, using the apparent magnitude difference of 1.62 ± 0.03 mag between the two supernovae (Wells et al. 1994). These data give a modulus of $(m-M)_o = 31.84$ for NGC 1316 whereas Phillips used $(m-M)_o = 31.02$. Because of the spatial tightness of the E and S0 galaxy core to Fornax, it is likely that NGC 1380 is close to the same distance as NGC 1316. The modulus used for NGC 1380 by Phillips was 30.65.

These five overlaps with Phillips' (1993) Table 1 show differences in the distance moduli (Phillips' moduli are smaller) that range from 0.60 mag for NGC 4536 to 1.19 mag for NGC 1380. These differences are evident in Fig. 9. Phillips (1993) recognized the problem in his footnote 2, commenting on the systematic difference in the distance scales based on Cepheids and on the TF plus SBF methods.

The next development was made by Hamuy et al. (1996) using the new and extensive results of the highly successful Tololo/Calan SNe discovery and photometric program. From these data Hamuy et al. derived much shallower magnitude-decay rate relations in B, V and I than the initial relation by Phillips. Their diagram shown here in Fig. 10 was based on relative redshift distances calculated with $H_0 = 65$. This central paper of the Tololo/Calan collaboration revised an earlier discussion (Hamuy et al. 1995, their Figs. 4 and 5) that was based on $H_0 = 85$, but which still espoused the steep slope of the decay rate relation made by keeping all the Branch abnormal examples.

Although the bulk of the data in Fig. 10 are the kinematic (redshift) distances, the nine original calibrators used by Phillips are also shown by the filled circles, plotted with his originally adopted absolute magnitudes using "absolute"(i.e. directly determined) RF and SBF distances. The differences in zero point and slope (scale error) between the "relative (i.e. depending on the assumed Hubble constant) redshift distances and the TF, SBF distances are now clearly evident.

Amidst these developments on the size of the second parameter correction for decay rate, a most remarkable paper was published by van den Bergh (1996a), titled "The

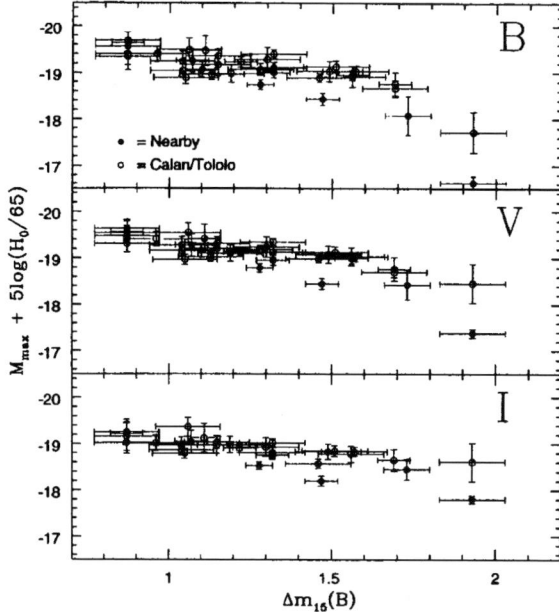

FIGURE 10. The calibration of the decay rate relation of Hamuy et al. (1996) using kinematic absolute magnitudes with $H_0 = 65$ and with the absolute distances used by Phillips from the TF and SFB methods shown as filled circles.

Luminosities of Supernovae of Type Ia Derived From Cepheid Calibrations." Van den Bergh again argued that the variation in $<M(max)>$ was a factor of 20 in intrinsic luminosity, implying that this range crippled the use of such SNe to determine H_0, consistent with his frequent reviews (van den Bergh 1992, 1993, 1994, 1996b) that defined his short distance scale.

His conclusion was based on his summary Table 1 which, however, has four debatable features.

(1) He includes the two known highly peculiar cases of SN 1991T and SN 1991bg which are Branch abnormal and therefore would not be part of our HST calibration experiments where the SNe for the Hubble diagram are restricted to Branch normal SNe Ia bluer than $B - V = 0.2$ at maximum.

(2) He uses $M(max) = -20.40$ for SN 1960F, whereas the value determined in the HST experiment is $M(max)_V = -19.62$ (Saha et al. 1996b).

(3) He lists $<M(max)> = -18.76$ for the mean of eight SNe Ia in Virgo-Cluster-complex galaxies, based on his assumed Virgo modulus of $(m - M) = 31.02$, whereas the modulus is undoubtedly larger for the reasons discussed earlier. Our mean Virgo cluster modulus is $(M - M) = 31.7$ (op cit.).

(4) He includes the problematical case of SN 1885 in M31 using $M(max) = -17.33$, assuming that it was a classical, bona fide SNe Ia.

Excluding SN 1885, SN 1991T, SN 1991bg, and his Virgo entry, each on the basis of our precept that we restrict the sample to Branch normal SNe Ia, the remaining five entries in his Table 1 give $<M(max)> = -19.51 \pm 0.04$ with an rms of 0.08 mag. Most of these five entries are in fact from our HST calibration program. The agreement of this mean with our current calibration of $<M(max)> = -19.48 \pm 0.07$ from eight calibrators (Saha et al. 1999), suggests that van den Bergh would now accept the long distance scale with $H_0 \sim 55$ when the supernova method is used with the Hubble diagrams such as

Figures 5 and 6, or the new, even tighter diagrams of Humay et al. (1995), Saha et al. (1997), Riess et al. (1996), Parodi et al. (2000), and Phillips et al. (2000).

Finally, an important paper by Maza et al. (1994) was influential in showing that the Pskovskii-Phillips is real and that the differences in $M(max)_{\text{SNe Ia}}$ are indeed correlated with their decay rates. Maza et al. set out the case of SN 1992bc and 1992bo that have closely the same redshifts (5940 km s^{-1} and 5445 km s^{-1}) but have very different decay rates (0.87 and 1.69 mag respectively). These SNe Ia differ by 0.9 mag in absolute magnitude (based on redshift ratios). SN 1992bc appeared in an Sab galaxy; SN 1992bo was in an E galaxy. The sense of the difference in luminosity is the same as the general rule (Fig. 6) that SNe Ia in E and S0 galaxies are fainter than those in late spiral types. The sense is also consistent with the Phillips relation; those with fast decay rates (large Δm_{15}) are fainter than those with smaller rates.

Hence it had become clear by 1996 that the luminosities of SNe Ia form a continuum with a range of at least 0.6 mag (cf. Fig. 9) over the range of decay rates that would still encompass Branch normal events and also the decay rates of our calibrators.

By 1995 a number of other groups had now entered the arena of collection and analysis of new SNe data. We were all faced now with the problem of how to use this development of diversity for the H_0 calibration problem.

5. Present status of the HST Cepheid-SNe Ia calibration program

Various approaches have been made to the problems of either restricting the sample or correcting the absolute luminosities of particular SNe Ia to increase the homogeneity of $<M(max)>$. The variety of the results for H_0 depend on the assumed severity of the second-parameter problem and the particular precepts used for the restrictions.

Results from the SN method to date range from $H_0 = 60 \pm 6$ (Saha et al. 1999; Parodi et al. 2000) to $H_0 = 68 \pm 7$ (Freedman 1997; Gibson et al. 2000, which however is incorrect in their treatment of the extant calibration data in the correction, or not, for reddening). Most others average between 60 and 64 (cf. Riess, Press, & Kirshner 1996; Tripp & Branch 2000, Phillips et al. 2000), based on a Cepheid zero point with $(m-M)_o = 18.50$ for the LMC.

We continue to assert that the correction for second-parameter effects to a raw value of H_0 is less than 10%, based on a non-linear Pskovskii-Phillips decay rate correlation that is nearly flat in the mid range of the decay rates events (Saha et al. 1999; Parodi et al. 2000), and on the eight Cepheid calibrators now available. The extremes of the effects of the correction range from a $\sim 5\%$ effect in Saha et al. (1999) to an $\sim 11\%$ effect in Phillips et al. (2000), discussed next.

5.1. A minimum correction

Fig. 11 illustrates the reason why we initially adopted the precept for no correction as long as the sample was "appropriately" restricted. The diagram is the same as Figure 10 but the top panel has been altered to exclude the extremes of the Δm_{15} distribution cut at 1.0 and 1.5 mag. The remaining range includes most of our absolute calibrators (Saha et al. 1999, Table 5).

Ignoring the five black dots from (Phillips 1993) for the reasons given in section 4 leaves a distribution that is fitted by a shallow correlation with a slope of $(dM/d\text{ decay rate})_B = 0.38$ in this restricted decay-rate range. Recall that the initial Phillips (1993) slope was $(dM/d\text{ decay rate})_B = 2.70$. Recall also that we derived $(dM/d\text{ decay rate})_B = 0.88$ in our 1995 dissension, but there we used a wider range of decay rate than in Fig. 11

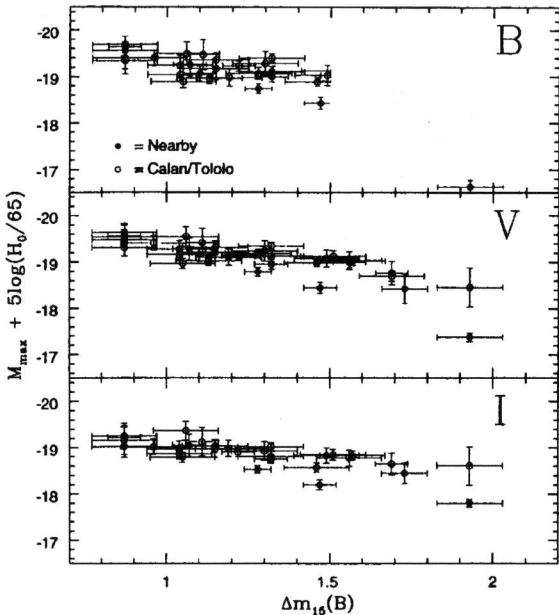

FIGURE 11. Same as Fig. 10 but with the data in the top panel restricted to decay rates between Δm_{15} of 1.0 and 1.5, showing the nearly flat relation in this interval, emphasizing the non-linearity of the correlation with absolute magnitude in Fig. 10 over the entire range of decay rates.

(top). From these differences, depending on the assumed range of decay rate, it is evident that the relation between absolute magnitude and decay is nonlinear.

From this fact, Saha et al. (1999) adopted the cubic relation shown in Fig. 12. The data are from the fiducial SNe Ia sample known to 1997 that met our restrictions on spectral normality and color (Table 6 of Saha et al. 1999). Note the difference between Fig. 12 and Fig. 8.

The Roman crosses in Fig. 12 are for the early type spirals in the fiducial sample of 1997. The open circles are for late type spirals. The skipping-jack crosses are for E and S0 galaxies in that sample.

A notable feature is that the open circles occupy a different part of the diagram than the skipping-jack crosses. This is the galaxy type dependence of the absolute magnitude between spirals and early type galaxies shown in the Hubble diagram of Fig. 6 (see Table 6 of Saha et al. 1999).

Fig. 12 was used by Saha et al. (1999) to correct the data for the decay rate effect for both the fiducial sample and the calibrators. After applying this correction we then searched the resulting modified absolute magnitudes for additional second parameter correlations. A much smaller second correlation with the colors was found, although the statistical significance was small (see below). Similar results had been reported earlier by Tripp (1998), by Riess (private comment), and again by Tripp & Branch (2000).

Our third-parameter color variation is shown in Fig. 13 where the decay-rate corrected absolute magnitudes, called now M^{15}, are plotted against the observed $B_o - V_o$ colors, at maximum. The subscript zero here denotes "as observed." They are not corrected for reddening (absorption); recall that the restricted sample has already excluded all SNe ia that had $B_{\max} - V_{\max} > 0.20$.

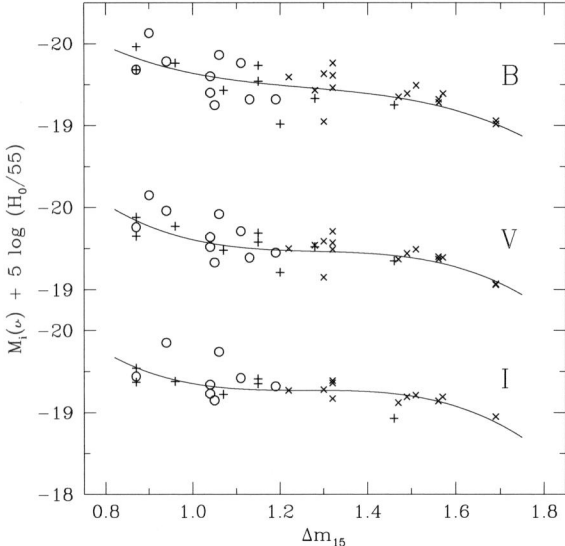

FIGURE 12. The data for the 1997 fiducial sample used by Saha et al. (1999, their Table 6) fitted with a cubic showing the nearly level correlation at intermediate decay rates. Open circles are late type spirals, Roman crosses are early type spirals. Skipping jacks are E and S0 galaxies. Note that the decay rates for the E plus S0 galaxies are systematically larger than for spirals in the mean. Absolute magnitudes are based on redshifts and a variable H_0 that changes from 60 for zero redshift to 55 at $v = 10,000$ km s^{-1}. Diagram from Saha et al. (1999).

We concluded from Fig. 13 (Saha et al. 1999) that the variation of color of this fiducial sample was not due to reddening but, if real, must be intrinsic. The slopes of the correlations in each panel are far from the reddening values.

As discussed below, we still assert the premise that reddening is not the cause of the apparent correlations in Fig. 13, but a later analysis makes problematical if even the apparent variation of color shown in the diagram is real, given the known random color errors quoted by Hamuy et al. of rms = 0.1 mag (Parodi et al. 2000). In any case, the variations shown in Fig. 13 are so small as to be within the statistical errors of "no effect."

With the corrections of Fig. 12 and 13 applied to the Hubble diagram defined by the fiducial sample, and with the nine Cepheid calibrators set out in Table 1, and by correcting the Cepheid P-L relation to a zero point such that $(m - M)_o = 18.58$ for the LMC (cf. Federspiel, Tammann, & Sandage 1998; Sandage, Bell, & Tripicco 1999) we have obtained $H_0 = 58 \pm 2$ (internal) km s^{-1} Mpc^{-1} (Saha et al. 1999).

5.2. A second method

The formulation in the last section was made using a variable local Hubble constant that decreases outward by 10% until it reaches the global value at $v = 10,000$ km s^{-1}. The evidence was from Tammann (1998a), Zehavi et al. (1998), and an early version of Parodi et al. (2000) where the slope of the Hubble diagram was derived to be $d \log v/dm = 0.192$ "locally" ($v < 10,000$ km s^{-1}) rather than the global value of 0.200.

However, we have also made a solution using a fixed value of H_0 for all distances (Parodi et al. 2000) to assess the sensitivity of the second-parameter corrections to the differences in the precepts used in Saha et al. (1999). The velocities were corrected to the frame of the microwave background. We only considered data for SNe events after

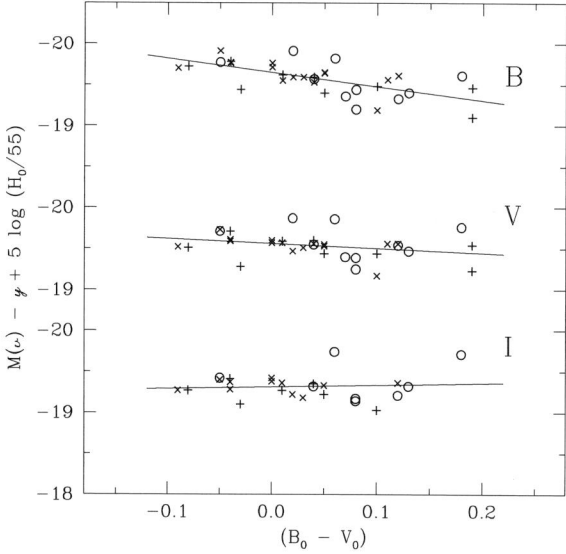

FIGURE 13. Correlation of observed color, $B_{max} - V_{max}$, with absolute magnitude after applying a first correction for decay rate using Fig. 12. Symbols are the same as in Fig. 12. Diagram from Saha et al. (1999).

SNe Ia	Galaxy Cepheids	$(m-M)_{AB}$	$(m-M)_{AV}$	$M_{Bo}(max)$	$M_{Vo}(max)$	Δm_{15}	Sources
1937C	IC 4182	28.36	28.36	−19.53	−19.48	0.87	1,2
	(39)	0.09	0.12	0.15	0.17		
1895B	NGC 5253	28.13	28.10	−19.87	—		3,4
	(15)	0.08	0.07	0.22			
1972E	NGC 5253	28.13	28.10	−19.52	−19.49	0.87	3,5
	(15)	0.08	0.07	0.22	0.14		
1981B	NGC 4536	31.10[a]	31.10[a]	−19.46	−19.44	1.10	6,7
	(74)	0.05	0.05	0.21	0.18		
1960F	NGC 4496A	31.16	31.13	−19.56	−19.62	1.06	8,9
	(95)	0.10	0.10	0.14	0.18		
1990N	NGC 4639	32.03[a]	32.03[a]	−19.33	−19.42	1.07	10,11
	(20)	0.22	0.22	0.23	0.23		
1989B	NGC 3627	30.22[a]	30.22[a]	−19.36	−19.34	1.31	12,13
	(58)	0.12	0.12	0.18	0.16		
1974G	NGC 4414	31.41	31.41	−19.59	−19.61	1.11	14
	(9)	0.17	0.17	0.35	0.30		
1998bu	NGC 3368	30.37[a]	30.37[a]	−19.53	−19.51	0.95	15,16
	(7)			0.16	0.30		
Straight mean (neglects 1895B)				−19.49	−19.48		
				0.03	0.03		
Weighted mean (neglects 1895B)				−19.49	−19.48		
				0.07	0.07		

[a] The true modulus is listed.

Sources: (1) Sandage et al. 1992; Saha et al. 1994; (2) Schaefer 1996; Jacoby & Pierce 1996; (3) Sandage et al. 1994; Saha et al. 1995; (4) Schaefer 1995a; (5) Hamuy et al. 1995; (6) Saha et al. 1996a; (7) Schaefer 1995a,b; Phillips 1993; (8) Saha et al. 1996b; (9) Leibundgut et al. 1991b; Schaefer 1995c; (10) Sandage et al. 1996; (11) Leibundgut et al. 1991a; (12) Wells et al. 1994; (13) Saha et al. 1999; (14) Schafer 1998; (15) Tanvir et al. 1995; (16) Suntzeff et al. 1999.

TABLE 1. Data for nine Cepheid-calibrated SNe Ia

1985 and with $v < 30,000$ km s^{-1} in the Parodi et al. sample. Details are in Parodi (2000).

The conclusions are:

(1) There is a narrow distribution of observed color that is well fitted by a Gaussian with the remarkably small σ of ~ 0.05 mag. In view of the claimed accuracy of the observations of color at maximum of no better than a σ of between 0.05 and 0.10 mag (Hamuy et al. 1996), there is now no room for intrinsic variation of color. Hence Fig. 13 may not in fact show a significant intrinsic variation. This is a current problem awaiting solution using increased samples with precision photometry.

(2) The Hubble diagrams in B, V, and I in Fig. 14, using the decay rate corrections similar to Fig. 10, but over the restricted range similar to Fig. 11 (top), are very tight showing the power of the supernova method.

(3) The slope of the Hubble diagram is 0.200 only for $v < 10,000$ km s^{-1}. The slight systematic deviation of the points for $v > 10,000$ is not presently understood but may be due to a systematic difference in *mean* galaxy type as a function of redshift that is not completely accounted for by the decay-rate corrections that are also related to Fig. 6. It is this deviation that gave the overall slope of 0.192 used in the calculations of section 5.1.

(4) Calculations using the calibrators in Table 1, corrected for decay rate in the same way as in Fig. 14, and corrected to the Cepheid zero point for the P-L relation using $(m-M)_o = 18.58$ (Panagia 1999) for LMC gives H_0 (global) $= 58 \pm 2$. This is identical with the result of section 5.2, but there are many subtleties not yet understood at the $\sim 5\%$ level.

5.3. Other formulations

As mentioned several times earlier, many groups have discussed the second-parameter corrections, deriving their own versions of Hubble constants using various manipulations of the calibrator Table 1 as that table has evolved with the progressive adding of new data. The analyses of Riess et al. (1996) and of Tripp (1994) and Tripp & Branch (2000, $H_0 = 62$) have already been mentioned. We now discuss here two others where we question several of the precepts.

5.3.1. Phillips et al. (2000)

This paper has often been cited above. The bottom line of their analysis is contained in three conclusions.

(a) The Hubble diagram corrected only for Galactic reddening and using the calibrators as set out in their equivalent of Table 1 gives $H_0 = 57.7 \pm 3.5$ if no second-parameter corrections are applied, either to the calibrators or to the Hubble-diagram sample.

(b) When the additional second-parameter corrections due to decay rate alone are applied as formulated by them (again these differ from ours), their result is $H_0 = 64.6 \pm 3.2$.

(c) They introduce a third correction for "host galaxy reddening" of the SN themselves, which, when applied to (a) and (b) gives $H_0 = 64.0 \pm 2.8$.

Because their calibrators are based on their version of Table 1, these values for H_0 are based on the Cepheid zero point that gives 18.50 for the LMC modulus. The quoted values of H_0 must then be reduced by 4% for an LMC modulus of 18.58.

Our approach differs from that of Phillips et al. (2000). They assert that the color range at a given decline rate (i.e. after correction for the Δm_{15} dependence) is due to reddening, and proceed accordingly, correction for internal "absorption." The philosophy is similar to that of van den Bergh and Pandzer, but their "absorption" corrections are of course smaller. Our conclusion from the analysis leading to Fig. 13 (and a similar diagram

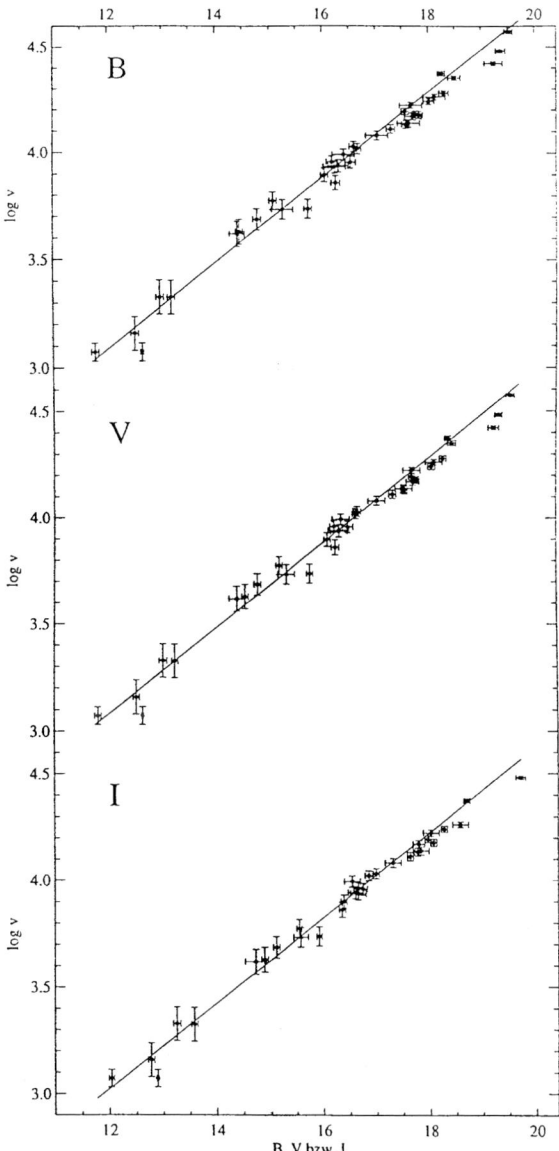

FIGURE 14. Hubble diagrams in B, V, and I from the fiducial sample of Parodi et al. (2000) as corrected by them with their version of the decay rate-absolute magnitude relation. The slope of the relation for $v < 10,000$ km s^{-1} is 0.2. The deviation of the data at larger velocities may be due to a systematic second-parameter morphological-type effect rather than a real property of the expansion. Diagram from Parodi et al. (2000).

using $V - I$ colors; Saha et al. 1999) is that whatever spread is in our color-restricted fiducial sample, the apparent color spread at a given decline rate must be intrinsic, not due to reddening; the color/absorption slopes are not normal. But, as mentioned in the last section, there may not be a color range at all, given the claimed random errors in the observed colors with a σ as large as 0.05 mag (Parodi et al. 2000).

Further, we believe that (1) the decay rate correction in Fig. 12 in the relevant range of decay rates is smaller than that of Phillips et al. (2000) by about a factor of two, and (2) no correction for "absorption" should be applied.

5.3.2. The "Key Project" formulation

Kennicutt, Mould, & Freedman (1998, KMF), and Freedman (1997) discard several of our calibrators for reasons that we find debatable.† They then add three of their own that are not based on direct Cepheid distances and, in addition, that are ~ 0.5 mag fainter than can be justified by external evidence. Consequently, they have derived fainter mean absolute magnitudes than in Table 1. Their version of the calibration also gives a steeper decay rate-absolute magnitude relation than either Fig. 11 (top) or Fig. 12.

Specifically, as was discussed in section 4 in connection to the distances used by Phillips, KMF assume that the distances of the early type galaxies NGC 1316 and NGC 1380 in the Fornax Cluster, parent galaxies to SN 1980N, SN 1981D, and SN 1992A, are identical with that of the spiral NGC 1365 for which they have obtained a Cepheid distance. (The same assumption is suggested by Suntzeff et al. 1999 for the Tololo/Calan discussion of H_0. They question the Cepheid calibration rather than the debatable precept that the distance to the core of the Fornax cluster is given by the distance to the single spiral NGC 1365, which for us is clearly in the foreground).

Moreover, there is direct evidence that the distance of NGC 1316 is ~ 0.5 mag larger than that of NGC 1365 by the discussion in section 4 based on the data by Wells (1994) for SN 1989B in NGC 3627 and those for SN 1980N in NGC 1316. Our Cepheid modulus of NGC 3627 is $(m-M)_o = 30.22 \pm 0.12$. Hence the inferred modulus of NGC 1316 is $(m-M)_o = 31.84 \pm 0.21$ based on the observed magnitude difference between the two SNe of 1.62 ± 0.03 mag given by Wells et al. (Note that we must combine with the 0.17 mag uncertainty to allow for scatter in the difference in peak of two SNe Ia with the same decay rate). This modulus is 0.51 mag fainter than that of NGC 1365 (Madore et al. 1998) but is close to the mean value for the early-type galaxies in the Fornax cluster core by independent methods that give $(m-M)_o = 31.80$ (Tammann & Federspiel 1997; Sandage & Tammann 1997; Tammann 1998b). The implication of Kennicutt et al. (1998) and Freedman (1998) that (a) the luminosities of two SNe Ia of 1989B in NGC 3627 and 1980N in NGC 1316 differ by ~ 0.5 mag whereas they are identical in their observed light curves and spectra, and (b) that SN 1980N, SN 1981D, and SN 1992A in the Fornax cluster are all three fainter again by the same (but now for a different reason) ~ 0.5 mag than the mean in Table 1, is not credible. For us, the simplest solution is that NGC 1365 is in the foreground and that it does not determine the distance to the compact core of E and S0 galaxies in the Fornax cluster.

The resulting fainter calibration of $<M(max)>_{\rm SNe\ Ia}$ and the steeper decay rate derived thereby by KMF compromises their conclusions concerning H_0, explaining their high value of $H_0 \sim 73$.

---

† Their assumptions for discarding these calibrators (SN 1960F and 1974G) are that the photometry of the light curves are the most uncertain of the list. We debate this on the basis that the original light curves have been corrected to the modern magnitude scales by photoelectric redeterminations of the comparison stars used by the original observers. It is also important to point out that the net worth of a calibrator is not just the uncertainty in the photometry of the supernova. Rather, it is the combined uncertainty with that of the Cepheid distance of the parent galaxy. Note from Table 1 that the Cepheid modulus of NGC 4496A is among the most accurate of the calibrating galaxies.

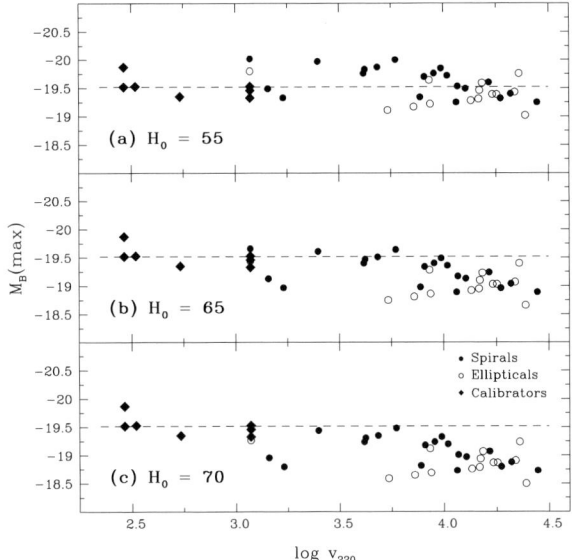

FIGURE 15. Kinematic absolute magnitudes for three different values of H_0 using the fiducial sample for SNe Ia after 1985 but with no second-parameter corrections. The seven Cepheid-based calibrators known in 1998 are closed diamonds. Unless the SNe Ia that are more distant than $v = 1600$ km s^{-1} are significantly fainter than the nearby calibrators, H_0 must be smaller than 65 from these data. The calibrators (diamonds) are from Table 1. Diagram from Saha et al. (1997).

5.4. Summary of the use of SNe Ia for H_0

As soon as the case was made by Phillips (1993) for diversity in the absolute magnitudes of even Branch normal SNe Ia, critics of the long distance scale suggested abandoning the SNe method altogether. The claim was that as long as there is a decay-rate absolute magnitude correlation at all, the supernova method loses its beauty and cannot, therefore, be used with any certainty.

Clearly this is false. Witness the tightness of the Hubble diagrams even without second-parameter corrections (cf. Figs. 5 and 6) if the samples are restricted by the spectroscopic and color criteria that we are discussing here.

All groups that have used the SNe Ia method (with the exception of the "Key Project" consortium) have obtained Hubble constants in the range of 55 to 64 depending on how they make the restrictive cuts to their samples and how they structure their decay rate and color corrections.

Nevertheless, the diversity of arguments on the size of the corrections and on how to treat the calibrators show that any final firm conclusion will come only by understanding the details of the systematic and statistical errors at the $\sim 15\%$ level (e.g. compare $H_0 = 55$ with 64). There *is* a Pskovskii-Phillips effect and perhaps a color correction as well.

The one conclusion that is clear from the available Hubble diagrams is that H_0 must be smaller than 65 km s^{-1} Mpc^{-1} if the calibration in Table 1 is used as given. The evidence is from Fig. 15 (Saha et al. 1997, their Fig. 11). Plotted are the kinematic absolute magnitudes for SNe Ia in our fiducial sample of 1997. Open and closed circles are from that fiducial sample. The seven calibrators known at the time (Saha et al. 1997, Table 6) are closed diamonds. The kinematic absolute magnitudes for the open and closed circles were calculated using the three values of H_0 shown.

Clearly, neither the values of 65 and 70 fit the calibrators (the contrary conclusion by Gibson et al. 2000 based on certain manipulations of the Table 1 calibrators can be shown to be contradictory to our original papers; Saha et al. 2000), whereas $H_0 = 55$ gives no difference between the calibrator (Cepheid) absolute magnitudes and the kinematic magnitudes for the SNe Ia with $1,600 < v_{220} < 30,000$ km s^{-1}.

6. Other ways to H_o

In view of the present 5% to 15% differences in H_0 determined by different groups with the SNe Ia method, it is important to use other methods to H_0 (global) to test the method externally. We have used three such independent astronomical methods, each of which give $H_0 \sim 55$. A fourth set of physical methods (gravitational lenses and the Sunyaev-Zeldovich effect) has been used by many groups, also favoring the long distance scale with $H_0 < 60$. Detailed reviews are in many of the sources cited earlier (cf. Tammann 1996, 1997, 1998a,b; Tammann & Federspiel 1997; Sandage & Tammann 1997; Sandage, Tammann, & Saha 1998). In this final section we simply summarize the three other methods.

6.1. The route through the Virgo and Fornax clusters

At various times we have used six independent methods to derive the distance to the Virgo cluster core. Details are in the reviews just cited. The methods include Tully-Fisher corrected for the Teerikorpi (1987, 1990) cluster incompleteness bias (Federspiel et al. 1998), globular clusters (Sandage & Tammann 1995), and normal novae (Pritchet & van den Bergh 1987). The mean distance modulus from the six methods is $(m-M)_o = 31.66 \pm 0.09$ ($D = 21.5$ Mpc). Contrast this with the value adopted by the "Key Project" consortium of $(m-M)_o = 31.25 \pm 0.45$ ($D = 17.8$ Mpc).

Similar differences exist for the distance to the core of the Fornax cluster. We obtained $(m-M)_o = 31.84$ for NGC 1316 in the last section. The value used by Madore et al. (1998) is $(m-M)_o = 31.32 \pm 0.19$.

We also derive smaller global recession velocities, corrected to the microwave background kinematic frame. Our values are 175 ± 70 km s^{-1} for Virgo and 1259 ± 155 for Fornax, based on the method of tying to the remote frame in Sandage & Tammann (1990) and Jerjen & Tammann (1993), (cf. Sandage & Tammann 1998 eq. 5). The "global velocity of Virgo" initially used by Freedman et al. (1994) was 1404 km s^{-1} based a version of the mapping of the local velocity field as perturbed by Virgo. Their Fornax velocity is 1300 km s^{-1} (Freedman 1998). The route using Jerjen & Tammann avoids all local perturbations. The local velocity field need not be mapped.

The effect of our larger distances and smaller velocities is shown in Fig. 16 from Freedman (1997) but with our Virgo and Fornax points superposed and connected by arrows to those of Freedman. H_0 cannot be even as large as 70 (which is the latest value from the "Key Project") using the larger distances and smaller global velocities advocated here.

6.2. The route through Tully-Fisher using local field galaxies

This is a much more complicated route than it seems at first because the answer is dominated by the effect of observational selection bias which must be identified and accounted for. Previous very high values of H_0 between 80 and 100 obtained using the method are not correct because of the failure to account for the bias in flux-limited samples compared with the calibrators that are more nearly drawn from a distance-limited sample.

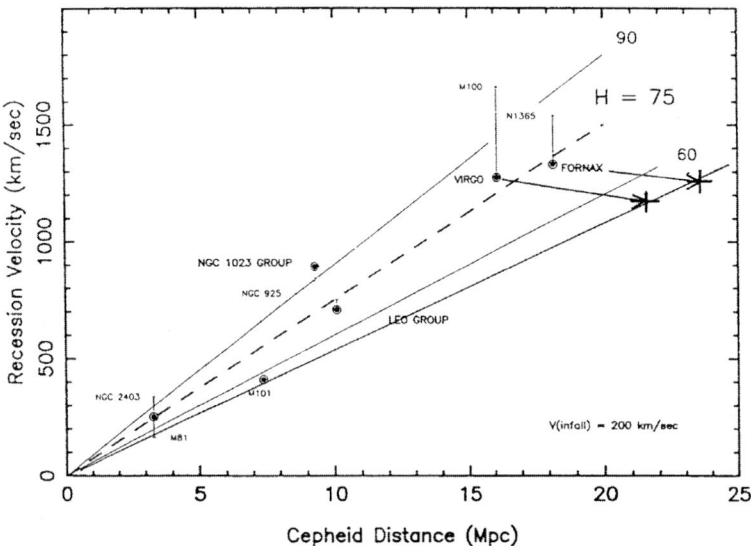

FIGURE 16. Diagram from Freedman (1997) but with our values of the distances and velocities of the Virgo and Fornax clusters shown at the ends of the arrows. Our data are $D = 21.5$ Mpc and v(global) = 1175 km s^{-1} for Virgo and $D = 23.3$ Mpc and v(global) = 1259 km s^{-1} for Fornax. The unmarked line is for $H_0 = 55$ km s^{-1} Mpc^{-1}.

A large literature exists, both for methods to correct for the bias (cf. Bottinelli et al. 1986a,b; Teerikorpi 1984, 1993 1997; Sandage 1988a,b, 1994a,b; Hendry & Simmons 1990, 1994) and the results of such correction (cf. Federspiel et al. 1994; Sandage, Tammann, & Federspiel 1995; Theureau et al. 1997; Goodwin Gribbin, & Hendry 1987 using diameters; Sandage 1999a). All results, corrected for bias, are consistent with $H_0 < 60$.

It should also be noted that the "inverse" TF method applied to field galaxies, said to be free of the simple Malmquist bias, is not in fact free from a bias of a different kind. Correction of the inverse calculation for this "calibration" bias also gives $H_0 \sim 55$ (Teerikorpi et al. 2000).

The results from the TF method via field galaxies gives, of course, only the "local" value of H_0. However, it is known that the expansion rate within the "local bubble" is same as the global rate to better than 10% (reviews in Sandage 1999b; Tammann 1999).

6.3. Luminosity function of galaxies corrected to a fixed morphological type and luminosity class

The first determinations of H_0 were by Lemaitre (1927, 1931), Robertson (1928), and Hubble & Humason (1931). Their method was to calibrating the mean absolute magnitude, $<M>$, of local galaxies of known distance and to apply that calibration to a general field sample where individual redshifts are known. The method was improved fundamentally by van den Bergh's development of luminosity classes for Sb and Sc galaxies with which the type-specific luminosity function of spirals could be narrowed significantly (Sandage 1996, 1999a,b).

The result for the *distance limited* subset of field Sc spirals in the complete sample from the Revised Shapley Ames Catalog, and calibrated using the 25 galaxies for which HST Cepheid distances are available, gives $H_0 = 55 \pm 5$ (Sandage 1999b). Use of the distance-limited sample avoids correction for selection bias. Had we used the complete (flux-limited) sample of Sc galaxies in the RSA and had we not corrected for observational selection bias, the method would have given $H_0 = 73$.

REFERENCES

AARONSON, M. & MOULD, J. 1986 *ApJ*, **303**, 1.

BAADE, W. 1938 *ApJ*, **88**, 285.

BAADE, W. & ZWICKY, F. 1938 *ApJ*, **88**, 411.

BARBON, R., CIATTI, F., & ROSINO, L. 1973 *A&A*, **25**, 241.

BERTOLA, F. 1964 *A&A*, **27**, 319.

BERTOLA, F., MAMMANO, A., & PERINOTTO, M. 1965 *Contr. Asiago Obs.*, **174**, 51.

BERTOLA, F. & SUSSI, M. G. 1965 *Contr. Oss. Astron. Univ. Padua*, **No. 176**

BOTTINELLI, L., GOUGUENHEIM, L., PATUREL, G. H., & TEERIKORPI, P. 1986a *A&A*, **156**, 157.

BOTTINELLI, L., GOUGUENHEIM, L., PATUREL, G. H., & TEERIKORPI, P. 1986b *A&A*, **166**, 39.

BRANCH, D. 1981 *ApJ*, **248**, 1076.

BRANCH, D. 1982 *ApJ*, **258**, 35.

BRANCH, D. 1986 *ApJ*, **300**, L51.

BRANCH, D., FISHER, A., & NUGENT, P. 1993 *AJ*, **106**, 2383.

BRANCH, D., NOMOTO, K., & FILIPPENKO, A. V. 1991 *Comments Astrophys.*, **15**, 221.

BRANCH, D. & TAMMANN, G. A. 1992 *Ann. Rev. A&A*, **30**, 352.

CADONAU R. 1987 *Ph.D. Thesis*, Univ. Basel.

CADONAU, R. & LEIBUNDGUT, B. 1990 *A&AS*, **82**, 145.

CADONAU, R., SANDAGE, A., & TAMMANN, G. A. 1985 in *Supernovae As Distance Indicators* (ed. N. Bartel). Lecture Notes in Physics No. 224, p. 151. Springer.

DE VAUCOULEURS, G. 1979 *ApJ*, **227**, 729.

ELIAS, J. H., FROGEL, J. A., HACKWELL, J., & PERSSON, S. E. 1981 *ApJ*, **251**, L13.

ELIAS, T. H., MATHEWS, K., NEUGEBAUER, G., & PERSSON, S. E. 1985 *ApJ*, **296**, 378.

FEDERSPIEL, M., SANDAGE, A., & TAMMANN, G. A. 1994 *ApJ*, **430**, 29 (paper III of the Bias series).

FEDERSPIEL, M., TAMMANN, G. A., & SANDAGE, A. 1998 *ApJ*, **495**, 115.

FILIPPENKO, A. V. 1997 *Ann. Rev. A&A*, **35**, 309.

FREEDMAN, W. L. 1997 in *Critical Dialogues in Cosmology*, (ed. N. Turok. p. 92. World Scientific.

FREEDMAN, W. L. 1998 in *Sources and Detection of Dark Matter in the Universe* (ed. D. Cline). p. 45. Elsevier.

FREEDMAN, W. L., ET AL. 1994 *Nature*, **371**, 757.

GOODWIN, S. P., GRIBBIN, J., & HENDRY, M. A. 1997 *AJ*, **114**, 2212.

GIBSON, B. K., ET AL. 2000 for the "Key Project" consortium, *ApJ*, in press.

HAMUY, M. 1992 *Lecture at Viña del Mar on SNe Ia homogeneity*, unpublished.

HAMUY, M., ET AL. 1993 *Rev. Mex. A&A*, **26**, 114.

HAMUY, M., PHILLIPS, M. M., MAZA, J., SUNTZEFF, N. B., SCHOMMER, R. A., & AVILES, R. 1995 *AJ*, **109**, 1.

HAMUY, M., PHILLIPS, M. M., SCHOMMER, R. A., SUNTZEFF, N. B., MAZA, J., & AVILES, R. 1996 *AJ*, **112**, 2391.

HAMUY, M., ET AL. 1996 *AJ*, **112**, 2408.

HARKNESS, R. P. & WHEELER, J. C. 1990 in *Supernovae*, (ed. A. G. Petschek), p. 1. Springer.

HARKNESS, R. P., WHEELER, J. C., MARGON, B., DOWNES, R. A., KIRSHNER, R. P., ET AL. 1987 *ApJ*, **317**, 355.

HENDRY, M. A. & SIMMONS, J. F. L. 1990 *A&A*, **237**, 275.

HENDRY, M. A. & SIMMONS, J. F. L. 1994 *ApJ*, **435**, 515.

HUBBLE, E. & HUMASON, M. L. 1931 *ApJ*, **74**, 43.

JACOBY, G., ET AL. 1992 *PASP*, **104**, 599.

JERJEN, H. & TAMMANN, G. A. 1993 *A&A*, **273**, 354.

KENNICUTT, R. C., MOULD, J., & FREEDMAN, W. L. 1998, preprint.

KOWAL, C. T. 1968 *AJ*, **73**, 1021.

KRAAN-KORTEWEG, R. C. 1986 *A&AS*, **66**, 255 (Basel Preprint 18)

LEIBUNDGUT, B. 1988 *Ph.D. Thesis*, Univ. Basel.

LEIBUNDGUT, B. 1990 *A&A*, **229**, 1.

LEIBUNDGUT, B., ET AL. 1991a *ApJ*, **371**, L23.

LEIBUNDGUT, B. 1991b *A&A*, **229**, 1.

LEIBUNDGUT, B., TAMMANN, G. A., CADONAU, R., & CERRITO, D. 1991b *A&AS*, **89**, 537.

LEMAITRE, G. 1927 *Ann. de la Soc. Sci. de Bruxelles*, **Tome 47A**, 49B.

LEMAITRE, G. 1931 *MNRAS*, **91**, 483, a translation of the above.

MINKOWSKI, R. 1964 *Ann. Rev. A&A*, **2**, 247.

MADORE, B. F., ET AL. 1998 *Nature*, **395**, 47.

MAZA, J., HAMUY, M., PHILLIPS, M. M., SUNTZEFF, N. B., & AVILES, R. 1944 *ApJ*, **424**, L107.

PANAGIA, N. 1985 in *Supernovae as Distance Indicators* (ed. N. Bartel). Lecture Notes in Physics No. 224, p. 14. Springer.

PANAGIA, N. 1999 in *New Views of the Magellanic Clouds*, IAU Symp. #190 (eds. Y.-H. Chu, N. Suntzeff, J. Hesser & D. Bohlender), p. 549–556. ASP.

PARODI, B. R., SAHA, A., SANDAGE, A., & TAMMANN, G. A. 2000 *AJ*, in press.

PATAT, F., BARBON, R., CAPPELLARO, E., & TURATTO, M. 1997 *A&A*, **317**, 423.

PHILLIPS, M. M. 1992 *unpublished paper at the Viña del Mar meeting in which he emphasized diversity*.

PHILLIPS, M. M. 1993 *ApJ*, **413**, L105.

PHILLIPS, M. M., ET AL. 2000, preprint.

PIERCE, M. J. 1994 *ApJ*, **430**, 53.

PIERCE, M. J. & TULLY, R. B. 1992 *ApJ*, **387**, 47.

PORTER, A. C. & FILIPPENKO, A. V. 1987 *AJ*, **93**, 1372.

PRITCHET, C. J. & VAN DEN BERGH, S. 1987 *ApJ*, **318**, 507.

PSKOVSKII, Y. P. 1967 *Sov. Astron.*, **11**, 63.

PSKOVSKII, Y. P. 1971 *Sov.Astron.*, **14**, 798.

PSKOVSKII, Y. P. 1984 *Sov. Astron.*, **28**, 658.

RIESS, A. G. 1996 *Ph.D. thesis*, Harvard.

RIESS, A. G., PRESS, W. H., & KIRSHNER, R. P. 1996 *ApJ*, **473**, 88.

ROBERTSON, H. P. 1928 *Phil. Mag.*, **5**, 845.

SAHA, A., LABHARDT, L., SCHWENGELER, H., MACCHETTO, F. D., PANAGIA, N., SANDAGE, A., & TAMMANN, G. A. 1994 *ApJ*, **425**, 14 (IC 4182).

SAHA, A., SANDAGE, A., LABHARDT, L., SCHWENGELER, H., TAMMANN, G. A., PANAGIA, N., & MACCHETTO, F. D. 1995 *ApJ*, **438**, 8 (NGC 5253).

SAHA, A., SANDAGE, A., LABHARDT, L., TAMMANN, G. A., MACCHETTO, F. D., & PANAGIA, N. 1996a *ApJ*, **466**, 55 (NGC 4536).

SAHA, A., SANDAGE, A., LABHARDT, L., TAMMANN, G. A., MACCHETTO, F. D., & PANAGIA, N. 1996b, *ApJS*, **107**, 693 (NGC 4496A).

SAHA, A., SANDAGE, A., LABHARDT, L., TAMMANN, G. A., MACCHETTO, F. D., & PANAGIA, N. 1997 *ApJ*, **486**, 1 (NGC 4639).

SAHA, A., SANDAGE, A., TAMMANN, G. A., LABHARDT, L., MACCHETTO, D. F., & PANAGIA, N. 1999 *ApJ*, in press (NGC 3627).

SAHA, A., TAMMANN, G. A. & SANDAGE, A. 2000 *ApJ*, in preparation; reply to the incorrect reconstruction of the calibrator Table 1 by Gibson et al. 2000.

SANDAGE, A. 1988a *ApJ*, **331**, 583.

SANDAGE, A. 1988b *ApJ*, **331**, 605.

SANDAGE, A. 1994a *ApJ*, **430**, 1 (paper I of the Bias series).

SANDAGE, A. 1994b *ApJ*, **430**, 13 (paper II of the Bias series).

SANDAGE, A. 1996 *AJ*, **111**, 1 (paper V of the Bias series).

SANDAGE, A. 1999a *AJ*, **117**, 157 (paper VII of the Bias series).

SANDAGE, A. 1999b *ApJ*, in press, (paper VIII of the Bias series).

SANDAGE, A., BELL, R. A., & TRIPICCO, M. 1999 *ApJ*, in press.

SANDAGE, A., SAHA, A., TAMMANN, G. A., LABHARDT, L., SCHWENGELER, H., PANAGIA, N., & MACCHETTO, F. D. 1994 *ApJ*, **423**, L13 (NGC 5253).

SANDAGE, A., SAHA, A., TAMMANN, G. A., PANAGIA, N., MACCHETTO, D. F. 1992 *ApJ*, **401**, L7 (NGC 4182).

SANDAGE, A. & TAMMANN, G. A. 1982 *ApJ*, **256**, 339 (Steps VII)

SANDAGE, A. & TAMMANN, G. A. 1990 *ApJ*, **365**, 1.

SANDAGE, A. & TAMMANN, G. A. 1993 *ApJ*, **415**, 1.

SANDAGE, A. & TAMMANN, G. A. 1995 *ApJ*, **446**, 1.

SANDAGE, A. & TAMMANN, G. A. 1997 in *Critical Dialogues in Cosmology* (ed. N. Turok), p. 130. World Scientific.

SANDAGE, A., TAMMANN, G. A., & FEDERSPIEL, M. 1995 *ApJ*, **452**, 1 (paper IV of the Bias series).

SANDAGE, A., TAMMANN, G. A., & SAHA, A. 1998 in *Sources and Detection of Dark Matter in the Universe* (ed. D. Cline), p. 1. Elsevier.

SCHAEFER, B. E. 1995a *ApJ*, **447**, L13.

SCHAEFER, B. E. 1995b *ApJ*, **449**, L9.

SCHAEFER, B. E. 1995c *ApJ*, **450**, L5.

SCHAEFER, B. E. 1996 *AJ*, **111**, 1668.

SCHAEFER, B. E. 1998 *ApJ*, **509**, 80.

SCHMIDT, T. 1957 *Z. f. Ap.*, **41**, 182.

SUNTZEFF, N. B., ET AL. 1999 *AJ*, **117**, 1175.

TAMMANN, G. A. 1978 in *Astronomical Uses of the Space Telescope* (eds. F. Macchetto, F. Pacini, & M. Tarenghi). p. 329. ESO Proceedings.

TAMMANN, G. A. 1982 in *Supernovae: A Survey of Current Research* (eds. M. J. Rees & R. J. Stoneham). p. 371. Reidel.

TAMMANN, G. A. 1996 *Rev. Mod. Astronomy*, **9**, 139.

TAMMANN, G. A. 1997 in *Relativistic Astrophysics* (eds. B. J. T. Jones & D. Markovic). p. 17. Cambridge Univ. Press.

TAMMANN, G. A. 1998a in *General Relativity, 8th Marcel Grossmann Symposium* (ed. T. Piran). World Scientific, in press.

TAMMANN, G. A. 1998b in *Harmonizing Cosmic Distance Scales in a Post Hipparcos Era* (eds. D. Egret & A. Heck), in press.

TAMMANN, G. A. 1999 in *IAU 183* (ed. Sato). p. 30. Kluwer Academic.

TAMMANN, G. A. & FEDERSPIEL, M. 1997 in *The Extragalactic Distance Scale* (eds. M. Livio, M. Donahue, & N. Panagia). p. 137. Cambridge Univ. Press.

TAMMANN, G. A. & LEIBUNDGUT, B. 1990 *A&A*, **236**, 9.

TAMMANN, G. A. & SANDAGE, A. 1995 *ApJ*, **452**, 16.

TANVIR, N. R., SHANKS, T., FERGUSON, H. C., & ROBINSON, D. T. R. 1995 *Nature*, **377**, 27.

TEERIKORPI, P. 1984 *A&A*, **141**, 407.

TEERIKORPI, P. 1987 *Ann. Rev. A&A*, **35**, 101.

TEERIKORPI, P. 1990 *A&A*, **234**, 1.

TEERIKORPI, P. 1993 *A&A*, **280**, 443.

TEERIKORPI, P., ET AL. 2000, preprint.

THEUREAU, G., HANSKI, M., EKHOLM, T., BOTTINELLI, L., GOUGUENHEIM, L., PATUREL, G., & TEERIKORPI, P. 1997 *A&A*, **322**, 730.

TONRY, J. L. 1991 *ApJ*, **373**, L1.

TRIPP, R. 1998 *A&A*, **331**, 815.

TRIPP, R. & BRANCH, D. 2000 *ApJ*, in press.

UOMOTO, A. & KIRSHNER, R. P. 1985 *A&A*, **149**, L7.

VAN DEN BERGH, S. 1960 *Zs. f. Ap.*, **49**, 201.

VAN DEN BERGH, S. 1992 *PASP*, **104**, 861.

VAN DEN BERGH, S. 1993 *Rev. Mex. A&A*, **26**, 73.

VAN DEN BERGH, S. 1994 *PASP*, **106**, 705.

VAN DEN BERGH, S. 1996a *ApJ*, **472**, 431.

VAN DEN BERGH, S. 1996b *PASP*, **108**, 1091.

VAN DEN BERGH, S. & PAZDER, J. 1992 *ApJ*, **390**, 34.

WELLS, L. A., ET AL. 1994 *AJ*, **108**, 2233.

WEILER, K. & SRAMEK, R. A. 1988 *Ann. Rev. A&A*, **26**, 295.

WHEELER, J. C. & LEVREAULT, R. 1985 *ApJ*, **313**, L69.

ZEHAVI, I., RIESS, A. G., KIRSHNER, R. P., & DEKEL, A. 1998 *ApJ*, **503**, 483.

ZWICKY, F. 1958 in *Handbuch der Physik* (ed. S. Fluge), Vol. 51. p. 766. Springer.

ZWICKY, F. 1962 in *Problems of Extra-Galactic Research, IAU Symposium 15* (ed. G. C. McVittie). pp. 357, 378. Macmillan Co.

Type Ia Supernovae and their implications for cosmology

By MARIO LIVIO

Space Telescope Science Institute, 3700 San Martin Drive, Baltimore, MD 21218

Models for Type Ia Supernovae (SNe Ia) are reviewed. It is shown that there are strong reasons to believe that most SNe Ia represent thermonuclear disruptions of C–O white dwarfs, when these white dwarfs reach the Chandrasekhar limit and ignite carbon at their centers.

Different progenitor scenarios are reviewed critically and the strengths and weaknesses of each scenario are presented in detail. It is argued that theoretical considerations currently favor single-degenerate models, in which the white dwarf accretes from a subgiant or giant companion. However, it is still possible that more than one progenitor class contributes to the observed sample. The relation of the different models to the use of SNe Ia for the determination of cosmological parameters is discussed. It is shown that while the observed diversity of SNe Ia may argue for the existence of different progenitor classes, this does not affect the interpretation of an accelerating expansion of the universe.

Crucial observational tests of the conclusions are suggested.

1. Introduction

During the past three years two groups (Perlmutter et al. 1997; Schmidt et al. 1998) have presented strong evidence that the expansion of the universe is accelerating rather than decelerating (Riess et al. 1998; Perlmutter et al. 1998, 1999; and see Livio 1999 for a perspective). This surprising result comes from distance measurements to more than fifty supernovae Type Ia in the redshift range $z = 0.1$ to $z = 1$. The results are consistent with the cosmological constant (or vacuum energy) contributing to the total energy density about 60–70% of the critical density, which in turn, is consistent with recent measurements of the anisotropy of the cosmic microwave background (e.g. Miller et al. 1999; Wilson et al. 1999; Mauskopf et al. 1999).

This unexpected finding, as well as the use of supernovae Type Ia to measure the Hubble constant (e.g. Sandage et al. 1996; Saha et al. 1997), have focused the attention again on the frustrating fact that in spite of decades of research, the exact nature of the progenitors of supernovae Type Ia remains unknown. Until this problem is solved, one cannot be fully confident that supernovae at higher redshifts are not somehow different from their low redshift counterparts. In the present review I therefore examine critically models for supernovae Type Ia and their progenitors. Other recent reviews include Branch et al. (1995), Livio (1996a; 2000), Renzini (1996), Iben (1997), and see Höflich & Dominguez, these procedings.

2. SNe Ia characteristics and the basic model

The *defining* characteristics of supernovae Type Ia (SNe Ia) are both spectral: (i) the *lack* of lines of hydrogen, and (ii) the *presence* of a strong red Si II absorption feature ($\lambda 6355$ shifted to ~ 6100 Å).

Once defined as SNe Ia, the following are several of the important *observational characteristics* of the class which may help in the search for progenitors:

(1) *Homogeneity*: Until very recently, it has generally been claimed that more than 80% of all SNe Ia form a homogeneous class (see however (2) below) in terms of their *spectra* (e.g. Branch, Fisher, & Nugent 1993), *light curves*, and *peak absolute magnitudes*. The latter are given by

$$M_B \simeq M_V \simeq -19.30(\pm 0.03) + 5\log(H_0/60 \text{ km s}^{-1} \text{ Mpc}^{-1}) \quad (2.1)$$

with a dispersion of $\sigma(M_B) \sim \sigma(M_V) \sim 0.2$–$0.3$ (Hamuy et al. 1996a; Tamman & Sandage 1995; and see Branch 1998 for a review).

(2) *Inhomogeneity*: Some differences in the spectra and light curves have been known to exist for a while (e.g. Hamuy et al. 1996b). In terms of explosion strength, SNe Ia have traditionally been roughly ordered as follows: SNe Ia like SN 1991bg and SN 1992K represent the weakest events, followed by weak events like 1986G, followed by about 80% of all SNe Ia which are called "normals" (or sometimes "Branch normals"), to the stronger than normal events like SN 1991T. In a very recent work, however, Li et al. (2000) find indications for a considerably higher *peculiarity* rate, a total of $(39\pm 10)\%$; of which $(19\pm 7)\%$ and $(21\pm 7)\%$ are SN 1991bg-like and SN 1991T-like objects respectively.

(3) The *luminosity function* of SNe Ia was found in earlier studies to decline very steeply on the bright side (e.g. Vaughan et al. 1995). Since selection effects cannot prevent the discovery of SNe which are brighter than the "normals" (unless they occur preferentially in high extinction regions), this is usually taken to imply that *the normals are essentially the brightest*. The recent study of Li et al. (2000) seems to show, however, that the luminosity function is relatively flat at both the overluminous and underluminous ends.

(4) Near maximum light, the spectra are characterized by *high velocity* (8000–30,000 km s^{-1}) *intermediate mass elements* (O–Ca). In the late, nebular phase, the spectra are dominated by forbidden lines of iron (e.g. Kirshner et al. 1993; Wheeler et al. 1995; Ruiz-Lapuente et al. 1995; Gómez et al. 1996; Filippenko 1997).

(5) Fairly young populations appear to be very efficient at producing SNe Ia (e.g. they tend to be associated with spiral arms in spirals; Della Valle & Livio 1994; Bartunov, Tsvetkov & Filimonova 1994), but relatively old populations ($\tau \gtrsim 4 \times 10^9$ yr) can also produce them. In particular, *SNe Ia do occur in ellipticals* (e.g. Turatto, Cappellaro & Benetti 1994). In fact, the rates of SNe Ia in ellipticals appear to be similar to those in spirals, $\sim 0.18 SNu$ (where $SNu = 1 SN (100 \text{ yr})^{-1} \left(10^{10} L_\odot^B\right)^{-1}$; Turatto, Cappellaro & Petrosian 1999). This immediately implies that *SNe Ia are not caused by the core collapse of stars more massive than 8 $M_\odot$*.

(6) There exist a number of correlations between different pairs of observables (see e.g. Branch 1998 for a review). Of these, the most frequently used in the context of determinations of cosmological parameters is the correlation between the *absolute magnitude and the shape of the light curve*. Basically, brighter SNe Ia decline more slowly. A parameter commonly used to quantify the light curve shape is Δm_{15} (Phillips 1993), the decline in magnitudes in the B band during the first 15 days after maximum light. Hamuy et al. (1996a) find slopes $dM_B/d\Delta m_{15} = 0.78\pm 0.17$, $dM_V/d\Delta m_{15} = 0.71\pm 0.14$, and $dM_I/d\Delta m_{15} = 0.58 \pm 0.13$. Using a stretch-factor s (Perlmutter et al. 1997), one can write $M_B = M_B(s=1) - \alpha * (s-1)$, with $M_B(s=1) = -19.46$ (e.g. Sandage et al. 1996), and $\alpha = 1.74$ (Perlmutter et al. 1999). Sophisticated techniques for using the different correlations in distance determinations have been developed (e.g. Riess et al. 1996, 1998).

The above characteristics can be augmented by the following suggestive facts:

(1) The *energy* per unit mass, $1/2(\sim 10^4 \text{ km s}^{-1})^2$, is of the order of the one obtained from the conversion of carbon and oxygen to iron.

(2) The fact that the event is explosive suggests that *degeneracy* may play a role.
(3) The spectrum appears to contain no hydrogen.
(4) The explosions can occur with long delays, after the cessation of star formation.

All the properties above have led to one agreed upon model: *SNe Ia represent thermonuclear disruptions of mass accreting white dwarfs.*

It is interesting that there exists a unanimous consensus on this model in spite of the fact that the essence of flame physics, burning front propagation, and the details of the (presumed) transition from deflagration to detonation (in particular the density at which the transition occurs), which are at the heart of the model, remain as major unsolved problems (e.g. Khokhlov, Oran & Wheeler 1997; Woosley 1997; Reinecke, Hillebrandt & Niemeyer 1998; and see Höflich & Dominguez, and Khokhlov, these proceedings). In fact, given these uncertainties, it is almost difficult to understand how the entire family of SNe Ia light curves can be fitted essentially with one parameter (e.g. Perlmutter et al. 1997), although it is possible that all SNe Ia explode at the same WD mass (see §4), and that the entire observed diversity stems from different ^{56}Ni masses.

3. Why is identifying the progenitors important?

The fact that we do not know yet what are the progenitor systems of some of the most dramatic explosions in the universe has become a major embarrassment and one of the key unsolved problems in stellar and binary star evolution. There are several important reasons why identifying the progenitors has become more crucial than ever:

(i) The use of SNe Ia as one of the main ways to determine key cosmological parameters like H_0, and the contributions to the energy density (by matter and by the cosmological constant) Ω_M, Ω_Λ requires an understanding of the evolution of the luminosity, and the SN rate with cosmic epoch. Both of these depend directly on the nature of the progenitors.

(ii) Galaxy evolution depends on the radiative, kinetic energy, and nucleosynthetic output of SNe Ia (e.g. Kauffmann, White & Guiderdoni 1993).

(iii) Due to the uncertainties that still exist in the explosion mechanism itself, a knowledge of the initial conditions and of the distribution of matter in the environment of the exploding star are essential for the understanding of the explosion.

(iv) An unambiguous identification of the progenitors, coupled with observationally determined SNe Ia rates can help to place meaningful constraints on the theory of binary star evolution (e.g. Livio 1996b; Li & van den Heuvel 1997; Yungelson & Livio 1998; Hachisu, Kato & Nomoto 1999). In particular, a semi-empirical determination of the elusive common-envelope-ejection efficiency parameter, α_{CE}, may be possible (e.g. Iben & Livio 1993).

4. Refinements to the basic model

The basic model for SNe Ia (that essentially all researchers in the field agree upon) is that of a thermonuclear disruption of an accreting white dwarf (WD). However, additional refinements to the model are possible on the basis of existing observational data and theoretical models. These refinements still do not involve the question of the *progenitor systems*. Rather, they address the question of the WD *composition*, and of its *mass* at the instant of explosion.

4.1. *The composition of the exploding WD*

In principle, the WD that accretes to the point of explosion could be composed of He, of C–O, or of O–Ne. Let us examine these possibilities one by one.

(i) *He WDs*: Helium WDs have typical masses that are smaller than ~ 0.45 M$_\odot$ (e.g. Iben & Tutukov 1985). While if accreting, these He WDs can explode following central He ignition at ~ 0.7 M$_\odot$, the composition of the ejected matter in this case will be that of He, ^{56}Ni and decay products (e.g. Nomoto & Sugimoto 1977; Woosley, Taam & Weaver 1986). This is entirely inconsistent with observations (observational characteristic (4) in §2). Therefore, *He WDs certainly do not produce the bulk of SNe Ia*.

(ii) *O–Ne WDs*: Oxygen–Neon WDs form in binaries from main sequence stars of ~ 10 M$_\odot$, although the precise range which allows formation is somewhat uncertain (e.g. Iben & Tutukov 1985; Canal, Isern & Labay 1990; Dominguez, Tornambé & Isern 1993). These systems are probably not numerous enough to constitute the main channel of SNe Ia (e.g. Livio & Truran 1992; Livio 1993). It is also generally expected that O–Ne WDs that manage to accrete enough material to reach the Chandrasekhar limit will produce (via electron capture) preferentially accretion-induced collapses (to form neutron stars) rather than SNe Ia (e.g. Nomoto & Kondo 1991; Gutierrez et al. 1996). Accretion induced collapses do not eject enough nickel to match the light curves of normal SNe Ia, although they may be able to explain very subluminous events like SN 1991bg (e.g. Fryer et al. 1999). I should note that the existing calculations have been performed for WDs of O–Ne–Mg composition, while some recent calculations of the evolution of a 10 M$_\odot$ star produce degenerate cores which are almost devoid of magnesium (Ritossa, Garcia-Berro & Iben 1996). Nevertheless, because of the above two points *it is unlikely that O–Ne WDs produce the bulk of SNe Ia*.

(iii) *C–O WDs*: Carbon–Oxygen WDs are formed in binaries from main sequence stars of up to ~ 10 M$_\odot$. They are therefore both relatively numerous, and they provide a significant "phase space volume" (masses in the range 0.8–1.2 M$_\odot$; accretion rates in the range 10^{-8}–10^{-6} M$_\odot$/yr) in which they are expected to produce SNe Ia (upon reaching the Chandrasekhar limit; e.g. Nomoto & Kondo 1991). Consequently, *the accreting WDs that produce most of the SNe Ia are very probably of C–O composition*!

4.2. *At what mass does the WD explode and where and in what fuel does the ignition take place?*

While there is virtually unanimous agreement about everything I said up to now, namely, that: *SNe Ia are thermonuclear disruptions of accreting C–O WDs*, the next step in the refinement to the model is more controversial. Two major classes of models have been considered, and they suggest entirely different answers to the questions posed by the title of this subsection. In one class, the WD explodes upon reaching the *Chandresekhar mass*, as *carbon* ignites at its *center*. In the second, the WD explodes at a *sub-Chandrasekhar mass*, as *helium* ignites *off-center*. I will now review briefly each of these classes and point out their strengths and weaknesses.

4.2.1. *Chandrasekhar mass carbon ignitors*

In this model, considered 'standard,' the WD accretes until it approaches the Chandrasekhar mass. Carbon ignition (triggered by compressional heating) occurs at or very near the center and the burning front propagates outwards. Three types of flame propagation models have been considered in the past three decades: (i) detonation (e.g. Arnett 1969; Hansen & Wheeler 1969), (ii) deflagration (e.g. Nomoto, Sugimoto & Neo 1976) and iii) delayed detonation, in which the flame starts as a deflagration which transitions into a detonation at some transition density (e.g. Khokhlov 1991; Woosley & Weaver

1994). Models of the latter two types ((iii) in particular) have generally been quite successful in explaining the observations (see e.g. Höflich & Dominguez, these proceedings). The main *strengths* of this model (central carbon ignition at the Chandrasekhar mass) are (see e.g. Höflich & Khokhlov 1996; Nugent et al. 1997; Höflich & Dominguez these proceedings, for detailed modeling):

(1) Some 10^{51} *ergs of kinetic energy* are deposited into the ejecta by nuclear energy.

(2) ^{56}Ni decay powers the *lightcurve*.

(3) The density and composition as a function of the ejection of velocity ($X_i(V_{ej})$) are consistent with the observed *spectra*.

(4) The fact that the explosion occurs at the Chandrasekhar mass may explain the broad-brush *homogeneity*.

(5) Spectra (e.g. of SNe 1994D, 1992A) can be fitted in great detail by theoretical models (e.g. Nugent et al. 1997).

The main *weaknesses* of the Chandrasekhar mass models are:

(1) It has proven more difficult than originally thought for WDs to accrete *up to the Chandrasekhar mass* in sufficient numbers to account for the SNe Ia rate. The difficulty is associated with mass loss episodes in nova explosions, in helium shell flashes and in massive winds or common envelope phases. I will return to some of these problems when I discuss specific progenitor models.

(2) For initial WD masses larger than ~ 1.2 M$_\odot$, (which can more easily, in principle, reach the Chandrasekhar mass) *accretion-induced collapse* is a more likely outcome than a SN Ia (e.g. Nomoto & Kondo 1991).

(3) *The late-time spectrum* (~ 300 days), and in particular the Fe III feature at ~ 4700 Å does not agree well with Chandrasekhar mass models (Liu, Jeffrey & Schultz 1998).

(4) The 'standard' model has some difficulty in reproducing the observed (e.g. Riess et al. 1999a) $\gtrsim 20$ days rise times.

My overall assessment of Chandrasekhar mass models is that the strengths significantly outweigh the weaknesses. The calculations of late-time, nebular spectra involve many uncertainties, and hence I do not regard weakness (3) above as fatal (although clearly more work will be required to explain it away). Both weaknesses (1) and (2) can be overcome if it can be demonstrated that SNe Ia statistics can be reproduced within the uncertainties that still plague the theoretical population synthesis models. As I will show in §5, this appears indeed to be the case. Weakness (4) can be overcome (in principle at least) by lower values of the C/O ratio, or by the presence of a 0.2–0.4 M$_\odot$ envelope (see e.g. Höflich, Wheeler & Thielemann 1998). This suggests to me that this is not a fundamental difficulty for the model.

4.2.2. *Sub-Chandrasekhar mass helium ignitors*

In these models a C–O WD accumulates a helium layer of ~ 0.15 M$_\odot$ while the total mass is sub-Chandrasekhar. The helium ignites off-center (at the bottom of the layer), resulting in an event known as "Indirect Double Detonation" (IDD) or "Edge Lit Detonation" (ELD). Basically, one detonation propagates outward (through the helium), while an inward propagating pressure wave compresses the C–O core which ignites off-center, followed by an outward detonation (e.g. Livne 1990; Livne & Glasner 1991; Woosley & Weaver 1994; Livne & Arnett 1995; Höflich & Khokhlov 1996; and Ruiz-Lapuente, talk presented at the Chicago meeting on Type Ia Supernovae: Theory and Cosmology, October 1998).

The main *strengths* of ELD (sub-Chandrasekhar) models are:

(1) It is easier to achieve the required *statistics*, since less mass needs to be accreted, and the WD does not need to be extremely massive (e.g. Ruiz-Lapuente, Canal & Burkert 1997; Di Stefano et al. 1997; Yungelson & Livio 1998).

(2) The *late-time spectrum* (in particular the Fe III feature at ~ 4700 Å) agrees better with ELD models.

(3) SNe Ia *light curves* can be reproduced adequately by ELD models (although the light curves rise somewhat faster than observed, due to ^{56}Ni heating; Höflich et al. 1997).

The main *weaknesses* of ELD models are:

(1) The *spectra* that are produced by ELD models generally do not agree with observations (e.g. of SN 1994D; Nugent et al. 1997). In particular, the spectra are very blue (due to heating by radioactive Ni), and are dominated by Ni lines, while not showing a strong Si line. The agreement is somewhat better for the subluminous SNe Ia (e.g. SN 1991bg; Nugent et al. 1997; Ruiz-Lapuente, talk presented at the Chicago meeting on Supernovae, October 1998), but even there it is not very good.

(2) The *highest velocity ejecta have the wrong composition* (^{56}Ni and He moving at 11,000 to 14,000 km s^{-1}, not intermediate mass elements; also no high velocity C; e.g. Livne & Arnett 1995). This is due to the fact that in these models, essentially by construction, the intermediate mass elements are sandwiched by Ni and He/Ni rich layers, at the inner and outer sides, respectively.

(3) Since ELD models allow for a range of WD masses, and since more massive WDs produce brighter SNe, one might expect this model to produce a more gradual decline on the bright side of the *luminosity function*. While this is in contradiction to the observed sharp decline obtained for some of the earlier samples, it may not be in contradiction with the more recent observations of a relatively flat luminosity function (see §2 characteristic (3)).

My overall assessment of the sub-Chandrasekhar mass model is that the weaknesses (and in particular weaknesses (1) and (2) which appear almost inevitable) greatly outweigh the strengths in terms of this being a model for the bulk of SNe Ia. It is still possible that ELDs may correctly represent some subluminous SNe Ia (e.g. Ruiz-Lapuente, Canal, & Burkert 1997; Pinto, private communication).

4.3. *The favored model*

On the basis of the above discussion the basic model can now be further refined, and I tentatively conclude that: *Most SNe Ia represent thermonuclear disruptions of mass accreting C–O white dwarfs, when these white dwarfs reach the Chandrasekhar limit and ignite carbon at their centers*!

5. The two possible scenarios

The next step, in which we search for the progenitor systems of SNe Ia is even more controversial. Two possible scenarios have been proposed: (i) The *double-degenerate* scenario, in which two CO WDs in a binary system are brought together by the emission of gravitational radiation and coalesce (Webbink 1984; Iben & Tutukov 1984). (ii) The *single-degenerate* scenario, in which a CO WD accretes hydrogen-rich or helium-rich material from a non-degenerate companion (Whelan & Iben 1973; Nomoto 1982).

In the first scenario the progenitor systems are necessarily *binary WD systems* in which the total mass exceeds the Chandrasekhar mass, and which have binary periods shorter than about thirteen hours (to allow merger within a Hubble time).

In the second scenario the progenitors could be systems like: (i) *Recurrent novae* (both of the type in which the WD accretes hydrogen from a giant like T CrB, RS Oph, and of the type in which the WD accretes helium rich material from a subgiant like U Sco, V394 CrA, and Nova LMC 1990#2), (ii) *Symbiotic Systems* (in which the WD accretes hydrogen-rich material from a low mass red giant or a Mira variable), or (iii) persistent *Supersoft X-ray Sources* (in which the WD accretes at a high rate $\gtrsim 10^{-7}$ $M_\odot$/yr from a subgiant companion).

I will now examine the strengths and weaknesses of each one of these scenarios.

5.1. *The double-degenerate scenario*

There is no question that close binary white dwarf systems in which the total mass exceeds the Chandrasekhar mass are an expected outcome of binary star evolution (e.g. Iben & Tutukov 1984; Iben & Livio 1993). Once the lighter WD (which has a larger radius) fills its Roche lobe, it is entirely dissipated within a few orbital periods, to form a massive disk around the primary (e.g. Rasio & Shapiro 1994; Benz, Thielemann & Hills 1989). The subsequent evolution of the system depends largely on the accretion rate through this disk (e.g. Mochkovitch & Livio 1990; see discussion below).

The main *strengths* of this scenario are the following:

(1) The *absence of hydrogen* in the spectrum is naturally explained in a model which involves the merger of two C–O WDs. In fact, if hydrogen is ever detected in the spectrum of a SN Ia, this would deal a fatal blow to this model. Tentative evidence for circumstellar Hα absorption is SN 1990M was presented by Polcaro and Viotti (1991). However, Della Valle, Benetti & Panagia (1996) demonstrated convincingly that the absorption was caused by the parent galaxy, rather than by the SN environment.

(2) In spite of some impressions to the contrary, *many double WD systems do exist*. In a sample of 153 field WDs and subdwarf B stars, Saffer, Livio & Yungelson (1998) found 18 new double-degenerate candidates. Maxted & Marsh (1999) showed (from a radial velocity survey of 46 WDs) that there is a 95% probability that the fraction of double degenerates among DA WDs lies in the range 0.017–0.19. There are currently eight known systems with orbital periods of less than half a day (and the subdwarf B stars PG 1432+159 and PG 2345+318, with orbital periods of 5.4 hr and 5.8 hr respectively may also have WD companions; Moran et al. 1999). While only one of all of these systems (KPD 0422+5421; Koen, Orosz & Wade (1998)) has a total mass which within the errors could be higher than the Chandrasekhar mass, the sample of confirmed short-period double-degenerates is still smaller than the number predicted to contain a massive system.

(3) Population synthesis calculations predict the *right statistics* for mergers, about 10^{-3} yr^{-1} events for populations that are $\sim 10^8$ yr old and 10^{-4} yr^{-1} for populations that are $\sim 10^{10}$ yr old.

(4) Since double WD systems were found to exist, *mergers* with some "interesting" consequences (either a SN Ia or an accretion-induced collapse) appear inevitable.

(5) The explosion or collapse is expected to occur at (or near) the Chandrasekhar mass, which as I noted in §4.3, I regard as a property of the favored model.

The main *weaknesses* of the double-degenerate scenario are the following:

(1) There are strong indications that WD mergers may lead to off-center carbon ignition, accompanied by the conversion of the C–O WD to an O–Ne–Mg composition, followed by an accretion-induced collapse rather than a SN Ia (e.g. Mochkovitch & Livio 1990; Saio & Nomoto 1985, 1998; Woosley & Weaver 1986).

(2) Galactic chemical evolution results, and in particular the behavior of the [O/Fe] ratio as a function of metallicity ([Fe/H]) have been claimed to be inconsistent with WD mergers as the mechanism for SNe Ia (Kobayashi et al. 1998).

(3) While the unusually high luminosity of SN 1991T and some of its other features have been tentatively attributed to a super-Chandrasekhar product of the merger of two WDs (Fisher et al. 1999), there is little evidence for example for the presence of unburned carbon (as might be expected from the disk formed in the merger process) in most SNe Ia.

Since we are now getting to the final stages in the identification of the progenitors, it is important to assess critically the severity of the above weaknesses. I will therefore discuss now each one of them in some detail.

5.1.1. Constraints from Galactic chemical evolution

Supernovae Type II (SNe II) are explosions resulting from the core collapse of massive ($\gtrsim 8$ $M_\odot$) stars. These supernovae produce relatively more oxygen and magnesium than iron ([O/Fe] > 0). On the other hand, SNe Ia produce mostly iron and little oxygen. Generally, the impression is that metal poor stars ([Fe/H] ≤ -1) have a nearly flat relation of [O/Fe] vs. [Fe/H], with a value of [O/Fe] ~ 0.45 (e.g. Nissen et al. 1994), while disk stars ([Fe/H] $\gtrsim -1$) show a linearly decreasing [O/Fe] with increasing metallicity (e.g. Edvardsson et al. 1993; McWilliam 1997). The "observed" (but see below) break (from flat to linearly decreasing) near [Fe/H] ~ -1 is traditionally explained by the fact that the early heavy element production was done exclusively by SNe II, with the break occurring when the larger Fe production by SNe Ia kicks in (e.g. Matteucci & Greggio 1986).

Recently, Kobayashi et al. (1998) performed chemical evolution calculations for both the double-degenerate scenario and for the single-degenerate scenario. For the latter they used two types of progenitor systems: one with a red giant companion and an orbital period of tens to hundreds of days, and the other with a near main sequence companion and a period of a few tenths of a day to a few days.

They obtained for the double-degenerate scenario (for which they took a time delay to the explosion of ~ 0.1–0.3 Gyr) a break at [Fe/H] ~ -2. For the single-degenerate scenario (with a delay caused by the main sequence lifetime of $\gtrsim 1$ Gyr; including metallicity effects), they obtained a break at [Fe/H] ~ -1. Kobayashi et al. (1998) thus concluded that the Galactic chemical evolution that results from the double-degenerate scenario is inconsistent with observations.

Personally, I am not too convinced by this apparent discrepancy, since Galactic chemical evolution calculations and observations are notoriously uncertain. For example, a recent determination of [Ba/Fe] as a function of [Fe/H] shows a break near [Fe/H] ~ -2, which would be consistent with the double-degenerate scenario prediction (Burris et al. 1999). In addition, recent Keck observations of oxygen in unevolved metal-poor stars appear to show no break in the [O/Fe] vs. [Fe/H] relation. Rather, oxygen is enhanced relative to iron over three orders of magnitude in [Fe/H] in a linear relation (Boesgaard et al. 1999; see also Israelian, Garcia Lopez & Rebolo 1998). While some reservations about these findings have been raised, in particular, a re-analysis of two of the stars of Israelian et al. shows [O/Fe] ratios which are discrepant with the results of Israelian et al. and of Boesgaard et al. (Fulbright & Kraft 1999), this in fact demonstrates the uncertainties involved in such determinations (see also Stephens 2000).

5.1.2. Merger only applicable to relatively rare events?

As I noted above, it has been shown that if SN 1991T is at the same distance as SNe 1981B and 1960F, then its luminosity is too high to be explained in terms of a

Chandrasekhar mass ejection (Fisher et al. 1999). Thus, it has been suggested that this SN resulted from the explosion of a super-Chandrasekhar object, indicating perhaps that WD mergers may be responsible for at least some SNe Ia.

However, events like SN 1991T, which seem to be associated with regions of active star formation, represent at most $\sim 20\%$ of the SNe Ia (Li et al. 2000), and therefore, even if they are the results of mergers this still does not mean that WD mergers are the main class of progenitors of SNe Ia. In addition, it is still far from clear whether mergers can lead to explosions at all (see §5.1.3 below). Incidentally, data for a cepheid distance to NGC 4527 (which will help determine the true intrinsic luminosity of SN 1991T) have been obtained with HST, and the analysis is in progress (Saha et al. 2000).

5.1.3. *SN Ia or accretion induced collapse?*

Potentially the most serious (and possibly even fatal) weakness of the double-degenerate scenario comes from the fact that some estimates and calculations indicate that the coalescence of two C–O WDs may lead to an accretion-induced collapse rather than to a SN explosion (e.g. Mochkovitch & Livio 1990; Saio & Nomoto 1985, 1998; Kawai, Saio & Nomoto 1987; Timmes, Woosley & Taam 1994; Mochkovitch, Guerrero & Segretain 1997).

The point is the following: once the lighter WD fills its Roche lobe, it is dissipated within a few orbital periods (Benz et al. 1990; Rasio & Shapiro 1995; Guerrero 1994), and it forms a hot thick disk configuration around the more massive white dwarf. This disk is mainly rotationally supported and hence central carbon ignition does not take place immediately, but rather the subsequent evolution depends largely on the rate of angular momentum transport and removal, since they determine the accretion rate onto the primary WD. As long as the accretion rate is higher than about $\dot{M} \gtrsim 2.7 \times 10^{-6}$ M$_\odot$ yr^{-1}, *carbon is ignited off-center* (at the core-disk boundary; this may happen during the merger itself; e.g. Segretain 1994). Under such conditions, the flame was found (in spherically symmetric calculations) to propagate all the way to the center within a few thousand years, thus burning the C–O into an O–Ne–Mg mixture with *no explosion* (i.e. before carbon is centrally ignited; e.g. Saio & Nomoto 1998). Such configurations are expected to collapse (following electron captures on ^{24}Mg) to form neutron stars (Nomoto & Kondo 1991; Canal 1997). The main questions are therefore:

(i) What accretion rates can be expected from the initial WD-thick disk configuration?

(ii) May some aspects of the flame propagation be different given the fact that the real problem is three-dimensional while most of the existing calculations were performed using a spherically symmetric code? In particular, *could the carbon burning be quenched before the transformation to O–Ne–Mg composition occurs?*

(iii) Could the WDs ignite even prior to the merger due to tidal heating, and what would be the outcome of such pre-merger ignition?

The answers to all of these questions involve uncertainties, however some possibilities appear more likely than others. First, it appears *very difficult to avoid high accretion rates*. If the MHD turbulence that is expected to develop in accretion disks (e.g. Balbus & Hawley 1998) is operative, with a corresponding viscosity parameter of $\alpha \sim 0.01$ (where the viscosity is given by $\nu \sim \alpha c_s H$, with H being a vertical scaleheight in the disk and c_s the speed of sound; e.g. Balbus, Hawley & Stone 1996), then angular momentum can be removed in a matter of days! In such a case, even if the accretion rate is Eddington limited (at $\sim 10^{-5}$ M$_\odot$/yr), off-center carbon ignition should still occur, with an eventual collapse rather than an explosion. Deviations from spherical symmetry can only hurt, since they may allow accretion to proceed at a super-Eddington rate. It is difficult to see why the dynamo-generated viscosity would be suppressed for the kind of shear and temperatures expected in the disk.

Concerning the burning itself, recent attempts at multi-dimensional calculations of the flame propagation and a more detailed analysis of some of the processes involved (Garcia-Senz, Bravo & Serichol 1998; Bravo & Garcia-Senz 1999) indicate that if anything, accretion induced collapses are an even more likely outcome than previously thought. This is due to the effects of electron captures in Nuclear Statistical Equilibrium which tend to stabilize the thermonuclear flame, and to Coulomb corrections to the equation of state. The latter has the effect of reducing the flame velocities and the electronic and ionic pressures, all of which result in a reduction in the critical density which separates explosions from collapses.

As two WDs approach merger, their interiors can be spun up by tidal forces. *If* these tides can bring (at least one of) the WDs into quasi synchronism between the spin period and the orbital period, high dissipation rates and heating will ensue (Rieutord & Bonazolla 1987). The obtained luminosities due to this tidal heating can reach values as high as $\gtrsim 10^{37}$ erg s^{-1} (Rieutord & Bonazzola 1987; Iben, Tutukov & Federova 1998). If such heating indeed occurs, it could have (in principle at least) two important effects: (i) it would increase the probability of detection of pre-merger WDs, due to the increased luminosity and the expected periodic variability (due to mutual occultations), (ii) heating could lead to carbon ignition prior to or during the merger.

From the point of view of the present discussion it is important to assess whether the latter possibility makes the merging WDs more viable progenitor systems. This does not appear to be the case for the following reasons:

(i) It is not obvious that a WD can be brought to synchronous rotation. The normal viscosity of WD matter is very low (e.g. Durisen 1973), which can make the viscous timescale even longer than the system's lifetime. This problem however may be overcome if turbulence develops due to the strong shear.

(ii) It is not clear if carbon will be ignited even if tidal heating occurs. In fact, in the calculations of Iben et al. (1998) carbon failed to be ignited (although only by a relatively narrow margin).

(iii) Even if carbon is ignited, it is very likely that the ignition will occur off-center, making an accretion induced collapse a more likely outcome than a SN Ia (as explained above).

Finally, on the observational side there are also two points which argue at some level against WD mergers as the main SNe Ia progenitors.

(i) Even if MHD viscosity could somehow be suppressed in the disk, and the disk surrounding the primary WD could cool down, so that angular momentum would be transported only via the viscosity of (partially) degenerate electrons, this would result in an accretion timescale of $\sim 10^9$ yrs (Mochkovitch & Livio 1990; Mochkovitch et al. 1997). The system prior to the explosion would have an absolute magnitude of $M_V \lesssim 10$ (with much of the emission occurring in the UV). There is no evidence for the existence of some $\sim 10^7$ such objects in the Galaxy.

(ii) The existence of planets around the pulsars PSR 1257+12 and PSR 1620–26 (Wolszczan 1997; Backer 1993; Thorsett, Arzoumanian & Taylor 1993) could be taken to mean (this is a model dependent statement) that mergers tend to produce accretion induced collapses rather than SNe Ia. In one of the leading models for the formation of such planets (Podsiadlowski; Pringle & Rees 1991; Livio, Pringle & Saffer 1992), the planets form in the following sequence of events. The lighter WD is dissipated (upon Roche lobe overflow) to form a disk around the primary. As material from this disk is accreted, matter at the outer edge of the disk has to absorb the angular momentum, thereby expanding the disk to a large radius. The planets form from this disk in the same way that they did in the solar system, while the central object collapses to form a neutron star.

5.1.4. Overall assessment of the double-degenerate scenario

It has now been observationally demonstrated that many double-degenerate systems exist. The general agreement between the distribution of the observed properties (e.g. orbital periods, masses) and those predicted by population synthesis calculations (Saffer, Livio & Yungelson 1998), suggests that the fact that no clear candidate (short period) system with a total mass exceeding the Chandrasekhar mass has been found yet, may merely reflect the insufficient size of the observational sample. Thus, there is very little doubt in my mind that statistics is not a serious problem. The most disturbing uncertainty is related to the outcome of the merger process itself. The discussion in §5.1.3 suggests that *collapse to a neutron star is more likely than a SN Ia* (see also Mochovitch et al. 1997).

5.2. The single-degenerate scenario

The main *strengths* of the single degenerate scenario are:

(1) A class of objects in which hydrogen is being transferred at such high rates that it *burns steadily* on the surface of the WD has been identified—the Supersoft X-ray Sources (e.g. Greiner, Hasinger & Kahabka 1991; van den Heuvel et al. 1992; Southwell et al. 1996; Kahabka and van den Heuvel 1997). If the accreted matter can indeed be retained, this provides a natural path to an increase in the WD mass towards the Chandrasekhar mass (e.g. Di Stefano & Rappaport 1994; Livio 1995, 1996a; Yungelson et al. 1996).

(2) Other candidate progenitor systems are known to exist, like symbiotic systems (e.g. Munari & Renzini 1992; Kenyon et al. 1993; Hachisu, Kato & Nomoto 1999) and recurrent novae (Hachisu et al. 1999a).

(3) There have been claims that the single degenerate scenario fits better the results of Galactic chemical evolution (e.g. Kobayashi et al. 1998). However, as I have shown in §5.1.1, recent observations cast doubt on this assertion. Similarly, nucleosynthesis results show that in order to avoid unacceptably large ratios of $^{54}Cr/^{56}Fe$ and $^{50}Ti/^{56}Fe$, the central density of the WD at the moment of thermonuclear runaway must be lower than $\sim 2 \times 10^9$ g cm^{-3} (Nomoto et al. 1997). Such low densities are realized for high accretion rates ($\gtrsim 10^{-7}$ M$_\odot$ yr^{-1}), which are typical for the Supersoft X-ray Sources. Nucleosynthesis results suffer too, however, from considerable uncertainties (e.g. Nagataki, Hashimoto & Sato 1998).

The main *weaknesses* of the single degenerate scenario are:

(1) The upper limits on *radio detection* of hydrogen at 2 and 6 cm in SN 1986G, taken approximately one week before optical maximum (Eck et al. 1995), rule out a symbiotic system progenitor for this system with a wind mass loss rate of $10^{-7} \lesssim \dot{M}_W \lesssim 10^{-6}$ M$_\odot$ yr^{-1} (Boffi & Branch 1995). This in itself is not fatal, since SN 1986G is somewhat peculiar (e.g. Branch and van den Bergh 1993), and the upper limit on the mass loss rate is at the high end of observed symbiotic winds. An even less stringent upper limit from x-ray and Hα observations exists for SN 1994D (Cumming et al. 1996).

(2) There exists some uncertainty whether WDs can even reach the Chandrasekhar mass *at all* by the accretion of hydrogen (e.g. Cassisi, Iben & Tornambe 1998). Furthermore, even if they can, the question of whether they can produce the required SNe Ia

statistics is highly controversial (e.g. Yungelson et al. 1995, 1996; Yungelson & Livio 1998, 1999; Hachisu, Kato & Nomoto 1999; Hachisu et al. 1999a).

I will now examine these weaknesses in some detail.

5.2.1. Observational detection of hydrogen

Ultimately, the presence or total absence of hydrogen in SNe Ia will distinguish unambiguously between single-degenerate and double-degenerate models. To date, hydrogen has not been convincingly detected in *any* SN Ia. It is interesting to note that narrow $\lambda 6300$, $\lambda 6363$ [OI] lines were observed only in one SN Ia (SN 1937C; Minkowski 1939), but even in that case there was no hint of a narrow Hα line. Hachisu, Kato & Nomoto (1999) estimate in one of their models (which involves stripping of material from the red giant; see below) a density measure of $\dot{M}/v_{10} \sim 10^{-8}$ M$_\odot$ yr^{-1} (where v_{10} is the wind velocity in units of 10 km s^{-1}), while the most stringent radio upper limit existing currently (for SN 1986G) is $\dot{M}/v_{10} \sim 10^{-7}$ M$_\odot$ yr^{-1} (Eck et al. 1995; for SN 1994D Cumming et al. (1996) find from Hα an upper limit of $\dot{M} \sim 1.5 \times 10^{-5}$ M$_\odot$ yr^{-1} for a wind speed of 10 km s^{-1}; for SN 1992A Schlegel & Petre (1993) find from X-ray observations an upper limit of $\dot{M}/v_{10} = (2-3) \times 10^{-6}$ M$_\odot$ yr^{-1}). Thus, while it is impossible at present to rule out single-degenerate models on the basis of the apparent absence of hydrogen, the hope is that near future observations will be able to determine definitively whether this absence is real or if it merely represents the limitations of existing observations (an improvement by two orders of magnitude in the sensitivity will give a definitive answer). I should note that a narrow emission feature possibly corresponding to Hα was detected in SN 1981b (in NGC 4536), however no trace of the emission was seen 5 days later (Branch et al. 1983).

5.2.2. Statistics

Growing the WD to the Chandrasekhar mass is not easy. At accretion rates below $\sim 10^{-8}$ M$_\odot$/yr WDs undergo repeated nova outbursts (e.g. Prialnik & Kovetz 1995), in which the WDs lose more mass than they accrete between outbursts (e.g. Livio & Truran 1992). For accretion rates in the range 10^{-8}–a few $\times 10^{-7}$ M$_\odot$/yr, while helium can accumulate, the WDs experience mass loss due to helium shell flashes and due to the common envelope phase which results from the engulfing of the secondary star in the expanding envelope (with mass loss occurring due to drag energy deposition). At accretion rates above a few $\times 10^{-7}$ M$_\odot$/yr, the WDs expand to red giant configurations and lose mass due to drag in the common envelope and due to winds (e.g. Cassisi et al. 1998). The net result of these constraints has been that population synthesis calculations which follow the evolution of all the binary systems in the Galaxy, tended until recently to conclude that single degenerate channels manage to bring WDs to the Chandrasekhar mass only at about 10% of the inferred SNe Ia frequency of 4×10^{-3} yr^{-1} (e.g. Yungelson et al. 1995, 1996; Yungelson & Livio 1998; Di Stefano et al. 1997; although see Li & van den Heuvel 1997).

Very recently, a few serious attempts have been made to investigate whether the statistics could be improved by increasing the "phase space" for single degenerate scenarios, given the fact that population synthesis calculations involve many assumptions. These attempts resulted in the identification of three directions in which the phase space could (potentially) be increased.

(i) The accumulation efficiency of helium has been recalculated using OPAL opacities (Kato & Hachisu 1999). These authors concluded that helium can accumulate much more efficiently than found by Cassisi et al. (1998), mainly because the latter authors used relatively low WD masses (0.516 M$_\odot$ and 0.8 M$_\odot$) and old opacities in their calculations.

(ii) Hachisu et al. (1999a,b) claimed to have identified two evolutionary channels for single-degenerate systems previously overlooked in population synthesis calculations. In the first of these channels, the C–O WD is formed from a red giant with a helium core of 0.8–2.0 $M_\odot$ (rather than from an asymptotic giant branch star with a C–O core). The immediate progenitors in this case are expected to be either helium-rich Supersoft X-ray Sources or recurrent novae of the U Sco subclass (where the accreted material appears to be helium rich).

In the second channel, Hachisu et al. (1999b) considered very wide (initial separations as large as $\sim 40,000$ $R_\odot$) symbiotic systems, in which the components are brought together by the inclusion of new physical effects (see (iii)–(3) below).

(iii) It has been suggested that the inclusion of a few additional physical effects, can increase substantially the phase space of the symbiotic channel (Hashisu, Kato & Nomoto 1996, 1999b). These new effects included:

(1) The WD loses much of the transferred mass in a massive wind. This has the effect that the mass transfer process is stabilized for a wider range of mass ratios, up to $q_{max} \equiv m_2/m_1 = 1.15$ instead of $q_{max} = 0.79$ without the massive wind.

(2) It has been suggested that the wind from the WD strips the outer layers of the red giant at a high rate. This increases the allowed mass ratios (for stability) even above 1.15, essentially indefinitely.

(3) It has been suggested that at large separations (up to $\sim 40,000$ $R_\odot$), when the orbital velocity is of the order of the wind velocity, the wind from the red giant acts like a common envelope to reduce the separation, thus allowing much wider initial separations to result in interaction.

There are many uncertainties associated with all of these attempts to increase the phase space. For example, the efficiency of mass stripping from the giant by the wind from the WD may be much smaller than assumed by Hachisu et al. (1999b), for the following reasons. At high accretion rates, much of the mass loss from the WD may be in the form of an outflow or a collimated jet, perpendicular to the accretion disk rather than in the direction of the giant. Evidence that this is the case is provided by the jet satellite lines to He II 4686, $H\beta$ and $H\alpha$ observed in the Supersoft X-ray Source RX J0513.9–6951 (Southwell et al. 1996). These jet lines are very similar to those seen in the prototypical jet source SS 433 (e.g. Vermeulen et al. 1992). Furthermore, even if some of the WD wind hits the surface of the giant, it is not clear how efficient it would be in stripping mass, since the rate of energy deposition per unit area by the wind is smaller by two orders of magnitude that the giant's own intrinsic flux.

Similarly, the efficiency of helium accumulation is still highly uncertain, as the differences between the results of Kato & Hachisu (1999) and Cassisi et al. (1998) have shown.

Also, the particular form of the specific angular momentum in the wind used by Hachisu, Kato & Nomoto (1999b; point (3) above) may be realized only in relatively rare cases (Yungelson & Livio 2000).

Finally, all the new suggestions for the increase in phase space rely very heavily on the results of the wind solutions of Kato (1990; 1991), which involve a treatment of the radiation and hydronamics not nearly as sophisticated as that of more state of the art radiative transfer codes (e.g. Hauschildt et al. 1995, 1996).

5.2.3. Overall assessment of the single-degenerate scenario

The above discussion suggests that probably not all the scenarios for increasing the "phase space" of the single-degenerate channels work (if they did, we might have had the opposite problem of too high a frequency of SNe Ia!). However, these attempts serve to demonstrate that *the input physics to population synthesis codes still involves many uncertainties*. My feeling is therefore that given the many potential channels leading to SNe Ia, statistics should not be regarded as a serious problem.

Single-degenerate scenarios therefore appear quite promising, since unlike the situation a decade ago, a class of objects in which the WDs accrete hydrogen steadily (the Supersoft X-Ray Sources) has actually been identified. The main problem with single-degenerate scenarios remains the non-detection of hydrogen so far. While a difficult observational problem (see §6), *the establishment of the presence or absence of hydrogen in SNe Ia should become a first priority for SNe observers*.

5.3. What if....?

Given the fact that there are still uncertainties involved in identifying the SNe Ia progenitors, and that WD mergers and some form of off-center helium ignitions almost certainly occur, it is instructive to pose a few "what if" questions. For example: *What if WD mergers with a total mass exceeding Chandrasekhar do not produce SNe Ia, what do they produce then?* The answer in this case will have to be that they almost certainly produce either neutron stars via accretion induced collapses, or single WDs, if the merger is accompanied by extensive mass loss from the system. Fryer et al. (1999) estimate (from nucleosynthesis constraints) that less than 0.1% of the total Galactic neutron star population is produced via accretion induced collapses.

What if off-center helium ignitions (ELDs) do not produce SNe Ia? In this case, if an explosive event indeed ensues, a population of "super novae" (with ~ 0.15 $M_\odot$ of ^{56}Ni and He) is yet to be detected (maybe SN 1885A in M31 was such an event?). *What if off-center helium ignitions do produce SNe Ia? What comes out of the systems with $M_{WD} \gtrsim 1$ $M_\odot$, which should be even brighter?* If indeed $\sim 20\%$ of SNe Ia are SN 1991T-like (Li et al. 2000), then maybe these could be represented by such events. This is far from certain, however, since an analysis of the properties of SN 1991T showed that these properties would be very difficult to reproduce even with a nickel mass approaching the Chandrasekhar mass (Fisher et al. 1999). Thus, we see that off-center helium ignitions seem to present an observational problem both if they *do* and if they *do not* produce SNe Ia. To me this suggests that the physics of these events is not yet well understood (for example, maybe off-center helium ignition fails to ignite the C–O core after all).

6. How can we hope to unambiguously identify the progenitors?

There are several ways in which observations of both nearby and distant supernovae could solve the mystery of SNe Ia progenitors:

(1) A combination of *early high resolution optical spectroscopy, x-ray observations* and *radio observations* of nearby SNe Ia can both provide limits on $\dot{M}/v$ from the progenitors and potentially detect the presence of circumstellar hydrogen (if it exists).

For example, narrow HI in emission or absorption could be detected either very early, or shortly after the ejecta become optically thin (~ 100 days). The latter is true because the SN ejecta probably engulfs the companion at early times (e.g. Chugai 1986; Livne, Tuchman & Wheeler 1992). The interaction of the ejecta with the circumstellar medium can be observed either in the radio (e.g. Boffi & Branch 1995) or in x-rays (e.g. Schlegel 1995). The collision of the ejecta (with circumstellar matter) can also set up a forward and a reverse shock (e.g. Chevalier 1984; Fransson, Lundqvist & Chevalier 1996), and radiation from the latter can ionize the wind and produce Hα emission (e.g. Cumming et al. 1996).

(2) Early observations (again of nearby SNe Ia) of the gamma-ray light curve (or gamma-ray line profiles) could distinguish between carbon ignitors and sub-Chandrasekhar helium ignitor models (see §4.2.2) since the latter can be expected to result in a quicker rise of the gamma-ray light curve due to the presence of ^{56}Ni in the outer layers (and different gamma-ray line profiles; because of the high velocity ^{56}Ni).

(3) Another important aspect of the single degenerate scenario which can be tested by observations of both nearby and very distant supernovae is the dependence on metallicity. The increase in the "phase space" of single degenerate progenitors, which is required to make the statistics more compatible with observations (§5.2.2), relies heavily on the existence of an optically thick wind from the WD. For a low metallicity of the accreted matter ([Fe/H] $\lesssim -1$), the wind from the WD is strongly suppressed (since the wind is driven by a peak in the opacity, which is due to iron lines; e.g. Kobayashi et al. 1998). Consequently, it is expected that SNe Ia rates will be significantly lower in low iron abundance environments. Thus, determinations of relative rates in dwarf galaxies (and in the very outskirts of spiral galaxies) can help determine the viability of a key ingredient in the single degenerate scenario.

Similarly, a significant drop may be expected in the rate of SNe Ia at redshift $z \sim 2$ (again due to the decrease in metallicity).

(4) In general, observations of very distant supernovae (at $z \sim 2$–4) with the Next Generation Space Telescope (NGST) can help significantly in identifying the progenitors (e.g. Yungelson & Livio 1999, 2000; Nomoto et al. 2000). For example, the progenitors can be identified from the observed frequency of SNe Ia as a function of redshift (e.g. Yungelson & Livio 1998, 1999, 2000; Ruiz-Lapuente & Canal 1998; Madau, Della Valle & Panagia 1998; Nomoto et al. 2000; and see §8), since different progenitor models produce different redshift distributions. Personally, I think that it would be quite pathetic to have to resort to this possibility. Rather, one would like to be able to identify the progenitors independently, and then use the observations of supernovae at high z to constrain models of cosmic star formation rates, and of cosmic evolution of SNe rates, luminosity, and input into galaxies.

7. Cosmological implications: Could we be fooled?

One of the key questions that result from the uncertainties in the theoretical models and the fact that we do not know with certainty which systems are the progenitors of SNe Ia is clearly: *is it possible that SNe Ia at higher redshifts are systematically dimmer than their low-redshift counterparts?* In this respect it is important to remember that a systematic decrease in the brightness by ~ 0.25 magnitudes is sufficient to explain away the need for a cosmological constant. This question became particularly relevant when an analysis of the rise times of SNe Ia (which was based on preliminary estimates for the high-z sample; by Goldhaber 1998 and Groom 1998) seemed to show that high-redshift SNe have shorter rise times by 2.5 days than the low-z SNe (Riess et al. 1999b). A more recent analysis (which used more realistic error estimates than those used by Goldhaber 1998), however, found a better agreement (within 2σ) between the rise times of the low- and high-redshift SNe Ia (Aldering, Knop & Nugent 2000).

In a recent work, Yungelson & Livio (1999) calculated the expected ratio of the rate of SNe Ia to the rate of SNe from massive stars (Types II, Ia, Ic) as a function of redshift for *several* progenitor models. *The possibility of having different classes of progenitors contributing to the total SNe Ia rate should definitely be considered*, especially in view of the tentative finding by Li et al. (2000) that there is a relatively high rate ($\sim 40\%$) of peculiar SNe Ia among the local sample. If confirmed, these findings suggest that homogeneity should no longer be considered a very strong constraint on progenitor models. I should note though that diversity among SNe Ia does not *necessarily* imply different progenitors, since even in the context of one progenitor model diversity may arise for example from changes in the carbon mass fraction of the WD, which in turn may depend on the environment (e.g. Nomoto et al. 2000). Yungelson and Livio (1999) showed (within the uncertainties of population synthesis models) that *if* different progenitor systems can contribute to the total SNe Ia rate (e.g. double-degenerates and single degenerates), then it is possible, in principle, that one class of progenitors (e.g. double degenerates) will dominate the rates of the local (low-z) sample, while a different progenitor class (e.g. single degenerate) will start to dominate at $z \lesssim 1$. This is a consequence of the fact that the SNe Ia rate from double degenerates is expected to decline quite steeply from $z = 0$ to $z \sim 1$, while the rate from single degenerates is expected to stay relatively flat in this redshift interval. However, at least within the assumptions of their model calculations, it appears that such a transition is not very likely, because of the following reason: If the contribution from physically different channels (like double degenerates and single degenerates, both at the Chandrasekhar mass) was indeed significant, with a transition from dominance by double degenerates to single degenerates occurring at $z \lesssim 1$, one would have expected to observe this division more clearly in the local and distant samples. For example, the local sample should be dominated by the double degenerate progenitors (with a ratio of double to single degenerates which should be consistent with the results of Li et al. 2000). At the same time, however, the high-z ($z \lesssim 1$) sample should be dominated by single degenerates, but with the contributions from the single and double degenerate channels not being vastly different (in particular, the contribution should be equal at the transition point). This is *not* consistent with the observations of the high-z sample, the latter appearing to be (within the observational uncertainties) *very homogeneous* (Li et al. 2000). Consequently, I do not think it likely that the observed universal acceleration is an artifact of the observed SNe Ia sample being dominated by different progenitor classes at high- and low-z (this view is supported by the measurements of the anisotropy of the microwave background).

I should note that the most surprising aspect in the results of Li et al. (2000) is the fact that although very bright SNe Ia (SN 1991T-like) constitute $(21 \pm 7)\%$ of the local sample, these bright objects appear to be totally absent from the high-z sample. One way in which one could (in principle) explain this fact is the following. Suppose that the SN 1991T-like events are caused by mergers (since a super-Chandrasekhar mass is possible in this case), while the "normals" are caused by single degenerates. In this case the local sample would be dominated by single degenerates ($\sim 60\%$ of the events to agree with Li et al., 2000), while double degenerates would contribute $\sim 20\%$ of the events (I ignore here the weak events since they cannot be seen at high-z). Now, since the rate of events from single degenerates stays quite flat till $z \sim 1$, while the rate of events from double degenerates declines quite steeply towards $z \sim 1$, the bright objects will be missing from the high-z sample. Note, however, that this potential explanation for the behavior of the diversity has no obvious implications for the finding of accelerating expansion, since the same class of progenitors dominates both the high-z and low-z samples. Nevertheless, a better understanding of the apparent diversity at low-z and apparent lack thereof at high-z is definitely needed.

Other evolutionary effects are still possible, in principle (e.g. Drell, Loredo & Wasserman, 1999; Hillebrandt 2000), however, as far as I am aware, only one that is physically meaningful, likely, and mimics accelerated expansion, has been identified so far.

This one potential evolutionary effect that certainly deserves more work is *the effect of metallicity on the density at the point of carbon ignition*. Generally, it is expected that a lower metallicity will result in a lower central density (e.g. Nomoto et al. 1997). This is because a lower metallicity results in a lower abundance of the Urca-active element ^{21}Ne, which in turn reduces the neutrino cooling and leads to an earlier ignition. A lower central density could (in principle at least) result in a more rapid light curve development (due to the lower WD binding energy), and a lower inferred maximum brightness.

It is important to note that in a recent work, Riess et al. (2000) have shown that it is highly unlikely that the dimming of distant SNe Ia is caused by dust opacity (Galactic-type dust was rejected at the 3.4σ confidence level, and "gray" dust with grain size > 0.1 μm was rejected at the 2.3 to 2.6σ confidence level).

8. Tentative conclusions and observational tests

On the basis of the analysis and discussion in the present work, the following tentative conclusions can be drawn:

(1) SNe Ia are almost certainly thermonuclear disruptions of mass accreting *C–O white dwarfs*.

(2) It is very likely that the explosion occurs *at the Chandrasekhar mass*, as *carbon is ignited at (or very near) the WD center*. The flame propagates either as a deflagration, or, more likely perhaps, starting as a deflagration which transitions into a detonation. Off-center ignition of helium at sub-Chandrasekhar masses may still be responsible for a subset of the SNe Ia which are subluminous, but this is less clear.

(3) The immediate progenitor systems are still not known with certainty. From the discussion in §5 (see in particular §5.1.3 and 5.2.3) however, I conclude that presently *single degenerate scenarios look more promising*, with hydrogen or helium rich material being transferred from a subgiant or giant companion (systems like Supersoft X-Ray Sources and Symbiotics). It is still possible, however, in view of the apparent diversity in the local sample, that more than one progenitor class contributes to the total SNe Ia rate. In particular, a scenario in which single degenerates contribute $\gtrsim 60\%$ of the events and double degenerates $\sim 20\%$, appears to be consistent with the diversity of the $z \sim 0$ sample and the lack thereof in the distant sample.

(4) Definitive answers concerning the nature of the progenitors can be obtained from observations taken as early as possible in: *x-rays, radio, and high resolution optical spectroscopy. The establishment of the presence or absence of hydrogen in SNe Ia should be regarded as an extremely high priority goal for supernovae observers*. If hydrogen will not be detected at interesting limits (corresponding to $\dot{M}/v_{10} \sim 10^{-8}$ M$_\odot$ yr^{-1}), this will point clearly towards the double-degenerate scenario.

(5) Observations of SNe Ia at high redshifts can help to test particular ingredients of the models which are directly related to the nature of the progenitors. For example, most of the models aiming at improving the statistics of the single-degenerate scenarios rely on a strong wind from the accreting WD. These models thus predict an "inhibition" of SNe Ia in low-metallicity environments, and in particular a significant decrease in the rate of SNe Ia in spirals at $z \sim 2$ (Kobayashi et al. 1998; Nomoto et al. 2000). Furthermore, if the inferred cosmic star formation rate is used (e.g. Pettini et al. 1998), then the SN Ia rate is expected to drop significantly at $z \sim 1.6$. At present, the detection of a very likely SN Ia at redshift $z = 1.32$ (SN 1997ff; Gilliland, Nugent & Phillips 1999) in the Hubble Deep Field, and two more at redshifts 1.20 and 1.23 (Perlmutter et al., private communication and Tonry et al., private communication) appear to be at least mildly inconsistent with this prediction, but more observations will be required to give a more definitive answer.

(6) The potential "inhibition" of SNe Ia due to low metallicity should also manifest itself in an absence (or at least a significant decline in the rate) of SNe Ia in dwarf galaxies and in the outer regions of spirals. The statistics necessary to test this prediction are starting to accumulate.

(7) It is possible, in principle, that the local and high-z SNe Ia samples are dominated by different progenitor classes, thus mimicking accelerated expansion, however, this is neither very likely nor consistent with the observations of diversity. *With more detections of SNe Ia at redshifts $z \gtrsim 1$ it will probably become possible to directly confirm the transition in the expansion of the universe from deceleration to acceleration.* Such a transition would be difficult to mimic by systematic or evolutionary effects, and it would therefore confirm the accelerated expansion.

This research has been supported in part by NASA Grant NAG5–6857.
I am grateful to Adam Riess for helpful discussions.

REFERENCES

ALDERING, G., KNOP, R., & NUGENT, P. 2000 astro-ph/0001049.
ARNETT, W. D. 1969 *ApJS*, **5**, 180.
BACKER, D. C. 1993. In *ASP. Conf. Ser. 36* (ed. J. A. Phillips, et al.). Planets around Pulsars, p. 11. ASP.
BALBUS, S. A. & HAWLEY, J. F. 1998 *Rev. Mod. Phys.*, **70**, 1.
BALBUS, S. A., HAWLEY, J. F., & STONE, J. M. 1996 *ApJ*, **467**, 76.
BARTUNOV, O. S., TSVETKOV, D. YU, & FILIMONOVA, I. V. 1994 *PASP*, **106**, 1276.
BENZ, W., THIELEMANN, F. K., & HILLS, J. G. 1989 *ApJ*, 342, 986.
BENZ, W., CAMERON, A. G. W., BOWERS, R. L., & PRESS, W. H. 1990 *ApJ*, **348**, 647.
BOESGAARD, A. M., KING, J. R., DELIGANNIS, C. P. & VOGT, S. S. 1999 *AJ*, in press.
BOFFI, F. R. & BRANCH, D. 1995 *PASP*, **107**, 347.
BRANCH, D. 1998 *ARAA*, **36**, 17.
BRANCH, D., FISHER, A., & NUGENT, P. 1993 *AJ*, **106**, 2383.
BRANCH, D., LACY, C. H., MCCALL, M. L., SUTHERLAND, P. G., UOMOTO, A., WHEELER, J. C., & WILLS, B. J. 1983 *ApJ*, **270**, 123.
BRANCH, D., LIVIO, M., YUNGELSON, L. R., BOFFI, F. R., & BARON, E. 1995 *PASP*, **107**, 717.
BRANCH, D. & VAN DEN BERGH, S. 1993 *AJ*, **105**, 2231.
BRAVO, E. & GARCIA-SENZ, D. 1999, *MNRAS*, submitted.
BURRIS, D. L., PILACHOWSKI, C. A., ARMANDROFF, T. A., SNEDEN, C., COWAN, J. J., & ROE, H. 1999 preprint.
CANAL, R. 1997. In *Thermonuclear Supernovae* (eds. P. Ruiz-Lapuente, R. Canal & J. Isern). p 257. Kluwer.
CANAL, R., ISERN, J. & LABAY, J. 1990 *ARAA*, **28**, 183.
CASSISI, S., IBEN, I. JR., & TORNANBE, A. 1998 *ApJ*, **496**, 376.
CHEVALIER, R. A. 1984 *ApJ*, **285**, L63.
CHUGAI, N. N. 1986 *SvA*, **30**, 563.
CUMMING, R. J., LUNDQVIST, P., SMITH, L. J., PETTINI, M., & KING, D. L. 1996 *MNRAS*, **283**, 1355.
DELLA VALLE, M., BENETTI, S., & PANAGIA, N. 1996 *ApJ*, **459**, 23.
DELLA VALLE, M. & LIVIO, M. 1994 *ApJ*, **423**, L31.

DI STEFANO, R., NELSON, L. A., LEE, W., WOOD, T. H., & RAPPAPORT, S. 1997. In *Thermonuclear Supernovae* (ed. R. Ruiz-Lapuente, R. Canal, & J. Isern). p. 147. Kluwer.

DI STEFANO, R. & RAPPAPORT, S. 1994 *ApJ*, **437**, 733.

DOMINGUEZ, I., TORNAMBÉ, A., & ISERN, J. 1993 *ApJ*, **419**, 268.

DRELL, P. S., LOREDO, T. J., & WASSERMAN, I. 1999, astro-ph/9905027.

DURISEN, R. H. 1973 *ApJ*, **183**, 205.

ECK, C., COWAN, J. J., ROBERTS, D., BOFFI, F. R. & BRANCH, D. 1995 *ApJ*, **451**, L53.

EDVARDSSON, B., ANDERSEN, J., GUSTOFSSON, B., LAMBERT, D. L., NISSEN, P. E., & TOMKIN, J. 1993 *A&A*, **275**, 101.

FILIPPENKO, A. V. 1997 *ARAA*, **35**, 309.

FISHER, A., BRANCH, D., HATANO, K., & BARON, E. 1999 *MNRAS*, **304**, 67.

FRANSSON, C., LUNDQVIST, P., & CHEVALIER, R. A. 1996 *ApJ*, **461**, 993.

FRYER, C., BENZ, W., HERANT, M., & COLGATE, S. A. 1999 *ApJ*, **516**, 892.

FULBRIGHT, J. P. & KRAFT, R. P. 1999 *AJ*, **118**, 527.

GARCIA-SENZ, D., BRAVO, E., & SERICHOL, N. 1998 *ApJS*, **115**, 119

GILLILAND, R. I., NUGENT, P. E., & PHILLIPS, M. M. 1999 *ApJ*, in press.

GOLDHABER, G. 1998, in *Gravity: From the Hubble Length to the Planck Length*. SLAC.

GÓMEZ, G., LÓPEZ, R., & SÁNCHEZ, F. 1996 *AJ*, **112**, 2094.

GREINER, J., HASINGER, G., & KAHABKA, P. 1991 *A&A*, **246**, L17.

GROOM, D. E. 1998 *BAAS*, **193**, 111.02.

GUERRERO, J. 1994 PhD Thesis, Univ. of Barcelona.

GUTIÉRREZ, J., GARCIA-BERRO, E., IBEN, I. JR., ISERN, J., LABAY, J. & CANAL, R. 1996 *ApJ*, **459**, 701.

HACHISU, I., KATO, M., & NOMOTO, K. 1996 *ApJ*, **470**, L97.

HACHISU, I., KATO, M., & NOMOTO, K. 1999b *ApJ*, **522**, 487.

HACHISU, I., KATO, M., NOMOTO, K., & UMEDA, H. 1999a *ApJ*, **519**, 314.

HAMUY, M., PHILLIPS, M. M., SUNTZEFF, N. B., SCHOMMER, R. A. MAZA, J. & AVILÉS, R. 1996a *AJ*, **112**, 2398.

HAMUY, M., PHILLIPS, M. M., SUNTZEFF, N. B., SCHOMMER, R. A., MAZA, J., ET AL. 1996b *AJ*, **112**, 2438.

HANSEN, C. J. & WHEELER, J. C. 1969 *ApJS*, **3**, 464.

HAUSCHILDT, P. H., BARON, E., STARRFIELD, S., & ALLARD, F. 1996 *ApJ*, **462**, 386.

HAUSCHILDT, P. H., STARRFIELD, S., SHORE, S. N., ALLARD, F., & BARON, E. 1995 *ApJ*, **447**, 829.

HILLEBRANDT, W. 2000. In *Type Ia Supernovae: Theory and Cosmology* (eds. J. C. Niemeyer & J. W. Truran), CUP, in press.

HÖFLICH, P. & KHOKHLOV, A. 1996 *ApJ*, **457**, 500.

HÖFLICH, P., WHEELER, J. C., & THIELEMANN, F. K. 1998 *ApJ*, **495**, 617.

IBEN, I. JR. & LIVIO, M. 1993 *PASP*, **105**, 1373.

IBEN, I. JR. & TUTUKOV, A. V. 1984 *ApJS*, **54**, 355.

IBEN, I. JR. & TUTUKOV, A. V. 1985 *ApJS*, **58**, 661.

IBEN, I. JR., TUTUKOV, A. V., & FEDOROVA, A. V. 1998 *ApJ*, **503**, 344.

ISRAELIAN, G., GARCIA LOPEZ, R. J., & REBOLO, R. 1998 *ApJ*, **507**, 805.

KAHABKA, P. & VAN DEN HEUVEL, E. P. J. 1997 *ARA&A*, **35**, 69.

KATO, M. 1990 *ApJ*, **355**, 277.

KATO, M. 1991 *ApJ*, **369**, 471.

KATO, M. & HACHISU, I. 1999 *ApJ*, submitted, astro-ph/991080.

KAUFFMANN, G., WHITE, S. D. M. & GUIDERDONI, B. 1993 *MNRAS*, **264**, 201.
KAWAI, Y., SAIO, H. & NOMOTO, K. 1987 *ApJ*, **315**, 229.
KENYON, S. J., LIVIO, M., MIKOLAJEWSKI, J. & TOUT, C. A. 1993 *ApJ*, **407**, L81.
KIRSHNER, R. P. ET AL. 1993 *ApJ*, **415**, 589.
KHOKHLOV, A. 1991 *A&A*, **245**, 114.
KHOKHLOV, A. M., ORAN, E. S. & WHEELER, I. C. 1997 *ApJ*, **478**, 678.
KOBAYASHI, C., TSUJIMATO, T., NOMOTO, K., HACHISU, I. & KATO, M. 1998 *ApJ*, **503**, L155.
KOEN, C., OROSZ, J. A. & WADE, R. A. 1998 *MNRAS*, **300**, 695.
LI, W., FILIPPENKO, A. V., RIESS, A. G., TREFFERS, R. R., HU, J., & QIU, Y. 2000, preprint.
LI, X.-D. & VAN DEN HEUVEL, E. P. J. 1997 *A&A*, **322**, L9
LIU, W., JEFFREY, D. D. & SCHULTZ, D. R. 1998 *ApJ*, **494**, 812.
LIVIO, M. 1993. In *Cataclysmic Variables and Related Physics* (eds. O. Regev & G. Shaviv), p. 57. Institute of Physics Publishing.
LIVIO, M. 1995. In *Millisecond Pulsars: A Decade of Surprise* (eds. A. S. Fruchter, M. Tavani & D. R. Backer) volume 72, p. 105. ASP.
LIVIO, M. 1996a. In *Supersoft X-Ray Sources* (ed. G. Greiner) p. 183. Springer.
LIVIO, M. 1996b. In *Evolutionary Processes in Binary Stars* (eds. R. A. M. J. Wijers, M. B. Davies & C. A. Tout) p. 141. Kluwer.
LIVIO, M. 1999 *Science*, **286**, 1689.
LIVIO, M. 2000. In *Type Ia Supernovae: Theory and Cosmology* (eds. J. C. Niemeyer & J. W. Truran), CUP, in press.
LIVIO, M., PRINGLE, J. E. & SAFFER, R. A. 1992 *MNRAS*, **257**, 15p.
LIVIO, M. & TRURAN, J. W. 1992 *ApJ*, **389**, 695.
LIVNE, E. 1990 *ApJ*, **354**, L53.
LIVNE, E. & ARNETT, D. 1995 *ApJ*, **454**, 62.
LIVNE, E. & GLASNER, A. 1991 *ApJ*, **370**, 272.
LIVNE, E., TUCHMAN, Y. & WHEELER, J. C. 1992, *ApJ*, **399**, 665.
MADAU, P., DELLA VALLE, M. & PANAGIA, N. 1998 *MNRAS*, **297**, L17
MATTEUCCI, F. & GREGGIO, L. 1986 *A&A*, **154**, 279.
MAUSKOPF, P. D., ET AL. 1999 astro-ph/9911444.
MAXTED, P. F. L. & MARSH, T. R. 1999, *MNRAS*, **307**, 122.
McWILLIAM, A. 1997 *ARA&A*, **35**, 503.
MILLER, A. D., ET AL. 1999, *ApJ*, **524**, L1.
MINKOWSKI, R. 1939 *ApJ*, **89**, 156.
MOCHKOVITCH, R., GUERRERO, J. & SEGRETAIN, L. 1997. In *Thermonuclear Supernovae* (eds. P. Ruiz-Lapuente, R. Canal & J. Isern). p 187. Kluwer.
MOCHKOVITCH, R. & LIVIO, M. 1990 *A&A*, **236**, 378.
MORAN, C., MAXTED, P., MARSH, T. R., SAFFER, R. A., & LIVIO, M. 1999 *MNRAS*, **304**, 535.
MUNARI, V. & RENZINI, A. 1992 *ApJ*, **397**, L87.
NAGATAKI, S., HASHIMOTO, M. & SATO, K. 1998 *PSAJ*, **50**, 75.
NISSEN, P. E., GUSTAFSSON, G., EDVARDSSON, B. & GILMORE, G. 1994 *A&A*, **285**, 440.
NOMOTO, K. 1982 *ApJ*, **253**, 798.
NOMOTO, K., IWAMOTO, K., NAKASATO, N., THIELEMANN, F.-K., BRACHWITZ, F., TSUJIMOTO, T., KUBO, Y., & KISHIMOTO, N. 1997 *Nuc. Phys.*, **A621**, 467c.
NOMOTO, K. & KONDO, Y. 1991 *ApJ*, **367**, L19.
NOMOTO, K. & SUGIMOTO, D. 1977 *PASJ*, **29**, 765.

NOMOTO, K., SUGIMOTO, D., & NEO, S. 1976 *ApJS*, **39**, 137.

NOMOTO, K., ET AL. 1997. In *Thermonuclear Supernovae* (eds. P. Ruiz-Lapuente et al.) p. 349. Kluwer.

NOMOTO, K., UMEDA, H., HACHISU, I., KATO, M., KOBAYASHI, C., & TSAJIMOTO, T. 2000. In *Type Ia Supernovae: Theory and Cosmology* (eds. J. Truran & J. Niemeyer). CUP, in press.

NUGENT, P., BARON, E., BRANCH, D., FISHER, A., & HAUSCHILDT, P. H. 1997 *ApJ*, **485**, 812

PERLMUTTER, S., ET AL. 1997 *ApJ*, **483**, 565.

PERLMUTTER, S., ET AL. 1998 *Nature*, **391**, 51.

PERLMUTTER, S., ET AL. 1999 *ApJ*, **517**, 565.

PODSIADLOWSKI, PH., PRINGLE, J. E., & REES, M. J. 1991 *Nature*, **352**, 783.

POLCARO, V. F. & VIOTTI, R. 1991 *A&A*, **242**, L9.

PRIALNIK, D. & KOVETZ, A. 1995 *ApJ*, **445**, 789.

RASIO, F. A. & SHAPIRO, S. L. 1995, *ApJ*, **438**, 887.

REINECKE, M., HILLEBRANDT, W., & NIEMEYER, J. C. 1998 *A&A*, in press, astro-ph/9812120.

RENZINI, A. 1996 In *Supernovae and Supernova Remnants* (ed. R. McCray & Z. Wang). p 77. CUP.

RIESS, A. G., ET AL. 1998 *AJ*, **116**, 1009.

RIESS, A. G., ET AL. 1999a *AJ*, **118**, 2675.

RIESS, A. G., ET AL. 2000, astro-ph/0001384.

RIESS, A. G., FILIPPENKO, A. V., LI, W., & SCHMIDT, B. P. 1999b *AJ*, **118**, 2668.

RIESS, A. G., NUGENT, P., FILIPPENKO, A. V., KIRSHNER, R. P. & PERLMUTTER, S. 1998 *ApJ*, **504**, 935.

RIESS, A. G., PRESS, W. H., & KIRSHNER, R. P. 1996 *ApJ*, **473**, 88.

RIEUTORD, M. & BONAZZOLA, S. 1987 *MNRAS*, **227**, 295.

RITOSSA, C., GARCIA-BERRO, E., & IBEN, I. JR. 1996 *ApJ*, **460**, 489.

RUIZ-LAPUENTE, P. & CANAL, R. 1998 *ApJ*, **497**, L57.

RUIZ-LAPUENTE, P., CANAL, R., & BURKERT, A. 1997. In *Thermonuclear Supernovae* (ed. R. Ruiz-Lapuente, R. Canal, & J. Isern). p. 205. Kluwer.

RUIZ-LAPUENTE, P., KIRSHNER, R. P., PHILLIPS, M. M., CHALLIS, P. M., SCHMIDT, B. P., FILIPPENKO, A. V., & WHEELER, J. C. 1995 *ApJ*, **439**, 60.

SAFFER, R. A., LIVIO, M. & YUNGELSON, L. R. 1998 *ApJ*, **502**, 394.

SAHA, A., SANDAGE, A., LABHARDT, L., TAMMANN, G. A., MACCHETTO, F. D. & PANAGIA, N. 1997 *ApJ*, **486**, 1.

SAHA, A., ET AL. 2000, in preparation.

SAIO, H. & NOMOTO, K. 1985 *A&A*, **150**, L21.

SAIO, H. & NOMOTO, K. 1998 *ApJ*, **500**, 388.

SANDAGE, A., SAHA, A., TAMMANN, G. A., LABHARDT, L., PANAGIA, N. & MACCHETTO, F. D. 1996 *ApJ*, **460**, L15.

SCHLEGEL, E. M. 1995 *Rep. Prog. Phys.* **58**, 1375.

SCHMIDT, B., ET AL. 1998 *ApJ*, **507**, 46.

SEGRETAIN, L. 1994 PhD Thesis, University of Lyon.

SOUTHWELL, K. A., LIVIO, M., CHARLES, P. A., O'DONOGHUE, D., & SUTHERLAND, W. J. 1996 *ApJ*, **470**, 1065.

STEPHENS, A. 2000 *Ph.D. thesis*, University of Hawaii.

TAMMAN, G. A. & SANDAGE, A. 1995 *ApJ*, **452**, 16.

THORSETT, S. E., ARZOWMANIAN, Z., & TAYLOR, J. H. 1993 *ApJ*, **412**, L33.

TIMMES, F. X., WOOSLEY, S. E., & TAAM, R. E. 1994 *ApJ*, **420**, 348.

TURATTO, M., CAPPELLARO, E., & BENETTI, S. 1994 *AJ*, **108**, 202.

TURATTO, M., CAPPELLARO, E., & PETROSIAN, A. R. 1999. In *Active Galactic Nuclei & Related Phenomena*, IAU Symp. 194 (eds. Y. Terzian, D. Weedman, & E. Khachikian). p. 364. ASP.

VAUGHAN, T. E., BRANCH, D., MILLER, D. L., & PERLMUTTER, S. 1995 *ApJ*, **439**, 558.

WEBBINK, R. F. 1984 *ApJ*, **227**, 355.

WHEELER, J. C., HARKNESS, R. P., KHOKHLOV, A. V., & HOEFLICH, P. 1995 *Phys. Rep.* **256**, 211.

WHELAN, J. & IBEN, I. JR. 1973 *ApJ*, **186**, 1007.

WILSON, G. W., ET AL. 1999 astro-ph/9902047.

WOLSZCZAN, A. 1997 *Cel. Mech. & Dyn. Aston.*, **68**, 13.

WOOSLEY, S. E. 1997. In *Thermonuclear Supernovae* (eds. P. Ruiz-Lapuente, R. Canal & J. Isern). p 313. Kluwer.

WOOSLEY, S. E., TAAM, R. E., & WEAVER, T. A. 1986 *ApJ*, **301**, 601.

WOOSLEY, S. E. & WEAVER, T. A. 1994 *ApJ*, **423**, 371.

VAN DEN HEUVEL, E. P. J., BHATTACHARYA, D., NOMOTO, K., & RAPPAPORT, S. A. 1992 *A&A*, **262**, 97.

VERMEULEN, R. C., ET AL. 1992 *A&A*, **270**, 204.

YUNGELSON, L. & LIVIO, M. 1998 *ApJ*, **497**, 168.

YUNGELSON, L. R. & LIVIO, M. 1999, *ApJ*, in press, astro-ph/9907359.

YUNGELSON, L. R. & LIVIO, M. 2000, in preparation.

YUNGELSON, L., LIVIO, M., TRURAN, J. W., TUTUKOV, A., & FEDOROVA, A. V. 1996 *ApJ*, **466**, 890.

YUNGELSON, L., LIVIO, M., TUTUKOV, A., & KENYON, S. J. 1995 *ApJ*, **447**, 656.

Conference summary: Supernovae and Gamma-Ray Bursts

By J. CRAIG WHEELER

Department of Astronomy, University of Texas, Austin, TX 78712

There are hints that nearby Type Ia supernovae may be a little different than those at large redshift. Confidence in the conclusion that there is a cosmological constant and an accelerating Universe thus still requires the hard work of sorting out potential systematic effects. Polarization data show that core-collapse supernovae (Type II and Ib/c) probably depart strongly from spherical symmetry. Evidence for exceedingly energetic supernovae must be considered self-consistently with evidence that they are asymmetric, a condition that affects energy estimates. Jets arising near the compact object can produce such asymmetries. There is growing conviction that gamma-ray bursts intrinsically involve collimated or jet-like flow and hence that they are also strongly asymmetric. SN 1998bw is a potential rosetta stone that will help to sort out the physics of explosive events. Are events like SN 1998bw more closely related to "ordinary" supernovae or "hypernovae?" Do they leave behind neutron stars as "ordinary" pulsars or "magnetars" or is the remnant a black hole? Are any of these events associated with classic cosmic gamma-ray bursts as suggested by the supernova-like modulation of the afterglows of GRB 970228, GRB 980326 and GRB 990712?

New data is driving both supernova and γ-ray burst research and suggesting that these subjects may be related. A central issue in both areas is the breaking of spherical symmetry. This conference celebrated as much as any single thing the emergence of evidence and argument for strong breakdown of spherical symmetry for both supernovae and γ-ray bursts and especially the resulting potential for links between them, whether those links are supernovae or "hypernovae." Supernova studies have also revolutionized the study of cosmology. There spherical symmetry is not at issue, but the prospect of an accelerating Universe presents many challenges.

I cannot synthesize all the energetic work presented in oral talks and poster presentations, never mind over coffee and dinner, at this symposium that mixed two "exploding" fields. I will attempt to give a summary of certain highlights and connective themes that I think will set the course for future research. In §2, I give a brief summary of the exciting work on SN Ia, their associated physics, and their application to cosmology. Some perspectives on γ-ray bursts and their possible link to supernovae or "hypernovae" are presented in §3. New results on the propagation of a jet through a stellar core are given in §4. Some perspectives and conclusions are presented in §5.

1. Type Ia supernovae and cosmology

There is a general concensus that the strong majority, if not all, Type Ia supernovae arise in carbon/oxygen white dwarfs of very near the Chandrasekhar mass. The evidence in favor of this was given by Livio (2000) who also summarized the problems of understanding the binary evolution that allows a sufficient number of white dwarfs to grow to carbon ignition at the mass limit. Most of the observed Type Ia are "Branch normal," (Branch, Fisher & Nugent, 1993) and allowance for deviations from "standard candles" can be made rather successfully with one-parameter brightness/decline rate relations (Phillips 1993; Riess, Press & Kirshner 1996; Perlmutter et al. 1999; Sandage, Tammann & Saha 2000).

A dichotomy of thinking arises at this point. The theorists say that a one-parameter brightness/decline rate relation cannot be the whole story. Theory suggests appreciable variation with input parameters (Khokhlov 2000; Höflich & Dominguez 2000) and the observations themselves suggest departures from one-parameter relations. Observers, on the other hand, point out that utilizing these one-parameter relations works remarkably well in practice in correcting for deviations (Schmidt 2000; Perlmutter 2000; Sandage 2000). The conclusion, as emphasized by Höflich & Dominguez (2000) seems to be that some of the input parameters that are varied independently in the evolutionary and dynamic models—the carbon ignition density, the density of transition from subsonic deflagration to supersonic detonation (Khokhlov, Oran & Wheeler 1997a,b; Niemeyer & Woosley 1997; Khokhlov 2000), rotation, progenitor mass, progenitor metallicity—must be correlated in ways we have yet to elucidate.

To make progress, we must understand the combustion physics (Niemeyer 1999; Khokhlov 2000), and we need to better understand the progenitor evolution (Livio 2000; Höflich & Dominguez 2000). I would be delighted to witness even a shred of direct observational evidence that Type Ia are in binary systems, a conclusion to which I hold firmly despite vivid understanding that there is no proof.

Despite the uncertainties that still plague work on SN Ia, they have been used with great effect to explore cosmological issues. Theory has been used to compare models with individual supernovae in a way that does not require secondary distance calibration. The result is that the value of the Hubble constant is estimated to be 67 ± 9 km s^{-1} Mpc^{-1} (Höflich & Khokhlov 1996). Supernova observations calibrated with Cepheid variables give values in the range 60 ± 9 km s^{-1} Mpc^{-1} (Sandage et al. 2000) to 65.2 ± 1.3 km s^{-1} Mpc^{-1} (Riess et al. 1998).

The application to cosmology has been even more startling (Riess et al. 1998; Perlmutter et al. 1999; Perlmutter 2000; Schmidt 2000). The value of the normalized cosmological matter density derived from Type Ia supernovae is $\Omega_m \sim 1/3$ and that of the cosmological constant, $\Omega_\Lambda \sim 2/3$. This raises two sets of issues. One is a possible new view of the Universe. With a cosmological constant that might not be "constant," there is the potential for closed Universes that expand forever or open Universes that collapse. There must be a consideration of new physics to understand the microscopic origin of the vacuum energy that poses as a cosmological constant and why it is so small, but not zero, at just this epoch.

The other issue is, to use Brian Schmidt's (2000) phrase, the "mundane." There may be subtle systematic effects that bias the estimates of brightness of supernovae as a function of redshift and which masquerade as the effect of a cosmological constant. Howell, Wang & Wheeler (1999) have noted that the nearby Type Ia that are used to calibrate the light curve decline relation are primarily discovered photographically so that they are susceptible to the "Shaw effect" of being lost near the centers of galaxies due to saturation, whereas the deep searches are done with CCDs that are less susceptible to this effect. The properties of the Type Ia vary with galactic radius (Wang, Höflich & Wheeler 1997; Riess et al. 1999a) and the radial distributions of the calibration sample and the distant sample are distinctly different. This difference may be removed by light curve decline corrections, but such systematic effects need more study. Riess et al. (1999b,c) report that nearby Type Ia might have systematically slower rise times than the cosmological events for similar decay times. Suntzeff (1999) notes that the sample of events that show distinct departures from a one-parameter brightness/decline relation is real and growing and that while 6 of 40 Type Ia in a nearby sample are of the very bright kind, no events like SN 1991T have been observed in the much larger deep sample.

970111	970828	980326?	980613	990123	990704
970228	971024	980329	980703	990506	990705
970402	971214	980425	980923	990510	990712
970508	971227?	980519	981226	990520	990806
970815				990627	

TABLE 1. Gamma-Ray Bursts with X-ray afterglows
(http://www.aip.de/People/JGreiner/grbgen.html)

All these developments are hints of systematic effects that must be better understood, both physically and observationally. Resolving this issue of the cosmological versus the mundane one way or the other will take several year's hard, slogging work by both theorists and observers in the supernova community. The necessary program of careful comparison of the spectra and light curves of near, intermediate, and far supernovae is underway.

2. Gamma-Ray Bursts, hypernovae, and supernovae

2.1. *Gamma-Ray Bursts, collimation, and jets*

The past year has seen discussions of extreme energies, ranging up to 3×10^{54} ergs (Kulkarni et al. 1999), and extreme degrees of collimation (Wang & Wheeler 1998). The community seems to have gotten those excesses out of its system and is now buckling down to the hard work of figuring out the true nature of the γ-ray bursts and their afterglows. Table 1 gives a list of the γ-ray bursts with observed X-ray afterglows. Table 2 gives a compilation of some of the relevant properties of γ-ray bursts in the afterglow era (see also http://astro.uchicago.edu/home/web/reichart/grb/grb.html; http://www.aip.de/People/JGreiner/grbgen.html). Excellent reviews of γ-ray bursts and afterglows were given by Paczyński (2000), Fishman (2000), Piro (2000), Fruchter (2000), Kulkarni (2000), Rees (2000), and Piran (2000).

Prior to BeppoSAX and the discovery of the afterglows, there was some discussion in the literature of the possibility of collimation (Woosley 1993; Rhoads 1997), but by and large the general community paid only lip service to collimation, if any mention was made of it at all. In the last year, the phrase "isotropic equivalent" has become a common and even mandatory part of the vocabulary of papers on γ-ray bursts and afterglows. Even more recently, judging by postings to "astro-ph," the specific phrase "jets" has become common parlance. Over this year there has been a maturity from musing about collimation to wide-spread general acceptance that collimation is a critical aspect of some, if not all, γ-ray bursts.

Judging from presentations at this meeting, I have already lost this battle, but I would like to plead with the community to use the words "collimation" or "jets" or their equivalent when non-spherical flow is implied rather than the phrase "beaming." The latter is often clear in context, but is prone to confusion with the Lorentz beaming that is purely a kinematic effect of relativistic motion. Authors who want to discuss both collimation and Lorentz beaming in the same paper are inviting confusion if they do not clearly discriminate. The total energy can be determined straightforwardly in principle by multiplying the isotropic equivalent energy with the collimation factor, $\Delta\Omega/4\pi$, but the discussion of luminosity that involves Doppler factors in a more complex way can be hopelessly muddled if the effects of collimation and Lorentz beaming are not clearly dilineated.

GRB	radio afterglow	redshift	spectral slope[a] α	temporal slope[a] β	γ-ray energy[b] erg	collimation $\Delta\Omega/4\pi$	host galaxy R^e	comment
970228	—	0.695^c 1	—	1.58 ± 0.28 2	5×10^{51} 3	little?	24.7 4	red, SN-like excess at 20 d[2,5]; no jet[6,7,8];
970508	yes?	0.835^c 9,10	0.61 ± 0.32 2	1.23 ± 0.04 11	8×10^{51} 3	>0.11 7	25.0 4	peak flux not single power law[6] no break for 9 months[11]; no jet[6,7] X-ray resurgence at $10^5-4\times10^5$ s[12]; Fe lines??[12]
970828	yes?	—	—	—	—	—	—	rapid decline? jet?[6] X-ray resurgence; Fe lines??[13]
971214	no	3.418^c 14,15	$0.93^{+0.04}_{-0.07}$ 16	1.22 ± 0.13 16	3×10^{53} 17	little?	26.2 4	no jet[6]
971227	—	1?? 20	?	2.5 19	—	yes?	>22 4	no X-ray resurgence[18], no Fe lines[18]
980326	—	5?? 21	0.8 ± 0.4 20	2.0 ± 0.1 20	3×10^{51} 20 if z=1	yes?	>27.1 20,4	host confusion[6] like SN1998bw at 20 days at z=1 20
980329	yes	0.0085^c 23	—	$1.21^{+0.13}_{-0.12}$ 22	—	little?	>25.5 4	no X-ray afterglow
980425	yes	—	—	—	$\sim10^{48}$ 23	little?	14.1 4	SN1998bw
980519	yes	—	1.20 ± 0.25 24	2.05 ± 0.4 24	6×10^{52} 24 if z=1	<0.003 7,24	24.8 4	changed slope $\sim$ hours, after GRB, before afterglow[24] BeppoSAX fluence rank 2 7
980613	—	1.096^c 25	—	—	—	—	—	no X-ray resurgence[18]; no Fe line[18]
980703	yes	0.966^c 26	~0.78 27	1.22 ± 0.35 27	1×10^{53} 19	little?	23.4 4	brightest host; no jet?
980923	—	—	—	—	—	—	22.6 4	abrupt onset of smooth γ-ray "afterglow"[28]
990123	yes	1.600^d 8,29	~0 (B-r) 17 0.79 ± 0.05 (r-K) 17	$1.08\pm0.03 < 2.04$ d then steeper 17	3×10^{54} 17	<0.01 30,17	23.7 4	decay after 40s of strong variability[28] optical 9^m↓! 31; polarization <2.3% 32 $z\lesssim2.2$ 17; BeppoSAX fluence rank 1 7
990308	—	>1.2? 33	0.3? 33	1.2 ± 0.1 33	—	—	>25.7 33	no host galaxy observed[33]
990510	—	$>1.619^d$ 34	0.61 ± 0.12 35	$-0.76\pm0.1 < 1$ day 35 $-2.40\pm0.02 \gg 1$d 35	3×10^{53} 3	<0.003 3	>26.6 36	polarized 1.7%, t=18.5 hours[37,38] BeppoSAX fluence rank 4 3
990705	?	—	—	<1.4 39	—	—	—	near LMC X-2
990712	—	0.430 40	—	0.81 41,42	—	—	22? 42	SN-like contribution?[43]

TABLE 2. Information on Gamma-ray Bursts with afterglows

$^a F_{opt}\propto\nu_{opt}^{-\alpha}t^{-\beta}$ ^{b}isotropic equivalent ^{c}host galaxy ^{d}afterglow ^{e}corrected for reddening

REFERENCES—(1) Djorgovski et al. 1999b. (2) Reichart 1999. (3) Harrison et al. 1999. (4) Hogg & Fruchter 1999. (5) Galama et al. 1999a. (6) Rhoads 1999. (7) Sari, Piran & Halpern 1999. (8) Fruchter et al. 1999. (9) Metzger et al. 1997. (10) Bloom et al. 1998a. (11) Zharikov et al. 1998. (12) Piro et al. 1998. (13) Yoshida et al. 1999. (14) Kulkarni et al. 1998a. (15) Odewahn et al. 1998. (16) Reichart 1998. (17) Kulkarni et al. 1999. (18) Piro et al. 1999. (19) Djorgovski et al. 1998a. (20) Bloom et al. 1999b. (21) Fruchter 1999. (22) Reichart et al. 1999. (23) Galama et al. 1998. (24) Halpern et al. 1999 (980519). (25) Djorgovski et al. 1999a. (26) Djorgovski et al. 1998b. (27) Bloom et al. 1998b. (28) Giblin et al. 1999. (29) Bloom et al. 1999a. (30) Anderson et al. 1999. (31) Akerlof & McKay 1999. (32) Hjorth, J. et al. 1999a. (33) Schaefer et al. 1999. (34) Vreeswijk et al. 1999. (35) Stanek et al. 1999. (36) Israel et al. 1999. (37) Covino et al. 1999. (38) Wijers et al. 1999. (39) Palazzi et al. 1999. (40) Galama et al. 1999b. (41) Bakos et al. 1999. (42) Hjorth, J. et al. 1999b. (43) Hjorth, J. et al. 1999c.

At this conference, we heard about energies associated with γ-ray bursts ranging from 4×10^{50} to 3×10^{54} ergs (Frail 2000; Kulkarni 2000). In fact, energies of 4×10^{50} to 10^{52} ergs were associated with the same object! Frail (2000) made a clear case for the need to be careful in how energies are assigned to afterglows. The energies are sensitive to model parameters (e.g. break frequencies), the estimates of which are in turn subject to observational uncertainties. The lower energies represent circumstantial evidence that the highest energies are, indeed, only isotropic equivalents and must be scaled down by a substantial collimation factor.

One expects the decay slopes to become steeper as a collimated flow slows and spreads laterally (Rhoads 1997,1999), and evidence for steeper slopes and even breaks in the slope have now been observed (Kulkarni et al. 1999; Sari, Piran & Halpern 1999; Harrison et al. 1999; Table 2). This evidence for collimation is only circumstantial, and the rate of change of the slope may be a gradual rather than sudden process (Panaitescu & Mészáros 1998; Moderski, Sikora & Bulik 1999). The slope can also be affected by density gradients in the surrounding medium (Chevalier & Li 1999) and by variations in power output in the underlying "machine" (Li & Chevalier 1999; Dai & Lu 1999a,b). With these caveats, the strongest evidence for collimation is in the brightest sources, with GRB 990123 with an isotropic equivalent γ-ray energy of 3×10^{54} ergs having an estimated collimation factor of ~ 0.01 for an actual γ-ray energy of $\sim 3 \times 10^{52}$ ergs (Kulkarni et al. 1999) and GRB 980519 and GRB 990510 (which went off just after this conference) having a beaming factor of less than ~ 0.003 (Sari, Piran & Halpern 1999; Harrison et al. 1999). This suggests that three of the γ-ray bursts with the highest isotropic equivalent energy are also the ones with strongest evidence for collimation.

Another recent argument in favor of some form of collimation comes from observations of polarization of γ-ray bursts. The synchrotron radiation from γ-ray bursts and their afterglows should be strongly polarized with the degree of polarization from a single patch ranging as high as 70% (Sari 1999), but if the source is spherically symmetric and the field tangled, then the degree of polarization will be reduced. In this context, the upper limit to the polarization of 2.3% for GRB 990123 (Hjorth et al. 1999a) and the measured polarization of 1.7% in GRB 990510 (Covino et al. 1999; Wijers et al. 1999) are very significant for being so small. This may suggest that in addition to a tangled magnetic field, we are observing collimated flow and emission (Gruzinov 1999; Ghisellini & Lazzati 1999; Sari 1999). Note that, as Gruzinov stresses, the origin of the magnetic field, assumed to be in near equipartition, remains a major stumbling block for the relativistic blast wave synchrotron emission model.

With the evidence for collimation, there is a suggestion that the true γ-ray energies of all the γ-ray bursts are in the range 10^{50}–10^{52} ergs, from 0.1 to 10 times the canonical energy of a supernova, but still substantially less than the binding energy of a single neutron star. It is also true that as the evidence for collimation has grown, so have the highest recorded isotropic equivalent energies, leaving most people with the feeling that, at best, the target energy is near the upper end of this range, $\sim 10^{52}$ ergs. The issue becomes the nature of supernovae, hypernovae, and their link to γ-ray bursts.

2.2. Polarization of supernovae

Like many people in the supernova community, we at the University of Texas got actively involved in the supernova/soft-gamma-ray repeater/magnetar/γ-ray burst topic with the advent of SN 1998bw and its possible connection to GRB 980425. We brought a different perspective to this issue because of work we have done over the last four years on supernova spectropolarimetry.

We have been making spectropolarimetric observations of all accessible supernovae at McDonald Observatory (Wang et al. 1996; Wheeler, Wang & Höflich 1999; Wheeler, Höflich & Wang 1999). The result has been that most Type Ia have low polarization and hence are substantially spherically symmetric. Many have only upper limits of order 0.1–0.2%. A few have detected, but low polarization, of order 0.2%. The polarization observed is consistent with theoretical models of delayed detonation models (Wang, Wheeler & Höflich 1997) and may be a useful probe of the combustion physics. We have detected one exception, SN 1997bp, which was observed a week before maximum light to have a polarization of about 1%. The polarization was low in post-maximum spectra, but this event remains a challenge to understand. It is important to establish whether such events are common, the physical reason for the large polarization, and whether or not there could be an asymmetric luminosity distribution that could affect estimates of cosmological parameters.

More importantly in the current context are our observations of presumed core-collapse events, Type II and Type Ib/c. We have found that all such events are polarized at about the 1% level. So far there have been no exceptions in about a dozen events. There could be a myriad reasons for polarization, but our data suggest a very important trend: the smaller the hydrogen envelope, the larger the observed polarization. As an example, SN 1987A with a 10 $M_\odot$ envelope had a polarization of about 0.5% (Méndez et al. 1988), SN 1993J and a very similar object, SN 1996cb, with small hydrogen envelopes, ~ 0.1 $M_\odot$, were polarized at the 1–2% level (Trammell, Hines & Wheeler 1993; Tran et al. 1997), and Type Ic SN 1997X which showed no substantial hydrogen nor helium was polarized at perhaps greater than 3% (Wheeler, Höflich & Wang 1999). These are difficult observations requiring special care in the reduction to remove the effects of the ISM (the latter greatly aided by wavelength and temporal coverage), and there is a pressing need to expand the statistical sample. Nevertheless, this trend suggests that the core-collapse process itself is strongly asymmetric and that evidence for that asymmetry is damped by the addition of outer envelope material.

The level of polarization we have observed for core collapse events, $\sim 1\%$, requires a substantial asymmetry with axis ratios of order 2 to 1 (Höflich 1995). Asymmetric explosions tend to turn spherical as they expand, so to leave an imprint of an asymmetry of this level in the homologously expanding matter requires a substantially larger asymmetric input of energy or momentum in the explosion process itself. These factors led us to the hypothesis that the core collapse process is intrinsically strongly asymmetric, much more so than current collapse calculations involving convectively unstable neutron stars. It was in this context that we greeted the news of SN 1998bw a year ago.

2.3. SN 1998bw and GRB 980425

SN 1998bw was a shock to both the supernova and γ-ray burst communities. For the supernova community, it was clearly an odd and exciting event, even without the possible association with GRB 980425. Although it resembled a Type Ic in the sense that there was no obvious evidence for H and He, it was different than the canonical Type Ic. It was very bright, it showed very large velocities, and it had a very bright radio source. While not the brightest supernova radio event on record (Weiler et al. 2000), the radio source associated with SN 1998bw was undoubtedly very luminous and arguments based on the brightness temperature alone suggested relativistic motion (Kulkarni et al. 1998b) depending on whether or not one assumed magnetic field equipartition (Waxman & Loeb 1999). On the other hand, the γ-ray community had just gone through the catharsis of proof that γ-ray bursts were, indeed, at cosmological distances as strongly suggested by the isotropy of the BATSE sources (Fishman 1995, 2000) and confirmed by redshifts measured for a number of the events with detected afterglows. The identification of SN 1998bw with GRB 980425 immediately confused the issue by raising the prospect of substantially different sources of γ-ray bursts that could not be easily differentiated by their γ-ray flux, fluence, time history, or spectra.

I took the opportunity of this meeting to test a rift I had suspected since the Texas Symposium in Paris in December, 1999. I inquired as to the people in the audience who primarily identified themselves with the supernova community and those who identified themselves primarily with the γ-ray burst community. I then asked for a show of hands of those who had reservations about the identification of SN 1998bw with GRB 980425 and of those who were convinced that the two objects were one and the same. While there were some cross-over votes, this excercise basically confirmed my suspicion. The supernova community has bought this identification hook, line, and sinker, based on the odd properties of the supernova. The γ-ray burst community remains substantially suspicious despite the low *a priori* probability, $\sim 10^{-4}$ (Galama et al. 1998), of an accidental alignment. One does not do science by democratic vote, and this is more an excercise of sociology than science, so I leave the reader to contemplate the meaning, if any, of the result (and to criticize the unscientific sampling method).

The issues raised by SN 1998bw are these. If one believes that SN 1998bw produced GRB 980425, then there are more than one type of γ-ray burst that cannot be easily distinguished by γ-ray properties. If one does not believe that these two events are identical, there is still a weird supernova to explain and the standard mystery of the nature of the cosmic γ-ray bursts remains. All these possibilities bring with them the strong suggestion that, for both the supernova and for γ-ray bursts, asymmetries and collimation are the rule, not the exception.

Even if one believes the identification of SN 1998bw with the γ-ray burst there are substantially different interpretations of the event. Was it a version of an "ordinary" supernova or was it something better labeled a "hypernova?" The phrase "hypernova" was coined in this context by Paczyński (1998, 2000), who meant by it any event that produced an isotropic equivalent luminosity substantially larger than a supernova, whatever the source of that luminosity, core collapse or merging compact stars. The phrase "hypernova" has taken on a somewhat more specific meaning in the supernova community in the context of attempts to understand the nature of SN 1998bw and possibly related events.

The reason for this evolution of the definition of "hypernova" is that attempts to construct models for SN 1998bw have led to the suggestion that the energy must be substantially in excess of 10^{52} ergs (Iwamoto et al. 1998; Woosley, Eastman & Schmidt 1999; Branch 2000; Nomoto 2000; Woosley 2000) and perhaps also that it had very large ejecta mass and radioactive nickel mass, the latter in excess of 0.5 $M_\odot$ (Iwamoto et al. 1998; Woosley, Eastman & Schmidt 1999). These models were all, by assumption, spherically symmetric, and, if nothing else, ignore the observations that SN 1998bw was polarized. On the other hand, Höflich, Wheeler & Wang (1999) have argued that a sufficiently asymmetric model to account for the polarization (an ellipsoid of axis ratio of 2 to 1) could account for the bolometric and multi-color light curves near maximum light. With the proper viewing angle, within about 35° of the symmetry axis for a model with oblate isodensity contours, the luminosity could be substantially higher than the mean spherical equivalent and the peak of the light curve reproduced in a model with an ejecta kinetic energy of 2×10^{51} ergs, 2 $M_\odot$ of ejecta, and 0.2 $M_\odot$ of ^{56}Ni. The kinetic energy of this model is a little higher than average, but well in the range of energies deduced for standard core-collapse events, and the other parameters are quite nominal.

SN 1998bw may have been a "hypernova" requiring expansion energy more than 10 times that normally associated with supernovae, but that is certainly not necessarily so.

Given its important role, SN 1998bw received a great deal of attention at the conference. It is poignant to note at a conference hosted by ST ScI that the request by the author and Lifan Wang for Director's Discretionary Time was turned down in the press of other observational priorities. The result is that there were no UV observations of SN 1998bw. This lack becomes more important as one ponders similar events at large redshift where the UV spectrum is shifted to the visible. We can guess what SN 1998bw would have revealed in such observations, but we will never know.

One of the principle controversies over the identification of SN 1998bw with GRB 980425 is the nature of the afterglow or the lack thereof. This issue was addressed anew by Pian (2000). The first narrow-field instrument BeppoSAX observations revealed two X-ray sources in the original wide-field detection image. Neither corresponded to the location of the supernova. One source (Source 1) was at first thought to be constant and the other (Source 2) to decay in a few days in a manner consistent with other observed X-ray afterglows. This observation has been the origin of understandable suspicion by many that the association of the supernova and the γ-ray burst were accidental despite the low probability.

Recalibration of positions (Pian et al. 1998) revealed that Source 1 was coincident within the errors with SN 1998bw, but that Source 2 was definitely not associated with the supernova. Further observations showed that Source 1 did vary, but slowly, perhaps a factor of 2 in 10 days. At this conference, Pian presented evidence that Source 2 also declined slowly or was even substantially constant over an interval of more than 100 days. Taken at face value, the data presented by Pian suggest that both Source 1 and Source 2 declined rather slowly after the first NFI observations beginning about 1 day after detection. The flux for both is about 10^5 less than that first detected in the WFC. It is possible that neither Source 1 nor Source 2 represent an afterglow. It is possible that the afterglow decayed so rapidly from the WFC detection that only background sources were detected at Source 1 and Source 2. The interpretation of the data depend substantially on the confidence placed in the detection of Source 2 at the 3σ level after day 100. If this detection is true, then it seems unlikely that either Source 1 or Source 2 are an afterglow. If this detection is only an upper limit, then it is still conceivable that Source 2 is an afterglow and the identification of GRB 980425 with SN 1998bw is still open to question. Observations of these sources by ASCA in April may have yielded only upper limits (Harrison 1999), but these may help to resolve this issue.

Danziger et al. (2000) presented photometric, spectroscopic and spectropolarimetric observations of SN 1998bw from ESO. Danziger concluded that the object was asymmetric, in substantial agreement with Höflich, Wheeler & Wang (1999). Nomoto (2000) presented model light curve calculations. He noted that the spherically symmetric models required "hypernova" energies to fit the peak, but that the same models declined too rapidly to fit the tail. He pointed out that one way to account for the tail was to invoke a smaller expansion energy to get greater trapping of γ-rays at later times. He concluded that this implicit contradiction was evidence for asymmetry. This underlines the basic point of Höflich, Wheeler & Wang (1999) that the energy cannot be determined independently of considerations of asymmetry for the dynamics and radiative transfer.

Danziger et al. (2000) also presented nebular spectra of SN 1998bw. These spectra revealed futher peculiarities. The line of [OI] $\lambda\lambda$ 6300,6364 was broader than the lines of [Fe II]. The Fe lines were comparable in width to those of normal Type Ic in contrast to the expectation from the basic hypernova models that, with their very high energies, have very high velocities, $\gtrsim 4000$ km s^{-1}, in the inner, iron-rich, regions. Discussion with

Nomoto and Mazzali suggested that the spectra could be fit with models, but only by adding *ad hoc* inner regions of slower moving matter. Such slow matter may be produced in a more realistic, multi-dimensional model, but it is not predicted in the spherically symmetric "hypernova" models.

Branch (2000) presented simple atmosphere models that illustrated the systematic differences between SN 1994I, SN 1997ef, and SN 1998bw. SN 1994I was a canonical, well-studied Type Ic supernova. SN 1997ef was also labeled a Type Ic, but while showing a normal peak luminosity, it had higher velocity at the photosphere than SN 1994I. Branch illustrated how the increased broadening of the lines carved out the red continuum and led to a rather steep decline from the blue. With the even higher photospheric velocities of SN 1998bw, Branch provided convincing evidence that the unprecedentedly steep decline from about 4000 to 5000 Angstroms in the continuum of SN 1998bw near maximum light could be explained. Coupling his atmosphere models with the assumption of spherical symmetry, Branch made estimates of the kinetic energy of each event, concluding that SN 1994I was consistent with $\sim 10^{51}$ ergs, but that both SN 1997ef and SN 1998bw could require "hypernovae" energies of $\gtrsim 10^{52}$ ergs. Branch pointed out that the ball was in the court of advocates of asymmetric models to show that these observations could be explained self-consistently with asymmetries and modest energy. That is completely correct.

An excellent light curve of SN 1998bw is presented by McKenzie & Schaeffer (1999). They have an well time-sampled data set that shows that after an early steep decline from maximum for about 25 days, B, V, and I have declined in a precisely exponential manner. The slopes in the three bands are slightly different and all three are steeper than that expected for ^{56}Co decay and full trapping of γ-rays. McKenzie & Schaeffer note that because there must be some leakage of γ-rays they can only set a lower limit to the amount of ^{56}Ni produced in SN 1998bw which they determine to be 0.22 ± 0.09 M$_\odot$. This lower limit is close to that estimated for the asymmetric models of Höflich, Wheeler & Wang (1999) and substantially less than the hypernova models of Iwamoto et al. (1998) and Woosley, Eastman & Schmidt (1999). Since the light curves are so precisely exponential, the γ-ray trapping fraction cannot be changing substantially. This suggests that while less than unity, the trapping fraction is substantially greater than 50% or the light curves would be steeply declining to the limit set by positron trapping. This argument suggests that the lower limit set by McKenzie & Schaeffer may be near the actual nickel mass produced by SN 1998bw and in contradiction to the hypernova models.

2.4. *Other possible supernova/Gamma-Ray Burst connections*

SN 1998bw/GRB 980425 is the most famous and best established supernova/γ-ray burst connection (despite or because of its debated reality), but other arguments have accumulated for such a connection. Some candidates as of this writing are given in Table 3.

Germany et al. have discussed the case of SN 1997cy. This supernova was odd in its own way and unlike SN 1998bw in many substantial ways. The supernova occurred in a low surface brightness galaxy at a redshift of $z = 0.063$. The date of the explosion is uncertain by a few months. The spectrum is characterized by a very strong line of Hα, unlike SN 1998bw which showed no evidence for hydrogen. The Hα line showed both broad and narrow components reminiscent of Type IIn supernovae (Schlegel 1990). SN 1997cy also showed lines of Fe II and [Fe III] that are more characteristic of the nebular phase of Type Ia events. The light curve of SN 1997cy followed the decay slope of ^{56}Co for about 60 days after discovery. The light curve then flattened for 200 days and then proceeded to a more steep decline. Assuming the early part of the light curve to be due to the trapping of γ-rays from ^{56}Co decay, Germany et al. deduce that SN 1997cy

SN	GRB	SN properties		GRB properties			REFERENCES
		spectra	light curve	flux^a	fluence^b	duration(s)	
—	970228^c	~98bw?	~98bw?	—	—	80	1, 2
1997cy	970514?	IIn			4×10^{-7}	1.3	3
1997ef	970125?? 971115??	~98bw?					4
—	980326	~98bw?	~98bw?	8×10^{-7}	7×10^{-7}	0.2	5, 6
1998bw	980425	~Ic			4×10^{-6}	35	7
1999E	980910?	~97cy	$M_V < -19.4$		2×10^{-7}	—	8, 9, 10, 11
—	990712	~98bw?	~98bw?	—	—	—	12

^{a} erg cm^{-2} s^{-1} ^{b} erg cm^{-2} ^{c} BATSE behind Earth

REFERENCES–(1) Reichart (1999). (2) Galama et al. (1999). (3) Germany et al. (1999). (4) Wang & Wheeler (1998). (5) Bloom et al. (1999). (6) Briggs et al. (1998). (7) Galama et al. (1998). (8) Filippenko, Leonard & Riess (1999). (9) Jha et al. (1999). (10) Cappellaro, Turatto & Mazzali (1999). (11) Thorsett & Hogg (1999). (12) Hjorth et al. (1999c).

TABLE 3. SN/GRB candidates

ejected 2 M$_\odot$ of ^{56}Ni. They tentatively ascribe the subsequent flattening and decline of the light curve to circumstellar interaction. Germany et al. note that SN 1997cy was about a factor of 50% brighter at discovery than SN 1998bw was at maximum. Late-time radio observations, 16 months after the explosion, revealed no detectable source. Germany et al. argue for a possible connection of SN 1997cy with GRB 970514. This burst lasted $\lesssim 1$ s and was classified as a high-energy event. They estimate the chance association of the two events to be about 1%. If the events were associated, then the γ-ray energy in the burst was about 4×10^{48} ergs, comparable to, but somewhat larger than, that ascribed to SN 1998bw/GRB 980425. Germany et al. also note that the decay of GRB 970508 was similar in slope to SN 1997cy and ^{56}Co. They also note that SN 1999E had a spectrum similar to SN 1997cy, that SN 1999E was especially bright, and that it might be temporally linked to GRB 980910. One must be somewhat cautious in interpreting the data from SN 1997cy, especially the key observation that the light curve traced cobalt decay. There is no question that the supernova was bright at its redshift. On the other hand, the light curve was observed to fall at the rate of ^{56}Co for only about 60 days, barely half a ^{56}Co e-fold time. If the association with GRB 970514 is questioned, then the time of explosion is uncertain and this also impacts the amount of ^{56}Co one would attribute to the event, even accepting that ^{56}Co decay is observed.

Another class of association of γ-ray bursts and supernovae comes from the discovery of transient brightening or modulation of afterglows. Bloom et al. (1999b) argue that a brightening of the light curve of GRB 980326 can be interpreted as the addition of the light of an event like SN 1998bw about 20 days after the explosion if the supernova were at about a redshift of 1. The optical transient became about 60 times brighter than expected from an extrapolation of the decline of flux at earlier times. In addition, Keck spectra showed that the continuum spectrum changed from being blue to being red. The latter is roughly consistent with the radiation from a supernova photosphere, but inconsistent with synchrotron radiation as might have occurred if there were delayed energy input or the blast wave ran into a dense cloud, possible alternative models (Panaitescu, Mészáros & Rees 1998; Dai & Lu 1998a,b; Piro et al. 1999a). Bloom et al. note that special circumstances might be necessary to reveal such a late-time rebrightening: a rapid afterglow decay, and a low surface-brightness host galaxy.

Reichart (1999) has advanced similar arguments in a study of an earlier event, GRB 970228. While there is no spectroscopy, Reichart notes that there is much more thorough photometry for GRB 970228 as compared to GRB 980326 and that there is a measured redshift of $z = 0.695$ (Djorgovsky et al. 1999). GRB 970228 showed strong reddening with time, a characteristic not explained by standard relativistic blast wave models. Reichart argues that the afterglow data is not consistent with a single power law spectrum nor a single power law temporal decay, but that it is consisent with the U-band light curve of SN 1998bw appropriately redshifted to the frame of GRB 970228. The spectral energy distribution is also consistent with the convolution of a SN 1998bw-like event and the power-law decline of a relativistic blastwave. Similar conclusions have been reached by Galama et al. (1999). Interestingly, this manifestation of a supernova-like resurgence is seen despite a relatively slow decline in the early afterglow $\propto t^{-1.58}$. The host galaxy was relatively dim compared to the early afterglow and the "supernova" contribution.

The most recent suggestion of such a supernova-like modulation has been given by Hjorth et al. (1999c) for GRB 990712. They argue that the R-band light curve is consistent with a temporal decay of the afterglow like $^{-1}$, a host galaxy of R = 21.76, and a "supernova" like SN 1998bw at the known redshift of $z = 0.430$ (Galama et al. 1999b).

Although it is not at all clear that it should be presented in this context, for completeness I will add the possibility that SN 1987A produced a jet of some sort, if not a γ-ray burst. Wang & Wheeler (1998), Cen (1998), and Nakamura (1998) all noted that if supernovae make jets that have some connection to γ-ray bursts there might be some relevance to the "mystery spot" of SN 1987A. Motivated by Cen's comments in this connection, Nisenson & Papaliolios (1999) re-examined their speckle data on SN 1987A and argued in favor of both a jet and counter jet from SN 1987A. From the kinematics they concluded that the counter jet must have moved at relativistic speeds. Some sort of jets might be produced in many core collapse events (see §3), but there is still debate and doubt on the reality of these jets. There is, of course, no direct link to a γ-ray burst. On the other hand, Nagataki (1999) makes a convincing case that a jet-like explosion in SN 1987A can resolve many of the issues of outward mixing of radioactive elements and line profiles.

3. Jet-induced supernovae

The goal of producing robustly asymmetric supernovae, weak γ-ray bursts of the sort observed in SN 1998bw/GRB 980425, and perhaps collimated high-energy bursts of γ-rays that could contribute to the cosmological γ-ray bursts suggests the following general picture. The best chance of imprinting the asymmetry and producing some sort of γ-ray burst is in the absence of an extended hydrogen envelope which could slow, delay, or disperse the propagation of an asymmetric flow of energy from a newly-formed neutron star. This makes a Type Ib/c, and especially a Type Ic configuration, a likely site for study, independent of the similarities of SN 1998bw to Type Ic.

The progenitor of a Type Ic is envisaged to be the core of a massive star, perhaps in excess of 15 $M_\odot$ on the main sequence, which has shed its hydrogen and helium by a winds or binary mass transfer. The iron core in such a progenitor collapses to form a neutron star, and the outer layers of Si, O, and C with longer free-fall times hover momentarily. The neutron star bounces and produces a standing shock that stalls without inducing an explosion. If the neutron star is a pulsar, then it is possible to create an MHD jet up the rotational axis at the time of the formation of the neutron star as in the old calculation of Leblanc & Wilson (1970). Such a jet could induce the requisite asymmetry in the ejecta and accelerate in the density gradient to produce a weak γ-ray burst. If the pulsar were very highly magnetized, a magnetar (Duncan & Thompson 1992; Kouveliotou et al. 1998; Harding 2000), then the pulsar could also potentially produce an intense flux of Poynting radiation (Usov 1992,1994; Thompson 1994; Mészáros & Rees 1997; Blackman & Yi 1998; see also Ostriker & Gunn 1971; Bisnovatyi-Kogan 1971). This Poynting flux would tend to flow out through the weakest part of the hovering mantle, namely the wound punched by the MHD jet. For very large magnetic fields, the collimation and energetics could be sufficient to produce a cosmic γ-ray burst.

This outline of a possible scenario illustrates a host of places where more rigorous physics is needed to determine whether the hypothesized outcome is reasonable. One issue is the propagation of the initial MHD jet out through the mantle and its effect on the star. This issue has been recently addressed by Khokhlov et al. (1999).

Khokhlov et al. adopt a progenitor Type Ib/c model consisting of a spherical helium star of radius $R_{\text{star}} = 1.88 \times 10^{10}$ cm and mass $M_{\text{star}} \simeq 4.1$ M$_\odot$. The inner Fe/Si core with mass $M_{\text{core}} \simeq 1.6 M_\odot$ and radius $R_{\text{core}} = 3.82 \times 10^8$ cm is assumed to have collapsed on a timescale much faster than the outer, lower-density material. This core is replaced by a point gravitational source with mass M_{core} representing the newly formed neutron star. The remaining mass, $\simeq 2.5$ M$_\odot$, consists of an O-Ne-Mg inner layer surrounded by the C-O and He mantles.

The jets are assumed to enter the mantle at two polar locations at R_{core}. An inflow with velocity v_j, density ρ_j and pressure P_j is imposed. The jet parameters are chosen to represent the results of LeBlanc & Wilson (1970). At R_{core}, the jet density and pressure are the same as those of the background material, $\rho_j = 6.5 \times 10^5$ gm cm^{-3} and $P_j = 1.0 \times 10^{23}$ ergs cm^{-3}, respectively. The radii of the cylindrical jets entering the computational domain are approximately $r_j = 1.2 \times 10^8$ cm. The jet velocity at R_{core} was held constant at $v_j = 3.22 \times 10^9$ cm^{-1} for 0.5 s. This results in a mass flux rate of $\sim 9.5 \times 10^{31}$ gm s^{-1} with an energy deposition rate $dE/dt = 5 \times 10^{50}$ erg s^{-1} for each jet. After 0.5 s, the velocity of the jets at R_{core} was gradually decreased to zero at approximately 1 s. The total energy deposited by the jets was $E_j \simeq 9 \times 10^{50}$ ergs and the total mass ejected is $M_j \simeq 2 \times 10^{32}$ grams or $\simeq 0.1$ M$_\odot$. These parameters are consistent with, but somewhat less than, those of the LeBlanc-Wilson model.

As the jets move outward, they remain collimated and do not develop much internal structure. A bow shock forms at the head of the jet and spreads in all directions, roughly cylindrically around each jet. The jet characteristic time, $\tau_j \simeq 1$ s, is much shorter than the sound crossing time of the star, $\tau(R_{\text{star}}) \simeq 10^3$ s. The jets stay collimated enough to reach the surface as strong jets. The stellar matter is shocked by the bow shock, and acts as a high-pressure confining medium by forming a cocoon around the jet. The sound crossing time of the dense O-Ne-Mg envelope, $\tau(\sim 10^9 \text{ cm}) \simeq 10$ s, is only ten times longer than τ_j, and the jets are capable of penetrating this dense inner part of the star in ~ 2 s. By the time the jets penetrate into the less dense C-O and He layers, the inflow of material into the jets has been turned off. By this time, however, the jets have become long bullets of high-density material moving through the background low-density material almost ballistically. The higher pressures in these jets cause them to spread laterally. This spreading is limited by a secondary shock that forms around each jet between the jet and the material already shocked by the bow shock. The radius of the jets, $\sim 3 \times 10^9$ cm as they emerge from the star, is larger than the initial radius, $\sim 10^8$ cm, but it is still significantly less than the radius of the star. After about 5.9 s, the bow shock reaches the edge of the star and breaks through. Figure 1 shows the subsequent evolution of the star after the breakthrough. By $\simeq 20$ s, most of the material in the jets has left the star and will propagate into the interstellar medium ballistically.

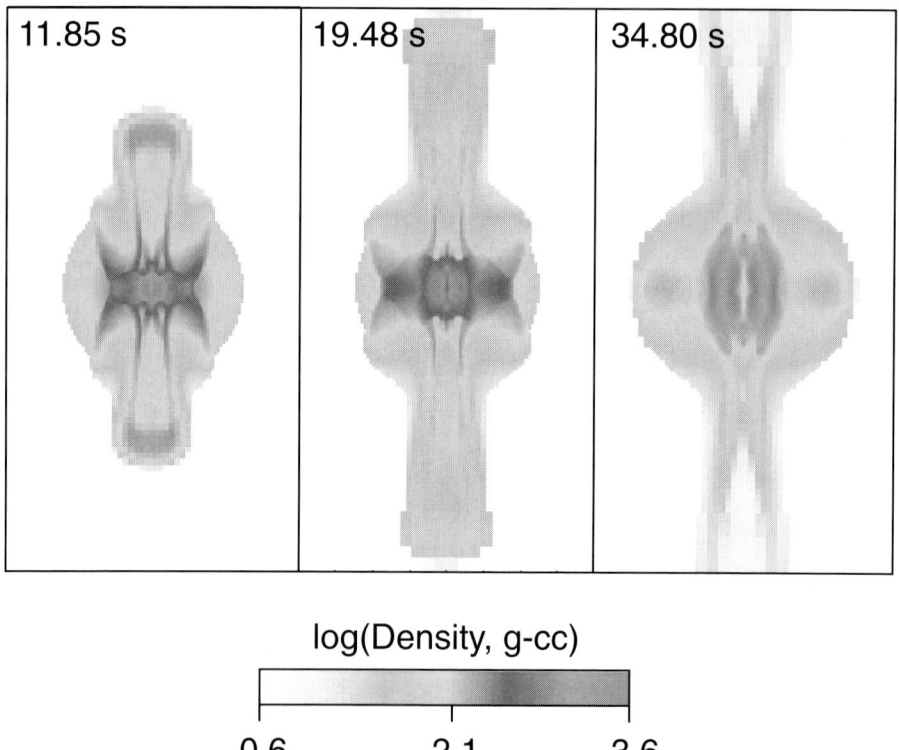

FIGURE 1. Jet evolution after breakout for the case of a progenitor which has lost its hydrogen rich-envelope (from Khokhlov et al. 1999). The frames show the density in the x-z plane passing through the center of the computational domain. The time since the beginning of the simulation is given in the upper left. The sizes of the frames are $\Delta x = 6.1 \times 10^{10}$ cm and $\Delta z = 1.125 \times 10^{11}$ cm.

The laterally expanding bow shocks generated by the jets move toward the equator where they collide with each other. The result is that the material in the equatorial plane is compressed and accelerated more than material in other directions (excluding the jet material). At $t \simeq 29$ s, the equatorial flow reaches the outer edge of the star, and the star begins to settle into the free expansion regime. The computation was terminated at $\simeq 35$ s, before free expansion was attained. The stellar ejecta at this time is highly asymmetric. The density contour of 50 gm cm^{-3}, which is the average density of the ejecta at this time, forms an oblate configuration with the equator-to-polar velocity and density ratios $\simeq 2/1$ and $4/1$, respectively. Complex shock and rarefaction interactions inside the expanding envelope will continue to change the distribution of the parameters inside the ejecta. Nonetheless, we expect that the resulting configuration will resemble an oblate ellipsoid with a very high degree of asymmetry ≥ 2.

The asymmetric explosion generated in this calculation provides ejection velocities that are comparable to those observed in supernovae, but with especially high velocities near the jet axis. For this particular calculation, an energy of 9×10^{50} ergs is input at the base of the jets. This energy is divided roughly equally between the emerging jets and the bulk of the asymmetric ejecta. The total mass in the two jets is $M_j \simeq 0.1$ M$_\odot$ and the total kinetic energy is $E_j \simeq 5 \times 10^{50}$ ergs. The average velocity of the jets is about

25,000 km s^{-1}. The outer 2.5 M$_\odot$ of mantle material is ejected with kinetic energy of 5×10^{50} ergs and average velocity 3,000–4,000 km s^{-1}.

The jet-induced explosion is entirely due to the action of the jet on the surrounding star. The mechanism that determines the energy of such an explosion is related to the shut-off of the accretion onto the neutron star by the lateral shocks that accelerate the material outwards. The explosion thus does not depend on neutrino transport or re-acceleration of the stalled shock.

The result of this calculation is a highly nonspherical supernova explosion with two high-velocity jets of material moving in polar directions and oblate, highly distorted ejecta containing most of the supernova material. This jet-induced explosion thus provides a satisfactory account of the degree of polarization and asymmetry observed for typical core-collapse supernovae. The luminosity and photospheric velocities will be a function of the aspect angle. This model gives at least the possibility of reproducing a standard Type Ic like SN 1994I by observations along the equator where the velocity and luminosity (and perhaps the γ-ray flux) will be minimum, but the polarization will be maximum, of reproducing SN 1997ef by an observation at intermediate angles, and reproducing SN 1998bw by observation near the poles (within 35 degrees; Höflich, Wheeler & Wang 1999) where the velocity and luminosity will be maximum and the polarization a minimum, albeit with perhaps somewhat higher explosion energy. This scheme might thus account for many of the properties of these three events as outlined by Branch (2000) without requiring a "hypernova" for any of them.

The jets provide a large kinetic energy per unit solid angle. When the jets break through the stellar photosphere, a small amount of mass will be accelerated down the density gradient to high velocities. Khokhlov et al. (1999) did not have sufficient resolution to make quantitative predictions; however, a small fraction of the material at the stellar surface had a velocity of up to $\sim 90,000$ km s^{-1}. There is thus a good liklihood of producing a weak γ-ray burst and a radio outburst of the type seen in SN 1998bw/GRB 980425 by the relativistic shock ejection mechanism of Colgate (1975).

Khokhlov et al. (1999) assumed that the jets were generated by a magneto-rotational mechanism during core collapse and neutron star formation (LeBlanc & Wilson 1970). The LeBlanc & Wilson calculation was criticized by Meier et al. (1976) as requiring extreme parameters of the progenitor star. These issues need to be re-examined in the current context, but several things are worth noting about the Meier et al. analysis. They argue that the MHD axial flow found by Leblanc & Wilson will not propagate to the stellar surface as a jet. The calculation of Khokhlov et al. shows this to be incorrect. Meier et al. based their analysis on stellar evolution calculations of the day, but they adopted a stellar core with central density of about 10^{10} gm cm^{-3} giving a binding energy of about 10^{52} ergs. This exaggerates the binding energy of the initial core by about a factor of 10 compared to modern calculations and gives an incorrectly small value of a key parameter of Meier et al., the ratio of the binding energy of the newly-formed neutron star to that of the initial core. Meier et al. also did not consider the possibility of an $\alpha - \Omega$ dynamo that could lead to exponential growth of the magnetic field (Duncan & Thompson 1992). The whole question of the initiation of MHD jets in association with neutron star formation needs to be considered anew.

A different mechanism of jet generation involving neutrino radiation or perhaps MHD jets during collapse of a very massive star into a black hole has been recently discussed by MacFadyen & Woosley (1999) and Woosley (2000) in the context of "collapsar" models of γ-ray bursts. The energy input rate and total energy of the jets in the calculation of MacFadyen & Woosley are similar to those of Khokhlov et al. (1999), but by choice of initial conditions, MacFadyen & Woosley inject energy into the jets as thermal energy whereas Khokhlov et al. assume the input as kinetic energy. Apparently this gives less mass in the jets for MacFadyen & Woosley and they find that their jets rapidly accelerate to relativistic speeds, whereas the jets of Khokhlov et al. remain sub-relativistic. This affects the dynamics of the jets. It is as if MacFadyen & Woosley were blowing a jet of air through water and Khokhlov et al. were blowing a jet of water through water. If there is to be a strong γ-ray burst due to Poynting flux from the neutron star as sketched earlier, there must be a subsequent phase where, to extend the analogy, a second jet of air blows out through the water jet. Clearly, an extensive amount of work is required to understand the origin of jets in this general context, their sensitivity to mode of initiation, and their propagation through various progenitor stars from compact cores through extended supergiants.

4. Perspectives and conclusions

A principle issue that confronts the subject of γ-ray bursts is that of diversity. Occams razor is a powerful tool, but sometimes it is not adequate for all of nature's handiwork. One measure of progress in this century was the development of our understanding of "novae." We now know that this apparently rather similar category of optical outbursts included a wide range of astrophysical phenomena: dwarf novae, classical novae, X-ray transients, and supernovae. A great amount of painstaking observational work and theoretical understanding was required to separate these categories, including the understanding that some of the "nebulae" hosting "novae" were galaxies giving rise to supernovae. It is sobering to recognize that the differences between X-ray transients involving black hole accretion and classical novae involving thermonuclear explosions on the surfaces of white dwarfs were only fully recognized in the last decade or so.

One must keep an open mind that something like this diversity of phenomena may occur in the γ-ray bursts despite obvious similarities and in the absence of other information. That information gap is now being filled in a rush in the age of the afterglow. The separation of γ-ray bursts into two morphological groups by the length of the outburst is well established and into different hardness categories is suspected (Lamb 1999; Fishman 2000). The γ-ray bursts with afterglows show a variety of light curve behaviors, some with monotonic power law declines, some with breaks or gradual changes of slope (Table 2). Some of the afterglows are seen in the X-ray, but not in the optical despite simple arguments that say Lorentz beaming should be more pronounced in higher energy bands. Time will tell whether this diversity in phenomenology is telling us about the complexity of a single category or a diversity of physical phenomena that share some properties.

SN 1998bw/GRB980425 plays a key role here. This event (and other apparently extreme supernovae) lies on the conceptual border between extreme classes of models. On one hand, one has asymmetric supernovae that appear to be commonly associated with the core collapse phenomenon, Type II and Type Ib/c. SN 1998bw could be an extreme version of that physics, requiring energies in the upper range for other observed supernovae, perhaps 2×10^{51} ergs versus $\simeq 10^{51}$ ergs for most core collapse events and 1.3×10^{51} ergs for the well-measured SN 1987A. In this case, SN 1998bw is predicted to leave behind a neutron star, a pulsar, perhaps even a magnetar. In the other extreme of some order parameter, we have the classic cosmological γ-ray bursts as revealed by BeppoSAX and the sterling follow-up work that has been done in all wavelengths. These cosmic γ-ray bursts could be "hypernovae" involving jets originating in "collapsars" as discussed by Woosley (2000) or neutron star-neutron star or neutron star-black hole collisions (or the many variations on that theme) or maybe both. In this picture, SN 1998bw might represent a mild, not so collimated or powerful version of the cosmological bursts. In this case, SN 1998bw is predicted to leave behind a black hole. Clearly, SN 1998bw remains a potential rosetta stone, one of fading, but still detectable brilliance at the time of the conference.

Too many issues arise in the total sweep of supernova and γ-ray burst research to touch on here. There are some topics that fall in the interstices that may illuminate both, and these are worth comment.

One interesting issue is the amplitude of the Lorentz factor. The Lorentz factor is typically limited to $\lesssim 10$ in both AGNs and the stellar mass black hole sources with superluminal jets, the microquasars (Mirabel & Rodriguez 1994; Fender 1999), whereas it is understood that that the Lorentz factors in γ-ray bursts must be $\gtrsim 100$ (Baring & Harding 1997). What is the difference in the physics of these situations? One factor that is thought to limit the speed of black hole jets is radiation drag by photons emitted from the surrounding disk (Sikora et al. 1996; Luo & Protheroe 1999). Does this phenomenon, as well as baryon loading, affect the models based on black hole accretion? Has nature, through the AGN and microquasars, already told us what black hole accretion does, or is the massive accretion rate postulated for "collapsar" models sufficient to account for the difference? Do γ-ray bursts with their exceedingly large Lorentz factors require some physical basis that is very different than accreting black holes?

In principle, either newly-born neutron stars or black holes could generate jets. How are those cases to be differentiated as we explore SN 1998bw and the other candidate "hypernova" events. One interesting possibility is to look at the iron abundances in X-ray spectra, as discussed by Piro (2000; see also Mészáros & Rees 1998; Piro et al. 1999a; Lazzati, Campana & Ghisellini 1999; Vietri et al. 1999; Table 2. Note that a work widely referred to by Yoshida et al. 1999, based on ASCA observations of GRB 879828, does not seem to have been submitted). Another important issue to explore is the nature of the birth of a "magnetar" with a magnetic field ranging up to perhaps 10^{16} G compared to the birth of a "normal" pulsar with a dipole field of 10^{12} G.

The recognition that γ-ray bursts may involve collimated flows has become widespread. There is already a debate about when and at what speed, lateral expansion of jets will occur and with what effect on the spectrum and temporal behavior of the afterglow. The question of different collimation and Lorentz beaming of the γ-ray burst and afterglow must be addressed. Chevalier (2000; Li & Chevalier 1999; Chevalier & Li 1999) has emphasized the difference of afterglows propagating into density gradients as opposed to uniform density environments. In general, a spherical relativistic blast wave propagating into a density gradient will give a light curve that declines more steeply than the case of a constant density. The result is that the effects of density gradients, for instance winds with density profile $\rho \propto r^{-2}$, can mimic the effects of a collimated jet that slows and spreads in a constant density environment. In addition, prolonged energy input into the blast wave can result in a shallower light curve decline in the case with a density gradient in a manner that can mimic the effect of a spherical blast wave in a constant density medium. The luminosity of a jet could fluctuate in space and time for dynamical and kinematic reasons. If one asks about collimated flow into an environment with density gradients, even clumpiness, then the range of phenomenology could be quite great (Mészáros, Rees & Wijers 1998). These issues could be related. While relativistic blast waves propagating in a steady state wind will decelerate, those propagating through steeper profiles ($\rho \propto r^{-n}$ with n > 3; Shapiro 1980) will accelerate (Colgate 1975; Shapiro 1980) and the gradient itself can lead to collimation (Shapiro 1979). My guess is that the full potential complexity of the behavior of collimated flow has not yet been appreciated nor evaluated.

Once again, while it is fading, SN 1998bw could be an important resource for raising these issues. All models for this event presuppose a stripped core of a massive star. A strong wind is the most likely candidate for the mass loss and the radio emission has already given evidence for a relativistic blast wave interacting with a $\rho \propto r^{-2}$ density gradient (Li & Chevalier 1999). Follow-up of SN 1998bw in all possible wavelengths until it is completely unobservable is strongly encouraged!

Finally, one of the most exciting events in a string of recent revolutions was the recording of the contemporary optical outburst associated with GRB 990123 (Akerloff et al. 1999). This burst was seen at 9th magnitude and Kehoe (2000) emphasized that they cannot guarantee having detected the peak due to their sampling! Kehoe also emphasized that they have observed other events at lower limits so some γ-ray bursts, at least, are not this bright. Still, some obviously are and the lesson for the ROTSE group and the LOTIS and other groups who are trying to do automated contemporary detections is to just be patient. There must be more such events.

What an incredible event this single observation of GRB 990123 represented. At 9th magnitude, this optical flare was almost naked eye! One of the first things that occurred to me, somewhat facetiously, was encouraging hoards of private citizens to go out every night in their back yards with a decent pair of binoculars and look for these events. BATSE detects about one γ-ray burst per day. If every one were like GRB 990123, there should be a bright optical flash for a minute or so once a day somewhere on the sky that would be easily visible with a decent pair of binoculars. A pair of binoculars allows you to see about one part in a thousand of the total sky. If you looked every night for three years running, you just might get lucky.

As it so happens, Gerry Fishman and Janet Mattei of the AAVSO were way ahead of me. They are already talking about working with amateurs with decent sized telescopes and commercial CCDs to undertake a project akin to Joe Patterson's Backyard Astronomy, the program he coordinates from Columbia. With this kind of organization we may indeed see the era of Backyard Cosmology.

My thanks personally and on behalf of all the attendees at the meeting for a job well done to Mario Livio and the Scientific Organizing Committee, to Patrick Godon and Rosie Diaz-Miller for handling the expert and neophyte Power Point operators as well as all the other A-V tasks, and especially to Lorraine Garcia and Theresa Bailey for their excellent work on the symposium arrangements and the myriad issues and questions that always arise during such a conference. I am grateful for scientific discussions of supernovae and gamma-ray bursts with Lifan Wang, Peter Höflich, Rob Duncan, Alexei Khokhlov, Elaine Oran, Insu Yi, Brian Schmidt, Saul Perlmutter, Allan Sandage, David Branch, Eddie Baron, Don Lamb, Shri Kulkarni, Josh Bloom, Dave Meier, Peter Mészáros, Martin Rees, Stan Woosley and Andrew MacFadyen. Special thanks go to Howie Marion for helping with the data collection and to Martin Lang for LaTeX wrangling. This research was supported in part by NSF Grant 95-28110, NASA Grant NAG 5-2888, and a grant from the Texas Advanced Research Program.

REFERENCES

AKERLOFF, C. W., ET AL. 1999 *Nature*, **398**, 400.
ANDERSON, M. I. 1999 *Science*, **283**, 2075.
BAKOS, G., ET AL. 1999 *IAUC*, **7225**.
BARING, M. G. & HARDING, A. K. 1997 *ApJ,*, **491**, 663.
BISNOVATYI-KOGAN, G. S. 1971 *Soviet Astronomy AJ*, **14**, 652.

BLACKMAN, E. G. & YI, I. 1998 *ApJ*, **498**, L31.

BLOOM, J. S., ET AL. 1998 *ApJ*, **508**, L21.

BLOOM, J. S., ET AL. 1999a *ApJ*, **518**, L1.

BLOOM, J. S., ET AL. 1999b *Nature*,, submitted, astro-ph/9905301.

BRANCH, D. 2000, in *The Largest Explosions Since the Big Bang: Supernovae and Gamma-Ray Bursts* (eds. M. Livio, N. Panagia & K. Sahu), in press, astro-ph/9906168.

BOND, H. E. 1997 *IAUC*, **6654**.

BRANCH, D., FISHER, A., & NUGENT, P. 1993 *AJ*, **106**, 2383.

CAPPELLARO, E., TURATTO, M., & MAZZALI, P. 1999 *IAUC*, **7091**.

CEN, R. 1998 *ApJ*, **507**, L131.

CHEVALIER, R. A. 2000, in *The Largest Explosions Since the Big Bang: Supernovae and Gamma-Ray Bursts* (eds. M. Livio, N. Panagia & K. Sahu) in press.

CHEVALIER, R. A. & LI, Z.-Y. 1999 ApJ, **520**, L29.

COLGATE, S. A. 1975 *ApJ*, **198**, 439.

COVINO, S., ET AL. 1999 *A&A*, in press, astro-ph/9906319.

DAI, Z. G. & LU, T. 1998a *A&A*, **333**, L87.

DAI, Z. G. & LU, T. 1998b *Phys. Rev. Lett.*, **81**, 4301.

DAI, Z. G. & LU, T. 1999a *ApJ*, **519**, L155.

DAI, Z. G. & LU, T. 1999b *ApJ*, submitted, astro-ph/9906109.

DJORGOVSKI, S. G. 1998a *GCN*, **025**.

DJORGOVSKI, S. G., ET AL. 1998b *ApJ*, **508**, L17.

DJORGOVSKI, S. G., ET AL. 1999a *GCN*, **189**.

DJORGOVSKI, S. G., ET AL. 1999b *GCN*, **289**.

DANZIGER, J., ET AL. 2000, in *The Largest Explosions Since the Big Bang: Supernovae and Gamma-Ray Bursts* (eds. M. Livio, N. Panagia & K. Sahu) in press.

DUNCAN, R. C. & THOMPSON, C. 1992 *ApJ*, **392**, L9.

FENDER, R. P. 1999, in *Astrophysics and Cosmology*, in press. Springer-Verlag. astro-ph/9907050.

FILIPPENKO, A. V., LEONARD, D. C., & RIESS, A. G. 1999 *IAUC*, **7091**.

FISHMAN, G. J. 1995 *PASP*, **107**, 1145.

FISHMAN, G. J. 2000, in *The Largest Explosions Since the Big Bang: Supernovae and Gamma-Ray Bursts* (eds. M. Livio, N. Panagia & K. Sahu) in press.

FRAIL, D. 2000, in *The Largest Explosions Since the Big Bang: Supernovae and Gamma-Ray Bursts* (eds. M. Livio, N. Panagia & K. Sahu) in press.

FRUCHTER, A. S. 1999 *ApJ*, **512**, L1.

FRUCHTER, A. S. 2000, in *The Largest Explosions Since the Big Bang: Supernovae and Gamma-Ray Bursts* (eds. M. Livio, N. Panagia & K. Sahu) in press.

FRUCHTER, A. S., ET AL. 1999 *ApJ*, **516**, 689.

GALAMA, T. J., ET AL. 1998 *Nature*, **395**, 670.

GALAMA, T. J., ET AL. 1999a *ApJ*, submitted, astro-ph/9907264.

GALAMA, T. J., ET AL. 1999b *GCN*, **388**.

GHISELLINI, G. & LAZZATI, D. 1999 *MNRAS*, in press, astro-ph/9906471.

GERMANY, L. M., REISS, D. J., SADLER, E. S., SCHMIDT, B. P. & STUBBS, C. W. 1999 *ApJ*, submitted, astro-ph/9906096.

GIBLIN, T. W., ET AL. 1999 *ApJ*, submitted, astro-ph/9908139.

GRUZINOV, A. 1999 *ApJ* submitted, astro-ph/9905276

HALPERN, J. P., KEMP, J., PIRAN, T., & BERSHADY, M. A. 1999 *ApJ*, **517**, L105.

HARDING, A. 2000, in *The Largest Explosions Since the Big Bang: Supernovae and Gamma-Ray Bursts* (eds. M. Livio, N. Panagia & K. Sahu) in press.

HARRISON, F. A. 1999 private communication.

HARRISON, F. A., ET AL. 1999 *ApJ*, in press, astro-ph/9905306.

HJORTH, J., ET AL. 1999a *Science*, **283**, 2073.

HJORTH, J., ET AL. 1999b *GCN*, **389**.

HJORTH, J., ET AL. 1999c *GCN*, **403**.

HÖFLICH, P. 1995 *ApJ*, **440**, 821.

HÖFLICH, P. & DOMINGUEZ, I. 2000, in *The Largest Explosions Since the Big Bang: Supernovae and Gamma-Ray Bursts* (eds. M. Livio, N. Panagia & K. Sahu) in press.

HÖFLICH, P. & KHOKHLOV, A. 1996 *ApJ*, **457**, 500.

HÖFLICH, P., WHEELER, J. C., & WANG, L. 1999 *ApJ*, **521**, 179.

HOGG, D. W. & FRUCHTER, A. S. 1999 *ApJ*, **520**, 54.

HOWELL, D. A., WANG, L., & WHEELER, J. C. 1999 *ApJ*, in press, astro-ph/9908127.

ISRAEL, G. L., ET AL. 1999 *A&A*, in press, astro-ph/9906409.

IWAMOTO, K., ET AL. 1998 *Nature*, **395**, 672.

JHA, S., GARNAVICH, P., CHALLIS, P., & KIRSHNER, R. P. 1999 *IAUC*, **7090**.

KEHOE, R. 2000, in *The Largest Explosions Since the Big Bang: Supernovae and Gamma-Ray Bursts* (eds. M. Livio, N. Panagia & K. Sahu) in press.

KHOKHLOV, A. 2000, in *The Largest Explosions Since the Big Bang: Supernovae and Gamma-Ray Bursts* (eds. M. Livio, N. Panagia & K. Sahu, in press.

KHOKHLOV, A. M., ORAN, E. S., & WHEELER, J. C. 1997a *Combustion & Flame* **108**, 503.

KHOKHLOV, A. M. ORAN, E. S., WHEELER, J. C. 1997b *ApJ*, **478**, 678.

KHOKHLOV, A. M., HÖFLICH, P. A., ORAN, E. S., WHEELER, J. C., WANG, L., & CHTCHELKANOVA, A. YU. 1999 *ApJ*, in press, astro-ph/9904419.

KOUVELIOTOU, C., STROHMAYER, T., HURLEY, K., VAN PARADIJS, J., FINGER, M. H., DIETERS, S., WOODS, P., THOMPSON, C., & DUNCAN, R. C. 1998 *ApJ*, **510**, 115.

KULKARNI, S., ET AL. 1998a *Nature*, **393**, 35.

KULKARNI, S., ET AL. 1998b *Nature*, **395**, 663.

KULKARNI, S., ET AL. 1999 *Nature*, **398**, 389.

KULKARNI, S. 2000, in *The Largest Explosions Since the Big Bang: Supernovae and Gamma-Ray Bursts* (eds. M. Livio, N. Panagia & K. Sahu) in press.

LAMB, D. Q. 1999 *A&A*, in press, astro-ph/9909026.

LAZZATI, D., CAMPANA, S. & GHISELLINI, G. 1999 *MNRAS*, in press, astro-ph/9902058.

LEBLANC, J. M. & WILSON, J. R. 1970 *ApJ*, **161**, 541.

LI, Z.-Y. & CHEVALIER, R. A. 1999 *ApJ*, in press, astro-ph/9903483.

LIVIO, M. 2000, in *The Largest Explosions Since the Big Bang: Supernovae and Gamma-Ray Bursts* (eds. M. Livio, N. Panagia & K. Sahu) in press.

LUO, Q. & PROTHEROE, R. J. 1999 *MNRAS*, **304**, 800.

MACFADYEN, A. & WOOSLEY, S. E. 1999 *ApJ*, in press, astro-ph/9810274.

MCKENZIE, E. H. & SCHAEFER, B. E. 1999 *ApJ*, submitted.

MEIER, D., EPSTEIN, R. I., ARNETT, W. D. & SCHRAMM, D. N. 1976 *ApJ*, **204**, 869.

MÉNDEZ, M., CLOCCHIATTI, A., BENVENUTO, G., FEINSTEIN, C. & MARRACO, U. G. 1988 *ApJ*, **334**, 295.

METZGER, M. R., ET AL. 1997 *IAUC*, **6676**.

MÉSZÁROS, P. & REES, M. J. 1997 *ApJ*, **482**, L29.

MÉSZÁROS, P. & REES, M. J. 1998 *MNRAS*, **299**, L10.

MÉSZÁROS, P., REES, M. J., & WIJERS, R. A. M. J. 1998 *ApJ*, **499**, 301.

MIRABEL, I. F. & RODRIGUEZ, L. F. 1994 *Nature*, **371**, 46.
MODERSKI, R., SIKORA, M., & BUKIL, T. 1999 *ApJ*, submitted, astro-ph/9904310.
NAGATAKI, S. 1999 *ApJ*, in press, astro-ph/9907109.
NAKAMURA, T. 1998 *Prog. Theor. Phys.*, **100**, 921.
NIEMEYER, J. & WOOSLEY, S. E. 1997 *ApJ*, **475**, 740.
NIEMEYER, J. 1999 *ApJ*, in press, astro-ph/9906142.
NISENSON, P. & PAPALIOLIOS, C. 1999 *ApJ*, **518**, L29.
NOMOTO, K. 2000, in *The Largest Explosions Since the Big Bang: Supernovae and Gamma-Ray Bursts* (eds. M. Livio, N. Panagia & K. Sahu) in press.
ODEWAHN, S. C., ET AL. 1998 *ApJ*, **509**, L5.
OSTRIKER, J. P. & GUNN, J. E. 1971 *ApJ*, **164**, L95.
PACZYŃSKI, B. E. 1998 *ApJ*, **494**, L45.
PACZYŃSKI, B. E. 2000, in *The Largest Explosions Since the Big Bang: Supernovae and Gamma-Ray Bursts* (eds. M. Livio, N. Panagia & K. Sahu) in press, astro-ph/9909048.
PALAZZI, E., ET AL. 1999 *GCN*, **377**.
PANAITESCU, A. & MÉSZÁROS, P. 1998 *ApJ*, **492**, 683.
PANAITESCU, A., MÉSZÁROS, P., & REES, M. J. 1998 *ApJ*, **503**, 314.
PERLMUTTER, S., ET AL. 1999 *ApJ*, **517**, 565.
PERLMUTTER, S. 2000, in *The Largest Explosions Since the Big Bang: Supernovae and Gamma-Ray Bursts* (eds. M. Livio, N. Panagia & K. Sahu) in press.
PHILLIPS, M. M. 1993 *ApJ*, **413**, L108.
PIAN, E. 2000, in *The Largest Explosions Since the Big Bang: Supernovae and Gamma-Ray Bursts* (eds. M. Livio, N. Panagia & K. Sahu) in press.
PIAN, E., ET AL. 1998 *GCN*, **158**.
PIRAN, T. 2000, in *The Largest Explosions Since the Big Bang: Supernovae and Gamma-Ray Bursts* (eds. M. Livio, N. Panagia & K. Sahu) in press.
PIRO, L., ET AL. 1999a *ApJ*, **514**, L73.
PIRO, L., ET AL. 1999b *A&A*, in press, astro-ph/9906363.
PIRO, L. 2000, in *The Largest Explosions Since the Big Bang: Supernovae and Gamma-Ray Bursts* (eds. M. Livio, N. Panagia & K. Sahu) in press.
REES, M. J. 2000, in *The Largest Explosions Since the Big Bang: Supernovae and Gamma-Ray Bursts* (eds. M. Livio, N. Panagia & K. Sahu) in press.
REICHART, D. E. 1998 *ApJ*, submitted, astro-ph/9901139.
REICHART, D. E. 1999 *ApJ*, **521**, 111.
REICHART, D. E., ET AL. 1999 *ApJ*, **517**, 692.
RHOADS, J. E. 1997 *ApJ*, **487**, L1.
RHOADS, J. E. 1999 *ApJ*, submitted, astro-ph/9903399.
RIESS, A. G., PRESS, W. H., & KIRSHNER, R. P. 1996 *ApJ*, **473**, 588.
RIESS, A. G., ET AL. 1998 *AJ*, **116**, 1009.
RIESS, A. G., ET AL. 1999a *AJ*, in press.
RIESS, A. G., ET AL. 1999b *ApJ*, submitted, astro-ph/9907037.
RIESS, A. G., ET AL. 1999c *ApJ*, submitted, astro-ph/9907038.
SANDAGE, A., TAMMANN, G., & SAHA, A. 2000, in *The Largest Explosions Since the Big Bang: Supernovae and Gamma-Ray Bursts* (eds. M. Livio, N. Panagia & K. Sahu) in press.
SARI, R. 1999 *ApJ*, submitted, astro-ph/9906503.
SARI, R., PIRAN, T., & HALPERN, J. P. 1999 *ApJ*, **519**, L17.
SCHLEGEL, E. M. 1990 *MNRAS*, **244**, 269.
SCHAEFER, B. E., ET AL. 1999 *ApJ*, in press, astro-ph/9907235.

SCHMIDT, B. 2000, in *The Largest Explosions Since the Big Bang: Supernovae and Gamma-Ray Bursts* (eds. M. Livio, N. Panagia & K. Sahu) in press.

SHAPIRO, P. R. 1979 *ApJ*, **223** 831.

SHAPIRO, P. R. 1980 *ApJ*, **236** 958.

SIKORA, M., SOL, H., BEGELMAN, M. C., & MADEJSKI, G. M. 1996 *MNRAS*, **280**, 781.

STANEK, K. Z., ET AL. 1999 *ApJ*, **522**, L39.

SUNTZEFF, N. 1999, *Aspen Summer Workshop*, unpublished.

THOMPSON, C. 1994 *MNRAS*, **270**, 480.

THORSETT, S. E. & HOGG, D. W. 1999 *GCN*, **197**.

TRAMMELL, S. R., HINES, D. C., & WHEELER, J. C. 1993 *ApJ*, **414**, L21.

TRAN, H. D., FILIPPENKO, A. V., SCHMIDT, G. D., BJORKMAN, K. S., JANUZZI, B. J., & SMITH, P. S. 1997 *PASP*, **109**, 489.

USOV, V. V. 1992 *Nature*, **357**, 452.

USOV, V. V. 1994 *MNRAS*, **267**, 1035.

VIETRI, M., PEROLA, C., PIRO, L. & STELLA, L. 1999 *MNRAS*, in press, astro-ph/9906288.

VREESWIJK, P. M., ET AL. 1999 *GCN*, **310**, 324.

WANG, L., HÖFLICH, P., & WHEELER, J. C. 1997 *ApJ*, **483**, L29.

WANG, L., WHEELER, J. C., & HÖFLICH, P. 1997 *ApJ*, **476**, L27.

WANG, L. & WHEELER, J. C. 1998 *ApJ*, **584**, L87.

WANG, L., WHEELER, J. C., LI, Z. W., & CLOCCHIATTI, A. 1996 *ApJ*, **467**, 435.

WANG, L., WHEELER, J. C., & HÖFLICH, P. 1999, in *SN 1987A* (eds. Phillips, et al.) PASP.

WAXMAN, E. & LOEB, A. 1999 *ApJ*, **515**, 721.

WEILER, K. W., PANAGIA, N., SRAMEK, R. A., VAN DYK, S. D., MONTES, M. J., & LACEY, C. K. 2000, in *The Largest Explosions Since the Big Bang: Supernovae and Gamma-Ray Bursts* (eds. M. Livio, N. Panagia & K. Sahu) in press.

WHEELER, J. C., HÖFLICH, P., & WANG, L. 1999, in *Workshop on Future Directions of Supernova Research: Progenitors to Remnants* (eds. J. Danziger & S. Cassisi) in press.

WIJERS, R. A. M. J., ET AL. 1999 *ApJ*, in press, astro-ph/9906346.

WOOSLEY, S. 1993 *ApJ*, **405**, 273.

WOOSLEY, S., EASTMAN, R., & SCHMIDT, M. 1999 *ApJ*, **516**, 788.

WOOSLEY, S. E., MACFADYEN, A. I., & HEGER, A. 2000, in *The Largest Explosions Since the Big Bang: Supernovae and Gamma-Ray Bursts* (eds. M. Livio, N. Panagia & K. Sahu) in press, astro-ph/9909034.

YOSHIDA, A., ET AL. 1999, *preprint*.

ZHARIKOV, S. V., SOKOLOV, V. V., & BARYSHEV, YU. V. 1998 *A&A*, **337**, 356.